EFFICIENT PREPARATIONS OF FLUORINE COMPOUNDS

EFFICIENT PREPARATIONS OF FLUORINE COMPOUNDS

Edited By

Herbert W. Roesky

WILEY

A JOHN WILEY & SONS, INC., PUBLICATION

Published by John Wiley & Sons, Inc., Hoboken, New Jersey
Published simultaneously in Canada.

Library of Congress Cataloging-in-Publication Data:

Efficient preparations of fluorine compounds / edited by Herbert W. Roesky.
pages cm
Includes index.
ISBN 978-1-118-07856-3 (hardback)
1. Fluorine compounds. I. Roesky, H. W., editor of compilation.
QD181.F1E34 2012
546′.7312–dc23

2012025732

ISBN: 9781118078563

10 9 8 7 6 5 4 3 2 1

CONTENTS

Foreword xi

Preface xiii

Contributors xv

1 Preparation of Elemental Fluorine 1
Karl O. Christe

2 Preparation of Highly Active Cesium Fluoride 7
Konrad Seppelt

3 Preparation of Highly Active Silver Fluoride 9
Konrad Seppelt

4 A Room-Temperature Non-Irradiative Synthesis of XeF_2 11
Ulf Breddemann, John R. DeBackere, and Gary J. Schrobilgen

5 Efficient Perfluorination of $K_2B_{12}H_{12}$ in Neutral Acetonitrile 16
Dmitry V. Peryshkov, Eric V. Bukovsky, and Steven H. Strauss

6 Efficient Preparation of the Highly Soluble *ortho*- and *para*-$C_{60}(CF_2C_6F_5)_2$ Derivatives 19
Igor V. Kuvychko, Bryon W. Larson, Steven H. Strauss, and Olga V. Boltalina

7 Synthesis of $Cs[1\text{-}H\text{-}CB_{11}F_{11}]$ 22
Michal Valášek, Filip Šembera, Michael J. Hughes, Ivan Stibor, Zbyněk Janoušek, and Josef Michl

8 Synthesis of Zero-Valent Trifluoromethyl Chalcogenato Derivatives, $[NMe_4]ECF_3$ ($E = S, Se, Te$), and Related Compounds 26
Wieland Tyrra and Silke Kremer

9 Synthesis Methods for Exotic Inorganic Fluorides with Varied Applications 35
Dayal T. Meshri, N. C. Mathur, and Ritesh R. Jain

10 Tris(pentafluorophenyl)borane, $B(C_6F_5)_3$, a Powerful, Well-Soluble, Nonoxidizing Lewis Acid and the Weakly Coordinating Tetrakis(pentafluorophenyl)borate Anion, $[B(C_6F_5)_4]^-$ 60
Hermann-Josef Frohn

11 **Pentafluorophenyldifluoroborane, $C_6F_5BF_2$, and Pentafluorophenyltrifluorosilane, $C_6F_5SiF_3$, Versatile Reagents for Fluorine/Pentafluorophenyl Substitution Reactions in Strongly Oxidizing Hypervalent Nonmetal Fluorides** 65
Hermann-Josef Frohn

12 **Iodine(III and V) Fluorides: Interesting Fluorinating Agents** 72
Hermann-Josef Frohn

13 **Tetramethylammonium Fluoride, $[N(CH_3)_4]F$, a Widely Applicable Reagent to Introduce Fluoride Ions and a Suitable Nucleophile to Initiate the Transfer of Perfluoroorganyl Groups to Electrophiles** 76
Hermann-Josef Frohn

14 **Preparation of Transition Metal Sulfide Fluorides** 79
Michael Gerken

15 **Transition Metal Carbonyl Sulfur Dioxide and Thiazylfluoride Complexes: Reactions at the Metal Center and at the Ligand** 82
Rüdiger Mews

16 **Cesium, Mercury, and Silver Salts with Sulfur–Nitrogen–Fluorine Anions: Useful Transfer Reagents for NSF Building Blocks** 88
Rüdiger Mews

17 **Laboratory-Scale Synthesis of Gold Trifluoride and Uranium Hexafluoride** 94
Thomas M. Klapötke

18 **Preparation of Transition Metal Fluorides using ClF_3** 100
Ralf Haiges

19 **Preparation of Fluorine-Containing Molecular Halides and Heteropolar Salts with Elements of Group 15 and Niobium and Tantalum Halides** 108
Lothar Kolditz

20 **Fluoro and Fluorohydroxy Complexes of As, Sb, and Sn** 117
Lothar Kolditz

21 **Trifluoromethyl Compounds via Electrophilic and Nucleophilic Reactions** 121
Jean'ne M. Shreeve

22 **Introduction of Fluorine into Compounds via Electrophilic and Nucleophilic Reactions** 127
Jean'ne M. Shreeve

23 **The "Real" Iodine and Bromine Monofluorides** 133
Shlomo Rozen

24 **The Versatile Chemistry of Acetyl Hypofluorite: The First Hypofluorite Not Bonded to a Polyhaloalkyl Group** 139
Shlomo Rozen

25 **Direct Fluorination of Organic Compounds with Elemental Fluorine** 146
Shlomo Rozen

26 **The Surprising Chemistry of Bromine Trifluoride** 154
Shlomo Rozen

27 **Preparation of Silicon- and Sulfur-Based Fluorinated Methane Derivatives as Versatile Fluoromethylation Reagents** 173
G.K. Surya Prakash and Fang Wang

28 **Pentafluoroethyl Lithium: Reactions with Carbonyl Compounds and Epoxides** 205
Olesya Kazakova and Gerd-Volker Röschenthaler

29 **Synthesis of γ-Hydroxy-α,α-difluoromethylenephosphonates** 210
Olesya Kazakova and Gerd-Volker Röschenthaler

30 **Synthesis of Pentafluoro-λ^6-sulfanyl-Substituted Acetylenes for Liquid Crystals** 214
Olesya Kazakova, Maxim Ponomarenko, and Gerd-Volker Röschenthaler

31 **Delocalized Lipophilic Cations as a Source of Naked Fluoride and Phase-Transfer Catalysts** 220
Nataliya Kalinovich, Olesya Kazakova, and Gerd-Volker Röschenthaler

32 **Methyltrifluoropyruvate Imines Possessing *N*-oxalyl and *N*-Phosphonoformyl Groups: Precursors to a Variety of α-CF_3-α-Amino Acid Derivatives** 227
Olesya Kazakova and Gerd-Volker Röschenthaler

33 **Rhodium-Mediated Synthesis of (3,3,3-Trifluoropropyl)trimethoxysilane and (3,3,3-Trifluoropropyl)triphenylsilane** 232
Falk Wehmeier and Thomas Braun

34 **Rhodium-Mediated Synthesis of a Tetrafluoropyridyl-2-boronate Ester** 235
Michael Teltewskoi and Thomas Braun

35 **Palladium-Mediated Synthesis of 4-Vinyltetrafluoropyridine and 2,3,5,6-Tetrafluoropyridine** 237
David Breyer and Thomas Braun

36 **Preparation of Polyfluoroaryl-1,2-difluorovinylsilanes** 240
Donald J. Burton and Ba V. Nguyen

37 **Preparation of (Z)-2-iodo-1,2-difluorostyrenes** 243
Donald J. Burton and Ling Xue

38 **Preparation of 1,4-Bis(Z-2-iodo-1,2-difluoroethenyl)benzene** 246
Donald J. Burton and Craig A. Wesolowski

39 **Stereospecific Carboamidation Route to (Z)-*N*-Phenyl-2,3-difluoro-3-(triethylsilyl) acrylamide** 248
Donald J. Burton and Craig A. Wesolowski

40 **Preparation of 2-Trifluoromethyl-3,3,3-trifluoropropanal** 250
Donald J. Burton and Donald A. Wiebe

41 **Synthesis of Tetrafluorocatechol** 252
Sudharsanam Ramanathan, Dayong Sang, Vivek Kumar, and David M. Lemal

42 **Preparation of an Unsymmetrical Bis((perfluoroalkyl)sulfonyl)imide** 256
Darryl D. Desmarteau and Thomas D. Hickman

43 **Preparation of Perfluoroalkyl Sulfilimines and Sulfoximines** 262
Emmanuel Magnier

44 **Preparation of Trifluoromethylsulfonium Salts** 266
Emmanuel Magnier

45 **Preparation of Organometallic Fluorides of Main Group and Transition Elements** 270
Herbert W. Roesky

46 **Preparation of Pentafluorosulfanyl Carbonyl Compounds** 275
John T. Welch

47 **Preparation of Power-Variable Electrophilic Trifluoromethylating Agents, *S*-(Trifluoromethyl)dibenzothiophenium Salts Series** 279
Teruo Umemoto

48 **Synthesis of Fluorine-Containing Heterocycles from α,α-Dihydropolyfluoroalkylsulfides and Fluorinated Thiocarboxylic Acids Derivatives** 290
Vadim M. Timoshemko and Yuriy G. Shermolovich

49 **Synthesis of Octafluorocyclooctatetraene** 308
John M. Buzby, Aldos C. Barefoot, III, Michael W. Grayston, Evan D. Laganis, Roy F. Waldron, and David M. Lemal

50 **Preparation of Fluoroolefins** 315
Moritz F. Kühnel and Dieter Lentz

51 **Preparation of Ionic Liquids of Fluorocomplex and Oxofluorocomplex Anions by Fluoroacid–Base Reactions** 334
Kazuhiko Matsumoto and Rika Hagiwara

52 **Synthesis of Difluorocyclopropyl Building Blocks: 2,2-Difluorocyclopropylmethanol and 2-(Bromomethyl)-1,1-difluorocyclopropane** 338
Wei Xu and William R. Dolbier Jr.

53 **Preparation of 1,1,2,2,9,9,10,10-Octafluoro[2.2]paracyclophane and Perfluoro[2.2]paracyclophane** 343
William R. Dolbier Jr., Jian-Xi Duan, Alex J. Roche, and Lianhao Zhang

54 **Synthesis and Application of gem-Difluoromethylenated and Trifluoromethylated Building Blocks** 349
Feng-Ling Qing and Xiao-Long Qiu

55 **1,3-Dipolar Cycloaddition Reactions to Fluoroalkenes** 365
Hanna Wójtowicz-Rajchel and Henryk Koroniak

56 **Synthesis of Fluorinated Vinyl Derivatives of Nucleic Acid Bases** 370
Hanna Wójtowicz-Rajchel, Donata Pluskota-Karwatka, and Henryk Koroniak

57 **Synthesis of CF_3-Substituted Aziridine Ring by the Gabriel Reaction** 375
Tomasz Cytlak, Bartosz Marciniak, and Henryk Koroniak

58 **Preparation of α-Fluoro Amino and α-Fluoro Enamino Reagents** 379
Justyna Walkowiak and Henryk Koroniak

59 **Synthesis of Original Fluoromonomers and Their Radical Copolymerization with Vinylidene Fluoride** 385
Bruno Ameduri

60 **Convergent ^{18}F Radiosynthesis** 396
V. Gouverneur and O. Lozano

61 **Asymmetric Fluorocyclization Reactions** 400
V. Gouverneur and O. Lozano

62 **Preparation of Allylic Fluorides** 405
V. Gouverneur and O. Lozano

63 **Dehydroxyfluorinations of Primary or Secondary Alcohols Using Perfluoro *n*-Butylsulfonyl Fluoride (Nonaflyl Fluoride) in Combination with 1,8-Diaza-bicyclo[5.4.0]undec-7-ene** 412
Orlin Petrov, Matthias Schneider, Rolf Bohlmann, Stephan Vettel, and Helmut Vorbrüggen

64 **Preparation of Rare Earth Fluorosulfides and Oxyfluorosulfides** 415
A. Demourgues and A. Tressaud

65 **Preparation of Carbon–Fluorine Compounds and Fluoride or Oxide Fluoride-Intercalated Graphites** 425
Tsuyoshi Nakajima

66 Safe Synthesis of Superstoichiometric Mesoporous Fluorocarbons **433**
Valentin N. Mitkin

67 Preparation of Fluorinated γ-Alumina **442**
Erhard Kemnitz, Tomaž Skapin, and John M. Winfield

68 High-Yield Synthesis of a Single Asymmetric Isomer of $C_{70}(CF_3)_{10}$ by High-Temperature Radical Trifluoromethylation **447**
Natalia B. Shustova, Steven H. Strauss, and Olga V. Boltalina

Index **451**

FOREWORD

When Herbert W. Roesky asked if I might write a foreword for his book, I accepted instantly, without even glancing at the list of authors or topics! The reason is very simple and it goes back many years.

Herbert W. Roesky and I have only met in person on a few occasions, yet I feel I know him well, mainly because I read his early papers in the Chemistry Library at Stanford where I was a graduate student in the 1960s. In hindsight, I realize we both like the idea of mixing up simple, readily available chemicals from all around the periodic table and seeing what happens. There are vast *terrae incognitae* of descriptive chemistry throughout "the table" still waiting to be entered that could provide access to materials and functions not yet imagined.

Berzelius was the first to make the distinction between organic and inorganic compounds, and to strictly entail their origins as animate and inanimate, respectively. The first shock for believers in the *vis vitalis* origin was delivered by Berzelius's own student, Friedrich Wöhler, who, in an 1828 issue of *Annalen der Physik und Chemie* disclosed the synthesis of "organic" urea by simple union of two "inorganics": water and isocyanate. Organic chemists since Wöhler rationally know that natural products contain no 'magic' force or aura. Nevertheless, the synthesis of natural products has long been and remains a central research endeavor in organic chemistry. In retrospect, this bias was very likely ordained by the field's "birth" under vitalism.

Every chemist needs to know physical chemistry, just as every natural scientist needs to know the basic laws of physics, above all thermodynamics. But, beyond that, I prefer to just be a chemist. The idea of there being organic and inorganic chemists is bad enough, but today there are now so many qualifiers before *chemist* I suggest we need to start over. The only subtype of chemist I am not ashamed of being is a "reactivity" chemist or a 'process' chemist, which are nearly the same thing and which I will define as a *real* chemist for the moment. A *real* chemist should have a quick, hence intuitive sense, of the most likely chemical reactions occurring when any element, or simple compound thereof, is combined with others from around *The Table*.

In my view, process/reactivity chemists are a breed apart, and, in fact, are the only real chemists. Fluorine chemists, for example, are by definition good process chemists or they could not survive, literally! Many years of experience, together with deep and catholic knowledge of chemical reactivity principles, is the important coin in this realm. The specific backgrounds of process chemists matter little, but they are united by an intense focus on issues affecting chemical reactivity and have a sixth sense for the critical points in the overall sequence. The best of this breed take great pride in anticipating serious obstacles and avoiding them altogether. Of course, even the most carefully considered routes hit unknowable barriers in the real world. For a gifted process chemist, such unexpected encounters elicit excitement, not anxiety. I personally like the cases where the fix involves the realization that one or more of the fundamental reaction parameters is the culprit.

For example, it is not uncommon for the pH to take a damaging excursion when a process aims for production scales. On a small lab scale, with everything being added quickly the pH problem is barely noticeable. However at scale, where one of the crucial reactants must be added slowly for safety reasons, there will be predictable circumstances in which the pH will transit a wide range over the course of the addition stage. This problem is easily solved by having the right buffer system present. This may sound like trivial "reaction doctoring" but, when it works perfectly as planned, it makes my day. Reactions that proceed smoothly—as if gliding along the desired path—can leave the chemist in charge with the whimsical sense of wielding power over the molecules, which of course is just plain absurd!

The contributors to this volume are all "reactivity hounds," hence real chemists by definition. The level of experimental detail here is extraordinary and thus fully enabling for those less experienced in dealing with highly reactive species. Nearly all the famous names in fluorine chemistry were included by the editor. I imagine he is hard to turn down in any case, but his plan was also compelling: reach out to all chemists with favorite recipes and transformations from the best in the field.

When I am with fellow process chemists or following their recipes, I love to glean the "hints" for success that inevitably pop up. These "hints" are precious gems that need years in the trenches to reach crystalline perfection. I can see this book serving researchers and students for years to come in the many fields dependent on new compounds possessed of new or better properties. Fluorine is already famous as a giver of unique, even unimagined, properties. The practical value of fluorine substitution in molecules can only continue to grow. Fluorine is abundant in the Earth's crust as gem quality crystalline forms of CaF_2, so its uses in the future are limited only by existing reactivity constraints on crucial reaction paths.

Chemists' power/value to society derives almost entirely from our ability to manipulate reactivity. Some reactions are easily manipulated; at the other extreme are reactions that we may never succeed in manipulating. Fluorine chemistry is an area rich in reactivity constraints but also in compounds with valuable properties, and is thus a fertile hunting ground for chemists. As chemists, it falls to us to either lower the constraining barrier for the desired reaction, or discover workaround routes skirting the high passes.

In short, I see a very bright future for new fluorine-containing products. If you are the monovalent element at the end of the electronegativity universe, your presence in a molecule could prove absolutely crucial for function. A stand-in (e.g., C-Cl) for a C-F group might suffice in some applications but, wishful thinking aside, an **honest** surrogate for a C-F unit does not exist—nothing even comes close.

KARL BARRY SHARPLESS

PREFACE

The idea to publish a book with the title *Efficient Preparations of Fluorine Compounds* resulted on behalf of two observations. Firstly, about 20% of pharmaceutical products as well as 30% of agrochemical compounds contain fluorine, and the proportion is increasing. This indicates the importance of organofluorine compounds worldwide. Secondly, the interest in fluorine chemistry at the university level is steadily decreasing during the last two decades. Therefore authors working in the fluorine field were asked to write about their discoveries in a way that young scientists may reproduce their results, and use the fluorine-containing compounds to add new facets to their research.

This book brings together contributions written by leading researchers and covering a wide scope of fluorine chemistry. Karl O. Christe describes an easy laboratory method for the preparation of elemental fluorine as the first experiment. This method guarantees an easy access to fluorine when compared with that of Henri Moissan. He discovered fluorine in 1886 by electrolysis of potassium fluoride in anhydrous hydrogen fluoride. Moissan was awarded the Nobel Prize in Chemistry in 1906 for the discovery of "le fluor."

Fluorine is a poisonous diatomic gas at room temperature. It is a very versatile reagent and can form compounds with almost every element. A milestone in fluorine chemistry was achieved in 1962, when Neil Bartlett obtained the noble gas compound Xenon fluoride. The general accepted view that noble gases were inert ended with this experiment. It led to the preparation of a great number of noble gas fluoride and their derivatives.

Fluorine continues to fascinate chemists who overcome the fear of handling fluorine for the preparation of new compounds and materials. Fluorine compounds play a key role in electric cars, electronic devices, space technology, pharmaceuticals, and agrochemicals. However, especially in fluorine chemistry, the words of Winston Churchill are true: "Success is the ability to go from one failure to another with no less of enthusiasm."

Finally and most importantly, I am very thankful to the authors for their excellent contributions, and I hope that this book will inspire a young generation to do research in fluorine chemistry.

HERBERT W. ROESKY

CONTRIBUTORS

Bruno Ameduri, Engineering and Macromolecular Architectures, Institute Charles Gerhardt UMR (CNRS) 5253, Ecole Nationale Supérieure de Chimie de Montpellier, 8 Rue de l'Ecole Normale, 34296 Montpellier Cedex 5, France

Aldos C. Barefoot, III, Department of Chemistry, Dartmouth College, Hanover, NH 03755, USA

Rolf Bohlmann, Global Drug Discovery, Bayer Pharma AG, Müllerstr. 178, D-13353 Berlin, Germany

Olga V. Boltalina, Department of Chemistry, Colorado State University, Fort Collins, CO 80523, USA

Thomas Braun, Department of Chemistry, Humboldt-Universität zu Berlin, Brook-Taylor-Str. 2, 12489 Berlin, Germany

Ulf Breddemann, Department of Chemistry, McMaster University, Hamilton, Ontario L8S 4M1, Canada

David Breyer, Department of Chemistry, Humboldt-Universität zu Berlin, Brook-Taylor-Str. 2, 12489 Berlin, Germany

Donald J. Burton, Department of Chemistry, University of Iowa, Iowa City, IA 52242, USA

John M. Buzby, Department of Chemistry, Dartmouth College, Hanover, NH 03755, USA

Karl O. Christe, Loker Research Institute and Department of Chemistry, University of Southern California, Los Angeles, CA 90089-1661, USA

Tomasz Cytlak, Faculty of Chemistry, Adam Mickiewicz University, Grunwaldzka 6, 60-780 Poznań, Poland

John R. DeBackere, Department of Chemistry, McMaster University, Hamilton, Ontario L8S 4M1, Canada

A. Demourgues, ICMCB-CNRS, Université Bordeaux, 87 avenue du Dr A. Schweitzer, 33608 Pessac Cedex, France

Darryl D. DesMarteau, Department of Chemistry, Clemson University, Box 340973, Clemson, SC 29634-0973, USA

William R. Dolbier, Jr., Department of Chemistry, University of Florida, Gainesville, FL 32611-7200, USA

Jian-Xi Duan, Department of Chemistry, University of Florida, Gainesville, FL 32611-7200, USA

Hermann-Josef Frohn, Universität Duisburg-Essen, Anorganische Chemie (Campus Duisburg), Lotharstr. 1, D-47048 Duisburg, Germany

Michael Gerken, Department of Chemistry and Biochemistry, University of Lethbridge, 4401 University Drive, Lethbridge Alberta, T1K3M4, Canada

V. Gouverneur, Chemistry Research Laboratory, University of Oxford, 12 Mansfield Road, Oxford, OX1 3TA, UK

Michael W. Grayston, Department of Chemistry, Dartmouth College, Hanover, NH 03755, USA

Rika Hagiwara, Graduate School of Energy Science, Kyoto University, Yoshida Sakyo-ku, Kyoto 606-8501, Japan

Ralf Haiges, Loker Research Institute and Department of Chemistry, University of Southern California, Los Angeles, CA 90089, USA

Thomas D. Hickman, Department of Chemistry, Clemson University, Box 340973, Clemson, SC 29634-0973, USA

Michael J. Hughes, Department of Organic Chemistry and Biochemistry, University of Colorado, Boulder, CO 80309-0215, USA

Ritesh R. Jain, Advance Research Chemicals, Inc., Catoosa, OK 74015, USA

Zbyněk Janoušek, Institute of Organic Chemistry and Biochemistry, Academy of Sciences of the Czech Republic, Flemingovo nám. 2, 16610 Prague 6, Czech Republic

Olesya Kazakova, School of Engineering and Science, Jacobs University Bremen, Campus Ring 1, D-28779 Bremen, Germany

Nataliya Kalinovich, School of Engineering and Science, Jacobs University Bremen, Campus Ring 1, D-28779 Bremen, Germany

Erhard Kemnitz, Humboldt-Universität zu Berlin, Institut für Chemie, 12489 Berlin, Germany

Thomas M. Klapötke, Department of Chemistry, LMU Munich, Butenandtstr. 5-13 (D), 81377 Munich, Germany

Lothar Kolditz, Steinförde Steinerne Furth 16, 16798 Fürstenberg/Havel, Germany

Henryk Koroniak, Faculty of Chemistry, Adam Mickiewicz University, Grunwaldzka 6, 60-780 Poznań, Poland

Silke Kremer, Department für Chemie, Universität zu Köln, Institut für Anorganische Chemie, Greinstraße 6, D-50939 Köln, Germany

Moritz F. Kühnel, Freie Universität Berlin, Institut für Chemie und Biochemie, Fabeckstr. 34-36, 14195 Berlin, Germany

Vivek Kumar, Department of Chemistry, Dartmouth College, Hanover, NH 03755, USA

Igor V. Kuvychko, Department of Chemistry, Colorado State University, Fort Collins, CO 80523, USA

Evan D. Laganis, Department of Chemistry, Dartmouth College, Hanover, NH 03755, USA

Bryon W. Larson, Department of Chemistry, Colorado State University, Fort Collins, CO 80523, USA

David M. Lemal, Department of Chemistry, Dartmouth College, Hanover, NH 03755, USA

Dieter Lentz, Institut für Chemie und Biochemie, Freie Universität Berlin, Fabeckstr. 34-36, 14195 Berlin, Germany

O. Lozano, Chemistry Research Laboratory, University of Oxford, 12 Mansfield Road, Oxford, OX1 3TA, UK

Emmanuel Magnier, Institut Lavoisier de Versailles, UMR CNRS 8180, Université de Versailles-St-Quentin, Bâtiment Lavoisier, 45 Avenue des Etats-Unis, 78035 Versailles cedex, France

Bartosz Marciniak, Faculty of Chemistry, Adam Mickiewicz University, Grunwaldzka 6, 60-780 Poznań, Poland

N. C. Mathur, Advance Research Chemicals, Inc., Catoosa, OK 74015, USA

Kazuhiko Matsumoto, Graduate School of Energy Science, Kyoto University, Yoshida Sakyo-ku, Kyoto 606-8501, Japan

Dayal T. Meshri, Advance Research Chemicals, Inc., Catoosa, OK 74015, USA

Rüdiger Mews, Institute for Inorganic Chemistry, University of Bremen, Leobenerstr. NW2C, D-28359 Bremen, Germany

Josef Michl, Institute of Organic Chemistry and Biochemistry, Academy of Sciences of the Czech Republic, Flemingovo nám. 2, 16610 Prague 6, Czech Republic; Department of Organic Chemistry and Biochemistry, University of Colorado, Boulder, CO 80309-0215, USA

Valentin N. Mitkin, Nikolaev Institute of Inorganic Chemistry SB RAS, 3, Acad. Lavrentjeva Avenue, Novosibirsk, 630090, Russia

Tsuyoshi Nakajima, Department of Applied Chemistry, Aichi Institute of Technology, Yakusa, Toyota 470-0392, Japan

Ba V. Nguyen, Department of Chemistry, University of Iowa, Iowa City, IA 52242, USA

Dmitry V. Peryshkov, Department of Chemistry, Colorado State University, Fort Collins, CO 80523, USA

Orlin Petrov, Global Drug Discovery, Bayer Pharma AG, Müllerstr. 178, D-13353 Berlin, Germany

Donata Pluskota-Karwatka, Faculty of Chemistry, Adam Mickiewicz University, Grunwaldzka 6, 60-780 Poznań, Poland

Maxim Ponomarenko, School of Engineering and Science, Jacobs University Bremen, Campus Ring 1, D-28779 Bremen, Germany

G. K. Surya Prakash, Loker Hydrocarbon Research Institute, University of Southern California, Los Angeles, CA 90089, USA

Feng-Ling Qing, Laboratory of Organofluorine Chemistry, Shanghai Institute of Organic Chemistry, Chinese Academy of Sciences, 354 Fenglin Lu, Shanghai 200032, China

Xiao-Long Qiu, Wisdom Pharmaceutical Co., Ltd., 601 East, Xiu-shan Road, Haimen, Jiangsu Province, 226100, China

Sudharsanam Ramanathan, Department of Chemistry, Dartmouth College Hanover, NH 03755, USA

Alex J. Roche, Department of Chemistry, University of Florida, Gainesville, FL 32611-7200, USA

Herbert W. Roesky, Institut für Anorganische Chemie, Georg August Universität Göttingen, Tammannstrasse 4, 37077 Göttingen, Germany

Gerd-Volker Röschenthaler, School of Engineering and Science, Jacobs University Bremen, Campus Ring 1, 28759 Bremen, Germany

Shlomo Rozen, School of Chemistry, Tel-Aviv University, Tel-Aviv 69978, Israel

Dayong Sang, Department of Chemistry, Dartmouth College, Hanover, NH 03755, USA

Matthias Schneider, Global Drug Discovery, Bayer Pharma AG, Müllerstr. 178, D-13353 Berlin, Germany

Gary J. Schrobilgen, Department of Chemistry, McMaster University, Hamilton, Ontario L8S 4M1, Canada

Filip Šembera, Institute of Organic Chemistry and Biochemistry, Academy of Sciences of the Czech Republic, Flemingovo nám. 2, 16610 Prague 6, Czech Republic

Konrad Seppelt, Institut für Chemie und Biochemie, Freie Universität Berlin, Anorganische Chemie, Fabeckstraße 34-36, 14195 Berlin, Germany

Yuriy G. Shermolovich, Institute of Organic Chemistry, NAS of Ukraine, Murmanskaya str.5, 02094, Kiev, Ukraine

Jean'ne M. Shreeve, Department of Chemistry, University of Idaho, Moscow, ID 83844-2343, USA

Tomaž Skapin, Department of Inorganic Chemistry and Technology, Jožef Stefan Institute, Jamova 39, SI-1000 Ljubljana, Slovenia

Ivan Stibor, Institute of Organic Chemistry and Biochemistry, Academy of Sciences of the Czech Republic, Flemingovo nám. 2, 16610 Prague 6, Czech Republic

Steven H. Strauss, Department of Chemistry, Colorado State University, Fort Collins, CO 80523, USA

Michael Teltewskoi, Institut für Chemie, Humboldt-Universität zu Berlin, Institut für Chemie, Brook-Taylor-Str. 2, 12489 Berlin, Germany

Vadim M. Timoshemko, Institute of Organic Chemistry, NAS of Ukraine, Murmanskaya str.5, 02094, Kiev, Ukraine

A. Tressaud, ICMCB-CNRS, Université Bordeaux, 87 avenue du Dr A. Schweitzer, 33608 Pessac Cedex, France

Wieland Tyrra, Department für Chemie, Universität zu Köln, Institut für Anorganische Chemie, Greinstraße 6, D-50939 Köln, Germany

Teruo Umemoto, UBE America Inc., 6860 N. Broadway, Suite B, Denver, CO 80221, USA

Michal Valášek, Institute of Organic Chemistry and Biochemistry, Academy of Sciences of the Czech Republic, Flemingovo nám. 2, 16610 Prague 6, Czech Republic

Stephan Vettel, Global Drug Discovery, Bayer Pharma AG, Müllerstr. 178, D-13353 Berlin, Germany

Helmut Vorbrüggen, Institut für Organische Chemie und Biochemie, Freie Universität Berlin, Takustr. 3, D-14195 Berlin, Germany

Roy F. Waldron, Department of Chemistry, Dartmouth College Department of Chemistry, Hanover, NH 03755, USA

Justyna Walkowiak, Faculty of Chemistry, Adam Mickiewicz University, Grunwaldzka 6, 60-780 Poznań, Poland

Fang Wang, Loker Hydrocarbon Research Institute, University of Southern California, Los Angeles, CA 90089, USA

Falk Wehmeier, Department of Chemistry, Humboldt-Universität zu Berlin, Brook-Taylor-Str. 2, 12489 Berlin, Germany

John T. Welch, Department of Chemistry, University at Albany, 1400 Washington Ave, Albany, NY 12203, USA

Craig A. Wesolowski, Department of Chemistry, University of Iowa, Iowa City, IA 52242, USA

Donald A. Wiebe, Department of Chemistry, University of Iowa, Iowa City, IA 52242, USA

John M. Winfield, School of Chemistry, University of Glasgow, Joseph Black Building, Glasgow, G12 8QQ, Scotland, UK

Hanna Wójtowicz-Rajchel, Faculty of Chemistry, Adam Mickiewicz University, Grunwaldzka 6, 60-780 Poznań, Poland

Wei Xu, Department of Chemistry, University of Florida, Gainesville, FL 32611-7200, USA

Ling Xue, Department of Chemistry, University of Iowa, Iowa City, IA 52242, USA

Lianhao Zhang, Department of Chemistry, University of Florida, Gainesville, FL 32611-7200, USA

CHAPTER 1

Preparation of Elemental Fluorine

KARL O. CHRISTE

Although the syntheses of fluorinated compounds usually do not involve the use of elemental fluorine (F_2), F_2 can be considered to be the mother of all fluorine compounds. Because fluorine is the most electronegative element, its synthesis presented an enormous challenge and had been pursued unsuccessfully for almost a century, until finally in 1886 Moissan succeeded to prepare it electrochemically [1]. For the next 100 years, every major chemistry textbook stated that for the above reasons F_2 cannot be prepared by purely chemical means. This dogma was shattered in 1986 by Christe who prepared and isolated in a 3-day tour de force [2] F_2 in high yield from potassium hexafluoromanganate (K_2MnF_6) and antimony pentafluoride (SbF_5), two compounds that had already been known in the days of Moissan. He used a combination of two very simple and well-known principles for his synthesis: (1) that stronger acids can displace weaker acids from their salts and (2) that high oxidation states are stabilized by formal negative charges. Thus, a high oxidation state complex fluoro anion can be prepared with relative ease and when converted by the acid displacement reaction to its neutral parent molecule, the latter, if thermodynamically unstable, might spontaneously decompose to a lower oxidation state and thereby liberate F_2.

$$K_2MnF_6 + 2\,SbF_5 \rightarrow 2\,KSbF_6 + [MnF_4]$$

$$2\,[MnF_4] \rightarrow 2\,MnF_3 + F_2$$

In view of the relative ease, simplicity, and historical significance of this synthesis, it has been included in this book. It might be attractive for demonstration purposes or when only smaller amounts of fluorine are desired and the costs of either setting up an electrochemical cell or a compressed F_2 gas-handling system are prohibitive. Since SbF_5 is commercially readily available from at least 38 global and 14 U.S.

Efficient Preparations of Fluorine Compounds, First Edition. Edited by Herbert W. Roesky.

suppliers, its synthesis is not described in this chapter. However, if desired, it can be prepared in high yield by purely chemical means from $SbCl_5$ and HF [3]. The other starting material, K_2MnF_6, is more difficult to buy commercially and, therefore, its one-step, one-pot synthesis [2,4,5] from $KMnO_4$, KF, H_2O_2, and aqueous HF is described here.

$$2\,KMnO_4 + KF + 10\,HF + 3\,H_2O_2 \rightarrow 2\,K_2MnF_6 \downarrow + 8\,H_2O + 3\,O_2$$

1.1 PREPARATION OF K_2MNF_6

Apparatus Two 4-L polyethylene beakers, Teflon-coated magnetic stirrer, polyethylene Buchner funnel with Teflon filter paper, safety glasses, laboratory coat, and protective gloves.

Chemicals $KMnO_4$, KF, 48–50% reagent grade aqueous HF (caution: technical grade HF should be avoided because it contains a significant amount of H_2SiF_6, resulting in a product containing significant amounts of K_2SiF_6 as impurity), and 30% aqueous H_2O_2.

Attention! Safety glasses and protective gloves must be used at all times because HF and H_2O_2 can cause skin burns.

Experimental Procedure A 4-L polyethylene beaker, equipped with a Teflon-coated magnetic stirring bar, is charged with 50% aqueous HF (1 L) and cooled with an ice bath. Then KF (240 g, 4.13 mol) and finely powdered $KMnO_4$ (15 g, 94.9 mmol) are added and the mixture is vigorously stirred for 15 min. The stirring is stopped and any undissolved material is allowed to settle. The supernatant solution is decanted into a second beaker to assure the absence of any undissolved $KMnO_4$, which could make the endpoint recognition in the subsequent titration step difficult. To the cold dark purple solution, 30% aqueous H_2O_2 is slowly added with an eyedropper. After the addition of each 5–10 drops, further additions are halted until O_2 evolution has ceased. After the addition of about 20 mL of H_2O_2, the endpoint is being approached. A brownish golden precipitate is formed and the endpoint can be judged by stopping the stirring and observing the color of the supernatant solution. The reaction is complete when the color of the solution has changed from purple to medium reddish brown. The golden yellow K_2MnF_6 precipitate is collected using a plastic Buchner funnel with Teflon filter paper. The precipitate is washed twice with cold acetone (10 mL each) and pumped to dryness to yield 18.44 g (78.6% based on $KMnO_4$) of yellow K_2MnF_6.

Characterization Data Yellow, non-hygroscopic, crystalline solid. IR (AgCl, cm^{-1}): $\bar{\upsilon}$ 620 vs, 340 s. RA (glass melting point (mp) capillary): $\bar{\upsilon}$ 601 vs, 512 m, 307 ms. Crystal data: hexagonal, $P6_3mc$, $a = 5.719(1)$ Å, $c = 9.330(3)$ Å [6].

Application In addition to serving as convenient starting materials for the chemical synthesis of F_2, alkali metal hexafluoromanganates can be used in acidified HF solutions as fluorinating agents or as starting materials for the synthesis of $(NF_4)_2MnF_6$ for solid propellant NF_3/F_2 gas generators for chemical HF/DF lasers [7,8].

1.2 PREPARATION OF F_2

Apparatus The apparatus used in Christe's original synthesis of F_2 is shown in Figure 1.1, which shows a typical set up for the transfer of a compound of relatively low volatility, such as SbF_5, in a dynamic vacuum from a storage vessel into a reaction U-tube. The reactor can be a $^1/_2$-in. or $^3/_4$-in. o.d. Teflon-FEP (perfluoroethylene/perfluoropropylene copolymer) or metal (Monel, nickel, copper, or stainless steel) U-tube reactor, closed at both ends with Hoke stainless steel valves. Since Teflon-FEP starts softening and being attacked by the nascent fluorine at about 200 °C, the use of a Monel U-tube is preferred, unless the visual observation of the reaction is desired. The connections can be made with either Teflon or preferentially metal tubing. The T-piece connector between the SbF_5 vessel and the reactor allows evacuation and passivation of the connection. The

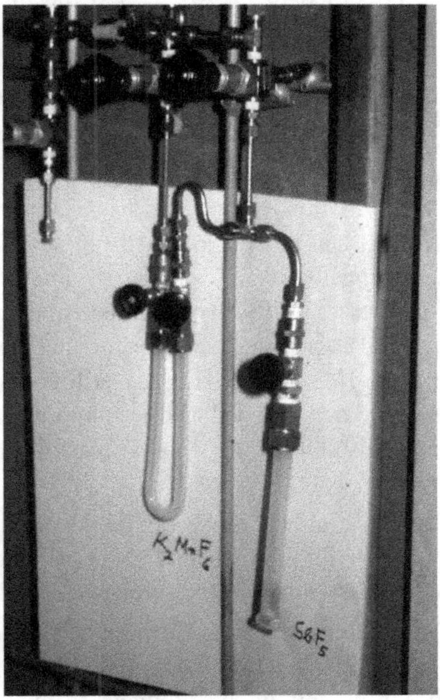

FIGURE 1.1 The original apparatus used by Christe in 1986 for the first chemical synthesis of F_2. (For a color version of the figure, please see color plates.)

exit side of the U-tube reactor is also connected to the vacuum manifold. The Teflon U-tube can be prepared by tightly packing a desired length of straight Teflon tubing with crystalline NaCl, closing both ends with rubber stoppers, heating the central part of the tube with a heat gun to the softening point of the Teflon, wrapping it 180° around an approximately 2-in. o.d. metal cylinder, allowing it to cool and removing the NaCl by pouring it out, and washing out any imbedded salt with water. The metal U-tubes are easily prepared with a tube bender. Safety requirements include face shield, safety glasses, laboratory coat, and protective gloves.

Chemicals Silicon-free $K_2[MnF_6]$, distilled SbF_5.

Attention! F_2 is a highly reactive gas with a very intense halogen odor and is easily detected already at very low concentrations (0.02 ppb) by its characteristic smell. Inhalation or contact with the skin must be strictly avoided. Laboratory coat, face shield, safety glasses, and protective gloves must be used at all times.

Experimental Procedure A passivated (with F_2 or preferentially ClF_3) $^3/_4$-in. o.d. Teflon-FEP ampoule, equipped with a valve, and a passivated $^1/_2$-in. o.d. Monel U-tube, closed at each end by a valve, are loaded in the dry box with distilled SbF_5 (~7 g or 32 mmol) and silicon-free K_2MnF_6 (1.912 g, 7.744 mmol), respectively, and are then connected to the vacuum manifold as shown in Figure 1.1. The connections are leak-checked and passivated. The Monel U-tube is cooled to −196 °C, and the SbF_5 is transferred in a dynamic vacuum from the Teflon ampoule to the Monel U-tube. After closing the valves, the Monel reactor is heated with an oil bath to 180 °C for 1 h and then cooled to −78 °C. The only product volatile at this temperature is the desired F_2 (56 mg, 1.47 mmol) in 38% yield, based on the limiting reagent K_2MnF_6.

Assay of the F_2 The formation, purity, and exact amount of fluorine formed in the above experiment can be verified easily by reacting the gas with mercury (Hg) and measuring the change in the volume of the gas by standard pressure–volume–temperature (PVT) techniques and monitoring the weight uptake of Hg. A typical experimental setup for this step is shown in Figure 1.2. Care must be taken to pump on Hg only at low temperatures (–78 °C or −196 °C), because Hg has some volatility at ambient temperature and even small losses will severely impact this analysis due to the high atomic weight of Hg.

Characterization Data Faint yellow-green, highly toxic, corrosive gas, mp −219.62 °C, boiling point (bp) −188.12 °C, standard atomic weight, 18.9984032 g/mol, first ionization energy, 1681.0 kJ/mol [9].

Waste Disposal The aqueous HF solution from the K_2MnF_6 preparation can be disposed of as NaF after neutralization with sodium bicarbonate. The $KSbF_6 \cdot nSbF_5$ and Mn-containing by-products from the chemical synthesis of F_2 have to be collected and properly deposited in a labeled container for toxic metal waste.

FIGURE 1.2 The original apparatus used by Christe in 1986 for the assay of F_2. On the left is the Teflon-FEP U-tube reactor containing the F_2 gas. It is connected to a glass bulb containing the mercury, and the whole system is connected to the Heise gauge for the pressure measurements. (For a color version of the figure, please see color plates.)

Application Due to its high reactivity and toxicity, F_2 is rarely used as a fluorinating agent in industrial processes. The major applications are the preparations of UF_6 for uranium isotope separation and SF_6 as a dielectric medium in transformers, and its use in the electronics industry for plasma etching and chamber cleaning.

REFERENCES

1. Moissan H. *C R Acad Sci* 1886;102:1543.
2. Christe KO. Chemical synthesis of elemental fluorine. *Inorg Chem* 1986;25:3721–3722.
3. Ruff O, Graf H, Heller W, Knoch M. *Ber Dtsch Chem Ges* 1906;39:4310–4327.
4. Weinland RF, Lauenstein O. Ueber fluormanganite. *Z Anorg Allg Chem* 1899;20:40–45.
5. Bode H, Jenssen H, Bandte F. Ueber eine neue Darstellung des Kalium-hexafluoromanganats(IV). *Angew Chem* 1953;65:304.
6. Bukovec P, Hoppe R. Zur Kenntnis von hexagonalem $K_2[MnF_6]$. *J Fluorine Chem* 1983;23:579–587.

7. Court TL, Dove MFA. J Chem Soc, Chem Commun 1971;726; Court TL, Dove MFA. Fluorine compounds of nickel(III). *J Chem Soc, Dalton Trans* 1973;1995–1997.

8. Christe KO, Wilson WW, Schack CJ, Wilson RD. Tetrafluoroammonium salts. *Inorg Synth* 1986;24:39–48.

9. For a summary of the properties of F_2, see: http://en.wikipedia.org/wiki/Fluorine

CHAPTER 2

Preparation of Highly Active Cesium Fluoride

KONRAD SEPPELT

Cesium fluoride (CsF) acts as a fluoride ion source in organic and inorganic chemistry. For example, it is used for C–Si cleavage reactions, halogen exchange reactions, or the isolation of highly coordinated anions AF_n^- or AF_n^{2-}. There are many sources for fluoride ions. But obviously the closest relatives to CsF, namely, NaF, KF, and RbF, are not so effective, although they are cheaper and easier to obtain in pure state. Alternatively, there is a good number of organic ammonium fluoride, the most prominent example is $(CH_3)_4NF$. Many of these are not easily prepared, and usually are thermally unstable, even $(CH_3)_4N^+F^-$ decomposes above 150 °C.

From a structural viewpoint, CsF is a very simple compound, having the NaCl structure. This is insofar surprising because the ionic radii of Cs^+ and F^- are such that a CsCl structure should be more stable, if only the ionic radii are considered. Therefore, CsF may be described as having an inverted NaCl structure, where the large Cs^+ ions form a cubic closest packing and the smaller F^- ions fill octahedral holes.

The unfortunate cation–anion size relation may explain the reactivity of CsF: It is extremely hygroscopic, similar to P_4O_{10}. Hydration would take place at the anionic sites to increase their sizes. Therefore, any CsF varies in reactivity, depending on the water content.

For many purposes, extreme dryness of CsF may not be necessary; in some cases, it may be too reactive, since it may catalyze side reactions. The reason why certain reactions of CsF are "source dependent" is certainly explained by its water content and surface area.

Efficient Preparations of Fluorine Compounds, First Edition. Edited by Herbert W. Roesky.

2.1 PREPARATION OF PURE CsF

Apparatus Platinum cup, heating furnace up to 800 °C, dry box, ball mill, and stainless steel plate.

Chemicals CsF, HF/H_2O.

Attention Safety glasses and gloves must be used at all times.

Caution All reactions should be carried out in a well-ventilated hood, if not done in closed systems (dry box, ball mill).

Experimental Procedure A small amount of CsF is dissolved in water and tested for neutrality. If it is basic, a few milliliters of HF/H_2O is given to the bulk amount. This is placed into the platinum cup and slowly heated up to 650 °C, until all solid has melted. As soon as this state is reached, the molten CsF is poured on the stainless steel plate, where it solidifies.

CsF is volatile at its melting point, so prolonged heating will make some or all CsF disappear! Mp 682 °C, bp 1251 °C. The molten CsF must be poured out of the platinum cup, because solid CsF, if it remains in there, will be difficult or impossible to get out without destroying the platinum vessel.

The solidified CsF chunks with the stainless steel plate are immediately, when they are still very hot, brought into the evacuation chamber of the dry box. The dry box should have a moisture content of 1 ppm or less. CsF is given into the ball mill capsule that needs to be kept free from moisture. The CsF is powdered in the ball mill. It is transferred back into the dry box, but even there it needs to be stored in airtight vessels.

The first traces of water will change the appearance of this CsF: it will start to become sticky, less powdery, very much like physical change of the uptake of water by P_4O_{10}.

Recycling of used CsF If larger amounts of used CsF have been accumulated, recycling may be considered. If it is without cationic impurities, the Cs salt mixture is dissolved in concentrated sulfuric acid in a platinum bowl. It is heated to red heat, until all volatiles have disappeared to about 800 °C. The solid, pure Cs_2SO_4 (mp 1019 °C) is weighted and reacted in water with the exact equivalent of $Ba(OH)_2$. Filtration of $BaSO_4$ and neutralization with HF/H_2O gives an aqueous solution of CsF, which then is dehydrated to CsF as described above. This recycled CsF will inevitably contain a small amount of a barium impurity.

CHAPTER 3

Preparation of Highly Active Silver Fluoride

KONRAD SEPPELT

Silver monofluoride (one of the five silver fluorides Ag_2F, AgF, AgF_2, Ag_3F_8, and AgF_3) is a very effective reagent for halogen exchange reactions. Driving force of this simple reaction is the formation of stable and insoluble AgCl, AgBr, or AgI.

There are several procedures for preparing AgF, and it is also commercially available. It has been observed that the AgF varies in its activity, which may be a consequence of residual water content, partial decomposition into elemental silver, or of its particle size.

Pure silver fluoride is a colorless, soft, crystalline material. It is extremely hydroscopic and light sensitive, so that it has usually a small water content and is brown or black, due to the presence of metallic silver.

To obtain pure AgF, the best preparation would be the reaction of Ag_2CO_3 with gaseous HF in a platinum tube under heating up to 310 °C. But this procedure is certainly demanding because of the need of a platinum tube and the handling of gaseous HF.

AgF has also been prepared by thermal decomposition of $Ag^+BrF_4^-$ [1] or $AgBF_4$, but these starting materials need to be prepared first.

The conventional way to prepare AgF, and possibly the way all commercial AgF is made, is reacting freshly precipitated Ag_2O or Ag_2CO_3 with excess of aqueous HF, and to free the solution from the water by heating.

In the following preparation, a variation of this method that gives AgF as a yellow solid is described.

3.1 PREPARATION OF PURE AgF

Apparatus Platinum cup.

Chemicals $AgNO_3$, Na_2CO_3, HF/H_2O, acetone.

Efficient Preparations of Fluorine Compounds, First Edition. Edited by Herbert W. Roesky.

Attention! Safety glasses and gloves must be used all the time.

Caution! All reactions should be carried out in a well-ventilated hood.

Experimental Procedure All preparation should be done under shelter against bright light. $AgNO_3$ is dissolved in water. A NaOH/water solution is added, until all brown Ag_2O is precipitated. (Alternatively, Ag_2CO_3 can be precipitated with a Na_2CO_3/H_2O solution. The disadvantage of the method is that in the following reaction with aqueous HF, large amounts of CO_2 are evolved that result in spraying and splashing.)

The Ag_2O precipitate is centrifuged and washed carefully with water. It should not be pumped to dryness, but reacted further immediately. The wet Ag_2O is placed into a platinum cup, and 40% HF/H_2O is added, just enough to dissolve all Ag_2O. At this part, the resulting solution is clear and off white.

Then acetone is added under stirring that initiates the precipitation of AgF. After cooling to slightly above 0 °C, the yellow AgF is filtered off and pumped to dryness by applying vacuum. The AgF is best stored in sealed vessels at low temperature (e.g., −30 °C) and in the dark to keep it from turning dark.

Recycling If larger amounts of reacted AgF have been accumulated, it may be worthwhile to regenerate AgF. Often the reaction product is a mixture of AgF, AgCl(Br,I), and/or Ag_2SO_4. Some 10% hydrochloric acid is given to the mixture, so that all water-soluble silver salts (AgF, Ag_2SO_4) are changed into AgCl. The solid silver halide mixture is filtered and washed carefully with water. Zinc bars are added to the wet silver salt. This transforms the silver halides into black, metallic silver. If all silver halide has disappeared, the remaining zinc bars are taken out and the silver is carefully washed with water. It can now be dissolved in HNO_3 forming $AgNO_3$ from which the AgF preparation can be restarted.

REFERENCE

1. Sharpe AG. The preparation of silver tetrafluoroborate and silver(I)fluoride. *J Chem Soc* 1952:4538–4539.

CHAPTER 4

A Room-Temperature Non-Irradiative Synthesis of XeF_2

ULF BREDDEMANN, JOHN R. DEBACKERE, AND GARY J. SCHROBILGEN

Xenon difluoride is one of the most accessible and easy to handle noble-gas compounds. Consequently, its chemistry and synthetic applications are the most extensive. Included among the synthetic applications of XeF_2 are its use as a gas-phase etchant for microelectromechanical systems (MEMS) [1] and electrophilic fluorination of biologically active compounds for use in ^{18}F positron emission tomography (PET) [2,3]. In addition, it has been widely used as a mild and convenient fluorinating/oxidizing agent in main group chemistry [4–6], transition metal chemistry [7–9], and organic chemistry [10–12].

A recent review provides a survey of the syntheses, properties, and chemistry of XeF_2 [13]. Xenon difluoride can be prepared from a gaseous mixture of Xe and F_2 using a variety of energy sources to dissociate F_2 into F· radicals. This can be accomplished by thermal dissociation at high temperatures, UV light from a mercury arc lamp or natural sunlight, electric discharge, high-intensity γ-radiation, and by irradiation in particle accelerators, for example, with 10-MeV protons from a cyclotron and electrons from a van de Graff accelerator. In order to prevent the formation of the higher fluorides of xenon, XeF_4 and XeF_6, the thermal synthesis of XeF_2 requires an excess of Xe. The use of a 2:1 molar ratio of Xe:F_2 for thermal synthesis at 400 °C [14] is prone, however, to give XeF_2 that contains a small quantity of XeF_4 contaminant. For cost effectiveness, this procedure also mandates xenon recovery for large-scale syntheses.

A little known and studied alternative synthesis of XeF_2 that requires neither thermal nor irradiative dissociation of F_2 is described in this paper. In a very brief account by Bartlett et al. [15], the reaction of Xe (1.26 mmol) with F_2 (1.65 mmol) in anhydrous HF (aHF; 2 mL) in a 42 mL FEP (perfluoroethylene/perfluoropropylene copolymer) reactor under dark conditions at 20 °C for 12 h was described and shown to give a 63% yield of XeF_2 (0.134 g, 0.792 mmol). It is clear from this and related

Efficient Preparations of Fluorine Compounds, First Edition. Edited by Herbert W. Roesky.

studies with AsF_5/Xe/F_2/aHF systems using $^{19}F_2$ [15] and [^{18}F]F_2 [16] that liquid aHF and strong Lewis acid fluoride ion acceptors such as AsF_5 serve to activate homolytic F–F bond dissociation.

The synthetic procedure outlined in this Chapter utilizes liquid aHF to activate F_2 and provides optimized conditions for the syntheses of gram quantities of very pure XeF_2 at ambient temperature in near quantitative yields.

Apparatus All manipulations are carried out under strictly anhydrous conditions on a metal vacuum line fabricated from nickel, 316 stainless steel, and FEP (Figure 4.1), which is equipped with a soda lime column for disposal of F_2 and aHF. Reaction vessels are fabricated from FEP tubing and are joined

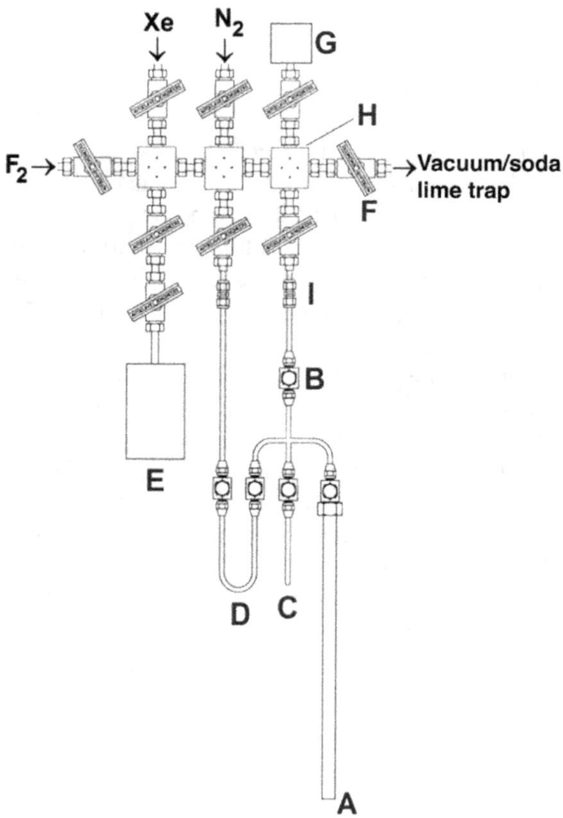

FIGURE 4.1 A segment of the metal vacuum line used for the synthesis of XeF_2. (A) FEP reaction vessel, (B) Kel-F valve, (C) FEP tube and Kel-F valve assembly used to weigh aHF, (D) FEP U-trap, (E) nickel canister used to measure Xe and F_2 gases, (F) high-pressure 316 stainless steel valve (Autoclave Engineers), (G) MKS Model PDR-5B pressure transducer (0–1000 Torr), (H) 316 stainless steel cross (Autoclave Engineers) employing $^3/_8$-in. o.d., $^1/_8$-in. i.d. threaded nickel tube connectors, (I) $^1/_4$-in. Swagelok Teflon union connected to a $^1/_4$-in. o.d., $^1/_8$-in. i.d. nickel tube end.

to Kel-F (chlorotrifluoroethylene polymer) valves by means of compression fittings. All metal and fluoroplastic surfaces are rigorously dried under dynamic vacuum prior to passivation overnight with 1 atm of F_2 gas. Xenon difluoride is recovered by transfer of the solid inside a nitrogen atmosphere dry box (H_2O, <1 ppm) into a FEP Raman sample tube, to check its purity, or into a FEP storage vessel.

Chemicals Hydrogen fluoride is purified prior to use by treatment of commercial aHF (Harshaw Chemical Co.) with F_2 gas to remove residual water [17]. Technical grade F_2 gas (Air Products) and Xe (Air Products, 99.995%) are used.

Attention! Safety glasses and protective gloves must be worn at all times; all operations must be carried out inside a well-vented fumehood.

Caution! Hydrogen fluoride and F_2 are extremely corrosive and destructive to tissue and must be handled using appropriate protective gear with immediate access to specialized treatment procedures [18–20] in the event of contact with aHF, HF vapor, HF-containing solutions, or F_2 gas. Hydrogen fluoride may prove fatal, if inhaled, absorbed through the skin, or swallowed. Like HF, F_2 gas causes severe burns that can cause serious damage to eyes, skin, and the respiratory system.

Experimental Procedure Anhydrous HF and Xe gas are condensed into an evacuated FEP reaction tube (A: 19.0 mm o.d., 15.8 mm i.d., 46.5 cm long, volume 91.2 mL) at -196 °C. Xenon and F_2 gases are measured out in a nickel canister (E: 286.4 mL). Because F_2 has an appreciable vapor pressure at -196 °C (273 Torr), it is the last component of the reaction mixture to be condensed into the reactor (A). A preweighed FEP measuring vessel (C: 6.4 mm o.d., 4.5 mm i.d., 20.5 cm long, volume 3.3 mL) is cooled to -196 °C and aHF is condensed into it at -196 °C from a Kel-F storage container at room temperature. The measuring vessel is reweighed and reconnected to the vacuum line. The evacuated reactor (A) is cooled to -196 °C and aHF is condensed into A from C at room temperature. The valve of A is closed and a known pressure of Xe gas, measured by means of a pressure transducer (G), is bled into the calibrated vacuum manifold and vessel E (total volume, 314.4 mL). The total contents of the manifold and E are then condensed into A at -196 °C. The valve of A is again closed and F_2 gas is introduced into the vacuum manifold and E, which is condensed into A at -196 °C. The valve of A is closed and A is removed from the vacuum manifold to be agitated on a mechanical shaker in complete darkness at room temperature. Upon completion of the reaction, excess F_2 is removed from the reaction vessel at -196 °C by pumping through a soda lime column. Anhydrous HF is removed from the reaction vessel at -65 to -55 °C by pumping for ~2 h into a soda lime column. Upon complete removal of HF, colorless, microcrystalline XeF_2 remains in the reaction vessel. The yield of XeF_2 is determined by sublimation of the product under dynamic vacuum into a preweighed U-trap, D (fabricated from a 36.5 cm length of 6.4 mm o.d., 4.5 mm i.d. FEP tubing, volume 5.8 mL), cooled to -196 °C.

TABLE 4.1 Representative Amounts and Yields of XeF_2 Resulting from the Reaction of Xe and F_2 in Liquid aHF[a]

Amount				XeF_2		
				Amount		
Xe, mmol	F_2, mmol	aHF, g	Time, h	g	mmol	% yield[b]
8.03	13.48	2.141	21.0	1.261	7.45	92.8
8.25	14.63	1.151	20.0	1.300	7.68	93.1
8.02	14.89	0.945	8.0	0.786	4.64	57.9
8.21	16.23	0.424	20.0	0.949	5.61	68.3
8.63	15.22	0.143	67.5	0.594	3.51	40.7
8.27	14.87	0.054[c]	143.5	0.000	0.00	0.0

[a]All reactions were carried out at ambient temperature, 23.5–25.5 °C, under dark conditions.
[b]Yields are based on Xe as the limiting reagent.
[c]Hydrogen fluoride is in the gas phase.

Table 4.1 should be consulted for representative quantities of reagents and yields. It is noteworthy that no reaction occurs between Xe and F_2 under dark conditions when HF is entirely in the gas phase.

Characterization The Raman bands of solid XeF_2 (−150 °C, 1064-nm excitation) occur at 495.9(100) [ν_{sym} (XeF_2)] and 118.9(13) [lattice mode] cm^{-1}. Xenon tetrafluoride, a possible contaminant, has intense factor-group split Raman bands (−160 °C) at 504(70), 508(19) and 545(100), 554(18) cm^{-1}, but is not detected using the present synthetic procedure.

REFERENCES

1. Easter C, O'Neal CB. Characterization of high-pressure XeF_2 vapor-phase silicon etching for MEMS processing. *J Microelectromech Syst* 2009;18:1054–1061.
2. Schrobilgen GJ, Firnau G, Chirakal R, Garnett ES. Synthesis of [^{18}F]XeF_2, a novel agent for the preparation of ^{18}F-radiopharmaceuticals. *J Chem Soc, Chem Commun* 1981:198–199.
3. Vasdev N, Pointner BE, Chirakal R, Schrobilgen GJ. On the preparation of fluorine-18 labelled XeF_2 and chemical exchange between fluoride ion and XeF_2. *J Am Chem Soc* 2002;124:12863–12868.
4. Yamaguchi S, Shirasaka T, Tamao K. Synthesis, structures, and UV-visible absorption spectra of tri(9-anthryl)bismuthine derivatives. *Organometallics* 2002;21:2555–2558.
5. Yamaguchi S, Akiyama S, Tamao K. The coordination number–photophysical properties relationship of trianthrylphosphorus compounds: doubly locked fluorescence of anthryl groups. *J Organomet Chem* 2002;646:277–281.
6. Klapötke TM, Krumm B, Mayer P, Piotrowski H, Ruscitti OP, Schiller A. Novel organotellurium(IV) diazides and triazides. *Inorg Chem* 2002;41:1184–1193.

7. Yahav A, Goldberg I, Vigalok A. Synthesis of the elusive $(R_3P)_2MF_2$ (M = Pd, Pt) complexes. *J Am Chem Soc* 2003;125:13634–13635.
8. Bernhardt E, Bach C, Bley B, Wartchow R, Westphal U, Sham IHT, von Ahsen B, Wang C, Willner H, Thompson RC, Aubke F. Homoleptic, σ-bonded octahedral $[M(CO)_6]^{2+}$ cations of iron(II), ruthenium(II), and osmium(II): Part 1: syntheses, thermochemical and vibrational characterizations, and molecular structures as $[Sb_2F_{11}]^-$ and $[SbF_6]^-$ salts. A comprehensive, comparative study. *Inorg Chem* 2005;44:4189–4205.
9. Yahav A, Goldberg I, Vigalok A. Difluoro complexes of platinum(II) and -(IV) with monodentate phosphine ligands: an exceptional stability of d^6 octahedral organometallic fluorides. *Inorg Chem* 2005;44:1547–1553.
10. Jeon K-J, Lee Z, Pollak E, Moreschini L, Bostwick A, Park C-M, Mendelsberg R, Radmilovic V, Kostecki R, Richardson TJ, Rotenberg E. Fluorographene: a wide bandgap semiconductor with ultraviolet luminescence. *ACS Nano* 2011;5:1042–1046.
11. Patrick TB, Mortezania R. Synthesis of 16α- and 16β-fluoro-17β-estradiol by fluorination of estrone enols. *J Org Chem* 1988;53:5153–5155.
12. Halpern DF. Xenon(II) fluoride. In: Paquette LA, Crich D, Fuchs PL, Molander GA, editors. *Encyclopedia of Reagents for Organic Synthesis,* 2nd ed., Vol. 13 Chichester, UK: John Wiley & Sons; 2009: pp 10869–10871.
13. Tramšek M, Žemva B. Synthesis, properties and chemistry of xenon(II) fluoride. *Acta Chim Slov* 2006;53:105–116.
14. Mercier HPA, Sanders JCP, Schrobilgen GJ, Tsai SS. The oxotrifluoroxenon(VI) cation: X-ray crystal structure of $XeOF_3^+ SbF_6^-$ and a solution ^{17}O and ^{129}Xe nuclear magnetic resonance study of the $^{17,18}O$-enriched $XeOF_3^+$ cation. *Inorg Chem* 1993;32:386–393.
15. Shen C, Hagiwara R, Mallouk TE, Bartlett N. Thermodynamic aspects of the remarkable oxidizing capabilities of fluorine—Lewis-fluoroacid mixtures. In: Thrasher J, Strauss SH, editors. *Inorganic Fluorine Chemistry, Toward the 21st Century*, ACS Symposium Series 555. Washington, DC: American Chemical Society; 1994, Chapter 2, pp 26–39.
16. Adams RM. Fluorine-18 tracer studies of inorganic and organic electrophilic fluorine sources. MSc Thesis, McMaster University, Hamilton, ON, Canada; 1994.
17. Emara AAA, Schrobilgen GJ. The fluoro(hydrogen cyanide)xenon(II) cation. Preparation of $HC{\equiv}NXeF^+ AsF_6^-$: a multinuclear magnetic resonance and Raman spectroscopic study. *Inorg Chem* 1992;31:1323–1332.
18. Bertolini JC. Hydrofluoric acid: a review of toxicity. *J Emerg Med* 1992;10:163–168.
19. Peters D, Mietchen R. Symptoms and treatment of hydrogen fluoride injuries. *J Fluorine Chem* 1996;79:161–165.
20. Segal EB. First aid for a unique acid, HF: a sequel. *Chem Health Saf* 2000;18–23.

CHAPTER 5

Efficient Perfluorination of $K_2B_{12}H_{12}$ in Neutral Acetonitrile

DMITRY V. PERYSHKOV, ERIC V. BUKOVSKY, AND STEVEN H. STRAUSS

The synthesis of $K_2B_{12}F_{12}$ by fluorination of $K_2B_{12}H_{12}$ in supercritical anhydrous HF was first reported by Solntsev et al. [1]. Ivanov et al. [2] used elemental fluorine as a fluorinating agent and anhydrous HF as the solvent to obtain $K_2B_{12}F_{12}$ in a batch process on approximately 1 g scale with more than 99% purity in 72% yield; the procedure took a total of 6 days to complete. Other solvents have been explored by Casteel and Ivanov [3] for the fluorination to scale-up the reaction. It was claimed that acidic medium is necessary for the fluorination; however, the complete fluorination was found to be difficult to achieve.

The present procedure was a successful attempt to use a continuous flow apparatus to scale-up the reaction (~20 g scale) and replace hazardous HF for CH_3CN as a solvent (with an additional benefit of the use of ordinary Pyrex glassware instead of Monel reactors) without decline in purity (99+%) or yield (74%) [4]. The crucial finding is that the presence of hydrofluoric acid, the by-product of the fluorination, inhibits the reaction. It is of the utmost importance to remove HF from the reaction mixture. It can be done by addition of solid KF, which binds HF and forms KHF_2 that is insoluble in acetonitrile. The synthesis takes only 2 days, including workup and recrystallization.

5.1 PREPARATIONS OF $K_2B_{12}F_{12}$ AND $CS_2B_{12}F_{12}$

Apparatus Three-necked, 1-L, Pyrex round-bottom flask fitted with a magnetic stir bar and two Teflon 22/29 tube adaptors. One tube is connected to a fluorine gas cylinder (20:80 F_2/N_2 mixture) and another one to a flask containing an aqueous solution of KI. The third flask joint is sealed with a glass stopper and can be used for sampling the reaction mixture. Teflon sleeves are used to ensure

Efficient Preparations of Fluorine Compounds, First Edition. Edited by Herbert W. Roesky.

proper fitting of the adaptors. Safety glasses, laboratory coat, and protective gloves must be used at all times.

Chemicals $K_2B_{12}H_{12}$, KF, stock acetonitrile, anhydrous acetonitrile, distilled water, 6% hydrogen peroxide.

Attention! Safety glasses and protective gloves must be used at all times.

Caution! F_2 is an extremely hazardous material and should only be handled by trained personnel. All reactions should be carried out in a well-ventilated hood.

It is essential that no H_2O_2 remains before the colorless solution is evaporated to dryness, and this can take more than 2 h depending on conditions. On one occasion a dried solid containing both $K_2B_{12}F_{12}$ and unreacted H_2O_2 was found to be shock sensitive and resulted in an energetic deflagration.

Experimental Procedure The compound $K_2B_{12}H_{12}$ (10.1 g, 46 mmol) is dissolved in a mixture of stock CH_3CN (490 mL) solution and H_2O (12 mL) in a three-necked, 1-L, Pyrex round-bottom flask to give a colorless solution. Finely ground KF (22.7 g, 390 mmol) is added, most of which does not dissolve. The reaction mixture is cooled to 0 °C, and a 20:80 F_2/N_2 mixture is bubbled through it at the rate of 115–125 mL min^{-1} for 6 h with vigorous stirring. The reaction mixture turns pale yellow in 3 h of fluorination. The gas flow is stopped after 6 h and the reaction mixture is bubbled with N_2 for 30 min to remove traces of F_2. The mixture is filtered and the filtrate is evaporated to dryness under vacuum. The solid residue is redissolved in 500 mL of anhydrous CH_3CN and filtered. Finely ground KF (27.3 g, 470 mmol) is added to the filtrate. The 20:80 F_2/N_2 mixture is bubbled through the reaction mixture at a slightly lower rate, 100–110 mL min^{-1}, for 7 h, again with vigorous stirring. The reaction is stopped when an aqueous KI solution in a trap connected to the outlet of the reaction flask has changed from colorless or light yellow to dark brown. The reaction mixture is filtered and the filtrate is neutralized to pH 7 with aqueous $KHCO_3$. All volatiles are removed under vacuum. The pale yellow solid residue is mixed with 100 mL of stock CH_3CN solution and filtered. The filtrate is evaporated to dryness under vacuum. The resulting pale yellow solid is dissolved in 6% hydrogen peroxide (50 mL) and heated to 80 °C for 2 h, after which the solution turns colorless. To isolate $K_2B_{12}F_{12}$, the colorless solution is evaporated to dryness and the white solid residue is recrystallized from acetonitrile, washed with ethanol, and dried at 80 °C under vacuum. Yield: 14.84 g (34 mmol $K_2B_{12}F_{12}$; 74% based on $K_2B_{12}H_{12}$). To isolate $Cs_2B_{12}F_{12}$, the colorless solution is treated with aqueous CsCl (18.1 g, 106 mmol) and cooled to 0 °C. The colorless needle-shaped crystals of $Cs_2B_{12}F_{12}.H_2O$ are isolated by filtration, washed twice with ice-cold water (2 × 10 mL), and dried in air. Yield: 22.42 g (34.9 mmol $Cs_2B_{12}F_{12}\cdot H_2O$; 76% yield based on $K_2B_{12}H_{12}$).

Characterization Data ($K_2B_{12}F_{12}$) Decomposition temperature 550 °C (in He atmosphere). $^{19}F\{^{11}B\}$ NMR (282.2 MHz, acetone–d_6): δ −269.6 (s). ^{11}B NMR (96.2 MHz, acetone–d_6): δ −17.0 (s). ESI-MS: *m/z* 178.8 ($B_{12}F_{12}^{2-}$), 396.8 ($KB_{12}F_{12}^{-}$).

Application The outstanding chemical and thermal stability of salts of $B_{12}F_{12}^{2-}$ anion was exploited to prepare solvent-free $Ag_2B_{12}F_{12}$ by desolvation of $Ag_2(CH_3CN)_4B_{12}F_{12}$ at 280 °C under vacuum (D.V. Peryshkov and S.H. Strauss, in preparation). The superweak nature of $B_{12}F_{12}^{2-}$ anion was shown by the rapid reaction of $Ag_2B_{12}F_{12}$ with CO yielding a nonclassical carbonyl $[Ag(CO)]_2B_{12}F_{12}$ with ν (CO) = 2198 cm^{-1}. The compound $K_2(H_2O)_2B_{12}F_{12}$ exhibits rapid and reversible dehydration/rehydration and H_2O/D_2O exchange at 25 °C, expanding the crystal lattice "on demand" to accommodate reactive gases, showing the concept of "latent" porosity [5].

REFERENCES

1. Solntsev KA, Mebel AM, Votinova NA, Kuznetsov NT, Charkin OP. *Koord Khim* 1992;18:340–364.
2. Ivanov SV, Miller SM, Anderson OP, Solntsev KA, Strauss SH. *J Am Chem Soc* 2003;125:4694–4695.
3. Casteel WJ, Jr., Ivanov SV, inventors; Air Products and Chemicals, Inc., USA, assignee. US Patent 6,781,005 B1. 2004.
4. Peryshkov DV, Popov AA, Strauss SH. *J Am Chem Soc* 2009;131:18393–18403.
5. Peryshkov DV, Popov AA, Strauss SH. *J Am Chem Soc* 2010;132:13902–13913.

CHAPTER 6

Efficient Preparation of the Highly Soluble *ortho*- and *para*-$C_{60}(CF_2C_6F_5)_2$ Derivatives

IGOR V. KUVYCHKO, BRYON W. LARSON, STEVEN H. STRAUSS, AND OLGA V. BOLTALINA

The preparation of *para*-$C_{60}(CH_2Ph)_2$ was first reported by Subramanian et al. [1]. A minor reaction product was originally identified as the *ortho*-isomer of $C_{60}(CH_2Ph)_2$, but later proved to be a methanofullerene, $C_{61}HPh$ (the actual *ortho*-isomer of $C_{60}(CH_2Ph)_2$ was also isolated) [2]. Recently, a partially fluorinated derivative *para*-$C_{60}(CH_2C_6F_5)_2$ was reported by Li et al. [3]. All of these reports used the same general approach. A solution of C_{60} was chemically or electrochemically reduced to C_{60}^{2-} anion, which was subsequently reacted with the corresponding electrophile (either $PhCH_2Br$ or $C_6F_5CH_2Br$). The crude products were separated using high-performance liquid chromatography (HPLC); or flash chromatography followed by HPLC producing pure *para*-isomers of $C_{60}(CH_2Ph)_2$ and $C_{60}(CH_2C_6F_5)_2$ with 55–60 mol% [1, 2] and 55 mol% [3] yield correspondingly (*ortho*-$C_{60}(CH_2Ph)_2$ was prepared only as a minor component with a yield of 3 mol%) [2].

Our initial attempts to adapt these synthetic procedures for the preparation of the fully fluorinated analogs *ortho*- and *para*-$C_{60}(CF_2C_6F_5)_2$ (reaction of the chemically generated C_{60}^{2-} with $C_6F_5CF_2I$) were unsuccessful (ca. 95–99 mol% of the starting C_{60} was recovered unchanged).

The present procedure uses a single-step thermal reaction between C_{60} dissolved in *ortho*-dichlorobenzene and $C_6F_5CF_2I$ in the presence of an excess of copper metal powder. This eliminates the need for the oxygen-sensitive and reactive chemical reducing agents or the use of an electrochemical cell and the supporting equipment [1–3]. The regioselectivity of $C_6F_5CF_2\cdot$ radical addition (which is likely to be the reactive intermediate generated from $C_6F_5CF_2I$ on heating) to C_{60} cage was moderated, by the reaction temperature. A lower reaction temperature of 130 °C leads to the

Efficient Preparations of Fluorine Compounds, First Edition. Edited by Herbert W. Roesky.

formation of both *ortho-* and *para*-$C_{60}(CF_2C_6F_5)_2$ with ca. 1:1 molar ratio (10–12 mol% yield each). A higher reaction temperature of 180 °C leads to a selective synthesis of *para*-$C_{60}(CF_2C_6F_5)_2$ derivative with ca. 25–30 mol% yield. We also found that crude fullerene extract (containing ca. 75–85% of C_{60}, 10–20% of C_{70}, and a few percentages of higher fullerenes) can be used as an economical substitute for pure C_{60} without adverse effects on the purity or yields of the target compounds.

6.1 PREPARATIONS OF *ortho-* AND *para*-$C_{60}(CF_2C_6F_5)_2$

Apparatus A 250-mL airfree storage flask (Chemglass part number AF-0093-03 or similar) fitted with a suitable magnetic stir bar.

Chemicals Pure C_{60} or crude fullerene extract, $C_6F_5CF_2I$, copper metal powder (325 mesh), *ortho*-dichlorobenzene, toluene.

Attention! Safety glasses and protective gloves must be used at all times.

Caution! $C_6F_5CF_2I$ is a lachrymator. All reactions should be carried out in a well-ventilated hood.

Experimental Procedure The reactor was charged with pure C_{60} (1.1 g, 1.5 mmol; or 1.2 g of fullerene extract), copper powder (1.0 g, 16 mmol), *ortho*-dichlorobenzene (120 mL), and $C_6F_5CF_2I$ (0.56 g, 240 μL, 1.5 mmol; this volatile reagent was measured using a 500-μL gastight syringe). This reaction mixture was degassed using a freeze-pump-thaw technique (the cycle was performed three times). The reactor was heated in an oil bath at 180 °C for 24 h with continuous stirring. After the reaction mixture cooled down, it was evaporated to dryness using a rotary evaporator equipped with a rotary vane vacuum pump. The dry residue was dissolved in a minimal amount of toluene, filtered, and separated using preparative HPLC (20 mm ID × 250 mm l, Cosmosil Buckyprep column, 100% toluene eluent, 16 mL/min flow rate, 300 nm detection wavelength). The fraction eluting between 5.7 and 6.3 min retention time was 98% pure *para*-$C_{60}(CF_2C_6F_5)_2$ (the unreacted C_{60} was collected between 8.8 and 9.8 min and recycled). The yield of the purified material was 25 mol%; 55 mol% of the starting C_{60} was recovered (as well as ca. 100 mol% of C_{70} contained in the fullerene extract). The synthesis and isolation of *ortho*-$C_{60}(CF_2C_6F_5)_2$ is accomplished using a similar procedure except for a larger amount of $C_6F_5CF_2I$ used (0.63 g, 290 μL, 1.8 mmol) and a lower reaction temperature of 130 °C. In addition to the 10–12 mol% *para*-$C_{60}(CF_2C_6F_5)_2$ eluting as described above, the 98% pure *ortho*-$C_{60}(CF_2C_6F_5)_2$ is collected between 10.7 and 11.8 min retention time also with 10–12 mol% yield.

Characterization Data *para*-$C_{60}(CF_2C_6F_5)_2$: ^{19}F NMR (376 MHz, $CDCl_3$, C_6F_6 internal standard is used, δ −164.9 ppm): δ −93.49 (m, 2F), −138.82 (q, 2F), −150.58 (t, 1F), −162.27 (t, 2F). UV–Vis (toluene solution): 327, 445, 536, 572, 602, 635, 662, 694 nm (poorly resolved). APCI-MS: *m*/*z* 937.0 ($C_{60}(CF_2C_6F_5)^-$), 1154.0 ($C_{60}(CF_2C_6F_5)_2^-$). Cyclic voltammetry: 0.10 V (0/−), −0.33 (−/2−), (−0.85) (2−/3−), (−1.36) (3−/4−). (Relative $E_{1/2}$ [V] vs $C_{60}^{0/-}$ in *ortho*-dichlorobenzene; all experiments were carried out in *ortho*-dichlorobenzene at 100 mV/s with $TBABF_4$ supporting electrolyte; the values in parentheses were obtained from square wave voltammetry.) Solubility in toluene at 20 °C is 400 mg/mL, 0.346 M. *ortho*-$C_{60}(CF_2C_6F_5)_2$: ^{19}F NMR (376 MHz, $CDCl_3$, C_6F_6 internal standard is used, δ −164.9 ppm): δ −(90.0–94.5) (group of multiplets, 2F), −137.46 (q, 2F), −149.95 (t, 1F), −161.79 (q, 2F). UV–Vis (toluene solution): 330, 446, 544, 605, 634, 665, 700 nm (poorly resolved). APCI-MS: *m/z* 937.0 ($C_{60}(CF_2C_6F_5)^-$), 1154.0 ($C_{60}(CF_2C_6F_5)_2^-$). Cyclic voltammetry: 0.00 V (0/−), −0.39 (−/2−), −0.90 (2−/3−), (−1.34) (3−/4−). (Relative $E_{1/2}$ [V] vs $C_{60}^{0/-}$ in *ortho*-dichlorobenzene; all experiments were carried out in *ortho*-dichlorobenzene at 100 mV/s with $TBABF_4$ supporting electrolyte ([0.1 M]); the values in parentheses were obtained from square wave voltammetry.) Solubility in toluene at 20 °C is 62 mg/mL, 0.054 M. Pt electrodes were used as working and counter electrodes whereas Ag wire was the quasi-reference electrode.

Application The $E_{1/2}$ value of *para*-$C_{60}(CF_2C_6F_5)_2$ is 100 mV more positive than that of C_{60}, whereas *ortho*-$C_{60}(CF_2C_6F_5)_2$ exhibits the same $E_{1/2}$ value as that of C_{60}, which has been used as an electron acceptor in various photoelectrochemical devices (e.g. solar cells). At the same time, *ortho*- and *para*-$C_{60}(CF_2C_6F_5)_2$ display molar solubilities in toluene, that are correspondingly 16 and 104 times larger than the molar solubility of the bare-cage C_{60} [4]. Such high solubility combined with good electron-accepting properties makes both compounds promising replacements for C_{60} in various solution-processed microelectronic devices.

REFERENCES

1. Subramanian R, Kadish K, Vijayashree MN, Gao X, Jones MT, Miller MD, Krause KL, Suenobu T, Fukuzumi S. Chemical generation of C_{60}^{2-} and electron transfer mechanism for the reactions with alkyl bromides. *J Phys Chem* 1996;100:16327–16335.
2. Zheng M, Li F, Shi Z, Gao X, Kadish K. Electrosynthesis and Characterization of 1,2-Dibenzyl C_{60}: A Revisit. *J Org Chem* 2007;72:2538–2542.
3. Li C-Z, Matsuo Y, Niinomi T, Sato Y, Nakamura E. Face-to-face C_6F_5-[60] fullerene interaction for ordering fullerene molecules and application to thin-film organic photovoltaics. *Chem Commun* 2010;46:8582–8584.
4. Ruoff RS, Tse DS, Malhotra R, Lorents DC. Solubility of C_{60} in a Variety of Solvents. *J Phys Chem* 1993;97:3379–3383.

CHAPTER 7

Synthesis of $Cs[1\text{-}H\text{-}CB_{11}F_{11}]$

MICHAL VALÁŠEK, FILIP ŠEMBERA, MICHAEL J. HUGHES, IVAN STIBOR, ZBYNĚK JANOUŠEK, AND JOSEF MICHL

The highly stable icosahedral 1-carba-*closo*-dodecaborate anion, $CB_{11}H_{12}^{-}$ (**1**), and its derivatives have many interesting properties [1, 2]. The most striking are their highly positive redox potentials, low nucleophilicity, and low coordinating ability, which make them unusually difficult to oxidize and make their protic acids extremely strong. These properties have been exploited for the stabilization of unusual and/or highly reactive cations [2–5]. They are particularly strongly expressed in the undecahalogenated derivatives, especially the undecafluorinated anion, 1-H-$CB_{11}F_{11}^{-}$ (**2**) [6–8]. The reversible oxidation potential of **2** in hexafluoroisopropyl alcohol lies fully 2.42 V above that of ferrocene/ferricinium [A. Wahab, F. Šembera, M. Valášek, J. Ludvík, I. Stibor, Z. Janoušek, J. Michl, unpublished observations]. Anions containing both trifluoromethyl and fluoro substituents are even harder to oxidize [9].

The procedure for the conversion of a commercial salt of **1** to a salt **2** on a multigram scale that is described below is based on a variation of the original literature process [6] and has been found useful in our laboratory (Scheme 7.1). It proceeds in two steps without an isolation of the intermediate. In the first step, the cesium salt of **1** is monofluorinated with liquid HF at 50 °C in about 2 days to afford the Cs salt of 1-H-12-F-$CB_{11}H_{10}^{-}$ (**3**). This fluorination is known to proceed selectively only to position 12 [10]. This product is not isolated and in an immediately following second step, 20% fluorine in nitrogen is bubbled through the suspension of the Cs salt of **3** in anhydrous HF at −78 °C under vigorous stirring and the reaction mixture is then exposed to a moderate pressure of F_2/N_2. Repetition of the second step to completion of the reaction (7–14 times), followed by isolation of the trimethylammonium salt of **2**, affords 72–83% yields. We also describe its conversion into the cesium salt in 86% yield.

Efficient Preparations of Fluorine Compounds, First Edition. Edited by Herbert W. Roesky.

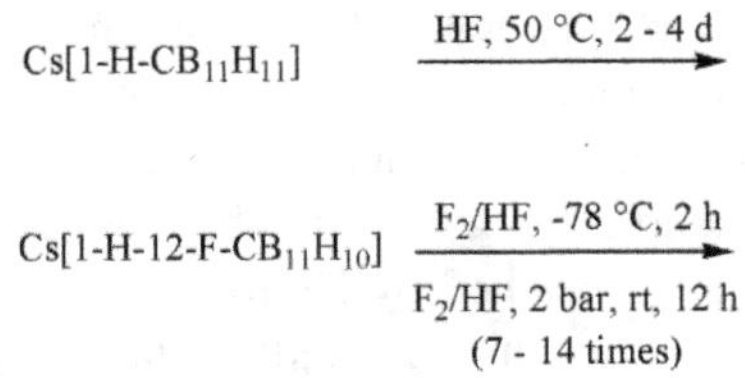

SCHEME 7.1 Fluorination of the 1-carba-*closo*-dodecaborate anion.

Chemicals Anhydrous HF (Solvay), 20% F_2/N_2 (Solvay), $Cs[CB_{11}H_{12}]$ (Katchem) [11], acetonitrile (HPLC grade), $N(CH_3)_3{\cdot}HCl$ (Sigma-Aldrich, 98%), water (HPLC grade), CsCl (Sigma-Aldrich, Grade II), NaOH (p.a.), NH_4OH (15% aqueous), HCl (35%, p.a.)

Attention! Gas masks and rubber gloves must be worn while handling HF and/or F_2.

Caution! HF and F_2/N_2 are extremely hazardous and should only be handled by trained personnel. All reactions should be carried out in a well ventilated hood.

Experimental Procedure *Trimethylammonium salt of 1-carba-(2-12)-undecafluoro-*closo*-dodecaborate* (**2**). [For a schematic representation of the fluorination line, see Figure 7.1.] A 120-mL "perfluoroalkoxide" (PFA) reactor (Savillex) equipped with a magnetic stirring bar was charged with dry $Cs[CB_{11}H_{12}]$ (1–6 g, 3.6–21 mmol) and flushed with nitrogen for 30 min. Then liquid anhydrous HF (ca. 50–80 mL) was distilled into the reactor at −78 °C. The reactor was capped and stirred for 2–4 days at 50 °C. The resulting white suspension was cooled to −78 °C and treated with 20% fluorine in nitrogen (the gas flow rate was 5–8 dm^3/h) for 1–2 h with vigorous stirring. Then the

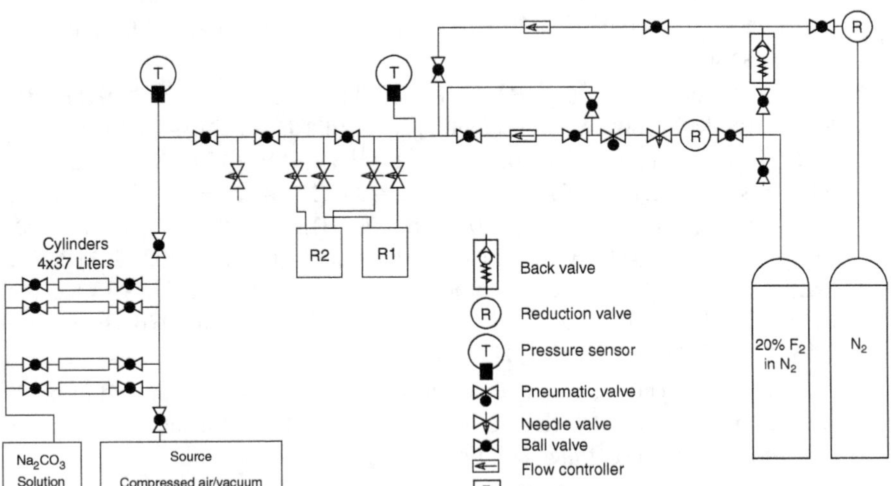

FIGURE 7.1 A schematic representation of the fluorination line.

reaction vessel was pressurized to 2 bar with the same gas mixture at −78 °C, closed, allowed to warm to room temperature, and stirred overnight. This procedure was repeated 7–14 times to ensure completion of the reaction, which was monitored by withdrawing a sample and measuring the mass spectrum (ESI-). *Sample withdrawal*: The reactor is flushed by N_2 for 15 min, pressurized to atmospheric pressure and the port for sample withdrawal is opened. If there is no port, the gas output tube is removed, samples are taken from the reactor with a PFA or PTFE (polytetrafluoroethylene) capillary at −78 °C, and the reactor is closed immediately. (*Caution*: Be extremely careful during this operation.)

The reaction mixture was then cooled to −78 °C and purged with nitrogen for 1 h to remove residual fluorine. After the removal of HF under reduced pressure, a white solid remained in the reactor. The residue was treated with excess acetonitrile (100–150 mL) and filtered to remove insoluble inorganic salts. The filtrate was neutralized with NH_4OH (15%) and evaporated to dryness under reduced pressure. The residue was dissolved in a minimal amount of water and treated with 3 equivalents of trimethylammonium chloride (30% solution in water), causing precipitation. The precipitate was filtered off, washed with water and then pentane, and dried under vacuum to afford $NHMe_3[1\text{-}H\text{-}CB_{11}F_{11}]$ as a white solid in 72–83% yield.

*Cesium salt of 1-carba-(2-12)-undecafluoro-*closo*-dodecaborate* (**2**). The $NHMe_3$ salt of the $1\text{-}H\text{-}CB_{11}F_{11}$ anion **2** (1.5 g, 3.7 mmol) was suspended in 30 mL of water and a solution of NaOH (330 mg, 8.3 mmol) in 10 mL of water was added. The addition caused dissolution of the suspension. This solution was stirred for 30 min. Water was then removed on a rotary evaporator and the solid was dissolved in a minimum amount of fresh water. The solution was neutralized with aqueous HCl (1 M) to pH ~5 and treated with CsCl (1.5 g, 8.9 mmol) dissolved in a minimum amount of water. The white solid was filtered off, washed with cold water (2 × 2 mL) and pentane (2 × 10 mL), and dried under vacuum (10 mbar/100 °C) to obtain 1.51 g of pure $Cs[1\text{-}H\text{-}CB_{11}F_{11}]$ as a white solid in 86% yield.

Characterization Data $^1H\{^{11}B\}$ NMR (499.8 MHz, acetone-d_6): 3.89 (bs, 1H, H-1). ^{11}B NMR (160.4 MHz, acetone-d_6, ext. std. $BF_3{\cdot}Et_2O$): −18.20 (bs, 5B); −16.54 (bs, 5B) ppm; −8.48 (bs, 1B). $^{19}F\ \{^{11}B\}$ NMR (470.3 MHz, acetone-d_6, ext. std. $CFCl_3$): −257.76 (bm, 5F); −257.12 (bm, 5F); −253.48 (bm, 1F) ppm. IR (KBr pellet, cm^{-1}) 3027 (m, ν(C–H)), 1308 (vs), 1285 (vs), 1265 (vs) and 1218 (s, ν(B–F)), 778 (m), 761 (m), 730 (m) and 702 (s, δ(B-F)). ESI(-) MS *m/z* 341, expected isotopic distribution. ESI(-) HRMS Calcd for $CHB_{11}F_{11}$, 343.0932. Found: *m/z* 343.0933. Anal. Calcd for $CHB_{11}F_{11}Cs$: C, 2.53; H, 0.21. Found: C, 2.57; H, 0.19.

Waste Disposal The fluorine line is a closed system. Gases are passed through the line into evacuated cylinders, where they are stored for later disposal by bubbling through a saturated solution of Na_2CO_3.

Notes An increase of the reaction temperature causes a degradation of the carborane and is to be avoided.

ACKNOWLEDGMENT

This work was supported by the Institute of Organic Chemistry and Biochemistry, Academy of Sciences of the Czech Republic (RVO:61388963), the Academy of Sciences of the Czech Republic (M200550906), the Grant Agency of the Czech Republic (203/09/J058), and the AFOSR (FA9550-10-1-0374).

REFERENCES

1. Körbe S, Schreiber PJ, Michl J. Chemistry of the carba-closo-dodecaborate(-) anion, $CB_{11}H_{12}^-$. *Chem Rev* 2006;106:5208–5249.
2. Reed CA. Carboranes: a new class of weakly coordinating anions for strong electrophiles, oxidants, and superacids. *Acc Chem Res* 1998;31:133–139.
3. Lupinetti AJ, Havighurst MD, Miller SM, Anderson OP, Strauss SH. Facile conversion of $[(^6\text{-}C_6H_6)Rh(CO)_2][1\text{-Et-}CB_{11}F_{11}]$ into the nonclassical rhodium (I) carbonyl $[Rh(CO)_4][1\text{-Et-}CB_{11}F_{11}]$. *J Am Chem Soc* 1999;121:11920–11921.
4. Ivanova SM, Ivanov SV, Miller SM, Anderson OP, Solntsev KA, Strauss SH. Mono-, di-, tri-, and tetracarbonyls of copper(I), including the structures of $Cu(CO)_2(1\text{-Bn-}CB_{11}F_{11})$ and $[Cu(CO)_4][1\text{-Et-}CB_{11}F_{11}]$. *Inorg Chem* 1999;38:3756–3757.
5. Kato T, Reed CA. Putting tert-butyl cation in a bottle. *Angew Chem Int Ed* 2004;43:2907–2911.
6. Ivanov SV, Rockwell JJ, Polyakov OG, Gaudinski CM, Anderson OP, Solntsev KA, Strauss SH. Highly fluorinated weakly coordinating monocarborane anions $1\text{-H-}CB_{11}F_{11}^-$, $1\text{-}CH_3\text{-}CB_{11}F_{11}^-$, and the structure of $[N(n\text{-Bu})_4]_2[CuCl(CB_{11}F_{11})]$. *J Am Chem Soc* 1998;120:4224–4225.
7. Finze M, Reiss GJ, Zähres M. $[1\text{-}H_2N\text{-}CB_{11}F_{11}](-)$ – Synthesis and reactions of a functionalized fluorinated carbadodecaborate anion. *Inorg Chem* 2007;46:9873–9883.
8. Küppers T, Bernhardt T, Eujen R, Willner H, Lehmann CW. $[Me_3Si][R\text{-}CB_{11}F_{11}]$-synthesis and properties. *Angew Chem Int Ed* 2007;46:6346–6349.
9. Fete MG, Havlas Z, Michl J. $HCB_{11}(CF_3)_nF_{(11-n)}^-$: inert anions with high anodic oxidation potentials. *J Am Chem Soc* 2011;133:4123–4131.
10. Ivanov SV, Lupinetti AJ, Miller SM, Anderson OP, Solntsev KA, Strauss SH. Regioselective fluorination of $CB_{11}H_{12}^-$. New weakly coordinating anions. *Inorg Chem* 1995;34:6419–6420.
11. Valášek M, Šlursa J, Pohl R, Michl J. Microwave-assisted alkylation of $[CB_{11}H_{12}]^-$ and related anions. *Inorg Chem*, 2010;49:10247–10254.

CHAPTER 8

Synthesis of Zero-Valent Trifluoromethyl Chalcogenato Derivatives, $[NMe_4]ECF_3$ (E = S, Se, Te), and Related Compounds

WIELAND TYRRA AND SILKE KREMER

The lipophilic electron-withdrawing nature especially of the SCF_3 group has been known for longer than half a century. As a result, numerous organic substrates of unique properties have been synthesized that are used mainly as agrochemicals or pharmaceuticals. Introduction of this group (and those of the heavier homologues) lacks from the fact that commonly used reagents such as $CuSCF_3$ or $AgSCF_3$ decompose more or less indiscriminately at temperatures required for a successful SCF_3 transfer and must be inconveniently prepared before use.

The investigations on SCF_3-substituted ($SeCF_3$ and possibly in the future times $TeCF_3$) pharmaceuticals, agrochemicals [1], and possible prospective precursors for metal and metal chalcogenide depositions [2] encouraged us to investigate the reactions of Me_3SiCF_3 (and related compounds) and the group 16 elements sulfur, selenium, and tellurium, suggesting the system Me_3SiCF_3/F^- as an equivalent to Grignard reagents [3–5].

Herein, we introduce metal-free alternatives of ECF_3 sources prepared via the interaction of commercially available Me_3SiCF_3, the elemental chalcogen in the presence of anhydrous $[NMe_4]F$ or a comparable fluoride source, which open a convenient access to the widely used copper and silver derivatives [3–6].

Experimental procedures are given in References 3–6. Explicit instructions are given later. For a better understanding of the saltlike nature of these compounds, the molecular structures of (*TDAE*)$[ECF_3]_2$ (*TDAE* = bis(dimethylamino)ethanediylidene, E = Se, Te) [7] are depicted in Figure 8.1.

Efficient Preparations of Fluorine Compounds, First Edition. Edited by Herbert W. Roesky.

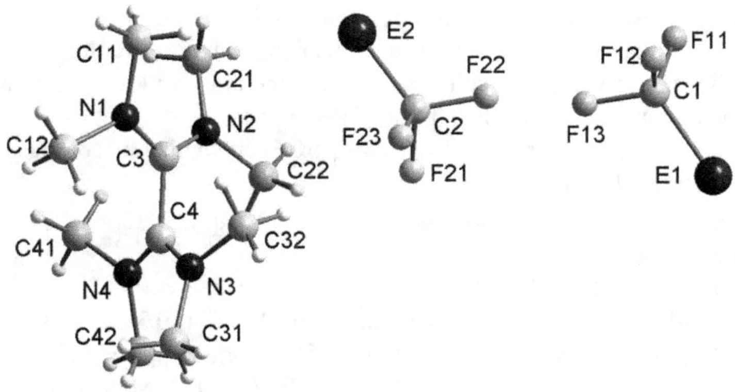

FIGURE 8.1 Molecular structure of (*TDAE*)$[ECF_3]_2$ [7].

8.1 PREPARATION OF $[NMe_4]SCF_3$

Apparatus A 50-mL Schlenk flask with stopcock (fitted with Glindemann ring), stopper, stirring bar, pipette (25 mL), Eppendorf pipette, Dewar vessel, Schlenk line, magnetic stirrer, funnel, filter, safety glasses, laboratory coat, protective gloves.

Chemicals Sulfur powder, Me_3SiCF_3, $[NMe_4]F$ [8], dimethoxyethane (glyme), ice bath (2-propanol, dry ice), liquid nitrogen.

Attention! Safety glasses must be used at all times and protective gloves at least while working with $[NMe_4]F$. The order of addition must be strictly followed; otherwise violent reactions between Me_3SiCF_3 and F^- may proceed even below room temperature. Intensive stirring is essential to obtain good yields and to avoid reactions of Me_3SiCF_3 with itself in the presence of F^- [9].

Caution! Because of its toxicity, care should be taken to avoid inhalation of $[NMe_4]F$ or organosulfur compounds. Also avoid contact of $[NMe_4]F$ with skin. All reactions should be carried out in a well-ventilated hood.

Experimental Procedure An oven-dried, 50-mL Schlenk flask equipped with a magnetic stirring bar and a stopcock was flushed with nitrogen. Sulfur (0.32 g, 10.0 mmol) was added and suspended in 25 mL glyme under nitrogen. To this suspension, Me_3SiCF_3 (1.647 mL, 11.0 mmol) was added using an Eppendorf pipette. The reaction mixture was cooled to −60 °C using a dry ice bath.

At this temperature and under a strong nitrogen flow, $[NMe_4]F$ (0.93 g, 10.0 mmol) was added in one single portion. Directly after addition, a change of color occurred from pale lemon via bright orange to pale yellow in most cases. The temperature should be kept at −60 °C for about 30 min and finally allowed to warm to room temperature overnight. The supernatant liquid may be colorless, in other cases yellow or brown.

The precipitated product ($[NMe_4]SCF_3$ is only very sparingly soluble in glyme) was isolated by decanting the supernatant. The raw material was washed several times with small portions of glyme until the supernatant was colorless. Finally, the product was dried in vacuum at ambient temperature.

The raw product is a colorless solid, sometimes yellow or tan, and was isolated in average yields better than 75%.

Moderately moisture-sensitive $[NMe_4]SCF_3$ crystallizes in colorless needles from saturated MeCN solutions at 0 °C.

Characterization Data Colorless needles, visible mp 153 °C (sintering); 210 °C (onset of decomposition).

^{19}F NMR (188.3 MHz, CD_3CN) −5.83 ppm ($^1J_{F,C}$ = 293 Hz), s. ^1H NMR (200.1 MHz, CD_3CN): δ 3.14 ppm, s. ^{13}C NMR (50.3 MHz, CD_3CN): δ 145.2 ppm, ($^1J_{F,C}$ = 293 Hz, SCF_3), q; 56.0 ppm ($^1J_{H,C}$ = 143 Hz, NMe_4) q. EIMS (20 eV): *m/z* 116 ($[MeSCF_3]^+$, 100%), 97 ($[MeSCF_2]^+$, 2%), 82 ($[SCF_2]^+$, 47%), 69 ($[CF_3]^+$, 8%), 63 ($[SCF]^+$, 4%), 59 ($[NMe_3]^+$, 88%), 58 ($[Me_2NCH_2]^+$, 90%), 47 ($[MeS]^+$, 15%).

Waste Disposal All residues as well as the supernatant and washing solutions have to be collected in a labeled container for toxic waste that has to be properly deposited.

Application $[NMe_4]SCF_3$ has already been successfully used to convert activated organics into trifluoromethyl sulfanes [3], acid chlorides into the corresponding thioesters [10] as well as metal halides, and related compounds into the corresponding trifluormethylthiolato derivatives [3]. The molecular structures of some $PtSCF_3$ compounds have been elucidated [11]. This concept together with very recent results using a Pd catalyst [12] may be considered as valuable approach for the convenient syntheses of pharmaceuticals and agrochemicals, while reactions with metal salts may open new access to metal sulfide or noble metal nanoparticles (NPs), that is by sol–gel or even by chemical vapor deposition (CVD) or atomic layer deposition (ALD) processes. For effective transformations, consult References 3, 10, and 11. The molecular structure of *cis*-$Pt(SCF_3)_2(PPh_3)_2$ is depicted in Figure 8.2.

8.2 PREPARATION OF $[NMe_4]SeCF_3$

Apparatus A 50-mL Schlenk flask with stopcock (fitted with Glindemann ring), stopper, stirring bar, pipette (25 mL), Eppendorf pipette, Dewar vessel, Schlenk line, magnetic stirrer, funnel, filter, safety glasses, laboratory coat, protective gloves.

Chemicals Selenium (preferred: red, nonmetallic modifications) powder, Me_3SiCF_3, $[NMe_4]F$ [8], dimethoxyethane (glyme), acetonitrile, *n*-pentane, ice bath (2-propanol, dry ice), liquid nitrogen

Attention! Safety glasses must be used at all times and protective gloves at least while working with $[NMe_4]F$. The order of addition must be strictly followed; otherwise violent reactions between Me_3SiCF_3 and F^- may proceed even below room temperature [9]. The product has a characteristic malodorous smell and should therefore be handled with gloves and under a ventilated hood.

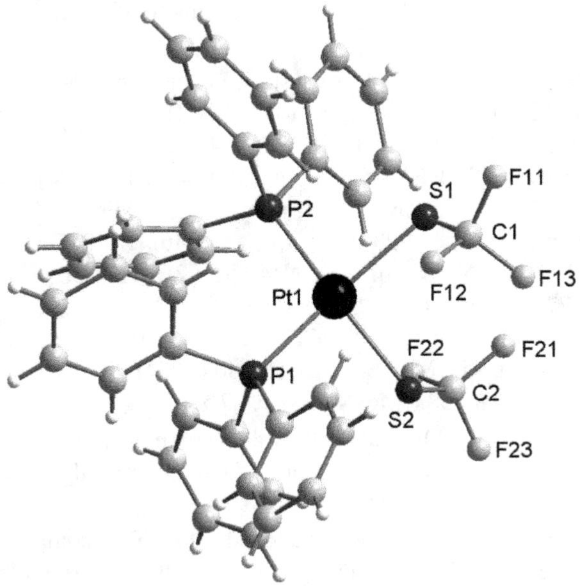

FIGURE 8.2 Molecular structure of *cis*-$Pt(SCF_3)_2(PPh_3)_2$ [11].

Caution! Because of its toxicity, care should be taken to avoid inhalation of elemental selenium, organoselenium compounds, and $[NMe_4]F$. Also avoid contact of all compounds with skin. All reactions should be carried out in a well-ventilated hood.

Experimental Procedure An oven-dried, 50-mL Schlenk flask equipped with a magnetic stirrer and a stopcock was flushed with nitrogen. To a suspension of (preferred) red selenium (1.58 g, 20.0 mmol) in 25 mL glyme, Me_3SiCF_3 (3.104 mL, 21.0 mmol) was added using an Eppendorf pipette. The reaction mixture was cooled to −78 °C using a dry ice bath. $[NMe_4]F$ (1.86 g, 20.0 mmol) was added in one portion under a strong nitrogen flow and with vigorous stirring. The temperature was held for approximately 1 h and then allowed to warm to room temperature overnight. The color of the suspension changed during this period from dark red to gray. After stirring ceased, a gray solid precipitated covered by a nearly colorless solution. The solution was decanted and the remaining solid was dried in vacuo. The gray residue was extracted with two portions of 5 mL MeCN. The extracts were combined and stored for 24 h at −30 °C. Crystalline $[NMe_4]SeCF_3$ precipitated from colorless, sometimes red or green, mother liquors. On removal of the cold MeCN solution, the remaining solid was washed with *n*-pentane and dried in vacuo giving $[NMe_4]SeCF_3$ in an average yield of 70%.

The colorless crystalline material can be handled at least for a short period in ambient atmosphere. Storing of the compound in Schlenk tubes or sealed glass tubes in daylight led to a red coloration (red selenium (?)). The compound is insoluble in alkanes, sparingly soluble in ethers, aromatics, and halogenated alkanes, but highly soluble in nitriles and dimethylformamide. The smell is intensively horseradish-like.

Characterization Data Colorless crystals, visible mp 215–217 °C (decomposition); 207 °C (onset of decomposition (DTA/TG)).

^{19}F NMR (188.3 MHz, DMF-D_7) −4.49 ppm ($^2J_{Se,F}$ = 79 Hz), s, $^1J_{F,C}$ = 323 Hz); (CD_3CN) −5.83 ppm ($^2J_{Se,F}$ = 77 Hz) s, $^1J_{F,C}$ = 323 Hz). ^{1}H NMR (200.1 MHz, CD_3CN): δ 3.14 ppm, s. ^{13}C NMR (75.4 MHz, CD_3CN): δ 132.2 ppm, ($^1J_{F,C}$ = 323 Hz, $^1J_{C,Se}$ = 246 Hz, $SeCF_3$), q; 55.6 ppm ($^1J_{H,C}$ = 144 Hz, NMe_4) q. ^{77}Se NMR (57.2 MHz, CD_3CN): δ 166 ppm ($^2J_{F,Se}$ = 77 Hz). EIMS (20 eV): *m/z* 164 ($[Me^{80}SeCF_3]^+$, 100%), 149 ($[^{80}SeCF_3]^+$, 6%), 145 ($[Me^{80}SeCF_2]^+$, 3%), 130 ($[^{80}SeCF_2]^+$, 21%), 111 ($[^{80}SeCF]^+$, 4%), 95 ($[Me^{80}Se]^+$, 37%), 80 ($[^{80}Se]^+$, 3%), 69 ($[CF_3]^+$, 7%), 59 ($[NMe_3]^+$, 63%), 58 ($[Me_2NCH_2]^+$, 87%).

Waste Disposal All residues as well as the supernatant and washing solutions have to be collected in a labeled container for toxic waste that has to be properly deposited.

Application The same as mentioned above for the SCF_3 compound may be applied to the $SeCF_3$ derivative, although the synthetic properties of this derivative have not yet been investigated in detail. Halide exchange reactions have been carried out successfully to obtain silver, gold [13], and platinum [14] derivatives. The molecular structure of $[PNP][Au(SeCF_3)_2]$ is shown in Figure 8.3.

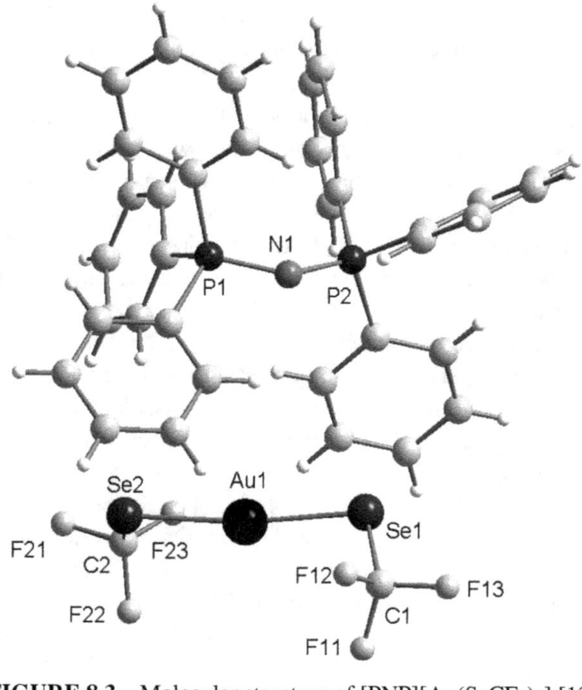

FIGURE 8.3 Molecular structure of $[PNP][Au(SeCF_3)_2]$ [13].

8.3 PREPARATION OF $[NMe_4]TeCF_3$

Apparatus Two 50-mL Schlenk flasks with stopcock (fitted with Glindemann ring), stopper, stirring bar, pipette (25 mL), Eppendorf pipette, Dewar vessel, Schlenk line, magnetic stirrer, funnel, filter, safety glasses, laboratory coat, protective gloves.

Chemicals Tellurium powder, Me_3SiCF_3, $[NMe_4]F$ [8], dimethoxyethane (glyme), acetonitrile, ice bath (2-propanol, dry ice), liquid nitrogen.

Attention! Safety glasses must be used at all times and protective gloves at least while working with $[NMe_4]F$. The order of addition must be strictly followed; otherwise violent reactions between Me_3SiCF_3 and F^- may proceed even below room temperature [9]. The product has a characteristic malodorous smell and should therefore be handled with gloves and under a ventilated hood.

Caution! Because of its toxicity, care should be taken to avoid inhalation of elemental tellurium, organotellurium compounds, and $[NMe_4]F$. Also avoid contact of all compounds with skin. All reactions should be carried out in a well-ventilated hood.

Experimental Procedure An oven-dried, 50-mL Schlenk flask equipped with a magnetic stirring bar and a stopcock was flushed with nitrogen. To a mixture of tellurium (1.91 g, 15.0 mmol) and Me_3SiCF_3 (1.647 mL, 11.0 mmol), added using an Eppendorf pipette, in 25 mL glyme at −60 °C (dry ice bath), $[NMe_4]F$ (0.93 g, 10.0 mmol) was added in one portion. The temperature was held for approximately 1 h and then allowed to warm to room temperature within approximately 2 h by removing the cooling bath. After stopping stirring, excess tellurium precipitated, covered by a yellow solution. In an inert atmosphere (argon may be used as protective gas), the solution was filtered through a funnel into another oven-dried Schlenk flask to remove excess tellurium. All volatile components were condensed off in vacuo at room temperature. $[NMe_4]TeCF_3$ was collected as a yellow solid in nearly quantitative yield. Recrystallization from MeCN (−28 °C) afforded a pale yellow, sometimes ochre, microcrystalline material in approximately 60% yield, which can be handled at least for a short period in ambient atmosphere. Storing of the compound in Schlenk tubes or sealed glass tubes in scattered daylight at room temperature effected slow elimination of elemental tellurium. In ambient atmosphere, bright yellow crystals grown from acetonitrile solutions became muddy red and finally gray over a period of 72 h with evolution of a malodorous smell.

$[NMe_4]TeCF_3$ is insoluble in alkanes, poorly soluble in aromatics (benzene, toluene) and ethers such as THF, glyme, and diethylether, highly soluble in dimethylformamide but it reacts with acetonitrile and dichloromethane.

Characterization Data Pale yellow crystals, visible mp 185 °C (onset of decomposition).

^{19}F NMR (282.3 MHz, DMF-d$_7$): δ +1.89 ppm ($^2J_{123Te,F}$ = 206 Hz, $^2J_{125Te,F}$ = 248 Hz; s, $^1J_{F,C}$ = 349 Hz, $^1\Delta(^{19}F\text{–}^{12/13}C)$ −0.1676 ppm), s. ^1H NMR (300.1 MHz, DMF-d$_7$) δ = 3.41 ppm, s. ^{13}C NMR (75.4 MHz, DMF-d$_7$)

δ 91.9 ppm, ($^1J_{F,C}$ = 349.0 Hz, ($^{13}C\{^{19}F\}$(−15 °C)) s, $^1J_{C,Te}$ = 677 Hz) $TeCF_3$), q; 55.3 ppm ($^1J_{H,C}$ = 143.9 Hz, $^3J_{H,C}$ = 3.5 Hz, NMe_4) q of decets. ^{125}Te NMR (126.2 MHz, DMF-d_7, −15 °C): δ 395 ppm, ($^2J_{Te,F}$ = 249 Hz), q. EIMS (20 eV) *m/z* 214 ($[Me^{130}TeCF_3]^+$, 100%), 199 ($[^{130}TeCF_3]^+$, 13%), 145 ($[Me^{130}Te]^+$, 48%), 130 ($[^{130}Te]^+$, 5%), 69 ($[CF_3]^+$, 6%), 59 ($[NMe_3]^+$, 58%), 58 ($[Me_2NCH_2]^+$, 76%). Negative ESI MS (MeCN) *m/z* 325 ($[Te_2CF_3])^-$, 100% most intense signal), 199 ($[^{130}TeCF_3]^-$, 92%).

The identical procedure may be applied to prepare $[NMe_4]TeC_2F_5$ using $Me_3SiC_2F_5$ instead of Me_3SiCF_3.

Characterization Data mp (decomposition) 115 °C.

^{19}F NMR (282.3 MHz, DMF-d_7) δ −66.6 (2F, $^2J_{Te,F}$ = 108 Hz, $^1J_{F,C}$ = 299 Hz, $^3J_{F,F}$ = 10 Hz), q, −83.5 (3F, $^3J_{Te,F}$ = 12 Hz; $^1J_{F,C}$ = 283 Hz, $^3J_{F,F}$ = 10 Hz), t ppm. ^{1}H NMR (300.1 MHz, DMF-d_7) δ_H 3.41 ppm. $^{13}C\{^{19}F\}$ NMR (75.4 MHz, DMF-d_7) δ 120.1 ($^2J_{Te,C}$ = 76 Hz), 95.1 ($^1J_{Te,C}$ = 574 Hz), 54.4 ppm (q of decet, $^1J_{C,H}$ = 144 Hz, $^3J_{C,H}$ = 4 Hz). $^{125}Te\{^{19}F\}$ NMR (126.2 MHz, DMF-d_7) δ 140 ppm.

Note An extension of the procedure to *n*-C_3F_7 and *n*-C_4F_9 derivatives failed due to the preferred formation of tetramethylammonium salts of perfluorinated tertiary carbanions.

Waste Disposal All residues as well as the supernatant and washing solutions have to be collected in a labeled container for toxic waste that has to be properly deposited.

Application Due to the very low solubility of $[NMe_4]X$ (X = Cl, Br, I, BF_4) in most common organic solvents, halide exchange reactions with activated organic and inorganic halides (or tetrafluoroborates) proceed quickly and selectively. This reagent together with its lighter homologues may be used as a $[ECF_3]^-$ transfer reagent to obtain organic and metal-organic species unknown so far.

Successful attempts in halide substitutions are limited to silver and gold [15] as well as platinum [14] derivatives.

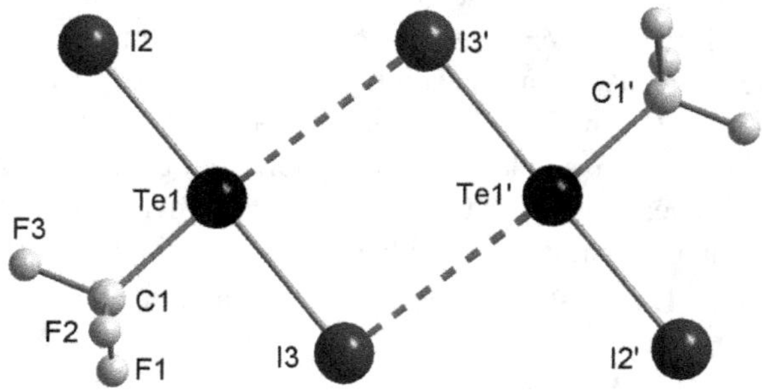

FIGURE 8.4 Molecular structure of the dimeric trifluoromethyldiiodotellurate(II) anion [17].

Salts with the anion $[TeCF_3]^-$ are easily oxidized by hexachloroethane, elemental bromine, and iodine to compounds with the anions $[Te(CF_3)X_2]^-$ (X = Cl, Br, I) [16, 17] or by disulfanes deriving from dithiocarbamic acids to derivatives with the anions $[Te(CF_3)(SC(S)NR_2)_2]^-$ (Figure 8.4) [6].

REFERENCES

1. Manteau L, Pazenok S, Vors JP, Leroux FR. New trends in the chemistry of α-fluorinated ethers, thioethers, amines and phosphines. *J Fluorine Chem* 2010;131:140–158.
2. Bochmann M. Metal chalcogenide materials: chalcogenolato complexes as "single-source" precursors. *Chem Vap Deposition* 1996;2:85–96.
3. Tyrra W, Naumann D, Hoge B, Yagupolskii YuL. A new synthesis of trifluoromethanethiolates – characterisation and properties of tetramethylammonium, cesium and di(benzo-15-crown-5)cesium trifluoromethanethiolates. *J Fluorine Chem* 2003;119:101–107.
4. Tyrra W, Naumann D, Yagupolskii YuL. Stable trifluoromethylselenates(0), $[A]SeCF_3$ – synthesis, characterizations and properties. *J Fluorine Chem* 2003;123:183–187.
5. Tyrra W, Kirij NV, Naumann D, Yagupolskii YuL. Tetramethylammonium trifluoromethyltellurate(0), $[NMe_4]TeCF_3$, synthesis, characterisation and properties. *J Fluorine Chem* 2004;125:1437–1440.
6. Tyrra W, Naumann D, Buslei S, Kremer S, Pantenburg I, Scherer H. Perfluoroalkyl(dithiocarbamato) tellurium(II) compounds. *Dalton Trans* 2007;1829–1837.
7. Tyrra W, Naumann D, Pohl H, Pantenburg I. Trifluoromethylselenate(0), $[SeCF_3]^-$, and trifluoromethyltellurate(0), $[TeCF_3]^-$—crystalline products of the reduction of the bis(trifluoromethyl)dichalkogen compounds, $E_2(CF_3)_2$ (E = Se, Te), with tetrakis(dimethylamino)ethene (TDAE). *Z Anorg Allg Chem* 2003;629:1039–1043.
8. Kolomeitsev AA, Seifert FU, Röschenthaler G-V. Simple preparation of difluorophosphoranes using anhydrous zinc and tetramethylammonium fluorides. *J Fluorine Chem* 1995;71:47–49.
9. Tyrra W, Kremlev MM, Naumann D, Scherer H, Schmidt H, Hoge B, Pantenburg I, Yagupolskii YuL. How trimethyl(trifluoromethyl)silane reacts with itself in the presence of naked fluoride—a one-pot synthesis of bis(15-crown-5)cesium 1,1,1,3,5,5,5-heptafluoro-2,4-bis(trifluoro methyl)-pentenide. *Chem Eur J* 2005;11:6514–6518.
10. Kremlev MM, Tyrra W, Naumann D, Yagupolskii YuL. *S*-Trifluoromethyl esters of thiocarboxylic acids, $RC(O)SCF_3$. *Tetrahedron Lett* 2004;45:6101–6104.
11. Kirij NV, Tyrra W, Naumann D, Pantenburg I, Yagupolskii YuL. Syntheses of trifluoromethanethiolato platinum(II) complexes—crystal structures of *cis*-$Pt(SCF_3)_2(PPh_3)_2$ and *trans*-$Pt(SCF_3)Cl(PPh_3)_2$. *Z Anorg Allg Chem* 2006;632:284–288.
12. Teverovskiy G, Surry DS, Buchwald SL. Pd-catalyzed synthesis of Ar-SCF_3 compounds under mild conditions. *Angew Chem Int Ed* 2011;50:7312–7314.
13. Naumann D, Tyrra W, Quadt S, Buslei S, Pantenburg I, Schäfer M. Bis(trifluoromethylselenato(0))metallates(I) of silver and gold, $[M(SeCF_3)_2]^-$ (M = Ag, Au). *Z Anorg Allg Chem* 2005;631:2733–2737.
14. Kirij NV, Tyrra W, Pantenburg I, Naumann D, Scherer H, Naumann D, Yagupolskii YuL. Trifluoromethylselenato(0) and trifluoromethyltellurato(0) complexes of platinum(II). *J Organomet Chem* 2006;691:2679–2685.

15. Kirij NV, Tyrra W, Naumann D, Yagupolskii YuL, Pantenburg I, Schäfer M. Bis(trifluoromethyltellurato(0))metallates(I) of copper, silver and gold, [$M(TeCF_3)_2$]$^-$ (M = Cu, Ag, Au). *J Fluorine Chem* 2004;125:1933–1938.

16. Fischer HTM. Synthesen und Strukturen neuer Trifluormethylchalkogen-Verbindungen, [*Dissertation*]. Universität zu Köln, Köln, Germany, 2007.

17. Fischer HTM, Naumann D, Tyrra W. A fluent transition from triiodide, I_3^-, to tris(trifluoromethyltellurate)(1-), [$(TeCF_3)_3$]$^-$—a structural study. *Chem Eur J* 2006;12:2515–2519.

CHAPTER 9

Synthesis Methods for Exotic Inorganic Fluorides with Varied Applications

DAYAL T. MESHRI, N.C. MATHUR, AND RITESH R. JAIN

The field of inorganic fluorides is vast, complex, and has far-reaching applications. A wide variety of inorganic fluorine compounds such as fluorides, triflates, tetrafluoroborates, and fluorophosphates of alkali, alkaline, rare earth, transition metal salts have been used extensively in the electronics, semiconductors, energy storage, and pharmaceutical industries. An emerging area of interest with enormous commercial potential for this important class of compounds involves optoelectronics, microlithography, and laser applications [1–4].

Metal triflates have been the subject of extensive academic and industrial research in the past two decades. These are used as catalyst in the polymerization of aromatic alkenes and aromatic monomers, in the electrophilic polymerization of 1,3-pentadiene, in the cationic ring-opening polymerization of tetrahydrofuran, and in the Michael reaction of O-silylated ketene acetals with α,β-unsaturated esters. Other areas of applications are aldol and Friedel–Crafts reactions [5]. Use of rare earth metal triflates in organic synthesis has been reviewed by Kobayashi et al. [6]. Recently, rare earth triflates have been used commercially in the synthesis of vitamin E [7].

Superacid systems are formed by mixing appropriate Lewis and Bronsted acids [8]. Trifluoromethane sulfonic acid or triflic acid, a superacid, forms water-soluble salts with nonhydrolyzable metals. Metal triflate salts used as catalysts, in general, have the following advantages:

a. fewer or no by-products, better yields, and higher selectivity than standard methods;
b. chiral forms can be highly diastereo- and enantioselective and hence good asymmetric catalysts;

Efficient Preparations of Fluorine Compounds, First Edition. Edited by Herbert W. Roesky.

c. greener, stable, fewer synthetic steps, use of non-chlorinated reagents (reactions can be performed in water), less toxic and corrosive, safer, easier to handle, milder reaction conditions, less energy requirements; and
d. less waste, metals are recoverable and usable.

The above properties make them an attractive alternative to Lewis acids such as $AlCl_3$.

The application and synthetic methods of inorganic fluorides and fluorinating reagents have been reviewed by Dayal Meshri in various monographs and articles [9, 10]. Some of these compounds are still not being produced in bulk scale and their synthesis methods are as much art as they are science. We have selected some important inorganic fluorides, including tetrafluoroborate, hexafluorophosphates, superacids, and triflates of aluminum, copper, lithium, potassium, etc. It has been attempted to showcase the variety of methods that may be used. These are just as varied in their applications as well.

9.1 PREPARATION OF COPPER FLUORIDE (CuF_2)

Apparatus Plastic beaker with Teflon-coated magnetic stirrer, tray oven.

Chemicals Copper carbonate and anhydrous hydrofluoric acid (AHF).

Attention! All personal protective gear such as safety glasses, full face acid respirator, full body alkyl suit with rubber gloves should be worn at all times.

Caution! AHF is extremely toxic and corrosive. Extreme care should be taken while handling. Make sure first aid treatment items in case of injury are readily available in close vicinity (eyewash station, water shower, calcium gluconate, etc.). The reaction must be carried out in a well-ventilated hood.

Experimental Procedure AHF (6000 g) is added to plastic beaker and stirred using Teflon-coated magnetic stirrer. Copper carbonate (4940 g, 40 gmol) is added slowly to AHF. The reaction is extremely exothermic liberating lot of AHF fumes and salt is formed immediately. After addition is complete, the mixture is allowed to stir for 4 h and then allowed to sit overnight without agitation. The salt is allowed to settle and the clear liquid decanted carefully. The wet salt is washed with AHF three times to remove the residual traces of water and dried under nitrogen inside a tray oven at 140 °C with gaseous AHF flowing over the trays at all times. After 1 day, the salt chunks are broken up under nitrogen in a dry glove box. The ground white powder is fluorinated again for 1 day with gaseous AHF at 140 °C. The AHF fumes are driven off by nitrogen purge and temperature of oven lowered. Sufficient flow of dry nitrogen is maintained in the ovens to prevent hydrolysis due to any moisture pickup. CuF_2 (3.74 kg, 92%) is yielded.

Characterization Data The product is characterized by elemental analysis. Copper content is typically from 61.3% to 63.8% and fluoride from 36.7% to 38.2%. Copper is analyzed by inductively coupled plasma (ICP) and fluoride content is determined using Willard–Winter distillation method. Both the methods are described in detail in Appendix A.

Application Its use has been explored as cathodic material in nonaqueous galvanic cells. Several intercalated compounds of CuF_2 have been explored recently [11–13].

9.2 PREPARATION OF POTASSIUM HEXAFLUORONICKELATE (K_2NiF_6)

Apparatus Monel tubular reactor with boat-shaped tray and radiant heaters.

Chemicals Potassium fluoride (anhydrous), nickel fluoride, elemental fluorine gas.

Attention! All personal protective gear such as safety glasses, full face acid respirator, and full body alkyl suit with rubber gloves should be worn at all times.

Caution! Fluorine gas is extremely toxic, strong oxidizer, and corrosive. Extreme care should be taken while handling. Make sure first aid treatment items in case of injury are readily available in close vicinity (eyewash station, water shower, calcium gluconate, etc.). The reaction must be carried out in a well-ventilated hood.

Experimental Procedure Potassium fluoride (580 g, 10 gmol) and nickel fluoride (485 g, 5 gmol) are ground and mixed intimately. The mixture is spread evenly in a Monel boat and placed in a tubular Monel reactor that is equipped with radiant heaters. The mixture is dried under vacuum at 200 °C for 4 h. Fluorine is then passed at very slow rates (0.25–0.50 L/m) overnight at 350 °C. The result is 95% converted chunky powder. The powder is ground and fluorinated again overnight. The resulting solid is K_2NiF_6 (1220 g) with almost 97% yield.

Characterization Data The product is characterized by elemental analysis. Fluoride content is typically from 44.5% to 46.3% and is determined using Willard–Winter distillation method. The method is described in detail in Appendix A.

Application K_2NiF_6 is an excellent source of elemental fluorine [14]. High valency complex fluorinating agent such as K_2NiF_6 converts aromatic and heterocyclic compounds to poly- and perfluorinated products, mostly unsaturated [15]. K_2NiF_6 with manganese fluoride has been successfully used for isolation of fluorooxyfluorofullerene ($C_{60}F_{17}OF$) for the first time [16]. Highly fluorinated semi-ionic graphites have been synthesized by combining elemental fluorine under pressure with K_2NiF_6 at much lower temperatures [17].

9.3 PREPARATION OF SILVER FLUORIDE (AgF)

Apparatus Five-gallon plastic jerry can with septum, Teflon-coated magnetic stirrer, and two nozzles at the top.

Chemicals Silver oxide powder, anhydrous hydrogen fluoride (AHF), anhydrous methanol, and anhydrous diethyl ether.

Attention! All personal protective gear such as safety glasses, full face acid respirator, and full body alkyl suit with rubber gloves should be worn at all times.

Caution! Please note hazards associated with AHF from the previous section. Make sure first aid treatment items in case of injury are readily available in close vicinity (eyewash station, water shower, calcium gluconate, etc.). The reaction must be carried out in a well-ventilated hood.

Experimental Procedure Silver oxide (Ag_2O, 3.5 kg, 15 mol) in a very fine powdered form is added to a clean 5-gallon plastic square jerry can. AHF (3.5 gallons) in liquid form is added to the container very slowly. The reaction is extremely exothermic and liquid AHF is allowed to react slowly with the Ag_2O powder in a stagnant manner for 2 days. The resulting solution is clear with very little unreacted Ag_2O impurities settled at the bottom. The impurities are probably Ag_2S due to trace amounts of H_2S in AHF and/or reduced silver metal. Please note that the solubility of AgF in AHF solution is about 83.2 g/100 g AHF at 11.9 °C [18]. The silver fluoride solution in AHF is decanted into another clean 5-gallon plastic square jerry can, equipped with a septum with two nozzles at the top for nitrogen gas inlet and HF vent. Nitrogen gas is bubbled through this solution for 2 days to drive off HF and yield white silver bifluoride (AgF·HF) salt. The solid chunks are broken and dried in nitrogen glove box for 6 h. Methanol (0.75 L) is added to solid and grinded to form slurry. The golden orange silver fluoride is precipitated by diethyl ether and washed several times with it (15 gallons total). The wet salt is dried under nitrogen for 2 days to yield 3.44 kg (90% yield) of AgF.

Characterization Data The product is characterized by elemental analysis. Silver content is typically from 83.7% to 86.3% and fluoride from 14.8% to 15.2%. Silver is analyzed by potassium thiocyanate titration and fluoride content is determined using Willard–Winter distillation method. The Willard–Winter distillation method is described in detail in Appendix A.

Application AgF has been used as fluorinating reagent for certain electrophilic olefins in the presence of proton donor. For example, CF_2CHSF_5 gives $CF_3(SF_5)CHAg$ in acetonitrile [19]. Treatment of siloxycyclopropane with allylic chlorides in the presence of silver fluoride results in effective formation of δ/η-unsaturated ketones [20]. Silver fluorides have also been used in batteries and as antimicrobial agents [21].

9.4 PREPARATION OF SILVER DIFLUORIDE (AgF_2)

Apparatus Monel tubular reactor with boat-shaped tray inside a copper tube and radiant heaters, safety glasses, full face acid respirator, and full body alkyl suit with rubber gloves.

Chemicals Silver fluoride, elemental fluorine gas.

Attention! All personal protective gear should be worn at all times.

Caution Fluorine gas is extremely toxic, strong oxidizer, and corrosive. Extreme care should be taken while handling. Make sure first aid treatment items in case of injury are readily available in close vicinity (eyewash station, water shower, calcium gluconate, etc.). The reaction must be carried out in a well-ventilated hood and away from organic solvents.

Experimental Procedure Silver fluoride (508 g, 4 mol) is spread evenly in a Monel boat and placed in a copper tube that is placed in a tubular Monel reactor that is equipped with radiant heaters. The fluorine is fed through an inlet tube made of copper or nickel. The powder is dried under vacuum at 150 °C for 4 h. Fluorine is then passed at very slow rates (0.25–0.50 L/min) overnight at 180 °C. The result is 95% conversion and formation of chunky powder. The chunks are unloaded from the reactor, ground to a powder, and refluorinated overnight under the same conditions. The resulting solid is AgF_2 (573 g, 98% yield). Similarly, silver chloride (AgCl) can be used to make AgF_2 by reaction with fluorine gas.

Characterization Data The product is characterized by elemental analysis. Silver content is typically from 72.1% to 75.8% and fluoride content from 25.4% to 26.7%. Silver is analyzed by potassium thiocyanate titration and fluoride content is determined using Willard–Winter distillation method. The Willard–Winter distillation method is described in detail in Appendix A.

Application AgF_2 is a strong fluorinating and oxidizing agent. It is capable of replacing aromatic ring hydrogen atoms at moderate temperatures. For example, conversion of benzene to fluorobenzene [22, 23]. It is used in the fluorination and preparation of organic perfluoro compounds [24]. It has been used as fluorinating agent for fluorination of C_{60} fullerenes. At 300 °C, $C_{60}F_{44}$ was produced with remarkable selectivity using AgF_2 [25]. AgF_2 oxidizes xenon to xenon difluoride in anhydrous HF solutions [26].

$$2AgF_2 + Xe \xrightarrow{HF} AgF + XeF_2$$

9.5 PREPARATION OF TRIETHYLAMINE TRIHYDROGEN FLUORIDE ($(C_2H_5)_3N{\cdot}3HF$)

Apparatus Four-liter high-density polyethylene (HDPE) beaker with Teflon-coated magnetic stirrer.

Chemicals Triethylamine (TEA), AHF.

Attention! All personal protective gear such as safety glasses, full face acid respirator, and full body alkyl suit with rubber gloves should be worn at all times.

Caution AHF is extremely toxic and corrosive. Extreme care should be taken while handling. Make sure first aid treatment items in case of injury are readily available in close vicinity (eyewash station, water shower, calcium gluconate, etc.). The reaction must be carried out in a well-ventilated hood.

Experimental Procedure Take calculated amount of AHF (600 g, 30 mol) in a 4-L HDPE beaker. Chill down the contents by placing the container in an ice bath. AHF boils at ~20 °C. Keep the temperature below it to prevent rapid loss of acid. Add a Teflon-coated magnetic stir bar and set the bath and the vessel on a stirring plate. Keep nitrogen purge around the neck of the container to limit exposure to moisture to a minimum. Slowly add TEA with stirring to AHF. The reaction is "extremely exothermic." The additions have to be drop by drop initially. Measure the temperature continuously. As more TEA is added, the boiling point of the reaction mixture increases. For the first 30% of the reaction, keep the temperature below 30 °C to avoid significant loss of AHF. The additions thus end up being very slow in this phase. For the next 30% addition, the temperature might be allowed to go up to 90 °C. Essentially, control the temperature to prevent rapid loss of reaction products. The second addition is faster than the first phase. The TEA addition can be rapid during the last stage. It is desirable to let the temperature increase to no more than 90 °C toward the end. Lot of solid is formed toward the end. The final amount has to be added slowly to allow for solids to dissolve. After the calculated TEA (1010 g, 10 mol) has been added and all solids have dissolved, send sample to lab for analysis. Minor adjustments may need to be made at the end to account for loss of AHF. The resulting pale yellow liquid may be filtered (1594 g, 99% yield).

Characterization Data The product is characterized by perchloric acid titration. Assay is typically greater than 97.5%.

Application TEA–HF complex has a pH close to neutral and is easier to handle than anhydrous HF and is now increasingly being used in organic synthesis. It is versatile fluorinating agent for synthesis of acid fluorides and alkyl fluorides [27–30]. It is an excellent selective reagent for cleavage of O-silyl groups in carbohydrates, nucleotides, and cyanohydrins [31–34]. Fluorination reactions using TEA–HF complex has been reviewed by Haufe [35].

9.6 PREPARATION OF LITHIUM TETRAFLUOROBORATE ($LiBF_4$)

Apparatus Five-gallon plastic jerry can with septum, Teflon-coated magnetic stirrer, and two nozzles at the top.

Chemicals Lithium fluoride, hydrofluoric acid, BF_3 gas.

Attention! All personal protective gear such as safety glasses, full face acid respirator, and full body alkyl suit with rubber gloves should be worn at all times.

Caution AHF and BF_3 gas are extremely toxic and corrosive. Extreme care should be taken while handling. Make sure first aid treatment items in case of injury are readily available in close vicinity (eyewash station, water shower, calcium gluconate, etc.). The reaction must be carried out in a well-ventilated hood.

Experimental Procedure High-purity lithium fluoride (LiF; 1040 g, 40 mol) in a very fine powdered form (−60 mesh) is added to a clean 5-gallon plastic square jerry can. AHF (2.5 gallons) in liquid form is added to the container very slowly. The resulting mixture is a clear solution that is transferred into another clean 5-gallon plastic square jerry can, equipped with a Teflon-coated magnetic stirrer, a septum with two nozzles at the top for BF_3 gas inlet and BF_3/HF vent. BF_3 gas is bubbled through this stirred solution to make $LiBF_4$ salt that precipitates out. Care must be taken to prevent plugging of the nozzle inlet inside the solution. The reaction is over when no further precipitation is observed. The salt is allowed to settle and the clear liquid decanted carefully. The wet salt is loaded in Teflon trays in tray oven at 55 °C for 2 days. Sufficient flow of nitrogen is maintained through the ovens to dry the salt and prevent hydrolysis due to moisture. $LiBF_4$ (2.25 kg, 75%) is yielded. The salt can also be dried in a vacuum dryer for very low moisture content, especially if meant for use in lithium ion battery electrolyte.

Characterization Data The product is characterized by elemental analysis. Boron content is typically from 11.2% to 11.9% and fluoride content from 78.6% to 83.5%. Boron is analyzed by ICP and fluoride content is determined using Willard–Winter distillation method. Both the methods are described in detail in Appendix A. Moisture is typically less than 1% by Karl Fischer.

Application $LiBF_4$ is extensively used as an electrolyte in lithium ion battery [36]. It has been investigated for potential use as Lewis acid in the intramolecular Diels–Alder reaction [37]. In acetonitrile, it efficiently catalyzes the aminolysis of trimethylene oxide and 2-octyl oxetane under mild conditions to give the corresponding γ-amino alcohol in good yields [38]. It efficiently catalyzes an unusual cyclization of *o*-hydroxybenzaldimines with 2,3-dihydrofuran and 3,4-dihydro-2*H*-pyran at ambient temperature to afford a class of new pyrano- and furanobenzopyran derivatives in excellent yields and high diastereoselectivity [39].

9.7 PREPARATION OF NITRONIUM TETRAFLUOROBORATE (NO_2BF_4)

Apparatus Five-gallon plastic jerry can with septum, Teflon-coated magnetic stirrer, ice bath, two nozzles at the top, safety glasses, full face acid respirator, and full body alkyl suit with rubber gloves.

Chemicals Fuming nitric acid (HNO_3), dry ice, anhydrous hydrogen fluoride (AHF), methanol, and BF_3 gas.

Attention! All personal protective equipment should be worn at all times.

Caution! AHF, fuming nitric acid, and BF_3 are extremely toxic and corrosive. Extreme care should be taken while handling. Make sure first aid treatment items in case of injury are readily available in close vicinity (eyewash station, water shower, calcium gluconate, etc.). The reaction must be carried out in a well-ventilated hood.

Experimental Procedure Fuming nitric acid (4.1 kg, 59 mol) is added to a clean 5-gallon plastic square jerry can, equipped with a Teflon-coated magnetic stirrer, placed in an ice bath. The bath is filled with dry ice/methanol. AHF (2.4 kg, 120 mol) in liquid form is added to the cold HNO_3 very slowly. BF_3 gas is bubbled slowly through this cold HNO_3/HF solution kept in the −78 °C dry ice bath. NO_2BF_4 salt will precipitate out gradually. Care must be taken to prevent plugging of the inlet nozzle inside the solution by NO_2BF_4 precipitate. The reaction is over when no further precipitation takes place. This is indicated by significant evolution of BF_3. The salt is allowed to settle and the clear liquid decanted carefully. The wet salt is loaded in plastic trays in tray oven at room temperature for 5 days. Sufficient flow of nitrogen is maintained through the ovens to dry the salt and prevent hydrolysis due to moisture. About 7.2 kg, 54 mol (92%) of NO_2BF_4 is obtained.

Note of caution: a rise in temperature of solution, blockage of BF_3 bubbling tube, or irregular stirring may cause an explosion.

Characterization Data The product is characterized by elemental analysis. Boron content is typically from 7.8% to 8.5% and fluoride content from 54.9% to 59.5%. Boron is analyzed by ICP and fluoride content is determined using Willard–Winter distillation method. Both the methods are described in detail in Appendix A.

Application NO_2BF_4 is extremely active nitration agent for aromatics. It typically requires mild conditions giving high yields. It also readily nitrates aliphatics. It has been used for selective and controllable nitration of *meso*-tetraphenylporphyrin to nitroporphyrins. Its versatility is seen in reactions with amines, esters, and amides [40–42].

9.8 PREPARATION OF SILVER TETRAFLUOROBORATE ($AgBF_4$)

Apparatus Five-gallon plastic jerry can with septum, Teflon-coated magnetic stirrer, two nozzles at the top, and BF_3 gas delivery system.

Chemicals Silver oxide powder (Ag_2O), anhydrous hydrogen fluoride (AHF), and boron trifluoride (BF_3) gas.

Attention! All personal protective gear such as safety glasses, full face acid respirator, and full body alkyl suit with rubber gloves should be worn at all times.

Caution! AHF and BF_3 are extremely toxic, volatile, and corrosive. Extreme care should be taken while handling. Make sure first aid treatment items in case of injury are readily available in close vicinity (eyewash station, water shower, calcium gluconate, etc.). The reaction must be carried out in a well-ventilated hood.

Experimental Procedure Silver oxide (Ag_2O; 3.5 kg, 15 mol) in a very fine powdered form is added to a clean 5-gallon plastic square jerry can. AHF (3.5 gallons) in liquid form is added to the container very slowly. The reaction is extremely exothermic and liquid AHF is allowed to react slowly with the Ag_2O powder in a stagnant manner for 2 days. The resulting solution is clear with very little unreacted silver oxide impurities settled at the bottom. The silver bifluoride solution in AHF is decanted into another clean 5-gallon plastic square jerry can, equipped with a Teflon-coated magnetic stirrer, a septum with two nozzles at the top for BF_3 gas inlet and BF_3/HF vent. BF_3 gas is bubbled through this stirred solution to make $AgBF_4$ salt, which is less soluble in AHF and precipitates out. Care must be taken to prevent plugging of the nozzle inlet inside the solution. The reaction is over when no further precipitation is observed. The salt is allowed to settle and the clear liquid decanted carefully. The wet salt is washed with AHF three times (1.5 gallons each) to remove the water generated from Ag_2O and AHF reaction. The wet salt is then dried under nitrogen inside the jerry can. Once the salt is devoid of liquid AHF, it is dried in plastic trays at room temperature for 5 days. Sufficient flow of nitrogen is maintained through the ovens to dry the salt to prevent hydrolysis and moisture. $AgBF_4$ (5.5 kg, 93%) is yielded.

Characterization Data The product is characterized by elemental analysis involving wet chemistry methods. Silver content is typically between 54% and 56%, boron content between 5.4% and 5.7%, and fluoride content between 38.3% and 39.5%. Silver is analyzed by potassium thiocyanate titration. Boron is determined by ICP and fluoride content is determined using Willard–Winter distillation method. The ICP and Willard–Winter distillation methods are described in detail in Appendix A because they are used commonly for other subsequent chemicals described in this chapter.

Application $AgBF_4$ is a powerful catalyst for promoting nitration, acylation, and sulfonation. It can displace less activated halogen group to form respective fluorides. For example, it can convert $C_6H_5CHCl_2$ to $C_6H_5CHF_2$, $C_6H_5CCl_2CH_3$ to $C_6H_5CF_2CH_3$, and $C_6H_5CCl_3$ to $C_6H_5CF_3$ [43]. Its ability to form complexes with olefins and aromatics has helped development of separation processes for olefins from mixtures with paraffin's and acetylenes and separation of aromatics from cyclohexane [44, 45].

Silver tetrafluoroborate was identified as an excellent promoter for the activation of various glycosyl donors, including glycosyl halides, trichloroacetimidates, and thioimidates. Easy handling and no requirement for azeotropic dehydration before application makes $AgBF_4$ especially beneficial in comparison to the commonly used AgOTf [46].

9.9 PREPARATION OF HEXAFLUOROPHOSPHORIC ACID (HPF_6)

Apparatus Five-gallon Hastelloy C-276 pressure vessel with mechanical stirrer, cooling jacket.

Chemicals Phosphorous pentoxide (P_2O_5), AHF, and phosphoric acid.

Attention! All personal protective gear such as safety glasses, full face acid respirator, and full body alkyl suit with rubber gloves should be worn at all times.

Caution! Hydrofluoric acid is extremely toxic and corrosive. Extreme care should be taken while handling. Make sure first aid treatment items in case of injury are readily available in close vicinity (eyewash station, water shower, calcium gluconate, etc.). The reaction must be carried out in a well-ventilated hood.

Experimental Procedure Hexafluorophosphoric acid (HPF_6) is synthesized by reacting AHF with solid P_2O_5 inside a Hastelloy C-276 pressure vessel equipped with a mixer and external cooling jacket. Because the reaction is initially very violent and highly exothermic, temperatures of 300 °C routinely occur during the early stages of the reaction. The reaction contents have high vapor pressures in this temperature range. The rate of AHF addition is dependent on the reactor cooling available to keep the reaction temperature(s) and pressure(s) limits of the setup, typically 30 psig. The vessel is protected against overpressure by appropriate safety devices such as relief valves and/or rupture disks.

The reactor vessel is first charged with the required amount of P_2O_5 (4.970 kg, 35 mol). It is common practice to compensate with phosphoric acid (1.153 kg, 10 mol). Mixer and cooling water are started and then AHF addition is initiated at a controlled rate via a feed tube (9.60 kg, 480 mol). Check all safety devices, engineering controls, and process control equipment during the course of reaction. When the theoretical amount of HF has been added, the reaction mixture is allowed to cool to ambient temperature. Clear liquid containing 73–75% of fluorophosphoric acids (at least 60% hexafluorophosphoric acid, and smaller percentages of monofluoro-, difluoro-, and phosphoric acid) is obtained on filtration (15.57 kg, 99% yield).

Characterization Data The product is characterized by elemental analysis. Phosphorus content is typically from 15.5% to 15.9% and fluoride content from 57.0% to 58.6%. Phosphorus is analyzed by ICP and fluoride content is determined using Willard–Winter distillation method. Both the methods are described in detail in Appendix A.

Application Use of HPF_6 in the Schiemann reaction has been reported [47]. 1-Bromo-2-fluorobenzene can efficiently be prepared from *o*-bromoaniline and hexafluorophosphoric acid [48]. Single largest use of HPF_6 appears to be in synthesizing COX-2 inhibitors for pain and inflammatory use [49].

9.10 PREPARATION OF LITHIUM HEXAFLUOROPHOSPHATE ($LiPF_6$)

Apparatus Five-gallon Teflon narrow neck flask with septum, Teflon-coated magnetic stirrer, and two nozzles at the top.

Chemicals Lithium fluoride, hydrofluoric acid, phosphorous pentafluoride (PF_5) gas.

Attention! All personal protective gear such as safety glasses, full face acid respirator, and full body alkyl suit with rubber gloves should be worn at all times.

Caution! Hydrofluoric acid and PF_5 gas are extremely toxic and corrosive. Extreme care should be taken while handling. Make sure first aid treatment items in case of injury are readily available in close vicinity (eyewash station, water shower, calcium gluconate, etc.). The reaction must be carried out in a well-ventilated hood.

Experimental Procedure Take calculated amount of LiF (260 g, 10 mol) and dissolve it into AHF (2.50 kg). The solution is filtered and PF_5 (1.26 kg, 10 mol) is passed through the solution. Note that PF_5 is a high-pressure gas and all the necessary setup involving pressure regulator, check valves, shut off valves need to be carefully selected. Transition is made from stainless steel fitting from the PF_5 cylinder to Teflon feed tube going into the LiF solution in AHF. White precipitate is formed, which is $LiPF_6$. The liquid is decanted off and solid is dried at room temperature under nitrogen or vacuum dried (1.370 kg, 90% yield).

Characterization Data The product is characterized by elemental analysis. Phosphorus content is typically from 19.8% to 21.0% and fluoride content from 72.8% to 77.3%. Phosphorus is analyzed by ICP and fluoride content is determined using Willard–Winter distillation method. Both the methods are described in detail in Appendix A.

Application Solutions of lithium hexafluorophosphates in organic solvents such as propylene carbonate/dimethoxyethane are used as an electrolyte in lithium batteries [50,51]. $LiPF_6$ also catalyzes the tetrahydropyranylation of tertiary alcohols [52].

9.11 PREPARATION OF POTASSIUM HEXAFLUOROPHOSPHATE (KPF_6)

Apparatus Five-gallon 316 Stainless Steel pressure vessel with mechanical stirrer, and cooling jacket.

Chemicals Hexafluorophosphoric acid (HPF_6), potassium hydroxide (KOH).

Attention! All personal protective gear such as safety glasses, full face acid respirator, and full body alkyl suit with rubber gloves should be worn at all times.

Caution HPF_6 and KOH are extremely toxic and corrosive. Extreme care should be taken while handling. Make sure first aid treatment items in case of injury are readily available in close vicinity (eyewash station, water shower, calcium gluconate, etc.). The reaction must be carried out in a well-ventilated hood.

Experimental Procedure Potassium hexafluorophosphate (KPF_6) is synthesized by reacting HPF_6 with KOH inside a 316 Stainless Steel pressure vessel equipped with a mixer and external cooling jacket. The vessel is protected against overpressure by appropriate safety devices such as relief valves and/or rupture disks.

Check all safety devices, engineering controls, and process control equipment during the course of reaction. Add calculated amount of both KOH 45% (1.245 kg, 10 mol) solution and KOH Flake 90% (622 g, 10 mol). Mix until uniform. Add HPF_6, ~74% fluorophosphoric acid content (1.60 kg, 6.6 mol minimum as HPF_6), through a dip tube in the vessel, start cooling water on jacket. Rate of addition must be controlled. After calculated amount of HPF_6 is added, the reaction mixture is stirred for 1 h. The reaction slurry is filtered and washed with water. The wet cake is dried at 100 °C for 2 days. After pulverizing, a white crystalline powder is obtained (1.0 kg, 82% yield on HPF_6 basis).

Characterization Data The product is characterized by elemental analysis. Phosphorus content is typically from 16.6% to 17.1% and fluoride content from 61.0% to 62.9%. Phosphorus is analyzed by ICP and fluoride content is determined using Willard–Winter distillation method. Both the methods are described in detail in Appendix A.

Application KPF_6 has been used as associated electrolyte [53] and useful salt in rechargeable battery [54].

It has also been used as antistatic additives in organic polymer compositions [55]. It is also used in metathesis reactions in preparation of PF_6^- room temperature ionic liquids [56, 57].

9.12 PREPARATION OF ALUMINUM TRIFLATE ($Al(CF_3SO_3)_3$)

Apparatus Round-bottom, 4-L flask with reflux condenser with Teflon-coated mechanical stirrer.

Chemicals Aluminum metal powder, triflic acid (CF_3SO_3H).

Attention! All personal protective gear such as safety glasses, full face acid respirator, and full body alkyl suit with rubber gloves should be worn at all times.

Caution! Triflic acid is extremely toxic and corrosive. Extreme care should be taken while handling. Make sure first aid treatment items in case of injury are readily available in close vicinity (eyewash station, water shower, calcium gluconate, etc.). The reaction must be carried out in a well-ventilated hood.

Experimental Procedure Take 880 g of demineralized water in a 4-L glass beaker and carefully add triflic acid (1.50 kg, 10 mol) to it. The reaction is exothermic. Please note that the acid should be added to the water. The flask is equipped with a reflux condenser and placed in a heating mantle with constant stirring with an overhead mechanical stirrer. The contents of the flask are maintained at 150 °C. Water has to be added periodically to maintain the level of liquid, which is lost due to heating. Aluminum powder (−100 mesh; 81 g,

3 mol) in a very fine powdered form is added slowly to the aqueous triflic acid solution. The reaction is slow giving off hydrogen bubbles and the resulting aqueous aluminum triflate solution. The heat has to be controlled to prevent a boilover due to massive liberation of hydrogen gas. After stirring for 3–4 h, the solution is decanted followed by further filtration and evaporated at 150 °C for 2 days to give white powder that is grinded and dried under nitrogen (1.337 kg, 94% yield).

Characterization Data The product is characterized by elemental analysis. Aluminum content is typically from 5.4% to 6.0%. Aluminum is analyzed by ICP. The method is described in detail in Appendix A.

Application Aluminum triflate readily replaces Bronsted acid cocatalyst in the palladium-catalyzed methoxycarbonylation reaction of styrene and 1-pentene, producing catalysts that are more stable and more active than those using traditional acids [58].

Aluminum triflate has also been found to be an efficient catalyst for various transformations, such as

a. epoxide ring openings [58,59,60];
b. thiol protections [61];
c. silylation of alcohols and phenols [62];
d. acylations of alcohols, phenols, and thiophenols [63];
e. tetrahydropyranylation of alcohols under solvent-free conditions [64];
f. esterification or transesterification catalyst [65]; and
g. polymerization of biodegradable composites of poly(lactic acid) with cellulose fibers [66].

9.13 PREPARATION OF COPPER TRIFLATE ($Cu(CF_3SO_3)_2$)

Apparatus Four-liter glass beaker with Teflon-coated magnetic stirrer.

Chemicals Copper carbonate powder, triflic acid (CF_3SO_3H).

Attention! All personal protective gear such as safety glasses, full face acid respirator, and full body alkyl suit with rubber gloves should be worn at all times.

Caution! Triflic acid is extremely toxic and corrosive. Extreme care should be taken while handling. Make sure first aid treatment items in case of injury are readily available in close vicinity (eyewash station, water shower, calcium gluconate, etc.). The reaction must be carried out in a well-ventilated hood.

Experimental Procedure Take 600 g of demineralized water in a 4-L glass beaker and carefully add triflic acid (1.50 kg, 10 mol) to it. The reaction is exothermic. Please note that the acid should be added to the water. The beaker is set on a hot plate with constant stirring with a Teflon-coated magnetic stirrer maintaining at 100 °C. Copper carbonate (568 g, 4.6 mol) in a very fine powdered form is added slowly to the aqueous triflic acid solution. The reaction is instantaneous giving aqueous copper triflate solution. After stirring for 2 h, the solution is decanted and evaporated at 180 °C for 2 days to give white powder that is grinded and dried under nitrogen (1.596 kg, 96% yield).

Characterization Data The product is characterized by elemental analysis. Copper content is typically from 17.3% to 17.8%. Copper is analyzed by ICP. The method is described in detail in Appendix A.

Application Synthetic use of copper triflate as a new reagent for mild dehydration of alcohols has been discussed by Laali et al. [67]. Copper triflate has been used in ring opening of epoxides and aziridines [68], asymmetric cycloadditions of aldol condensations [69], asymmetric Michael addition of enamides [70], and asymmetric O–H insertion reaction [71]. Behr and coworkers reported Diels–Alder reaction of conjugated linoleic acid ethyl ester with different quinines and with a variety of α/β-unsaturated aldehydes and ketones uses copper triflate as catalyst at room temperature with good yields. For the first time in oleochemistry, it was possible to prepare Diels–Alder cycloaddition with catalyst concentration of 10 mol % instead of stoichiometric amounts of Lewis acid [72].

9.14 PREPARATION OF LANTHANUM TRIFLATE ($La(CF_3SO_3)_3$)

Apparatus Four-liter glass beaker with Teflon-coated magnetic stirrer.

Chemicals Lanthanum carbonate powder, triflic acid (CF_3SO_3H).

Attention! All personal protective gear such as safety glasses, full face acid respirator, and full body alkyl suit with rubber gloves should be worn at all times.

Caution! Triflic acid is extremely toxic and corrosive. Extreme care should be taken while handling. Make sure first aid treatment items in case of injury are readily available in close vicinity (eyewash station, water shower, calcium gluconate, etc.). The reaction must be carried out in a well-ventilated hood.

Experimental Procedure Take 550 g of demineralized water in a 4-L glass beaker and carefully add triflic acid (1.50 kg, 10 mol) to it. The reaction is exothermic. Please note that the acid should be added to the water. The beaker is set on a hot plate with constant stirring with a Teflon-coated magnetic stirrer maintaining 100 °C. Lanthanum carbonate (733 g, 1.6 mol) in a very fine powdered form is added slowly to the aqueous triflic acid solution. The reaction is instantaneous giving aqueous lanthanum triflate solution. Add more water (2.00 kg) as reaction progresses for uniform mixing. After stirring for 4 h, the solution is decanted and evaporated at 140 °C for 3 days to give white powder that is grinded and dried under nitrogen at 140 °C overnight (1.782 kg, 95% yield).

Characterization Data The product is characterized by elemental analysis. Lanthanum content is typically from 23.0% to 24.4%. Lanthanum is analyzed by ICP. The method is described in detail in Appendix A.

Application Lanthanum triflate from lanthanide triflate family is one of the most promising green chemistry catalysts. Kobayashi and Manabe [73] described their catalytic effect in water and established the range of applications of lanthanum triflate catalytic systems in a number of important organic synthetic reactions. Lanthanide triflate as recyclable catalysts for atom economic aromatic nitration has been reported [74]. Lanthanum triflate has been found to be effective catalyst for the Baylis–Hillman reaction, using 1,4-diazabicyclo(2,2,2) octane, increasing the rate up to 40-fold [75]. It catalyzes Michael additions to methyl vinyl ketone in water [76]. Lanthanum triflate in combination of benzoic acid catalyzes the allylation at low catalyst loading without the need to activate the catalyst in advance [77, 78]. *p*-Methoxybenzyl (PMB) ethers of alcohols have been prepared in high yield and short reaction times using the trichloroacetimidate of PMB alcohol and lanthanum triflate. The mild conditions allow protection of acid-sensitive alcohols [79].

9.15 PREPARATION OF LITHIUM TRIFLATE ($LiCF_3SO_3$)

Apparatus Four-liter glass beaker with Teflon-coated magnetic stirrer.

Chemicals Lithium carbonate powder, triflic acid (CF_3SO_3H).

Attention! All personal protective gear such as safety glasses, full face acid respirator, and full body alkyl suit with rubber gloves should be worn at all times.

Caution! Triflic acid is extremely toxic and corrosive. Extreme care should be taken while handling. Make sure first aid treatment items in case of injury are readily available in close vicinity (eyewash station, water shower, calcium gluconate, etc.). The reaction must be carried out in a well-ventilated hood.

Experimental Procedure Take 300 g of demineralized water in a 4-L glass beaker and carefully add triflic acid (1.50 kg, 10 mol) to it. The reaction is exothermic. Please note that the acid should be added to the water. The beaker is set on a hot plate with constant stirring with a Teflon-coated magnetic stirrer maintaining 100 °C. Lithium carbonate (375 g, 5 mol) in a very fine powdered form is added slowly to the aqueous triflic acid solution. The reaction is instantaneous giving aqueous lithium triflate solution. After stirring for 0.5 h, the solution is decanted and evaporated at 120 °C for 1 day to obtain white powder that is grinded and dried under nitrogen at 120 °C for 1 day (1.376 kg, 96% yield).

Characterization Data The product is characterized by elemental analysis. Lithium content is typically from 4.3% to 4.6%. Lithium is analyzed by ICP. The method is described in detail in Appendix A.

Application Lithium triflate is used in some lithium ion batteries as a component of the electrolyte [80–82]. It fulfills the great demand of mild and neutral catalyst in various organic reactions. Various types of carbonyl compounds have been synthesized by Firouzabadi et al. [83] and chemoselectively converted to their corresponding dithioacetals in the presence of catalytic amounts of lithium triflate under solvent-free conditions. Lithium triflate has also been chemoselectively used for aminolysis of oxiranes and glycosylation of nucleophiles under mild and neutral conditions [84, 85]. A variety of alcohols and aldehydes acetylated and diacetylated in presence of recyclable catalyst lithium triflate at room temperature produce corresponding esters and 1,1-diacetates, respectively, in excellent yield [86].

9.16 PREPARATION OF SILVER TRIFLATE ($AgCF_3SO_3$)

Apparatus Four-liter glass beaker with Teflon-coated magnetic stirrer.

Chemicals Silver oxide powder, triflic acid (CF_3SO_3H).

Attention! All personal protective gear such as safety glasses, full face acid respirator, and full body alkyl suit with rubber gloves should be worn at all times.

Caution! Triflic acid is extremely toxic and corrosive. Extreme care should be taken while handling. Make sure first aid treatment items in case of injury are readily available in close vicinity (eyewash station, water shower, calcium gluconate, etc.). The reaction must be carried out in a well-ventilated hood.

Experimental Procedure Take 550 g of demineralized water in a 4-L glass beaker and carefully add triflic acid (1.50 kg, 10 mol) to it. The reaction is exothermic. Note that the acid should be added to the water. The beaker is set on a hot plate with constant stirring with a Teflon-coated magnetic stirrer maintaining 100 °C temperature. Silver oxide (1.070 kg, 4.6 mol) in a fine powdered form is added slowly to the aqueous triflic acid solution. The reaction is instantaneous giving aqueous silver triflate solution. After stirring for 2 h, the solution is decanted and evaporated at 120 °C for 2 days to give white powder that is grinded and dried under nitrogen (2.320 kg, 98% yield).

Characterization Data The product is characterized by analysis of silver by potassium thiocyanate titration. Silver content is typically from 41.1% to 42.8%.

Application Silver triflate is widely used as catalyst and in preparation of derivatives of triflic acid, both covalent esters and ionic salts [87,88]. It is a very useful catalyst in a number of organic synthetic reactions. New synthetic methods for olefins from secondary phosphates and thiophosphates in better yield by using small amount of silver triflate have been reported [89]. Burk et al. [90] developed a mild procedure for etherification of alcohols with primary alkyl halides in the presence of silver triflate with good yields. Koch carbonylation using silver triflate has been reported [91]. Iodine chloride and silver triflate provide a convenient promoter system for O-glycoside synthesis. This system is advantageous over conventional promoter system for thioglycoside activation because of convenient handling of the reagents and absence of by-products related to

N-succinimide [92]. An interesting use of silver triflate as a poly(ethylene oxide) complex in galvanic sensor for monitoring nitrogen oxide has also been reported in literature [93].

9.17 PREPARATION OF MAGIC ACID ($FSO_3H{\cdot}SbF_5$)

Apparatus Four-liter Teflon bottle with Teflon-coated magnetic stirrer.

Chemicals Fluorosulfonic acid (FSO_3H), antimony pentafluoride (SbF_5).

Attention! All personal protective gear such as safety glasses, full face acid respirator, and full body alkyl suit with rubber gloves should be worn at all times.

Caution! Fluorosulfonic acid and antimony pentafluoride are extremely toxic and corrosive. Extreme care should be taken while handling. Make sure first aid treatment items in case of injury are readily available in close vicinity (eyewash station, water shower, calcium gluconate, etc.). The reaction must be carried out in a well-ventilated hood.

Experimental Procedure Add fluorosulfonic acid (1.00 kg, 10 mol) to a narrow neck Teflon bottle placed in a dry ice/methanol bath in a dry glove box. The temperature need to be below −20 °C. Add antimony pentafluoride (2.170 kg, 10 mol) very slowly with constant stirring. The reaction is exothermic. Maintain the low temperature of the reaction mixture. Allow the final mixture to stir for 30 min and then allow warming to room temperature before analysis and packaging. Clear liquid is obtained as product (3.930 kg, 99% yield).

Characterization Data The product is characterized by elemental analysis. Antimony content is typically from 37.5% to 39.5% and fluoride content from 35.1% to 36.9%. Antimony is analyzed by ICP and fluoride content is determined using Willard–Winter distillation method. Both the methods are described in detail in Appendix A.

Application Magic acid has been used to stabilize carbocations and hypercoordinated carbonium ions in liquid media. Magic acid and other superacids are also used to catalyze isomerization of saturated hydrocarbons, and have been shown to protonate even weak bases, including methane, xenon, halogens, and molecular hydrogen [94]. Magic acid catalyzes cleavage–rearrangement reactions of tertiary hydroperoxides and tertiary alcohols [95] and electrophilic hydroxylation of aromatic compounds with hydrogen peroxide [96].

9.18 PREPARATION OF HEXAFLUOROANTIMONIC ACID ($HSbF_6$)

Apparatus Two-liter Teflon bottle with Teflon-coated stirrer.

Chemicals Hydrofluoric acid, antimony pentafluoride (SbF_5).

Attention! All personal protective gear such as safety glasses, full face acid respirator, and full body alkyl suit with rubber gloves should be worn at all times.

Caution! Hydrofluoric acid and antimony pentafluoride are extremely toxic and corrosive. Extreme care should be taken while handling. Make sure first aid treatment items in case of injury are readily available in close vicinity (eyewash station, water shower, calcium gluconate, etc.). The reaction must be carried out in a well-ventilated hood.

Experimental Procedure Add antimony pentafluoride (2.170 kg, 10 mol) to a narrow neck Teflon beaker placed in a dry ice/methanol bath in a dry glove box. The temperature does not need to be below 10 °C. Add AHF (200 g, 10 mol) very slowly with constant stirring to the SbF_5. The reaction is exothermic. Maintain the low temperature of the reaction mixture. Allow the final mixture to stir for 30 min and then allow warming to room temperature before analysis and packaging. Clear liquid is obtained as product (2.350 kg, 99% yield).

Characterization Data The product is characterized by elemental analysis. Antimony content is typically from 50.2% to 52.8% and fluoride content from 46.9% to 49.3%. Antimony is analyzed by ICP and fluoride content is determined using Willard–Winter distillation method. Both the methods are described in detail in Appendix A.

Application Fluoroantimonic acid protonates nearly all organic compounds. Bickel and coworkers [97, 98] showed that HF-SbF_5 removes hydrogen from isobutane and methane and neopentane. HF-SbF_5 is an excellent medium for the preparation of alkylidene oxonium salts [99].

APPENDIX A

Analytical Procedure for ICP

This procedure describes the methods used for the operation of the ICP (IRIS Intrepid II XSP, Thermo Electron Corporation). The ICP setup involves a suitable ventilation system, an argon cylinder, and a computer for data collection. The vent must be turned on with appropriate argon flow. The ICP torch has to be lit and system purged for appropriate time. A 10-ppm standard will be needed for each element and can be prepared by pipetting 1 mL of each 1000-ppm standard into a 100-mL volumetric flask and diluting to the mark with 2% nitric acid. Check the solubility of the sample from reference books such as the *CRC Handbook of Chemistry and Physics*. The theoretical ppm concentration of the sample solution should be close to 10 ppm. The sample weight and dilution factor can be manipulated to get the appropriate concentration of sample solution. Most samples can be dissolved using a 2% solution of OmniTrace nitric acid or other high-purity nitric acid. Some samples may require other methods to dissolve. Concentrated nitric acid, 50% hydrofluoric acid, perchloric acid, heat, and sonic bath can all be used in any combination to dissolve samples. Once it has been determined how to dissolve the sample, weigh out the necessary amount. Dissolve and dilute to 100 mL in a volumetric flask using 2% nitric acid. Using the proper dilution factor, extract the necessary number of milliliters from the volumetric flask and transfer them to a second 100 mL volumetric flask. Dilute to 100 mL mark with 2% nitric acid. This is the solution to be analyzed with the ICP. Appropriate methods are selected from the software menu and standards and blanks are ran. The ICP is now standardized and ready to run samples. Run the unknown sample and collect the scan. The peak height is compared against the standard peaks and the concentration of the unknown solution is calculated using the dilution factor and calibration constant. The samples are run in duplicates.

Analytical Procedure for Fluorine Determination by Willard–Winter Distillation Method

This Willard–Winter distillation is used in determination of fluorine concentration in inorganic as well as some organic compounds. It is not generally applicable to compounds containing carbon–fluorine bonds. Some organic compounds may require sodium metal and/or sodium carbonate and potassium carbonate fusion. Equipment used is shown in Figure 9.1. H_2SO_4 (200 mL, 50%) is added to the fluorine glass still containing 20–30 glass beads. A sample is weighed out (45–50 mg of F^- ions) and transferred into the sulfuric solution. A small amount of demineralized water is used to rinse the sample into the still, which is then closed tightly. The jacket temperature

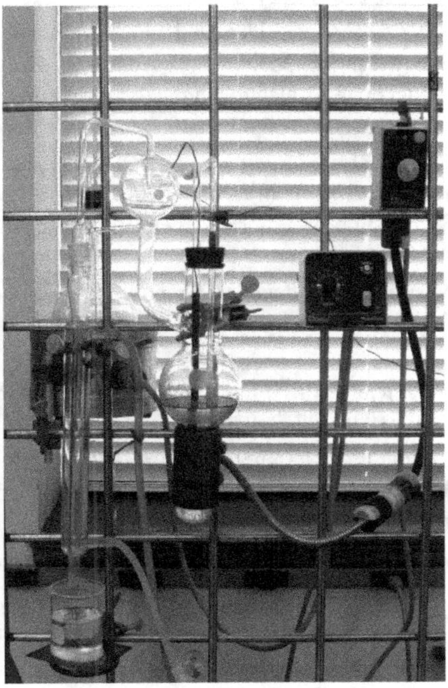

FIGURE 9.1 Setup for Willard–Winter distillation method. (For a color version of the figure, please see color plates.)

is set at high, boiling flask at 100 °C, and then filled with boiling water. The water to cool the condenser jacket is turned on and the distillate is received in a 250-mL volumetric flask. The temperature of the still is brought to 145 °C and temperature control set such that the temperature is maintained from 148–150 °C. Distillate (250 mL) is collected in a volumetric flask. An additional 150 mL of distillate for the tail is collected in a 250-mL beaker with a stirrer bar. 50 mL of the distillate is added to three different beakers each containing 50 mL of demineralized water and stirred. Six drops of sodium alizarin sulfonate indicator is added followed by a few drops of 2% NaOH until the solution turns pink. A few drops of HCl are added to turn it yellow. When the solution takes one drop to turn pink, carefully bring back to yellow. One milliliter of buffer is added and titrated with 0.001 N $Th(NO_3)_4$ to a faint pink endpoint. The titration is repeated for the other two beakers and the last 150-mL distillate fraction also. Average of first three titrations in conjunction with the last is used to compute the % of fluorine.

REFERENCES

1. Meshri DT, Meshri SD, Sridhar CN, Adams RL, Mathur NC. Exciting role of inorganic fluorides in 21st century. 15th European Symposium on Fluorine Chemistry; 2007 July 15–20; Prague, Czech Republic.
2. Meshri DT, Meshri S, Adams R, Jain R, Mathur NC, Pinnapareddy D. Inorganic fluorides: opportunities and challenges in energy sector. ACS 19th International Symposium on Fluorine Chemistry; 2009 Aug 23–28; Jackson Hole, WY.
3. Meshri DT, Shembel E, Meshri S, Mathur NC, Adams R, Bhagat S, Pinnapareddy D, Jain. R. Fluorine: the backbone of electrochemical conversion and storage. Fluorinated Materials & Energy Conversion Workshop; 2010 Apr 12–13; Bordeaux, France.
4. Meshri DT, Mathur NC, Meshri S, Adams R, Jain R, Pinnapareddy D, Shembel E. Applications of fluorine based materials in electrochemical conversion and storage devices. The 16th European Symposium on Fluorine Chemistry (16th ESFC); 2010 July 18–23; Ljubljana, Slovenia.
5. Finmans P, Hoell D, Vossler H-J, Heppner M, Green M, Roggenbuck R, inventors. Process for the production of metal salts of trifluoromethane sulphonic acid. EP 1 687 264 B1. 2010.
6. Kobayashi S, Sugiura M, Kitagawa H, Lam WWL. Rare-earth metal triflates in organic synthesis. *Chem Rev* 2002;102:2227–2302.
7. Bonrath W, Dittel C, Giraudi L, Netscher T, Pabst T. Rare earth triflate catalysts in the synthesis of vitamin E and its derivatives. *Catal Today* 2007;121:65–70.
8. Olah GA, Prakash GKS, Sommer J. *Superacids*. New York: Wiley; 1985.
9. Meshri D. *Advanced Inorganic Fluorides: Synthesis, Characterization and Applications*. Amsterdam: Elsevier Science S.A.; 2000. p 661–682.
10. Meshri D. *Chemistry of Organic Fluorine Compounds II*. ACS Monograph 187; American Chemical Society: Washington, DC, 1995; p 23–40.
11. Li H, Richter G, Maier J. Reversible formation and decomposition of LiF clusters using transition metal fluorides as precursors and their application in rechargeable Li batteries. *Adv Mater* 2003;15:736.

12. H. Li, Balaya P, Maier J. Li-storage via heterogeneous reaction in selected binary metal fluorides and oxides. *J Electrochem Soc* 2004;151:A1878.
13. Cui Y, Xue M, Zhou Y, Peng S, Wang X, Fu Z. The investigation on electrochemical reaction mechanism of CuF_2 thin film with lithium. *Electrochim Acta* 2011;56:2328.
14. Asprey LB. The preparation of very pure fluorine gas. *J Fluorine Chem* 1976;7:359–361.
15. Plevey RG, Rendell RW, Steward MP. Fluorination with complex metal fluorides Part III. The fluorination of benzene over potassium hexafluoronickelate(IV). *J Fluorine Chem* 1973/74;3:267–273.
16. Darwish AD, Abdul-Sada AK, Avent AG, Street JM, Taylor R. Isolation and characterisation of both the first fluoroxyfluorofullerene $C_{60}F_{17}OF$ and oxahomofluorofullerenol C60F17O.OH. *J Fluorine Chem* 2003;121:185–192.
17. Nakajima T, Koh M, Gupta V, Zemva B, Lutar K. Electrochemical behavior of graphite highly fluorinated by high oxidation state complex fluorides and elemental fluorine. *Electrochim Acta* 2000;45:1655–1661.
18. Jache AW, Cady GH. Solubility of fluorides of metals in liquid hydrogen fluoride. *J Phys Chem* 1952;56:1106–1109.
19. Efner HF, Kirk R, Noftle RE, Uhrig M. Preparation of some new pentafluorosulfur fluoroethanes; the vibrational spectrum of 1-hydro-1-pentafluorosulfur-F-ethane and its deuterium isotopomer. *Polyhedron* 1982;1:723–734.
20. Ryu I, Suzuki H, Ogawa A, Kambe M. Sonoda N. Allylative ring opening of siloxycyclopropanes by silver fluoride and allylic chlorides affording δ, ε-unsaturated ketones. *Tetrahedron Lett* 1988;29:6137–6140.
21. Meshri DT. *Kirk–Othmer Encyclopedia of Chemical Technology*. Volume 11, New York: John Wiley & Sons; 2001. p 422–426.
22. Fisher RG, Jr, Zweig A. US patent 4,394,527. 1983; Chem Abstr 99, 175351m, 1983.
23. Plevey RG, Steward MP, Tatlow JC. Fluorination with complex metal fluorides Part II. The fluorination of benzene, tetrahydrofuran and cyclohexanone with silver difluoride and potassium tetrafluoroargentate(III). *J Fluorine Chem* 1973/74;3:259–266.
24. Rausch DA, Davis RA, Osborne DW. The addition of fluorine to halogenated olefins by means of metal fluorides. *J Org Chem* 1963;28:494–497.
25. Goryunkov AA, Markov VY, Boltalina OV, Zemva B, Abdul-Sada AK, Taylor, R. Reaction of silver(I) and (II) fluorides with C_{60}: thermodynamic control over fluorination level. *J Fluorine Chem* 2001;112:191–196.
26. Levec J, Slivnik J, Zemva B. On the reaction between xenon and fluorine. *J Inorg Nuclear Chem* 1974;36:997.
27. Franz R. Ueber trishydrofluoride tertiaerer amine und ihren einsatz als fluorierungsmittel. *J Fluorine Chem* 1980;15:423–434.
28. Haufe G, Alvernhe G, Laurent A. Different selectivities in bromofluorination reactions using N-bromosuccinimide/triethylamine tris hydrofluoride or Olah's reagent. *Tetrahedron Lett* 1986;27:4449–4452.
29. Alvernhe G, Laurent A, Haufe G. Triethylamine tris-hydrofluoride [$(C_2H_5)_3N{\cdot}3HF$]: a highly versatile source of fluoride ion for the halofluorination of alkenes synthesis. *Synthesis* 1987;6:562.
30. Haufe G, Alvernhe G, Anker D, Laurent A, Saluzzo C. Formal addition of methanesulfenyl fluoride to unsaturated substrates. *Tetrahedron Lett* 1988;29:2311–2314.

31. Picq D, Anker D. Utilisation du complexe triéthylamine·acide fluorhydrique pour la synthèse de désoxyfluoropyranosides et la scission de groupes silylés substitués. *Carbohydr Res* 1987;166:309.

32. Fischer K, Huenig S. Trimethylsilyl cyanide-a reagent for umpolung. XIII. Furfural, a convenient precursor for intramolecular Diels-Alder reactions via umpolung with trimethylsilyl cyanide. *J Org Chem* 1987;52:564–569.

33. Pirrung MC, Shuey SW, Lever DC, Fallon L. A convenient procedure for the deprotection of silylated nucleosides and nucleotides using triethylamine trihydrofluoride. *Bioorg Med Chem Lett* 1994;4:1345.

34. Myers AG, Gin DY, Rogers DH. Synthetic studies of the tunicamycin antibiotics. Preparation of (+)-tunicaminyluracil, (+)-tunicamycin-V, and 5′-epi-tunicamycin-V. *J Am Chem Soc* 1994;116:4697–4718.

35. Haufe G. Triethylamine Trishydrofluoride in synthesis. *J Prakt Chem/Chem Ztg* 1996;338:99.

36. Zhang SS, Xu K, Jow TR. Study of $LiBF_4$ as an electrolyte salt for a Li-ion battery. *J Electrochem Soc* 2002;149:A586–A590.

37. Smith DA, Houk KN. $LiBF_4$: a mild lewis acid for intramolecular Diels-Alder reactions. *Tetrahedron Lett* 1991;32:1549–1552.

38. Chini M, Crotti P, Favero L, Macchia M. Mild $LiBF_4$-promoted aminolysis of oxetanes. *Tetrahedron Lett* 1994;35:761–764.

39. Yadav JS, Reddy BV, Madhuri CH, Sabitha G, Jagannadh B, Kumar SK, Kunwar AC. $LiBF_4$-catalyzed formation of fused pyrano- and furanobenzopyrans. *Tetrahedron Lett* 2001;42:6381–6384.

40. Olah GA, Narang SC, Olah JA, Lammertsma K. Recent aspects of nitration: new preparative methods and mechanistic studies (a review). *Proc Natl Acad Sci USA* 1982;79:4487.

41. Smith NW, Dzyuba SV. Efficient nitration of meso-tetraphenylporphyrin with nitronium tetrafluoroborate. *Arkivoc* 2010;7:10–18.

42. Olsen RE, Fish DW, Hamel EE. *Advanced Propellant Chemistry*, Sacramento, CA: Aerojet-General Corp.; 1966. Chapter 6 Synthetic Applications of Nitronium Tetrafluoroborate p 48–54.

43. Olah GA, Pavlath S, Kuhn S. *Chem Ind (London, U. K.)* 1957;50.

44. Baker BB, Knack DF, inventors; assignors to E. I. du Pont de Nemours and Company. Process for the separation of hydrocarbons. US patent 3,007,981. 1961 Nov 7.

45. Bloodworth AJ, Bowyer KJ, Mitchell JC. A mild, convenient, halogen-exchange route to gem-difluorides and trifluorides. *Tetrahedron Lett* 1987;28:5347–5350.

46. Kaeothip S, Pornsuriyasak P, Demchenko AV. Silver(I) tetrafluoroborate as a potent promoter for chemical glycosylation. *Tetrahedron Lett* 2008;49:1542–1545.

47. Rutherford KG, Redmond W, Rigamonti J. The use of hexafluorophosphoric acid in the schiemann reaction. *J Org Chem* 1961;26:5149–5152.

48. Rutherford KG, Redmond W. *Org Synth, Coll* 1973;5:133.

49. Corley EG, Davis IW, Larsen RD, Pye PJ, Rossen K, inventors; Merck & Co., Inc., assignee. Process for synthesizing COX-2 inhibitors. US patent 6,040,319. 2000 Mar 21.

50. Zhang SS, Jow TR, Amine K, Henriksen GL. $LiPF_6$–EC–EMC electrolyte for Li-ion battery. *J Power Sources* 2002;107:18–23.

51. Goodenough JB, Kim Y. Challenges for rechargeable Li batteries. *Chem Mater* 2010;22:587–603.

52. Hamada N, Tsuneo S. Lithium hexafluorophosphate-catalyzed efficient tetrahydropyranylation of tertiary alcohols under mild reaction conditions. *Synlett* 2004;10:1802.

53. Robinson RA, Stokes JM, Stokes RH. Potassium hexafluorophosphate-an associated electrolyte. *J Phys Chem* 1961;65:542–546.

54. Lyall AE, Seiger HN, inventors; US Industrial Chemicals, assignee. Negative lithium electrode and electrochemical battery containing the same. US patent 3,508,967. 1970 Apr 28.

55. Spicher DR, inventor; Cellular Technology International, Inc. and Chemfoan International, Inc., assignee. Antistatic additive for organic polymer compositions. US patent 5,955,526. 1999 Sept 21.

56. Dupont J, Consorti CS, Suarez PAZ, deSouza RF. Preparation of 1-butyl-3-methyl imidazolium-based room temperature ionic liquids. *Org Synth, Coll* 2004;10:184; 2002;79:236.

57. Homsi F, Robin S, Rousseau G. Preparation of bis(2,4,6-Trimethylpyridine)iodine(I) hexafluorophosphate and bis(2,4,6-trimethylpyridine) bromine(I) hexafluorophosphate. *Org Synth, Coll* 2004;10:122; 2000;77:206.

58. Williams DBG, Lawton M. Aluminium triflate: a remarkable Lewis acid catalyst for the ring opening of epoxides by alcohols. *Org Biomol Chem* 2005;3:3269.

59. Terblans YM, Huyser JJ, Huyser M, Green MJ, Young DA, Sibiya MS. Aluminium triflate as catalyst for epoxide ring-opening and esterification reactions—Mechanistic aspects. *Can J Chem* 2005;83:854.

60. Williams DBG, Lawton M. Aluminium triflate: an efficient recyclable Lewis acid catalyst for the aminolysis of epoxides. *Tetrahedron Lett* 2006;47:6557–6560.

61. Firouzabadi H, Iranpoor N, Kohmareh G. Highly efficient and chemoselective thioacetalization of carbonyl compounds catalyzed with aluminum trifluromethanesulfonate [$Al(OTf)_3$]. *Synth Commun* 2003;33:167.

62. Firouzabadi H, Iranpoor N, Sobhani S, Ghassamipour S. Aluminium triflate [$Al(OTf)_3$] as a recyclable catalyst for the conversion of α-hydroxyphosphonates, alcohols and phenols to their corresponding O-silylated products with hexamethyldisilazane (HMDS). *Synthesis* 2005;4: 595.

63. Kamal A, Khan MNA, Reddy KS, Srikant YVV, Krishnaji T. $Al(OTf)_3$ as a highly efficient catalyst for the rapid acetylation of alcohols, phenols and thiophenols under solvent-free conditions. *Tetrahedron Lett* 2007;48:3813–3818.

64. Kamal A, Khan MNA, Srikant YVV, Reddy KS. $Al(OTf)_3$—a highly efficient catalyst for the tetrahydropyranylation of alcohols under solvent-free conditions. *Can J Chem* 2008;86:1099.

65. Finmans P, Hoell D, Vossler H-J, Heppner M, Roggenbuck R, Green M, inventors; Müller Schupfner & Partner, assignee. Use of metal salts of trifuoromethane sulphonic acid as esterification catalyst and/or transesterification catalyst. EP 2 113 499 A1. 2009 Nov 4.

66. Funabashi M, Kunioka M. Macromolecular symposia special issue. Biobased Polymers, Recent Progress 2005; 224:309.

67. Laali K, Gerzina RJ, Flajnik CM, Geric CM, Dombroski AM. Copper(II) triflate, a new reagent for mild dehydration of alcohols: synthetic usefulness and mechanistic insight. *Helv Chim Acta* 1987;70:607–611.

68. Sekar G, Singh VK. An efficient method for cleavage of epoxides with aromatic amines. *J Org Chem* 1999;64:287–289.

69. Bolm C, Simić O. Highly enantioselective Hetero-Diels–Alder reactions catalyzed by a C2-symmetric bis(sulfoximine) copper(II) complex. *J Am Chem Soc* 2001;123:3830–3831.

70. Berthiol F, Matsubara R, Kawai N, Kobayashi S. Catalytic asymmetric Michael reactions with enamides as nucleophiles. *Angew. Chem, Int Ed* 2007;46:7803–7805.

71. Maier TC, Fu GC. Catalytic enantioselective O–H insertion reactions. *J Am Chem Soc* 2006;128:4594–4595.

72. Behr A, Fiene M, Naendrup F, Schurmann K. Transition-metal trifluoromethane-sulphonates-recyclable catalysts for the synthesis of branched fatty derivatives by Diels–Alder reactions at unsaturated fatty esters. *Eur J Lipid Sci Technol* 2000;102:342.

73. Kobayashi S, Manabe K. Green Lewis acid catalysis in organic synthesis. *Pure Appl Chem* 2000;72:1373.

74. Waller FJ, Barrett AGM, Braddock DC, Ramprasad D. Lanthanide(III) triflates as recyclable catalysts for atom economic aromatic nitration. Chem Commun 1997;6:613–614.

75. Aggarwal VK, Mereu A, Tarver GJ, McCague R. Metal- and ligand-accelerated catalysis of the Baylis–Hillman reaction. *J Org Chem* 1998;63:7183-7189.

76. Ding R, Katebzadeh K, Roman L, Bergquist K-E, Lindström UM. Expanding the scope of Lewis acid catalysis in water: remarkable ligand acceleration of aqueous ytterbium triflate catalyzed Michael addition reactions. *J Org Chem* 2006;71:352–355.

77. Aspinall HC, Bissett JS, Greeves N, Levin D. Lanthanum triflate-catalysed allylation of aldehydes: crucial activation by benzoic acid. *Tetrahedron Lett* 2002;43:319–321.

78. Aspinall HC, Bissett JS, Greeves N, Levin D. Three component synthesis of homoallylic amines. Efficient catalysis by lanthanum triflate/benzoic acid. *Tetrahedron Lett* 2002;43:323–325.

79. Rai AN, Basu A. An efficient method for para-methoxybenzyl ether formation with lanthanum triflate. *Tetrahedron Lett* 2003;44:2267–2269.

80. Crespi AM, Chen K, inventors; Medtronic, Inc. assignee. Electrolyte for LI/SVO batteries. US patent 5,766,797. 1998 June 16.

81. Kejha JB, Hope SF, inventors. Composite electrolytes for electrochemical devices. US patent 5,521,023. 1996 May 28.

82. Karan NK, Pradhan DK, Thomas R, Natesan B, Katiyar RS. Solid polymer electrolytes based on polyethylene oxide and lithium trifluoro- methane sulfonate (PEO–$LiCF_3SO_3$): ionic conductivity and dielectric relaxation. *Solid State Ionics* 2008;179:689.

83. Firouzabadi H, Karimi B, Eslami S. Lithium trifluoromethanesulfonate (LiOTf) as a highly efficient catalyst for chemoselective dithioacetalization of carbonyl compounds under neutral and solvent-free conditions. *Tetrahedron Lett* 1999;40:4055–4058.

84. Auge J, Leroy F. Lithium trifluoromethanesulfonate-catalysed aminolysis of oxiranes. *Tetrahedron Lett* 1996;37:7715–7716.

85. Lubineau A, Drouillat B. Lithium triflate as a new promoter of glycosylation under neutral conditions. *J Carbohydr Chem* 1997;16:1179.

86. Karimi B, Maleki J. Lithium trifluoromethanesulfonate (LiOTf) as a recyclable catalyst for highly efficient acetylation of alcohols and diacetylation of aldehydes under mild and neutral reaction conditions. *J Org Chem* 2003;68:4951–4954.

87. Stang PJ, Hanack M, Subramanian LR. Perfluoroalkanesulfonic esters: methods of preparation and applications in organic chemistry. *Synthesis* 1982;2:85.

88. Kobayashi Y, Kumadaki I, Yoshida T. *J Chem Res, Synop* 1977;215.

89. Shimagaki M, Fujieda Y, Kimura T, Nakata T. New synthetic methods for olefins from secondary phosphates and thiophosphates. *Tetrahedron Lett* 1995;36:719–722.

90. Burk RM, Gac TS, Roof MB. A mild procedure for etherification of alcohols with primary alkyl halides in the presence of silver triflate. *Tetrahedron Lett* 1994;35:8111–8112.

91. Mori H, Mori A, Xu Q, Souma Y. Koch carbonylation using silver trifluoromethanesulfonate. *Tetrahedron Lett* 2002;43:7871–7874.

92. Ercegovic T, Meijer A, Magnusson G, Ellervik U. Iodine monochloride/silver trifluoromethanesulfonate (ICl/AgOTf) as a convenient promoter system for O-glycoside synthesis. *Org Lett* 2001;3:913–915.

93. Nagashima K, Meguro K, Hobo T. A galvanic gas sensor using poly (ethylene oxide) complex of silver trifluoromethane sulphonate electrolyte. *Fresenius J Anal Chem* 1990;336:571–574.

94. Olah GA. Crossing conventional boundaries in half a century of research. *J Org Chem* 2005;70:2413–2429.

95. Olah GA, Parker DG, Yoneda Y, Pelizza F. Oxyfunctionalization of hydrocarbons. 1. Protolytic cleavage-rearrangement reactions of tertiary alkyl hydroperoxides with magic acid. *J Am Chem Soc* 1976;98:2245–2250.

96. Olah GA, Ohnishi R. Oxyfunctionalization of hydrocarbons. 8. Electrophilic hydroxylation of benzene, alkylbenzenes, and halobenzenes with hydrogen peroxide in superacids. *J Org Chem* 1978;43:865–867.

97. Bickel AF, Gaasbeek CJ, Hogeveen H, Oelderik JM, Platteeuw JC. Chemistry and spectroscopy in strongly acidic solutions: reversible reaction between aliphatic carbonium ions and hydrogen. *Chem Commun* 1967;13:634–635.

98. Hogeveen H, Bickel AF. Chemistry and spectroscopy in strongly acidic solutions: electrophilic substitution at alkane-carbon by protons. *Chem Commun* 1967;13:635–636.

99. Minkwitz R, Reinemann S, Blecher O, Hartl H, Brüdgam I. Halogenated carboxonium salts: preparation and structural analysis. *Inorg Chem* 1999;38:844–847.

CHAPTER 10

Tris(pentafluorophenyl)borane, $B(C_6F_5)_3$, a Powerful, Well-Soluble, Nonoxidizing Lewis Acid and the Weakly Coordinating Tetrakis(pentafluorophenyl)borate Anion, $[B(C_6F_5)_4]^-$

HERMANN-JOSEF FROHN

Tris(pentafluorophenyl)borane, ($B(C_6F_5)_3$) [1–3], is a Lewis acid [4] and as fluoride acceptor it is stronger than the inorganic parent molecule BF_3. In contrast to the prominent Lewis acids, the pentafluorides of the heavier elements of group 15, $B(C_6F_5)_3$ has no oxidizing properties. It forms adducts with neutral, even weak Lewis bases. An important application of $B(C_6F_5)_3$ is based on its capability to easily accept anionic ligands bonded to metals [5, 6].

$$B(C_6F_5)_3 + (C_5H_5)_2Zr(CH_3)_2 \rightarrow [(C_5H_5)_2ZrCH_3][B(C_6F_5)_3CH_3]$$

The resulting $[B(C_6F_5)_3CH_3]^-$ anion is weakly coordinating and enables the electrophilic site on Zr to interact with ethylene as a polymerization catalyst. The most popular weakly coordinating anion of the arylborate family is $[B(C_6F_5)_4]^-$ that is synthesized via $B(C_6F_5)_3$ [2, 7].

In synthetic chemistry, $B(C_6F_5)_3$ can be used as a reagent to introduce C_6F_5 groups into hypervalent nonmetal element fluorides EF_n. This property was first successfully applied for F/C_6F_5 substitution reactions in strongly oxidizing EF_n molecules such as noble gas fluorides [8] and halogen fluorides.

$$B(C_6F_5)_3 + XeF_2 \rightarrow [C_6F_5Xe][B(C_6F_5)_2F_2]$$

Efficient Preparations of Fluorine Compounds, First Edition. Edited by Herbert W. Roesky.

For the transfer of carbanions, it is unique that $B(C_6F_5)_3$ offers the possibility to perform F/C_6F_5 substitution reactions even in superacidic media such as anhydrous HF [9].

10.1 PREPARATION OF TRIS(PENTAFLUOROPHENYL)BORANE

Apparatus Three-necked, 250-mL, round-bottom flask, jacketed coil condenser with P_4O_{10} drying tube, dropping funnel, magnetic stirrer, sublimation apparatus, 50- and 500-mL round-bottom flasks, two Teflon stoppers with two drill holes for inserting Teflon tubes, safety glasses, laboratory coat, protective gloves.

Chemicals C_6F_5Br, Mg turnings, dry diethyl ether, freshly distilled $BF_3{\cdot}Et_2O$, dry *n*-hexane.

Attention and Caution Safety glasses and protective gloves must be used at all times. All reactions must be carried out in a well-ventilated hood. Contact of the skin with $BF_3{\cdot}Et_2O$ is dangerous (formation of HF by hydrolysis).

Experimental Procedure Pentafluorophenyl magnesium bromide, C_6F_5MgBr, is prepared under an atmosphere of dry Ar in dried glass equipment: a 250-mL, three-necked, round-bottom flask with a magnetic stirring bar, a dropping funnel, and a jacketed coil condenser topped with a P_4O_{10} drying tube. Bromopentafluorobenzene (25.46 g, 103 mmol) is diluted with Et_2O (40 mL) and added drop by drop to Mg turnings (3.40 g, 140 mmol) in dry Et_2O (60 mL) within 20 min at ~35 °C (beginning of reflux). The reaction mixture is stirred for 1 h. The completeness of the reaction can be proven by ^{19}F NMR spectroscopy (–113.4, –158.4, –162.3 ppm C_6F_5MgBr; –113.5, –158.8, –162.3 ppm $Mg(C_6F_5)_2$). In a second dried round-bottom flask (50 mL), freshly distilled $BF_3{\cdot}Et_2O$ (4.8 g, 34 mmol) is diluted in Et_2O (10 mL), cooled to –78 °C, and added to the cold (–78 °C) Grignard suspension within 5–10 min by means of a PTFE tube ($D_i = 0.7$ mm) and a slight pressure of dry Ar. The suspension is stirred in sequence for 15 min at –60 °C, for 45 min at 0 °C, and finally for 3.5 h at 20 °C. The completeness of the reaction is confirmed by ^{19}F NMR spectroscopy. Subsequently, the solvent is removed under dynamic vacuum. The residue is two times intensively stirred for 15 min with 75 mL portions of dry *n*-hexane, which are distilled-off (removal of embedded ether). The crude, brown, and sticky solid is treated in vacuum for 1 h and extracted three times with boiling dry *n*-hexane (75 mL portions). The hexane extracts are combined in a 500-mL flask and stored at 20 °C. Over time, transparent colorless crystals of $B(C_6F_5)_3$ grow that are isolated and dried (9.78 g, 19 mmol, 55.9%). $B(C_6F_5)_3$ is finally sublimed under dynamic vacuum at 100–110 °C onto a water-cooled cold-finger. After rigorous drying of the cold-finger, the sublimation apparatus is brought into a dry box where $B(C_6F_5)_3$ (6.73 g, 13 mmol, 38.2%) is isolated and stored.

Characterization Data mp (visual) 128 °C, DTA: 128 °C (T_{max}, endothermal), 363 °C (T_{max}, exothermal dec.). ^{11}B NMR (CH_2Cl_2) δ 59 (s) ppm. ^{13}C NMR

(CH_2Cl_2) δ 148.6 (m, $C^{2,6}$), 145.2 (br, C^4), 137.8 (m, $C^{3,5}$), 113.3 (br, C^1) ppm. ^{19}F NMR (CH_2Cl_2) δ –129.6 (m, $F^{2,6}$), –146.6 (s, F^4), –161.5 (m, $F^{3,5}$) ppm. IR (KBr, cm^{-1}): $\bar{\nu}$ 1652, 1524, 1380, 1293, 1108, 974. MS (EI): *m*/*z* 513 (M^+, 100%).

Waste Disposal Waste solvents (ether, *n*-hexane) have to be collected in the corresponding labeled containers for disposal.

Application Important applications of $B(C_6F_5)_3$ are the use as

a. strong Lewis acid, especially in catalyzed reactions (addition of silyl enol ethers to aldehydes, alkyl chlorides, α,β-unsaturated ketones; hydrosilylation of carbonyl functions)

b. starting material for adducts with bases of different donor strength (formation of tamed Lewis acids)

$$B(C_6F_5)_3 + P(C_6H_5)_3 \rightarrow B(C_6F_5)_3 \cdot P(C_6H_5)_3$$

With bulky ligands in the base, substitution of a *para*-fluorine atom in one C_6F_5 group becomes favored over adduct formation:

$$B(C_6F_5)_3 + P(Mes)_2H \rightarrow para\text{-}(^+P(Mes)_2H)C_6F_4B(C_6F_5)_2F^-$$

In the following step, the zwitterion can be converted into *para*-$(^+P(Mes)_2H)C_6F_4B(C_6F_5)_2H^-$, which principally allows hydrogen storage [10, 11].

c. acceptor of anionic groups from polar bonds under formation of electrophilic cations and weakly nucleophilic anions:

$$B(C_6F_5)_3 + (C_5H_5)_2Zr(CH_3)_2 \rightarrow [(C_5H_5)_2ZrCH_3][B(C_6F_5)_3CH_3]$$
$$B(C_6F_5)_3 + (C_6F_5)_2IF \rightarrow [(C_6F_5)_2I][B(C_6F_5)_3F]$$

d. reagent for F/C_6F_5 substitution in hypervalent element fluoride moieties:

$$B(C_6F_5)_3 + XeF_2 \rightarrow [C_6F_5Xe][B(C_6F_5)_2F_2]$$
$$B(C_6F_5)_3 + IF_3 \rightarrow [(C_6F_5)_2I][B(C_6F_5)_{4-n}F_n]$$

10.2 PREPARATION OF CESIUM TETRAKIS(PENTAFLUOROPHENYL)BORATE

Apparatus Two-necked, 100-mL, round-bottom flask, 250-mL round-bottom flask, two Teflon stoppers with two drill holes for inserting Teflon tubes, magnetic stirrer.

Chemicals C_6F_5Br, dry diethyl ether, butyllithium/*n*-pentane solution, $B(C_6F_5)_3$, dry *n*-pentane, $CsOH{\cdot}H_2O$.

Attention and Caution Safety glasses and protective gloves must be used at all times. All reactions must be carried out in a well-ventilated hood. C_6F_5Li must be handled at the given low temperature (≤ -25 °C) to avoid LiF elimination in an explosive manner.

Experimental Procedure *Preparation of the pentafluorophenyl lithium solution*: Under an atmosphere of dry Ar, bromopentafluorobenzene (1.783 g, 7.22 mmol) is dissolved in dry Et_2O (25 mL) and cooled to –78 °C. A butyllithium/*n*-pentane solution (2.5 M, 5 mL, 12.50 mmol, 20 °C) in a gastight syringe is added drop by drop on the cold upper part of the wall of the two-necked flask. The –78 °C cold mixture is stirred for 2 h. The completeness of the lithiation can be proven by ^{19}F NMR spectroscopy. In addition, a sample of defined volume (e.g. 250 μL) can be solvolyzed at –40 °C in an NMR tube with an excess of CH_3OH (~250 μL) and the amount of C_6F_5H can be determined by integration (^{19}F NMR) relative to the internal integral standard benzotrifluoride (10.0 μL, 0.082 mmol). *Caution*: Pentafluorophenyl lithium in Et_2O is unstable at temperatures higher than –25 °C and must be handled at lower temperatures! In a second round-bottom flask, $B(C_6F_5)_3$ (2.896 g, 5.66 mmol) is suspended in dry *n*-pentane (50 mL) and cooled to –78 °C. The –78 °C cold $C_6F_5Li/Et_2O/n$-pentane solution is added to the cold $B(C_6F_5)_3/n$-pentane suspension through a Teflon tube ($D_i = 0.7$ mm) by a slight pressure of dry Ar. The mixture is stirred in sequence for 15 min at –60 °C, for 60 min at –40 °C, and finally for 30 min at 20 °C. A white solid precipitates. The mother liquor is separated and the solid is washed three times (20 mL) with *n*-pentane. The white solid is dried for 8 h at 20 °C under vacuum, yielding $[Li(Et_2O)_x][B(C_6F_5)_4]$ ($x < 4$). The solid is dissolved in H_2O (30 mL) and H_2O is subsequently removed under vacuum. This treatment is repeated twice to remove coordinated Et_2O. A vacuum dried (20 °C, 2 h) white solid corresponding to $[Li(H_2O)]_x[B(C_6F_5)_4]$ ($x = \sim 1$) (3.42 g, 5 mmol; yield 88%, calculated for $x = 1$) is isolated. A portion of $[Li(H_2O)_x][B(C_6F_5)_4]$ (3.02 g, 4.3 mmol) is dissolved in water (50 mL). After addition of a saturated solution of $CsOH{\cdot}H_2O$ (1.854 g, 11.0 mmol) in H_2O (40 mL), a white solid precipitates immediately. The suspension is stirred for 2 h and finally the mother liquor is separated. The solid product is washed with H_2O until the pH becomes neutral. The white solid is dried for 6 h under vacuum at 20 °C, yielding 2.777 g (3.4 mmol, 79%) of $Cs[B(C_6F_5)_4]$.

Characterization Data ^{11}B NMR (acetone) δ –16.4 (s, br) ppm. ^{19}F NMR (acetone) δ –131.0 (s, br, $F^{2,6}$), –162.4 (t, $^3J = 21$ Hz, F^4), –166.3 (m, $F^{3,5}$) ppm. IR (AgCl, cm^{-1}): $\tilde{\nu}$ 1646, 1517, 1473, 1458, 1413, 1277, 1086, 1032, 983, 775, 758, 726, 685, 662, 611, 602, 573.

Waste Disposal Waste solvent mixtures and aqueous Li^+- and Cs^+-containing solutions have to be collected in the corresponding labeled containers for disposal.

Application $Cs[B(C_6F_5)_4]$ is a suitable reagent for metathesis reactions. Acetonitrile-soluble tetrafluoroborate salts can be converted with $Cs[B(C_6F_5)_4]$

into the corresponding $[B(C_6F_5)_4]^-$ salts accompanied by precipitation of $Cs[BF_4]$. In case of highly electrophilic cations, this procedure can be disadvantageous due to CH_3CN coordination on the cation. In such cases, an alternative reagent to $Cs[B(C_6F_5)_4]$ may be $[N(CH_3)_4][B(C_6F_5)_4]$ (in a less coordinating solvent), which can be obtained in an analogous procedure starting from $[Li(H_2O)_x][B(C_6F_5)_4]$ and $[N(CH_3)_4]Cl$.

REFERENCES

1. Massey AG, Park AJ, Stone FGA. Tris(pentafluorophenyl)boron. *Proc Chem Soc, London* 1963;212.
2. Massey AG, Parks AJ. Tris(pentafluorophenyl)boron. *J Organomet Chem* 1964;2:245–250.
3. Pohlmann JLW, Brinckmann FE. Preparation and characterization of group III A derivatives. *Z Naturforsch* 1965;20b:5–11.
4. Jacobsen H, Berke H, Döring S, Kehr G, Erker G, Fröhlich R, Meyer O. Lewis acid properties of tris(pentafluorophenyl)borane. Structure and bonding in L-$B(C_6F_5)_3$ complexes. *Organometallics* 1999;18:1724–1735.
5. Piers WE, Chivers T. Pentafluorophenylboranes: from obscurity to applications. *Chem Soc Rev* 1997;26:345–354.
6. Erker G. Tris(pentafluorophenyl)borane: a special boron Lewis acid for special reactions. *Dalton Trans* 2005;1883–1890.
7. Koppe K, Bilir V, Frohn H-J, Mercier HPA, Schrobilgen GJ. Syntheses, solution multi-NMR characterization, and reactivities of $[C_6F_5Xe]^+$ salts of weakly coordinating borate anions, $[BY_4]^-$ (Y=CF_3, C_6F_5, CN, or $OTeF_5$). *Inorg Chem* 2007;46:9425–9437.
8. Frohn H-J, Jakobs S, Henkel G. The acetonitrile(pentafluorophenyl)xenon(II) cation, $[MeCN–Xe–C_6F_5]^+$: the first structural characterization of a xenon-carbon bond. *Angew Chem Int Ed* 1989;28: 1506–1507.
9. Frohn H-J, Schroer T. The fluorine/pentafluorophenyl substitution in anhydrous hydrogen fluoride (aHF): a new interesting methodical approach to synthesize pentafluorophenylxenonium salts. *J Fluorine Chem* 2001;112:259–264.
10. Kubas GJ. Breaking the H_2 marriage and reuniting the couple. *Science* 2006;314:1096–1097.
11. Welch GC, San Juan RR, Masuda JD, Stephan DW. Reversible, metal-free hydrogen activation. *Science* 2006;314:1124–1126.

CHAPTER 11

Pentafluorophenyldifluoroborane, $C_6F_5BF_2$, and Pentafluorophenyltrifluorosilane, $C_6F_5SiF_3$, Versatile Reagents for Fluorine/Pentafluorophenyl Substitution Reactions in Strongly Oxidizing Hypervalent Nonmetal Fluorides

HERMANN-JOSEF FROHN

Perfluoroorganylmagnesium halides and perfluoroorganyllithium are the most frequently used reagents to introduce perfluoroorganyl groups into electrophilic sites of nonmetal and metal compounds. Because of the distinct carbanionic character of their organyl group, they do not resist to strong oxidizers. To perform fluorine substitution reactions on strongly oxidizing nonmetal molecules such as noble gas fluorides or halogen fluorides, alternative reagents are required. In such cases, perfluoroorganylfluoroboranes and -silanes are often the reagents of choice. The multistep syntheses of pentafluorophenyldifluoroborane and -trifluorosilane are described here. Pentafluorophenyldifluoroborane [1, 2] and its precursor potassium pentafluorophenyltrifluoroborate are prototypes of perfluoroorganyldifluoroboranes and -borates. The corresponding perfluorinated alkynyl, alkenyl, and alkyl boron compounds can be obtained in a similar manner [2, 3].

Pentafluorophenyldifluoroborane is a stronger Lewis acid than pentafluorophenyltrifluorosilane [4]. With fluorine atoms in hypervalent bonds, which carry a high partial negative charge, both reagents interact and enable acid-assisted nucleophilic fluorine/pentafluorophenyl substitutions. Examples are given next (see Application).

Efficient Preparations of Fluorine Compounds, First Edition. Edited by Herbert W. Roesky.

With the more acidic reagent, products of the onium type result. In principle, onium cations allow the electrophilic transfer of the perfluoroorganyl group to central atoms with an electron lone pair. With the reagent $C_6F_5SiF_3$, neutral molecules such as $C_6F_5IF_2$, $C_6F_5BrF_2$, or $C_6F_5IF_4$ are obtained, which are suitable, relatively mild fluorinating agents. They can add two fluorine atoms to central atoms with an electron lone pair and give a practical inert co-product (C_6F_5Br or C_6F_5I).

11.1 PREPARATION OF POTASSIUM PENTAFLUOROPHENYLTRIFLUOROBORATE

Apparatus Three-necked, 100-mL, round-bottom flask, jacketed coil condenser with P_4O_{10} drying tube, dropping funnel, magnetic stirrer, 100-mL flasks, 100-mL polypropylene (PP) beaker, safety glasses, laboratory coat, protective gloves.

Chemicals C_6F_5Br, Mg turnings, dry diethyl ether, freshly distilled $B(OCH_3)_3$, $K[HF_2]$.

Attention and Caution Safety glasses and protective gloves must be used at all times. All reactions must be carried out in a well-ventilated hood. Contact of the skin with $K[HF_2]$ is dangerous.

Experimental Procedure Potassium pentafluorophenyltrifluoroborate is prepared via pentafluorophenylmagnesium bromide, pentafluorophenyldi(methoxy)borane, and pentafluorophenylboronic acid.

C_6F_5MgBr: In a three-necked, 100-mL, round-bottom flask with a magnetic stirring bar, a dropping funnel, and a condenser with a drying tube, C_6F_5Br (4.9 g, 19.8 mmol) diluted with 15 mL ether is added drop by drop to Mg turnings (0.5 g, 20.6 mmol) in 25 mL ether under an atmosphere of dry Ar. To complete the reaction, the mixture is finally stirred under reflux for 1 h.

$C_6F_5B(OH)_2$: After cooling the C_6F_5MgBr reaction mixture to 0 °C, the mother liquor is transferred slowly to a cold stirred solution (0 °C) of freshly distilled $B(OCH_3)_3$ (4.2 g, 40 mmol) in 20 mL ether in a 100-mL flask under an atmosphere of dry Ar. After stirring for 1 h at 0 °C, the suspension is poured into a stirred 5% HCl_{aq} solution (50 mL). The ether phase is separated and the aqueous is extracted twice with 10 mL ether. The three ether phases are combined and dried with $MgSO_4$. After separation from $MgSO_4$, the solvent is removed. Finally, the product is treated under vacuum to yield crude $C_6F_5B(OH)_2$ (3.8 g) with small admixtures of $(C_6F_5)_2BOH$.

$K[C_6F_5BF_3]$: Crude $C_6F_5B(OH)_2$ is dissolved in methanol (5 mL) and added to a stirred solution of $K[HF_2]$ (4.7 g, 60.2 mmol) in 30 mL of water in a 100-mL PP beaker. A suspension is formed. After stirring for 4 h, the precipitate is filtered-off and washed with water (2 × 3 mL) and ether (2 × 3 mL). After drying in vacuum, $K[C_6F_5BF_3]$ is obtained as colorless solid (3.8 g, 13.9 mmol, 70.2%).

Characterization Data DSC: 327 °C (T_{onset}, exothermal dec.), ^{11}B NMR (CH_3CN) δ –1.6 (q, ^1J = 43 Hz) ppm, ^{13}C NMR (acetone) δ 148.9 ($C^{2,6}$), 139.9 (C^4), 137.4 ($C^{3,5}$), 119.2 (br, C^1) ppm, ^{19}F NMR (CH_3CN) δ –133.3 (qt, ^1J = 43 Hz, ^4J = 12 Hz, BF_3) –135.1 (m, $F^{2,6}$), –160.9 (t, ^3J = 19 Hz, F^4), –165.2 (m, $F^{3,5}$) ppm. IR (KBr, cm^{-1}): $\overline{\nu}$ 1656, 1532, 1466, 1388, 1312, 1144, 1004, 956, 816, 760, 636.

Waste Disposal Waste solvent (ether) has to be collected in the corresponding, labeled container for disposal. Aqueous F^--containing solutions are treated with a CaO suspension and CaF_2 is separated and stored for disposal.

Application $K[C_6F_5BF_3]$ is a well-storable source of the weakly nucleophilic C_6F_5 group. With substituted iodobenzenes p-IC_6H_4-R, it undergoes efficiently Pd-catalyzed cross-coupling reactions to polyfluorinated biphenyls [5].

$$K[C_6F_5BF_3] + p\text{-}IC_6H_4\text{-}R \rightarrow C_6F_5C_6H_4\text{-}R$$
$$R = NO_2, C(O)OC_2H_5, F, H, CH_3, OCH_3$$

Substituted benzenediazonium salts can also be used as electrophiles in catalyzed coupling reactions with $K[C_6F_5BF_3]$ [6].

11.2 PREPARATION OF PENTAFLUOROPHENYLDIFLUOROBORANE

Apparatus Two FEP traps (D_i = 23 mm, D_o = 25 mm), two Teflon stoppers with two drill holes for inserting Teflon tubes.

Chemicals $K[C_6F_5BF_3]$, dry CH_2Cl_2, BF_3 gas, NaF.

Attention and Caution Safety glasses and protective gloves must be used at all times. All reactions must be carried out in a well-ventilated hood. Contact of the skin with BF_3 gas or $C_6F_5BF_2$ (formation of HF by hydrolysis) is dangerous.

Experimental Procedure Under an atmosphere of dry Ar, $K[C_6F_5BF_3]$ (1.62 g, 5.9 mmol) is suspended in CH_2Cl_2 (5 mL) in an FEP trap (D_i = 23 mm) equipped with a magnetic stirring bar and a Teflon stopper with two drill holes for inserting Teflon tubes and cooled to –50 °C. Five equivalents of BF_3 from a stainless steel line with a reservoir are passed first a stirred NaF/CH_2Cl_2 suspension at 20 °C in an FEP trap before being slowly (ca. 1.5 h) bubbled into the $K[C_6F_5BF_3]$/CH_2Cl_2 suspension. At the end, the drilled stopper is exchanged against a Teflon stopper under protection of dry Ar, and the stirred reaction mixture is warmed to 20 °C. During this step, it is necessary to release the increased pressure of BF_3 several times. Subsequently, the suspension is stirred for further 1.5 h before still dissolved BF_3 gas is removed by pumping under dynamic vacuum at –78 °C for 15 min. The suspension is centrifuged at 20 °C and the mother liquor is separated under protection of dry Ar. The residue, $K[BF_4]$, is extracted three times with CH_2Cl_2 (1 mL). The amount of borane in the combined $C_6F_5BF_2$/CH_2Cl_2 phases is determined by ^{19}F NMR spectroscopy using the internal integral standard $C_6H_5CF_3$. The $C_6F_5BF_2/CH_2Cl_2$ solution can be used directly for reactions, for example, for F/C_6F_5 substitutions.

This protocol is also applicable to other types of perfluoroorganyldifluoroboranes.

Neat $C_6F_5BF_2$ can be obtained when instead of CH_2Cl_2 the more volatile solvent CCl_3F is used. In a flame-dried glass distillation apparatus, the solvent is distilled-off followed by the borane in a yield of 1.80 g (8.3 mmol, 83.4%).

Characterization Data Bp 104–105 °C. ^{11}B NMR (CH_2Cl_2) δ 21.8 (s, br) ppm. $^{13}C\{^{19}F\}$ NMR (CH_2Cl_2) δ 151.8 ($C^{2,6}$), 146.2 (C^4), 138.2 ($C^{3,5}$), 98.7 (br, C^1) ppm. ^{19}F NMR (CH_2Cl_2) δ −74.4 (s, br, BF_2), −129.0 (m, $F^{2,6}$), −143.9 (tt, $^3J = 20$ Hz, $^4J = 7$ Hz, F^4), −161.3 (m, $F^{3,5}$) ppm.

Waste Disposal Waste solutions of $C_6F_5BF_2$ have to be treated with a CaO/H_2O suspension in the hood. Precipitated CaF_2 and $K[BF_4]$ have to be collected in labeled containers for disposal.

Application $C_6F_5BF_2$ is a unique reagent to perform F/C_6F_5 substitution in hypervalent fluorine moieties of noble gas [7–9] and halogen fluorides [10–12]:

$$C_6F_5BF_2 + XeF_2 \rightarrow [C_6F_5Xe][BF_4]$$

$$C_6F_5BF_2 + XeF_4 \rightarrow [C_6F_5XeF_2][BF_4]$$

$$C_6F_5BF_2 + C_6F_5IF_2 \rightarrow [(C_6F_5)_2I][BF_4]$$

$$C_6F_5BF_2 + C_6F_5IF_4 \rightarrow [(C_6F_5)_2IF_2][BF_4]$$

$$2C_6F_5BF_2 + 2BrF_3 \rightarrow [(C_6F_5)_2Br][BF_4] + BrF_3 + BF_3$$

$$C_6F_5BF_2 + C_6F_5BrF_2 \rightarrow [(C_6F_5)_2Br][BF_4]$$

Formally, one fluorine of the hypervalent bond is substituted by a C_6F_5 group and the second fluorine is abstracted by the co-product BF_3. Thus, always electrophilic onium cations result in combination with the weakly coordinating anion $[BF_4]^-$.

11.3 PREPARATION OF PENTAFLUOROPHENYLTRIFLUOROSILANE

Apparatus Three-necked, 250-mL, round-bottom flask, 250-, 100-, and 50-mL, round-bottom flasks, jacketed coil condenser with a P_4O_{10} drying tube, dropping funnel, magnetic stirrer, short-path distillation apparatus with a Vigreux column.

Chemicals Mg turnings, C_6F_5Br, freshly distilled tetraethoxysilane, ($Si(OC_2H_5)_4$), dry diethyl ether, dry *n*-hexane, freshly distilled $SOCl_2$, Py·HCl, freshly sublimed SbF_3.

Attention and Caution Safety glasses and protective gloves must be used at all times. All reactions must be carried out in a well-ventilated hood. Contact of the skin with $C_6F_5SiF_3$ (formation of HF by hydrolysis) is dangerous.

Experimental Procedure Pentafluorophenyltrifluorosilane [13] is obtained in three steps via pentafluorophenyltri(ethoxy)silane and pentafluorophenyltrichlorosilane.

Pentafluorophenyltri(ethoxy)silane: In a dried, 250-mL, three-necked, round-bottom flask with a magnetic stirring bar, a jacketed coil condenser topped with a P_4O_{10} drying tube, Mg turnings (1.325 g, 54.5 mmol), C_6F_5Br (12.35 g, 50 mmol), and $Si(OC_2H_5)_4$ (freshly distilled, 41.6 g, 199.7 mmol) are mixed under an atmosphere of dry Ar. $(C_2H_5)_2O$ (dry, 30 mL) is added drop by drop under stirring. The start of the exothermic reaction is indicated by a change of the colorless suspension to gray. If the reaction does not start spontaneously, warming with a water bath (50 °C) is recommended. To complete the reaction, the mixture is refluxed for approximately 20 h. The Mg salts are precipitated by addition of dry *n*-hexane (70 mL) at 20 °C. After sedimentation, the mother liquor is separated and collected in a 250-mL flask under protection of dry Ar. A short-path distillation apparatus combined with a Vigreux column allows to separate ether, *n*-hexane, and $Si(OC_2H_5)_4$. Finally under vacuum (43 hPa), $C_6F_5Si(OC_2H_5)_3$ is distilled-off at 138 °C. The sticky residue contains mainly $(C_6F_5)_nSi(OC_2H_5)_{4-n}$ ($n \geq 2$). The yield of the colorless liquid $C_6F_5Si(OC_2H_5)_3$ is 7.37 g (22.3 mmol, 44.6%).

Pentafluorophenyltrichlorosilane: A 100-mL, round-bottom flask with a magnetic stirring bar and a jacketed coil condenser topped with a P_4O_{10} drying tube is charged in sequence with Py·HCl (1.6 g, 13.8 mmol) and freshly distilled $SOCl_2$ (48 mL, 658 mmol) under protection of dry Ar. $C_6F_5Si(OC_2H_5)_3$ (35.7 g, 108.1 mmol) is added in one portion under stirring. The exothermic reaction starts within the first hour. To complete the reaction, the mixture is stirred under reflux (bath ca. 100 °C) for 1 day. The excess of $SOCl_2$ is distilled-off and $C_6F_5SiCl_3$ is isolated by vacuum distillation (2 hPa; bp = 45 °C) in a yield of 27.2 g (90.2 mmol, 83.4%).

Pentafluorophenyltrifluorosilane: Freshly sublimed SbF_3 (12.3 g, 68.8 mmol) is added in one portion to $C_6F_5SiCl_3$ (17.02 g, 56.5 mmol) under protection of dry Ar in a 50-mL, round-bottom flask with a magnetic stirring bar and a jacketed coil condenser topped with a P_4O_{10} drying tube. The exothermic reaction starts after some minutes. The reaction mixture is stirred under reflux for 1 h. The product is distilled-off at slightly reduced pressure and redistilled to yield $C_6F_5SiF_3$ (13.47 g, 53.4 mmol, 94.5%).

Characterization Data bp 44 °C at 85 hPa. ^{13}C NMR (CH_2Cl_2): δ 151.2 ($C^{2,6}$), 146.9 (C^4), 139.1 ($C^{3,5}$), 96.1 (C^1) ppm. ^{19}F NMR ($CHCl_3$): δ –134.3 (dt, ^1J = 248 Hz, ^4J = 10 Hz SiF_3), –125.1 (m, $F^{2,6}$), –142.9 (tt, ^3J = 19 Hz, ^4J = 7 Hz, F^4), –159.0 (m, $F^{3,5}$) ppm. ^{29}Si NMR ($CHCl_3$): δ –77.8 (q, ^1J = 248 Hz, *Si*F_3) ppm.

Waste Disposal Waste solvents (ether, *n*-hexane) have to be collected in the corresponding labeled containers for disposal. Waste of $Si(OC_2H_5)_4$ and $SOCl_2$ has to be hydrolyzed in small portions in the hood and neutralized before depositing in labeled containers. Solutions of $C_6F_5SiF_3$ have to be treated with a CaO/H_2O suspension in the hood, and CaF_2 has to be filtered-off and stored in the labeled container for disposal. Residues that contain $SbF_3/SbCl_3$ mixture have to be collected in the corresponding labeled container for disposal.

Application $C_6F_5SiF_3$ is a weaker Lewis acid than $C_6F_5BF_2$. In acid-assisted reactions with halogen fluorides only one fluorine is substituted and $C_6F_5HalF_{n-1}$ molecules are formed [14–16]. The fluoride affinity of the co-product SiF_4 is too weak to abstract a fluorine atom from the molecule $C_6F_5HalF_{n-1}$ and form an onium fluorosilicate salt.

It is worth mentioning that $C_6F_5SiF_3$ does not facilitate F/C_6F_5 substitution in XeF_2.

$$C_6F_5SiF_3 + BrF_3 \rightarrow C_6F_5BrF_2 + SiF_4$$

$$C_6F_5SiF_3 + BrF_5 \rightarrow C_6F_5BrF_4 + SiF_4$$

$$C_6F_5SiF_3 + IF_5 \rightarrow C_6F_5IF_4 + SiF_4$$

REFERENCES

1. Frohn H-J, Franke H, Fritzen P, Bardin VV. Synthesis and spectroscopic characterization of potassium fluoroaryltrifluoroborates and fluoroarylboranes. *J Organomet Chem* 2000;598:127–135.
2. Bardin VV, Frohn H-J. (Polyfluoroorgano)haloboranes and (polyfluoroorgano)haloborate salts: preparation, NMR spectra and reactivity. *Main Group Met Chem* 2002;25:589–613.
3. Adonin NY, Bardin VV. Polyfluorinated compounds of boron. *Russ Chem Rev* 2010;79:757–785.
4. Abo-Amer A, Frohn H-J, Steinberg C, Westphal U. From hypervalent xenon difluoride and aryliodine(III) difluorides to onium salts: scope and limitation of acidic fluoroorganic reagents in the synthesis of fluoroorgano xenon(II) and iodine(III) onium salts. *J Fluorine Chem* 2006;127:1311–1323.
5. Frohn H-J, Adonin NY, Bardin VV, Starichenko VF. Highly efficient cross-coupling reactions with the perfluoroorganotrifluoroborate salts K $[R_FBF_3]$ ($R_F = C_6F_5$, CF_2=CF). *Tetrahedron Lett* 2002;43:8111–8114.
6. Frohn H-J, Adonin NY, Bardin VV, Starichenko VF. A new application of (polyfluoroorgano)trifluoroborate salts: the palladium-catalysed cross-coupling reaction with substituted benzenediazonium tetrafluoroborates. *J Fluorine Chem* 2002;117:115–120.
7. Frohn H-J, Franke H, Bardin VV. A simple and convenient route to arylxenon(II) tetrafluoroborates. *Z Naturforsch* 1999;54b:1495–1498.
8. Frohn H-J, Bardin VV. The reaction of R_FBF_2 and $K[R_FBF_3]$ (R_F = perfluorophenyl-, perfluoroalk-1-enyl-, and perfluoroalkyl) with xenon difluoride in anhydrous HF. *Z Anorg Allg Chem* 2002;628:1853–1856.
9. Frohn H-J, LeBlond N, Lutar K, Zemva B. The first organoxenon(IV)compound: pentafluorophenyldifluoroxenonium(IV) tetrafluoroborate. *Angew Chem Int Ed* 2000;39:391–393.
10. Bailly F, Barthen P, Frohn H-J, Köckerling M. Aryl(pentafluorophenyl)iodonium tetrafluoroborate: general method of synthesis, typical properties, and structural features. *Z Anorg Allg Chem* 2000;626:2419–2427.

11. Frohn H-J, Giesen M, Welting D, Bardin VV. Bis(perfluoroorganyl)bromonium salts $[(R_F)_2Br]Y$ (R_F = aryl, alkenyl, and alkynyl). *J Fluorine Chem* 2010;131:922–932.

12. Frohn H-J, Wenda A, Flörke U. $[(C_6F_5)_2IF_2][BF_4]$, the first salt with the electrophilic cation $[(C_6F_5)_2IF_2]^+$: synthesis, reactivity, and structure. *Z Anorg Allg Chem* 2008;634:764–770.

13. Frohn H-J, Giesen M, Klose A, Lewin A, Bardin VV. A convenient preparation of pentafluorophenyl(fluoro)silanes. Reactivity of pentafluorophenyltrifluorosilane. *J Organomet Chem* 1996;506:155–164.

14. Frohn H-J. Darstellung von Pentafluorphenyliodtetrafluorid und anderen Aryliodtetrafluoriden durch Reaktion von IF_5 mit Arylsilanen. *Chem-Ztg* 1984;108:146–147.

15. Frohn H-J, Giesen M. Pentafluorphenylbrom(III)difluorid und -bis(trifluoracetat). *J Fluorine Chem* 1984;24;9–15.

16. Breuer W, Frohn H-J. Pentafluorphenylbrom(V)tetrafluorid. *J Fluorine Chem* 1987;34:443–451.

CHAPTER 12

Iodine(III and V) Fluorides: Interesting Fluorinating Agents

HERMANN-JOSEF FROHN

Hypervalent fluorides are potential fluorinating agents. They allow the addition of two fluorine atoms to central atoms of molecules in a low valence state that possess an electron lone pair, for example, fluorination of phosphanes to difluorophosphoranes. More important than this fluorine addition are often other types of fluorinations, such as the selective substitution of one hydrogen in an organic moiety or the addition of fluorine to carbon–carbon multiple bonds. During the past two decades, Yoneda and Hara investigated the application of iodine(III and V) fluorides for such purposes. Following we offer a convenient preparative access to iodine(III and V) fluorides (without using elemental fluorine) by oxygen/fluorine substitution (without using chlorine and/or mercury salts [1,2] in a two-phase system involving hydrogen fluoride of different concentration [3].

12.1 PREPARATION OF $4\text{-}FC_6H_4IF_2$, $4\text{-}FC_6H_4IF_4$, AND IF_5 BY OXYGEN/FLUORINE SUBSTITUTION

Apparatus FEP traps (blockpolymer of perfluoroethylene and -propylene, D_i = 3.5 mm, D_o = 4.1 mm; D_i = 8 mm, D_o = 9 mm; or D_i = 23 mm, D_o = 25 mm) with Teflon stoppers, two Teflon stoppers with two drill holes for inserting Teflon tubes, magnetic stirrer, polyethylene (PE) syringe, safety glasses, laboratory coat, protective gloves suitable for working with HF.

Chemicals $4\text{-}FC_6H_4IO$, CH_2Cl_2, NaF, 48% HF_{aq} for $4\text{-}FC_6H_4IF_2$; $4\text{-}FC_6H_4IO_2$, CH_2Cl_2, NaF, anhydrous HF for $4\text{-}FC_6H_4IF_4$; I_2O_5 or $NaIO_3$, CH_2Cl_2, NaF, anhydrous HF for IF_5.

Attention and Caution Safety glasses and protective gloves must be used at all times. All reactions must be carried out in a well-ventilated hood. Contact of

---

Efficient Preparations of Fluorine Compounds, First Edition. Edited by Herbert W. Roesky.

the skin with HF_{aq} or anhydrous HF is very dangerous. Inhalation of HF vapor is extremely critical and must be avoided.

Experimental Procedure *Preparation of 4-fluoro(difluoroiodo)benzene from 4-fluoroiodosobenzene*: 4-FC_6H_4IO (11.0 g, 46.2 mmol) is suspended and intensively stirred in 15–20 mL CH_2Cl_2 in an FEP trap (D_i = 23 mm, D_o = 25 mm). HF (12 mL; 48%, 334 mmol) is added with a PE syringe. Solid iodosobenzene dissolved, and a two-phase system (a CH_2Cl_2 solution of 4-$FC_6H_4IF_2$ and a HF_{aq} phase) resulted that is stirred for further 15 min. The upper aqueous HF phase is separated and extracted two times with CH_2Cl_2 (2 × 5 mL). The solvent of the combined CH_2Cl_2 phases is distilled off. The white solid residue is dried in vacuum at 20 °C for 3 h. *p*-$FC_6H_4IF_2$ is obtained in a yield of 9.7 g (37.3 mmol, 80.7%).

Characterization Data Mp 101 °C. ^{1}H NMR (CH_2Cl_2): δ 8.0 (m, $H^{2,6}$), 7.4 (m, $H^{3,5}$) ppm. $^{13}C\{^1H\}$ NMR (CH_2Cl_2): δ 164.5 (d, C^4, ^1J = 253 Hz), 132.6 (dt, ^3J = 9 Hz, ^3J = 4 Hz, $C^{2,6}$), 118.7 (d, ^2J = 23 Hz, $C^{3,5}$), 117.9 (td, ^2J = 12 Hz, ^4J = 3 Hz, C^1) ppm. ^{19}F NMR (CH_2Cl_2): –108.9 (ttt, ^3J = 8 Hz, ^4J = 5 Hz, ^6J = 2 Hz, F^4), –174.4 (s, IF_2) ppm.

In analogy, the 2- and 3-$FC_6H_4IF_2$ isomers can be prepared.

Preparation of 4-fluoro(tetrafluoroiodo)benzene from 4-fluoro(iodyl) benzene: 4-$FC_6H_4IO_2$ (156 mg, 0.614 mmol) is suspended in 0.3 mL CH_2Cl_2 (–30 °C) and intensively stirred in an FEP trap (D_i = 8 mm, D_o = 9 mm). Cold anhydrous HF (0.15 mL, 7.61 mmol) is added drop by drop from a small FEP trap (D_i = 3.5 mm, D_o = 4.1 mm, –78 °C) by means of a Teflon tube (D_i = 0.7 mm) and a slight pressure of Ar. 4-$FC_6H_4IO_2$ dissolves and a two-phase system results. The temperature is raised in two steps to 0 °C and 20 °C. After cooling to –30 °C, the upper HF phase is separated and NaF (100 mg) is added to the lower 4-$FC_6H_4IF_4$/CH_2Cl_2 phase to remove traces of HF. After 15 min of stirring, the solution phase is separated from solid NaF. NaF is extracted twice with CH_2Cl_2 (0.15 mL). The solvent is evaporated from the combined CH_2Cl_2 phases in an FEP trap and the white solid is treated in vacuum at 20 °C for 3 h. 4-$FC_6H_4IF_4$ is isolated in a yield of 0.115 mg (0.386 mmol, 63%).

In analogy, the 2- and 3-$FC_6H_4IF_4$ isomers can be prepared.

Characterization Data mp 89 °C. ^{1}H NMR (CH_2Cl_2): δ 8.3 (td, ^3J = 9 Hz, ^4J = 4 Hz, $H^{2,6}$), 7.6 (td, ^3J = 9 Hz, ^3J = 9 Hz, $H^{3,5}$) ppm. $^{13}C\{^1H\}$ NMR (CH_2Cl_2): δ 165.6 (d, ^1J = 257 Hz, C^4), 154.8 (quin, ^2J = 9 Hz, C^1), 130.2 (d, ^3J = 10 Hz, $C^{2,6}$), 117.6 (d, ^2J = 24 Hz, $C^{3,5}$) ppm. ^{19}F NMR (CH_2Cl_2): –24.3 (s, IF_4), –102.8 (m, F^4) ppm.

Preparation of iodine pentafluoride from diiodine pentoxide: In an FEP trap (D_i = 8 mm, D_o = 9 mm) with a suitable magnetic stirrer, I_2O_5 (200 mg, 0.599 mmol) is suspended in CH_2Cl_2 (0.8 mL) and stirred at –78 °C. Cold anhydrous HF (0.15 mL, 7.61 mmol) is added drop by drop from a small FEP trap (D_i = 3.5 mm, D_o = 4.1 mm, –78 °C) by means of a Teflon tube (D_i = 0.7 mm) and a slight pressure of Ar. Within 0.5 h, the temperature is raised to –30 °C. The solid disappeared. The upper CH_2Cl_2 phase is separated at –30 °C from the lower HF phase and treated with NaF (400 mg) and stirred for 15 min in an FEP trap. The CH_2Cl_2 solution is separated from solid NaF. NaF is extracted twice with CH_2Cl_2 (0.15 mL, 20 °C). The CH_2Cl_2 phases are combined and the solvent is removed in vacuum at –78 °C. Colorless IF_5 (93.3 mg, 0.42 mmol, 70.1%) is obtained.

In an analogous procedure, IF_5 is obtained from $NaIO_3$ (213 mg, 1.076 mmol), CH_2Cl_2 (0.6 mL), and anhydrous HF (0.3 mL, 16.5 mmol) in a yield of 170.9 mg (0.77 mmol, 71.6%).

Characterization Data mp 9–10 °C. ^{19}F NMR (CH_2Cl_2): –52.9 (quin, ^2J = 89 Hz, IF_{ax}), 12.0 (d, ^2J = 89 Hz, IF_4) ppm.

Waste Disposal Inside an intensively ventilated hood, HF and solutions of HF (see Attention and Caution) are poured in small portions on ice followed by treatment with an aqueous CaO/H_2O suspension. CaF_2 is filtered off and collected in the labeled container for disposal. Waste solvent (CH_2Cl_2) has to be collected in the corresponding labeled container for disposal.

Application Difluoroiodobenzenes, such as $4\text{-}CH_3C_6H_4IF_2$, in combination with $N(C_2H_5)_3{\cdot}nHF$ have a wide range of applications as fluorinating agents for

a. fluorination of iodoalkanes [4]:

$$\mathrm{ArIF_2/N(C_2H_5)_3 \cdot 4HF + I-(CH_2)_{\mathit{n}} - X \rightarrow F-(CH_2)_{\mathit{n}} - X}$$
$$\mathrm{X = CH_3,\ CO_2CH_3,\ OAc,\ Cl,\ OTs}$$

b. [*E*]-β-fluoroalkenes from terminal alkynes [5]:

$$\mathrm{ArIF_2/N(C_2H_5)_3 \cdot 5HF + R - C \equiv CH \rightarrow [\mathit{E}]{-}F(R)C = CHI(F)Ar}$$

c. selective vicinal difluorination of terminal and cyclic alkenes [6]:

$$\mathrm{ArIF_2/N(C_2H_5)_3 \cdot 5HF + RCH = CH_2 \rightarrow RCHF - CH_2F}$$

d. selective α-fluorination of β-ketoesters [7]:

$$\mathrm{ArIF_2/N(C_2H_5)_3 \cdot \mathit{n}HF} + R^1 - \mathrm{C(O) - CHR^2 - C(O)OR^3}$$
$$\rightarrow \mathrm{R^1 - C(O) - CFR^2 - C(O)OR^3}$$

Even the combination $IF_5/N(C_2H_5)_3{\cdot}3HF$ offers a wide range of fluorination reactions [8]:

$$\mathrm{IF_5/N(C_2H_5)_3 \cdot 3HF + \mathit{n}\text{-}C_\mathit{n}H_{2\mathit{n}+1} - CO_2H \rightarrow \mathit{n}\text{-}C_\mathit{n}H_{2\mathit{n}+1} - COF}$$
$$\mathrm{IF_5/N(C_2H_5)_3 \cdot 3HF + \mathit{n}\text{-}C_\mathit{n}H_{2\mathit{n}+1} - OH \rightarrow \mathit{n}\text{-}C_\mathit{n}H_{2\mathit{n}+1} - F}$$
$$\mathrm{IF_5/N(C_2H_5)_3 \cdot 3HF + \mathit{p}\text{-}Tol - SCH_3 \rightarrow \mathit{p}\text{-}Tol - SCHF_2}$$
$$\mathrm{IF_5/N(C_2H_5)_3 \cdot 3HF + Ph - S - CH_2CO_2C_2H_5 \rightarrow Ph - S - CF_2CO_2C_2H_5}$$
$$\mathrm{IF_5/N(C_2H_5)_3 \cdot 3HF + Ph_2C = NNH_2 \rightarrow Ph_2CF_2}$$

REFERENCES

1. Carpenter W. Aryliodosodifluorides. *J Org Chem* 1966;31:2688–2689.
2. Sawaguchi M, Ayuba S, Hara S. A practical synthetic method of iodoarene difluorides without fluorine gas and mercury salts. *Synthesis* 2002;1802–1803.
3. Abo-Amer A, Frohn H-J, Steinberg C, Westphal U. From hypervalent xenon difluoride and aryliodine(III) difluorides to onium salts: scope and limitation of acidic fluoroorganic reagents in the synthesis of fluoroorgano xenon(II) and iodine(III) onium salts. *J Fluorine Chem* 2006;127:1311–1323.
4. Sawaguchi M, Hara S, Nakamura Y, Ayuba S, Fukuhara T, Yoneda N. Deiodinative fluorination of alkyl iodide with *p*-iodotoluene difluoride. *Tetrahedron* 2001;57:3315–3319.
5. Hara S, Yoshida M, Fukuhara T, Yoneda N. Stereo- and regio-selective addition of iodotoluene difluoride to alk-1-ynes. Selective synthesis of 2-fluoro-1-iodoalk-1-enes. *Chem Commun* 1998;965–966.
6. Hara S, Nakahigashi J, Ishi-i K, Sawaguchi M, Sakai H, Fukuhara T, Yoneda N. Difluorination of alkenes with iodotoluene difluoride. *Synlett* 1998; 495–496.
7. Yoneda N. Advances in the preparation of organofluorine compounds involving iodine and/or iodocompounds. *J Fluorine Chem* 2004;125:7–17.
8. Yoneda N, Fukuhara T. Novel fluorination reagent: IF_5/Et_3N–3HF. *Chem Lett* 2001;222–223.

CHAPTER 13

Tetramethylammonium Fluoride, $[N(CH_3)_4]F$, a Widely Applicable Reagent to Introduce Fluoride Ions and a Suitable Nucleophile to Initiate the Transfer of Perfluoroorganyl Groups to Electrophiles

HERMANN-JOSEF FROHN

Traditionally, alkali metal fluorides are often used as fluoride donor sources. Their disadvantage is the insolubility in most organic solvents. Thus, reactions with CsF, which possesses the lowest lattice energy of the alkali metal fluorides, proceed predominantly at the surface of the salt. In contrast to CsF, $[N(CH_3)_4]F$ [1] is soluble in some common organic solvents such as CH_2Cl_2 or CH_3CN [2]. The stability of such solutions depends strongly on the temperature. Already at –40 °C, $[N(CH_3)_4]F$ attacks CH_3CN in a noticeable quantity after 5 min. In the first step, abstraction of H^+ proceeds slowly: $F^- + CH_3CN \rightarrow HF + [CH_2CN]^-$, followed by the fast reaction of F^- with HF to $[HF_2]^-$. $[CH_2CN]^-$ ends with *trans*-amino-2-butenenitrile, $CH_3C(NH_2){=}CHCN$ [1]. CH_2Cl_2 solutions of $[N(CH_3)_4]F$ are stable over few hours at –40 °C. In the non-common solvent $CF_3CH_2CHF_2$, $[N(CH_3)_4]F$ shows a better stability even at 20 °C, but a lower solubility than in CH_2Cl_2. To avoid reactions with the solvent, it is often useful to prepare the $[N(CH_3)_4]F$ solution directly in the presence of the reactant, the fluoride acceptor.

Apparatus FEP traps (blockpolymer of perfluoroethylene and -propylene, $D_i = 8$ mm, $D_o = 9$ mm and $D_i = 23$ mm, $D_o = 25$ mm) with Teflon stoppers, glass trap with a stopcock adapter, safety glasses, laboratory coat, and protective gloves.

Efficient Preparations of Fluorine Compounds, First Edition. Edited by Herbert W. Roesky.

Chemicals $[N(CH_3)_4]F{\cdot}4H_2O$, dry *i*-propanol.

Attention and Caution! Safety glasses and protective gloves must be used at all times. All reactions should be carried out in a well-ventilated hood. Contact of the skin with fluoride ions must be avoided.

Experimental Procedure Six small FEP traps ($D_i = 8$ mm, $D_o = 9$ mm), each of them filled with $[N(CH_3)_4]F{\cdot}4H_2O$ till a height of ca. 20 mm, together with 6.686 g (40.5 mmol) $[N(CH_3)_4]F{\cdot}4H_2O$, are deposited in a glass trap with a 29/32 joint and a stopcock adapter on top. Under vacuum, the samples are heated for 1 h at 40–50 °C. Melting takes place and with the loss of water solidification proceeds. The first heating step is followed by further steps for 3 h at 80 °C, for 2 h at 120 °C, and finally 8 h at 140–150 °C. After cooling to 20 °C, the glass trap is pressurized by dry Ar and inside a glove box the solid product is transferred into a larger FEP trap ($D_i = 23$ mm, $D_o = 25$ mm). The salt is dissolved in dry *i*-propanol. Still present water is removed (*i*-propanol–water azeotrope) under reduced pressure (ca. 10 hPa), followed by a final treatment at 80 °C and 10^{-2} hPa for 8 h. $[N(CH_3)_4]F$ is obtained as a white, very electrostatic powder in yields of 2.5–3.2 g (26.8–34.4 mmol, 70–85%).

Instead of using $[N(CH_3)_4]F{\cdot}4H_2O$ as starting material, it is possible to neutralize an aqueous $[N(CH_3)_4]OH$ solution with an aqueous HF solution in a PFA flask and to remove water under reduced pressure till a solid $[N(CH_3)_4]F$ hydrate is obtained, which is further treated as described above.

An alternative preparation—a metathesis reaction—of potassium fluoride and tetramethylammonium tetrafluoroborate is based on the insolubility of potassium tetrafluoroborate in methanol.

Solid tetramethylammonium tetrafluoroborate (6.77 g, 42.1 mmol) is added under stirring and protection of dry Ar to a solution of potassium fluoride (2.56 g, 44.1 mmol) in 50 mL dry methanol in a 100 mL centrifuge trap. After some minutes, solid $K[BF_4]$ is precipitated. The suspension is centrifuged and the mother liquid separated. Methanol is removed from the solution and the solid residue is finally dried in vacuum at 140 °C for 5 h. 2.86 g (0.31 mmol, 73%) $[N(CH_3)_4]F$ is obtained.

Characterization Data ^{19}F NMR (CH_2Cl_2, –40 °C) –97.8 (s) ppm. {Traces of $[HF_2]^-$: ^{19}F NMR (CH_2Cl_2, –40 °C): –151.0 (d, $^1J = 121$ Hz) ppm.} ^1H NMR (CH_2Cl_2, –40 °C): 3.4 (s) ppm.

Waste Disposal Fluoride ions containing solutions can be treated with aqueous CaO/H_2O suspension. CaF_2 has to be collected in the labeled container for disposal. Waste solvent (i-C_3H_7OH) has to be collected in the corresponding labeled container for disposal.

Application Two main reactivities, which are important for applications, will be mentioned here:

a. The addition of fluoride ions to cationic [3], neutral, and anionic [4] fluoride acceptors:

$$[N(CH_3)_4]F + [C_6F_5Xe][BF_4] \rightarrow C_6F_5XeF + [N(CH_3)_4][BF_4]$$

$$[N(CH_3)_4]F + B(C_6F_5)_3 \rightarrow [N(CH_3)_4][B(C_6F_5)_3F]$$

$$[N(CH_3)_4]F + [N(CH_3)_4][IF_4] \rightarrow [N(CH_3)_4]_2[IF_5]$$

b. The interaction of catalytic amounts of $[N(CH_3)_4]F$ with fluoroorganyl element compounds, which form no stable fluoroanion, instead the intermediate fluoroanion releases a fluoroorganyl group—a weak nucleophile—as carbanion, which is trapped by a suitable electrophile [5–8]. This F^--catalyzed reaction is a methodical approach to introduce different types of perfluoroorganyl groups into molecules:

$$(CH_3)_3SiCF_3 + R^1R^2CO \xrightarrow{[F^-]} R^1R^2(CF_3)COSiMe_3$$

$$2(CH_3)_3SiC_6F_5 + XeF_2 \xrightarrow{[F^-]} Xe(C_6F_5)_2 + 2(CH_3)_3SiF$$

REFERENCES

1. Christe KO, Wilson WW, Wilson RD, Bau R, Feng J. Syntheses, properties, and structures of anhydrous tetramethylammonium fluoride and its 1:1 adduct with trans-3-amino-2-butenenitril. *J Am Chem Soc* 1990;112:7619–7625.
2. Christe KO, Wilson WW. Reaction of the fluoride anion with acetonitrile. Chloroform and methylene chloride. *J Fluorine Chem* 1990;47:117–120.
3. Frohn H-J, Theißen M. C_6F_5XeF, a key substrate in xenon-carbon chemistry: synthesis of symmetric and asymmetric pentafluorophenylxenon(II) derivatives. *Angew Chem Int Ed* 2000;39:4591–4593.
4. Christe KO, Wilson WW, Drake GW, Dixon DA, Boatz JA, Gnann RZ. Pentagonal planar AX_5 species: synthesis and characterization of the iodine(III) pentafluoride dianion, IF_5^{2-}. *J Am Chem Soc* 1998;120:4711–4716.
5. Denmark SE, Beutner GL. Lewis base catalysis in organic synthesis. *Angew Chem Int Ed* 2008;47:1560–1638.
6. Prakash GKS, Krishnamurti R, Olah GA. Fluoride-induced trifluoromethylation of carbonyl compounds with trifluoromethyltrimethylsilane (TMS-CF_3). A trifluoromethide equivalent. *J Am Chem Soc* 1989;111:393–395.
7. Maggiarosa N, Tyrra W, Naumann D, Kirij NV, Yagupolskii YL. $[Me_3Si(CF_3)F]^-$ and $[Me_3Si(CF_3)_2]^-$: reactive intermediates in fluoride-initiated trifluoromethylation with Me_3SiCF_3 – an NMR study. *Angew Chem Int Ed* 1999;38:2252–2253.
8. Kolomeitsev A, Bissky G, Lork E, Movchun V, Rusanov E, Kirsch P, Röschenthaler G-V. Different fluoride anion sources and (trifluoromethyl)trimethylsilane: molecular structure of tris(dimethylamino)sulfonium bis(trifluoromethyl)trimethylsiliconate, the first isolated pentacoordinate silicon species with five Si–C bonds. *Chem Commun* 1999;1017–1018.

CHAPTER 14

Preparation of Transition Metal Sulfide Fluorides

MICHAEL GERKEN

The chemistry of transition metal oxide fluorides has been studied extensively. A wide range of synthetic routes to such oxide fluorides is available, including the exchange of oxygen by fluorine, starting from the oxide precursor, or fluorine–oxygen exchange, starting from a fluoride precursor. The synthetic routes to sulfide fluorides are more limited. Few sulfide transfer agents have been used to introduce a sulfide in a controlled way. High-temperature reactions using Sb_2S_3, B_2S_3, or elemental sulfur in autoclaves were used for the preparation of WSF_4 [1, 2], $ReSF_4$ [2, 3], $ReSF_5$ [3], and $MoSF_4$ [4]. However, in some cases, significant amounts of side products were observed. In a solution synthesis, hexamethyldisilathiane, $[(CH_3)_3Si]_2S$, has been used in the preparation of WSF_4 [5] and $MoSF_4$ [6]. A different approach to transition metal sulfide fluorides uses chloride–fluoride exchange starting from the respective sulfide chloride. Salts of WSF_5^- have been synthesized and isolated by this route [7].

The use of Sb_2S_3 in anhydrous HF serves as a controlled way of introducing sulfide into WSF_4 [8], $ReSF_4$ [3], and $ReSF_5$ [3].

14.1 PREPARATIONS OF TRANSITION METAL SULFIDE FLUORIDES USING Sb_2S_3 IN ANHYDROUS HF AS A SULFIDE TRANSFER AGENT

Preparation of WSF_4

Apparatus T-shaped reactor built of 1/4-in. outer diameter FEP (copolymer of perfluoroethylene and perfluoropropylene) equipped with a valve, metal vacuum line system, safety glasses, laboratory coat, and protective gloves.

Chemicals WF_6, Sb_2S_3, anhydrous HF.

Efficient Preparations of Fluorine Compounds, First Edition. Edited by Herbert W. Roesky.

Attention! Safety glasses and protective gloves must be used at all times. The experiment has to be conducted under strictly anhydrous conditions in well-dried apparatus, because the reactant (WF_6), as well as the product (WSF_4), hydrolyzes on contact with traces of moisture.

Caution! Because of its toxicity, corrosive nature, and volatility, anhydrous HF has to be handled with extreme caution in closed, leak-checked systems. Skin contact with anhydrous HF and its inhalation have to be excluded. Calcium gluconate gel has to be available during this reaction for initial treatment of HF burns. Detailed information about the serious health effects of HF and the appropriate response to exposures are described in the literature, which has to be consulted before handling HF [9]. The apparatus has to be thoroughly dried, because traces of moisture will hydrolyze WF_6, generating HF.

Experimental Procedure Anhydrous HF is carefully dried using (a) the reaction with elemental fluorine [10], (b) the reaction with BiF_5 [11], or (c) the reaction with K_2NiF_6 [8]. In the latter method, K_2NiF_6 also serves as an indicator for the absence of water, because it dissolves in anhydrous HF giving a purple solution. A T-shaped FEP reactor equipped with a Kel-F or stainless steel valve is carefully dried and charged with Sb_2S_3 (0.136 g, 0.400 mmol) inside a dry box. On a vacuum manifold constructed from metal and fluoroplastic, anhydrous HF (1.76 g) and WF_6 (0.326 g, 1.09 mmol) are vacuum distilled onto the Sb_2S_3 at –196 °C. The mixture is warmed to room temperature and allowed to react for approximately 12 h. During this time, the solution slowly turns yellow. The bright yellow solution is decanted into the empty arm of the T-shaped reactor. Volatiles are removed under dynamic vacuum, yielding 0.258 g (79.3%) WSF_4.

Characterization Data Mp 89–90 °C. ^{19}F NMR: δ 77.4 (s) ppm, $^{1}J(^{183}\text{W-}^{19}\text{F}) =$ 42 Hz (anhydrous HF solvent); δ 85.4 (s) ppm, $^{1}J(^{183}\text{W-}^{19}\text{F}) = 33$ Hz (CH_3CN solvent). Raman (cm^{-1}): $\bar{\upsilon}$ 707, 687, 578, 568, 535, 513, 248, 236, 225. IR (KBr, cm^{-1}): $\bar{\upsilon}$ 695, 630, 576, 556, 514, 463.

Application Tungsten sulfide tetrafluoride acts as a Lewis acid toward fluoride ions and nitrogen bases. The reaction with acetonitrile yields the hexacoordinate $WSF_4{\cdot}CH_3CN$ adduct [8]. Depending on the ratio of WSF_4 toward F^-, the WSF_5^- and $W_2S_2F_9^-$ anions are obtained [12].

REFERENCES

1. Atherton MJ, Holloway JH. Transition-metal thiofluorides; preparation of WSF_4 and related anions. *J Chem Soc, Chem Commun* 1977;424–425.
2. Holloway JH, Kaučič V, Russell DR. Synthesis, chemistry, and crystal structures of high-valent transition-metal chalcogenide fluorides and their derivatives. *J Chem Soc, Chem Commun* 1983;1079–1081.
3. Holloway JH, Puddick DC, Staunton GM, Brown D. Hepta-, hexa- and quinquevalent rhenium chalcogenide fluorides: the preparation and characterization of ReF_5S, ReF_4S and ReF_3S. *Inorg Chim Acta* 1982;64:L209–L210.

4. Holloway JH, Puddick DC. Preparation and characterization of molybdenum thio- and seleno-tetrafluorides, $MoSF_4$ and $MoSeF_4$. *Inorg Nucl Chem Lett* 1979;15:85–87

5. Banger KK, Blackman CS, Brisdon AK. New synthetic route to WSF_4 and its solution-phase structure as determined by tungsten L(III)-edge extended X-ray absorption fine structure studies. *J Chem Soc, Dalton Trans* 1996;2975–2978.

6. Nieboer J, Mack JP, Mercier HPA, Gerken M. Synthesis, characterization, and computational study of $MoSF_4$. *Inorg Chem* 2010;49:6153–6159.

7. Hilbers M, Läge M, Mattes R. Preparation and structure of $(Ph_3P)_2N[WSF_5]$. *Inorg Chim Acta* 1992;201:1–3.

8. Nieboer J, Hillary W, Yu X, Mercier HPA, Gerken M. Syntheses, characterization, and computational study of WSF_4 and $WSF_4{\cdot}CH_3CN$. *Inorg Chem* 2009;48:11251–11258.

9. (a) Peters D, Miethchen R. Symptoms and treatment of hydrogen fluoride injuries. *J Fluorine Chem* 1996;79:161–165; (b) Segal EB. First aid for a unique acid, HF: a sequel. *J Chem Health Saf* 2000;7:18–23.

10. Emara AAA, Schrobilgen GJ. The fluoro(hydrogen cyanide)xenon(II) cation preparation of $HC{\equiv}NXeF^+ AsF^-$: a multinuclear magnetic resonance and Raman spectroscopic study. *Inorg Chem* 1992;31:1323–1332.

11. Christe KO, Wilson WW, Schack CJ. On the syntheses and properties of some hexafluorobismuthate(V) salts and their use in the metathetical synthesis of NF_4^+ salts. *J Fluorine Chem* 1978;11:71–85.

12. Nieboer J, Haiges R, Hillary W, Yu X, Richardet T, Mercier, HPA, Gerken M. Fluoride-ion acceptor properties of WSF_4: synthesis, characterization, and computational study of the WSF_5^-, and $W_2S_2F_9^-$ anions, and ^{19}F NMR spectroscopic characterization of the $W_2OSF_9^-$ anion. *Inorg Chem* 2012; 51: 6350–6359.

CHAPTER 15

Transition Metal Carbonyl Sulfur Dioxide and Thiazylfluoride Complexes: Reactions at the Metal Center and at the Ligand

RÜDIGER MEWS

Looking for possibilities to introduce thiazylfluoride, N≡SF, and thiazyltrifluoride, $N{\equiv}SF_3$, as ligands into organometallic chemistry, it seemed to be necessary to generate coordinatively unsaturated highly reactive species in a chemically very resistant, only weakly interacting solvent. Liquid SO_2 seemed to be suitable; it is chemically relatively inert and shows good solubility for many salts. Therefore, only small amounts of solvent are needed in these reactions.

Two different routes for the preparation of organometallic SO_2 complexes have been developed: The reaction of organometallic halides with silver salts of large, only weakly interacting anions (e.g., AsF_6^-, SbF_6^-) in liquid SO_2 at room temperature [1] and the oxidation of organometallic species containing metal–metal bonds by AsF_5 in liquid SO_2 at low temperature [2]. The safer route is the first method. $AgAsF_6$ or $AgSbF_6$ is readily prepared from the oxidation of silver by AsF_5 or SbF_5 [3], respectively. In the second approach, these steps are skipped, but the reaction is more difficult to handle. On the other hand, this method is carried out at low temperature and allows the preparation of thermally labile species. Starting materials might be all organometallic halide complexes, resistant toward SO_2, with halides (Cl^-, Br^-) suitable for abstraction by silver salts with almost not interacting counterions (e.g., $AgAsF_6$, $AgSbF_6$).

15.1 PREPARATION OF $[M(CO)_5(SO_2)]^+[AsF_6]^-$ (M = Mn, Re) [1]

$$M(CO)_5Br + AgAsF_6 \xrightarrow{SO_2} [(CO)_5M(SO_2)]^+[AsF_6]^- + AgBr \quad (M = Mn,\ Re)$$

Efficient Preparations of Fluorine Compounds, First Edition. Edited by Herbert W. Roesky.

Apparatus Special pressure-resistant Schlenk vessels [4] or Y-shaped glass vessels with Teflon valves, vacuum line, glove box, glass flask with Teflon valve (for SO_2 storage at room temperature).

Chemicals $Mn(CO)_5Br$, $Re(CO)_5Br$, $AgAsF_6$, SO_2 (dried over P_4O_{10} at room temperature).

Attention! Because the reaction vessel is under pressure, safety glasses must be used at all times.

Caution! SO_2 and the metal carbonyl compounds are very toxic, inhalation or contact with skin should be strictly avoided. Because the valves might leak or the vessel might break, all manipulations must be carried out in a well-ventilated hood.

Experimental Procedure In a glove box, equimolar amounts of $M(CO)_5Br$ and $AgAsF_6$ are filled separately in the two arms of the Y-shaped vessel, the silver salt in the straight arm and the carbonyl halide in the side arm. At a vacuum line, at −196 °C, SO_2 is condensed onto the silver salt (about 5 mL solvent for 10 mmol of each reactant). After warming to room temperature, the $AgAsF_6$ solution is transferred onto the carbonyl halide by tilting the vessel. By cooling the straight arm with cold water, SO_2 recondenses. This procedure is repeated until the silver salt is transferred quantitatively to the side arm. The reaction mixture is stirred at room temperature for 20 h. After the AgBr precipitated, the clear yellow brown (Mn) or yellow (Re) solutions are carefully decanted from the side arm to the straight arm. By recondensing SO_2 back to the side arm, the residue can be washed. These washings are repeated until a colorless solution in the side arm indicates quantitative transfer of the product. Finally, the SO_2 is recondensed to the side arm and the reaction vessel is connected to two condensation traps cooled to −80 °C. The solvent is evaporated at room temperature by opening the valve carefully, leaving behind brown yellow $[Mn(CO)_5SO_2]^+[AsF_6]^-$. When the yellow solution of $[Re(CO)_5(SO_2)]^+BF_4^-$ evaporates, a colorless residue is formed, indicating the loss of SO_2.

Characterization Data $[Mn(CO)_5(SO_2)]^+[AsF_6]^-$: IR ($cm^{-1}$) 2167 (vw), 2161 (vs), 2040 (vs), 1311/1305 (s), 1115 (m). $Re(CO)_5(SO_2)]^+[AsF_6]^-$: IR($cm^{-1}$) 2177 (w), 2059 (vs), 2075 (s), 1313/1307 (vs), 1148 (m).

15.2 PREPARATION OF $[CpFe(CO)_2(SO_2)]^+[AsF_6]^-$ [4]

$$[CpFe(CO)_2]_2 + 3AsF_5^- \xrightarrow{SO_2} 2[CpFe(CO)_2(SO_2)]^+[AsF_6]^- + AsF_3$$

Apparatus A special Schlenck flask [4], or more simple, a Y-shaped, pressure-resistant glass vessel with Teflon valve, a metered vacuum line, pressure-resistant glass flasks with Teflon valves for storage of SO_2 over P_4O_{10} at room temperature.

Attention! Because the reaction vessel is under high pressure, safety glasses must be used at all times.

Caution! SO_2 and the metal carbonyl compounds are very toxic, inhalation or contact with skin should be strictly avoided. Because the valve might leak or the vessel might break, all manipulations must be carried in a well-ventilated hood.

Experimental Procedure Onto 0.907 g (2.56 mmol) of $[CpFe(CO)_2]_2$ at −196 °C, 10 mL of SO_2 and a slight deficiency of AsF_5 (1.32 g, 7.77 mmol) are condensed. The reaction mixture is slowly warmed to room temperature and stirred for 5 h. The clear solution is decanted from a small residue from the side to the straight arm. By cooling the side arm with cold water, the solvent is recondensed. This procedure is repeated until the soluble product is quantitatively transferred. Then the solvent is completely condensed into the side arm. Then the flask is connected to two condensation traps. By slow opening of the valve, the solvent is evaporated, and the product remains at the bottom of the flask as green-black crystalline solid.

Characterization Data IR (Nujol, Kel-F) (cm^{-1}): 3140 (s), 2114 (vs), 2080 (vs), 2018 (s), 1430 (s), 1356 (s), 1143 (vs), 1122 (w), 882 (s), 700 (vs), 676 (m), 582 (m), 574 (m), 546 (s), 520 (s), 474 (w), 394 (s).

Application of SO_2 Complexes The synthesis of cationic organometallic sulfur dioxide complexes from the appropriate halides and silver salts with weakly coordinating anions is a general method for the preparation of this class of compounds. The applicability of the oxidative cleavage of metal–metal bonds by AsF_5 in liquid SO_2 depends on the resistance of the co-ligands toward oxidation.

The stability of the SO_2 complexes $[L_{x-1}M(SO_2)]^+ [MF_6]^-$ depends on the nucleophilicity of the anions $[MF_6]^-$, the Lewis acidity of the metal center M in the organometallic fragment, and the bonding mode. At room temperature, $[Mn(CO)_5(SO_2)]^+[AsF_6]^-$ is converted to $[(CO)_5MnFAsF_5]$ within 36 h, while $[Re(CO)_5FAsF_5]$ is quantitatively formed at 40–50 °C after 20 h [1]. The nucleophilicity of the anions increases with decreasing Lewis acidity of the acids from which the anions are formed. For example $[Re(CO)_5(SO_2)]^+[BF_4]^-$ is present only in solution, indicated by a yellow color. The precipitate from this solution is colorless and contains no SO_2. The SO_2 uptake and loss is reversible. Under the same conditions, $[CpFe(CO)_2(SO_2)]^+[AsF_6]^-$ is completely stable due to the high π-basicity of the $CpFe(CO)_2$ fragment, interacting strongly with a sulfur π-acceptor orbital. In $[(CO)_5MFAsF_5]$, the fluoro Lewis acid and the organometallic Lewis acid are bridged by a fluoride; in other words, the empty coordination site of the strong organometallic Lewis acid $[M(CO)_5]^+$ is occupied by the weakly interacting $[AsF_6]^-$ anion. Other pathways to strong organometallic Lewis acids with different weakly interacting anions as excellent leaving groups have been developed [5]. Besides their acceptor properties toward inorganic donors, their tremendous potential as acceptor for organic s- and p-donors, their use as catalysts, etc. have been reported [5].

15.3 $[Re(CO)_5(NSF)]^+[AsF_6]^-$ [6]

$$[Re(CO)_5SO_2]^+[AsF_6]^- + NSF \rightarrow [Re(CO)_5(NSF)]^+[AsF_6]^- + SO_2 \quad (15.1)$$

Apparatus Glass vessel with stirring bar and valve, vacuum line, glove box.

Chemicals $[Re(CO)_5(SO_2)]^+[AsF_6]^-$, NSF, dry SO_2.

Attention! Safety glasses and protective gloves must be used at all times.

Caution! SO_2 and metal carbonyls are toxic, inhalation or contact with skin has to be strictly avoided. The reaction must be carried out in a well-ventilated hood.

Experimental Procedure In a glove box, 2.64 g (4.56 mmol) of $[Re(CO)_5(SO_2]^+[AsF_6]^-$ is filled into a 200-mL glass flask with a Teflon valve. At −196 °C, 0.297 g (4.56 mmol) of NSF and 10 mL SO_2 are condensed to it. The reaction mixture is warmed to −30 °C to give a clear orange-yellow solution. After stirring for 1 h at this temperature, all volatile components are removed under vacuum. The orange-red crystalline solid remaining is analytically pure $[Re(CO)_5(NSF)]^+[AsF_6]^-$ in quantitative yield, mp 115 °C (dec).

Characterization Data ^{19}F NMR: δ (SF) 219.7; δ (AsF) 100.0 ppm, ^1J (AsF) = 930 Hz. IR(Nujol, (cm^{-1}): 2179 (m), 2046 (vs, br), 1390 (vw), 702 (vs), 675 (vw), 647 (m), 585 (s), 450 (w), 400 (s), 341 (w).

Application In the absence of moisture, $[Re(CO)_5(NSF)]^+[AsF_6]^-$ is stable at room temperature, while NSF itself—due to the highly polar $N(\delta^-)\equiv S(\delta^+)$ bond—rapidly decomposes (oligomerization, polymerization, etc.) under these conditions. Coordination to a metal center blocks the negative end of the dipole; the reactivity of the sulfur–fluorine bond is not affected. Nucleophilic exchange reactions lead to new thiazyl derivates, not observable in the free state. By fluoride abstraction with AsF_5, dicationic species are formed.

This chemistry is not restricted to the $[Re(CO)_5]^+$ fragment, further examples are the ions $[Mn(CO)_5]^+$, $[FeCp(CO)_2]^+$, and $[(CrCp(NO)_2]^+$. Instead of NSF, other labile donors might be stabilized at cationic metal centers.

Two different methods showing the different reaction possibilities of coordinated NSF are illustrated by the following equations:

Nucleophilic exchange [7, 8, 12]

$$[Re(CO)_5(NSF)]^+[AsF_6]^- + Me_2N - SiMe_3$$
$$\rightarrow [Re(CO)_5(NSNMe_2)]^+[AsF_6]^- + Me_3SiF \quad [7] \quad (15.2)$$

$$2[Re(CO)_5(NSF)]^+[AsF_6]^- + MeN(SiMe_3)_2$$
$$\rightarrow [(OC)_5ReNS - N(Me) - SNRe(CO)_5]^{2+}[AsF_6^-]_2 + 2\,Me_3SiF \quad [7] \quad (15.3)$$

$$4[Re(CO)_5(NSF)]^+[AsF_6]^- + SiX_4 \rightarrow 4[Re(CO)_5(NSX)]^2 + [AsF_6]^- + SiF_4 \quad [8, 12] \quad (X = Cl, Br) \tag{15.4}$$

F^- abstraction

$$[Re(CO)_5(NSF)]^+[AsF_6]^- + AsF_5 \rightarrow [Re(CO)_5(NS)]^{2+}[AsF_6^-]_2 \quad [6] \tag{15.5}$$

$[Re(CO)_5(NS)]^{2+}[AsF_6^-]_2$ was the first dicationic carbonyl complex [cf.9] reported in the literature. In a similar way, $[Mn(CO)_5(NS)]^{2+}[AsF_6]_2$, $[CpFe(CO)_2(NS)]^{2+}[AsF_6^-]_2$, and $[CpCr(NO)_2(NS)]^{2+}[AsF_6^-]_2$ were prepared [10].

Another route to the dications is the exchange of the SO_2 ligand in monocationic complexes by $[NS][AsF_6]^-$ [11].

REFERENCES

1. Mews R. Umsetzung von Übergangsmetallhalogeniden mit $AgAsF_6$ in flüssigem SO_2: Eine Schlüsselreaktion zur Darstellung neuer Komplex-Kationen. *Angew Chem* 1975;87:669–670; Reaction of transition metal carbonyl halides with $AgAsF_6$ in liquid SO_2: a key reaction for the preparation of novel complex cations. *Angew Chem, Int Ed Engl* 1975;14:640–641.
2. Hartmann G, Mews R. Eine einfache Methode zur Synthese kationischer metallorganischer Schwefeldioxid-Komplexe durch Metall-Metall-Bindungsspaltung. *Chem Ber* 1986;119:374–377.
3. Roesky HW, Witt M. Silver hexafluoroarsenate. *Inorg Synth* 1986;24:74–76.
4. Hartmann G, Froböse R, Mews R, Sheldrick GM. Synthese und Struktur von $[CpFe(CO)_2SO_2]^+AsF_6^-$. *Z Naturforsch* 1982;37b:1234–1240.
5. Beck W, Sünkel K. Metal complexes of weakly coordinating anions. Precursors of strong cationic organometallic Lewis acids. *Chem Rev* 1998;88:1405–1421.
6. Mews R, Liu CS. Synthese des Thionitrosyl-rhenium-dikations $[Re(CO)_5NS]^{2+}$ aus dem Thiazyfluoridkomplex $[Re(CO)_5NSF]^+$. *Angew Chem* 1983;95:156–157; Synthesis of the thionitrosylrhenium dication $[Re(CO)_5(NS)]^{2+}$ reactions and structure of the thiazylfluoride complex $[Re(CO)_5(NSF)]^+$. *Angew Chem, Int Ed Engl* 1983;22:158–159.
7. Hartmann G, Mews R, Sheldrick GM. Pentacarbonylrhenium-Komplexe mit Thiazylamid-Liganden. Angew Chem 1983;95:753–754; Pentacarbonylrhenium complexes with thiazylamide ligands. Angew Chem Int Ed Engl 1983;72:723–724; *Angew Chem Suppl* 1983;945–956.
8. Hartmann G, Mews R. Methoden zur Darstellung von Rheniumpentacarbonyl(thiazylhalogenid)Komplexen. *Z Naturforsch* 1985;40b:343–347.
9. Willner H, Aubke F. Homoleptische Carbonylkomplex-Kationen der elektronenreichen Metalle: Bildung in supersauren Medien sowie spektroskopische und strukturelle Charakterisierung. *Angew Chem* 1997;109:2506–2530; Homoleptic metal carbonyl cations of electron-rich metals: their generation in super-acid media together with their spectroscopic and structural characterisation. *Angew Chem, Int Ed Engl* 1997;36:2402–2425.

10. Beran K, Steinke G, Mews R. Metallorganische Thiazydfluorid- und Thionitrosyl-Komplexe. *Chem Ber* 1989;122:1613–1616.

11. Hartmann G, Mews R. Kationische Thionitrosyl- und Nitrosyl- Komplexe. *Angew Chem* 1985;97:218–219; Cationic thionitrosyl and nitrosyl complexes. *Angew Chem, Int Ed Engl* 1985;97:202–203.

12. Ruf C, Lork E, Behrens U, Mews R. Rheniumpentacarbonylthiazylhalogenid-Komplexe $[Re(CO)_5(NSX)]AsF_6$: Strukturen und Bindungsverhältnisse. *Z Anorg Allg Chem* 1997;623:444–448.

CHAPTER 16

Cesium, Mercury, and Silver Salts with Sulfur–Nitrogen–Fluorine Anions: Useful Transfer Reagents for NSF Building Blocks

RÜDIGER MEWS

CsF is the classical catalyst in fluorine chemistry for additions to multiple bonds. For the addition to element–nitrogen bonds, HgF_2 might be even more efficient. Primary step in this catalysis is the transfer of a fluoride with intermediate formation of anionic species. Only a few of them were isolated as salts, in many cases a stronger fluoride ion donor than CsF or HgF_2 (e.g., Tris(dimethylamino)sulfonium "fluoride (TASF)", $[(Me_2N)_3S]^+[Me_3SiF_2]^-$) is necessary.

Three different synthetic approaches have been used to generate Cs, Hg, or Ag salts of NSF anions:

a. Cleavage of an element–nitrogen bond by transfer of the fluoride to the element under formation of a leaving group, for example:

$$FC(O)N=SF_2 + F^- \rightarrow OCF_2 + NSF_2^- \quad [1,2] \tag{16.1}$$

$$Me_3SiN=S(O)F_2 + F^- \rightarrow Me_3SiF + NS(O)F_2^- \quad [3,4] \tag{16.2}$$

b. Addition of F^- to NS multiple bond systems, for example:

$$NSF + F^- \rightleftharpoons NSF_2^- \tag{16.3}$$

$$R-N=SF_4 + F^- \rightarrow RNSF_5^- \quad [5,6] \tag{16.4}$$

c. Reaction of persistent NSF acids with metal oxides or carbonates, for example:

$$Ag_2CO_3 + 2HN(SO_2F)_2 \rightarrow 2\,AgN(SO_2F)_2 + CO_2 + H_2O \quad [7] \tag{16.5}$$

Efficient Preparations of Fluorine Compounds, First Edition. Edited by Herbert W. Roesky.

Fluoride ion transfer and formation of stable salts occur, if three-coordinated (or pseudo-three) sulfur centers can be transformed to tetracoordinated systems (Eq. 16.3); or tetra- and pentacoordinated precursors to penta- and hexacoordinated systems, respectively (Eq. 16.4). With reagents containing an almost naked fluoride, for example, TASF ($[(Me_2N)_3S]^+[Me_3SiF_2]^-$) [8], transfer of the fluoride occurs, if the NS multiple bond system is a stronger acceptor than Me_3SiF [9]. By this transfer, several tetra- ($[RNS(O)F]^-$, $[(RN)_2SF]^-$) [10]), penta- ($[RNSF_3]^-$, $[RNSOF_3]^-$) [11], and hexacoordinated anions ($[R–N–SF_5]^-$) [5,6] have been prepared. Although it was shown that HgF_2 catalyzes addition to the NS triple bond of NSF_3 [12], all attempts to transfer NSF_3 to the pentacoordinated anion NSF_4^-, isoelectronic with OSF_4, failed.

The third method (c) is restricted to persistent NSF acids. These acids are persistent with sulfur(VI) substituents. Readily available are $HNS(O)F_2$ [13] and $HN(SO_2F)_2$ [14], while $(F_5S)_2NH$ [15] and $(F_5S)(FSO_2)NH$ [6] are difficult to prepare. For the generation of N metal bonds, the reaction of their precursors $F_5S–N=SF_4$ [16] and $FSO_2N=SF_4$ [17] with a fluoride donor according to (b) is the method of choice.

Examples for methods (a) and (b) will be given.

16.1 MERCURY SULFUR NITROGEN FLUORINE DERIVATIVES BY CLEAVAGE OF AN ELEMENT – NITROGEN – BOND BY HgF_2

Preparation of $Hg(NSF_2)_2$ [18a]

$$HgF_2 + 2FC(O)NSF_2 \rightarrow Hg(NSF_2)_2 + 2OCF_2$$

Apparatus A 250-mL glass flask, magnetic stirrer, 100-mL dropping funnel, reflux condenser, two condensation traps, 50-mL syringe, source of nitrogen, T-shaped N_2 inlet, safety glasses, laboratory coat, protective gloves.

Chemicals HgF_2, $FC(O)NSF_2$ [19] (prepared from $Si(NCO)_4$ [20] and SF_4).

Attention! Safety glasses and protective gloves must be used at all time.

Caution! Because of the high toxicity and volatility of $FC(O)NSF_2$, inhalation or contact with skin must be strictly avoided.

The reaction should be carried out in a well-ventilated hood. Decomposition products of NSF might react explosively with water.

Experimental Procedure In a stream of dry N_2, two oven-dried, hot condensation traps and a reflux condenser are cooled to room temperature. In a dry box a two-necked, 250-mL glass flask with magnetic stirring bar is filled with 11.0 g (0.046 mol) of HgF_2. The flask is connected to the condenser, and to the second neck a 100-mL dropping funnel is added. The funnel is closed by a stopcock, while simultaneously the N_2-T is opened. After cooling down the traps by liquid air, 35.4 g (0.27 mmol) of $FC(O)NSF_2$ is filled through a N_2-T with a syringe into the dropping funnel. Then, $FC(O)NSF_2$ is slowly added to the HgF_2 and stirred for 24 h at room temperature. After the reaction is finished, the reaction vessel is cooled to 0°C, excess of $FC(O)NSF_2$ is removed under vacuum, leaving behind 17.0 g of a colorless solid (80% yield rel. to $FC(O)NSF_2$). The purity of the product depends on the reaction conditions, higher temperature and lower pressure lead to the loss of NSF.

At room temperature, $Hg(NSF_2)_2$ decomposes slowly. The compound should be stored in a refrigerator at −30°C. With H_2O, it reacts vigorously to give SO_2, F^-, and NH_3.

Characterization Data IR (Nujol, cm^{-1}) 1313 (vs), 680 (vs), 574 (s), 550 (m).

$Hg(NSF_2)_2$ crystallizes in the space group *C2/c* with $a = 10.033$ Å, $b = 10.697$ Å, $c = 6.924$ Å, $\beta = 121.96°$, $V = 630.2$ Å^3 [21].

Applications

a. *Preparation of NSF* [18b]
$Hg(NSF_2)_2$ is the best source of pure NSF. At room temperature, the NSF pressure above the solid is small, but quantitative decomposition by heating to 110°C under high vacuum quickly occurs.

$$Hg(NSF_2)_2 \xrightarrow[\text{HV}]{110\ °\text{C}} HgF_2 + 2NSF$$

b. *Synthesis of N-halogen sulfur difluoride imides $XNSF_2$ (X = F* [22], *Cl, Br, I* [2,18c]*)*
$Hg(NSF_2)_2$ reacts at room temperature with Cl_2 in a flow system and with bromine in an autoclave to give the corresponding N-chloro derivative $ClNSF_2$ in 80% yield and the N-bromo derivative $BrNSF_2$ in 70% yield. The N-fluoro derivative $FNSF_2$, the isomer of $N{\equiv}SF_3$, was also obtained, but in very low yield.

$$Hg(NSF_2)_2 + X_2 \rightarrow HgX_2 + XNSF_2$$

c. *Cycloaddition of NSF to 1,3-dipolar reagents* [23]
$Hg(NSF_2)_2$ might be considered as "tamed" NSF. With $(CF_3)_2CN_2$ $Hg(NSF_2)_2$ reacts smoothly to give $(CF_3)_2C{=}N{-}SF$, an unusual stable sulfenyl fluoride. The direct reaction of the diazo compound with NSF leads to vigorous explosions.

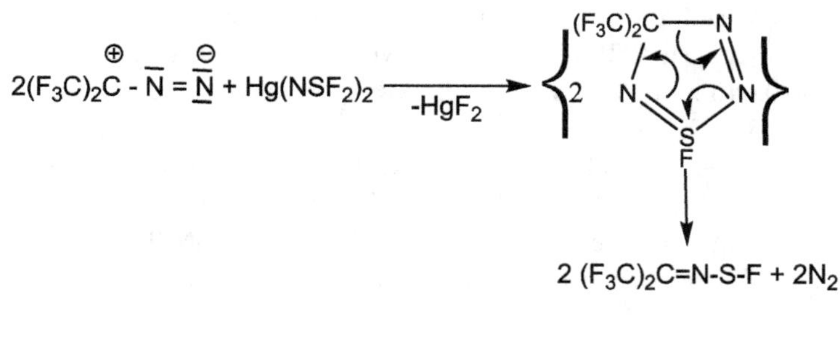

d. *Nucleophilic transfer of the NSF_2 group* [24]

Only a few attempts to transfer the NSF_2 group by nucleophilic exchange are reported in the literature, for example:

$$S_2Cl_2 + Hg(NSF_2)_2 \rightarrow HgCl_2 + F_2S = N - S - S - N = SF_2$$

Mostly, the resulting sulfur difluoride imide looses NSF. The result is a Cl–F exchange. NSF_2^- acts as a fluorinating agent similar to the isoelectronic fluorosulfinate SO_2F^-.

16.2 MERCURY AND CESIUM SULFUR NITROGEN FLUORINE DERIVATIVES BY ADDITION OF THEIR FLUORIDES TO S=N DOUBLE BONDS

Preparation of $Hg(N(SF_5)_2)_2$ [5]

$$HgF_2 + 2\,SF_5N = SF_4 \rightarrow Hg\,(N(SF_5)_2)_2$$

Apparatus A 50-mL Kel-F vessel with a valve, vacuum system, glove box, sublimation apparatus, glass bulb with Teflon valve.

Chemicals HgF_2, $SF_5N{=}SF_4$ (prepared from NSF_3 and fluorine [16]).

Attention! Safety glasses and protection gloves must be used at all times.

Caution! Mercury compounds are highly toxic, inhalation and contact with the skin must be strictly avoided. The toxicity of SF_5–$N{=}SF_4$ is not known, but one should assume, that it is highly toxic. Therefore, all manipulations should be carried out in a well-ventilated hood.

Experimental Procedure At a vacuum line onto 6.9 g of HgF_2 (28.9 mmol) (filled into a Kel-F vessel in a glove box) 11.2 g of SF_5–$N{=}SF_4$ (45.0 mmol) is condensed. The reaction mixture is heated to 50 °C for 14 days. Excess of SF_5NSF_4 is recondensed to a storage vessel. The remaining residue is transferred to a sublimation apparatus in a glove box. Sublimation under static vacuum gave 8.4 g (11.4 mmol) of $Hg(N(SF_5)_2)_2$ (yield 51%), mp 79°C.

Characterization Data IR (Nujol, cm^{-1}): 945 (vs), 893 (vs), 855 (s), 831 (s), 757 (s), 670 (s), 610 (m), 580 (m), 565 (m). ^{19}F NMR: δ_{FA} 83.14, δ_{FB4} ~ 81.95 ppm. MS (70 eV): m/z = 738 [M^+], 489 [$FHgN(SF_5)_2{}^+$], 343 [$FHgNSF_4{}^+$], 324 [$HgNSF_4{}^+$].

Preparation of $Cs(N(SO_2F)(SF_5))$ [6]

$$CsF + FSO_2N = SF_4 \rightarrow Cs(N(SO_2F)(SF_5))$$

Apparatus A 200-mL glass flask with Teflon valve, magnetic stirring bar, glove box, vacuum line.

Chemicals $FSO_2N{=}SF_4$, dried CsF, dry CH_3CN (solvent).

Attention! Safety glasses and protective gloves must be used at all times.

Caution! $FSO_2N{=}SF_4$ is moisture sensitive and hydrolyzes readily under formation of HF, $HN(SO_2F)_2$. Inhalation and contact with skin must be strictly avoided.

Experimental Procedure In a glove box, a 200-mL glass flask with Teflon valve is filled with 1.69 g (11.1 mmol) of rigorously dried CsF. To it by a vacuum line at −196°C, 2.52 g (18.3 mmol) of FSO_2NSF_4 and 5 mL of dry CH_3CN are condensed. Within 1 h, the reaction mixture is warmed to room temperature. After stirring for 24 h at this temperature, all volatiles are removed. About 3.93 g of analytically pure $Cs(N(SO_2F)(SF_5))$ (99% yield) remain as residue, mp 67°C.

Characterization Data IR (Kel-F/Nujol) (cm^{-1}): 1348 (vs), 1327 (sh), 1295 (m), 1210 (vs, br), 1124 (vs), 839 (vs,br), 818 (vs,br), 721 (vs), 704 (s), 613 (s), 598 (m), 589 (m), 578 (s), 565 (vs). ^{19}F NMR ($CD_3CN/CFCl_3$; 75 MHz; + 35°C): δ 100.59 (F_A), 83.15 (F_{B4}), 52.45 ppm (SO_2F); $^2J_{FA\text{-}FB} = 155.93$, $^4J_{FB4\text{-}SO2F} = 10.87$, $^4J_{FA\text{-}SO2F} = 0.84$ Hz.

Application $Hg[N(SF_5)_2]_2$ and $Cs[N(SF_5)(SO_2F)]$ and also $CsN(SF_5)_2$ might be used for introducing the NSF building blocks into inorganic and organic molecules, either directly or by the indirect way via N-halo amines.

$$Hg[N(SF_5)_2]_2 + 2CH_3I \rightarrow 2\,CH_3N(SF_5)_2 + HgI_2$$
$$Hg[N(SF_5)_2]_2 + 2\,CF_3SCl \rightarrow 2\,CF_3SN(SF_5)_2 + HgCl_2$$
$$CsN(SF_5)_2 + ClF \rightarrow ClN(SF_5)_2 + CsF$$
$$ClN(SO_2F)(SF_5) + H_2C = CH_2 \rightarrow ClCH_2CH_2N(SO_2F)(SF_5)$$

Salts of the $N(SO_2F)^-$ anion and their perfluoroalkyl analogs $[N(SO_2R_f)_2]$ have found broad application in electrochemistry (batteries, fuel cells) and as end groups in liquid crystal chemistry. Similar applications could find the $[N(SF_5)_2]^-$ and the $[N(SF_5)(FSO_2)]^-$ ions and their organo derivatives.

REFERENCES

1. Ruff JK. N-Haloiminosulfur derivates. *Inorg Chem* 1966;5:1787–1791.
2. Glemser O, Mews R, Roesky HW. Darstellung und Eigenschaften von Quecksilber-bis-schwefedifluoridimid, N-Chlor-schwefeldifluoridimid und N-Brom-schwefeldifluoridimid. *Chem Ber* 1969;102:1523–1528.
3. Glemser O, Saran H, Mews R. Über die Darstellung von N-Trimethylsilyl-schwefeloxiddifluoridimid und Quecksilber-bis-schwefeloxiddifluoridimid. *Chem Ber* 1971;104:696–699.
4. Seppelt K, Sundermeyer W. Neue Imidoschwefeloxiddifluoride. *Angew Chem* 1970;82:931; New Iminosulfur-oxydifluorides. *Angew Chem, Int Ed Engl* 1970;9:905.
5. Waterfeld A, Mews. R. Bis[bis(pentafluor-λ^6-sulfanyl)amino]quecksilber, $Hg[N(SF_5)_2]_2$. *Chem Ber* 1983;116:1674–1677.

6. Meier T, Mews R. Reaktionen des Fluorsulfonylimino-λ^6-schwefeltetrafluorid FSO_2NSF_4. *Chem Ber* 1993;126:2437–2439.
7. Ruff JK. The imidodisulfuryl Fluoride Ion. *Inorg Chem* 1965;4:1446–1449.
8. Middleton WJ. U.S. patent 3,940,402. 24.02.1976; Tris(Dimethylamino)Sulfonium Difluorotrimethyl Silicate, Middleton WJ., West FA., Vedejs E, *Org Synth* 1986;64:221.
9. Mews R, Borrmann T, Lork E, Stohrer W-D. Fluoride ion transfer and stabilisation of reactive ions. *J Fluorine Chem* 2004;125:903–916.
10. Heilemann W, Mews R. Imino- und Bis(imino)fluorosulfinates. *J Fluorine Chem* 1991;52:377–388.
11. Heilemann W, Mews R. Fluorinated sulfuramideanions II: Derivatives with penta-and ψ-pentacoordinated sulfur-centers. *Eur J Solid State Inorg Chem* 1992;29:799.
12. Waterfeld A, Mews R. HgF_2-assisted additions to the SN-and CN-Triple-Bond. *J Chem Soc, Chem Commun* 1982;839.
13. Parshall GW, Cramer R, Foster RE. Iminosulfur Oxydifluoride and Poly-(oxofluorosulfurnitride). *Inorg Chem* 1962;1:677–679.
14. Ruff JK, Lustig M. Imidodisulfuryl fluoride, cesium imidodisulfuryl fluoride, fluoroimidodisulfuryl fluoride. *Inorg Synth* 1968;11:138–147.
15. Waterfeld A, Oberhammer H, Mews R. Bis(pentafluorosulfanyl)amine. *Angew Chem* 1982;94:390–391; Bis(pentafluorosulfanyl)amines, *Angew Chem, Int Ed Engl* 1982;21: 355–356.
16. Waterfeld A, Mews R. N-Pentafluorsulfanyl Schwefeltetrafluoridimid aus der Reaktion von Thiazyltrifluorid mit elementarem Fluor. *Angew Chem* 1982;94:389–390: N-(Pentafluorosulfanyl)iminosulfur tetrafluoride by reaction of thiazyl trifluoride with fluorine. *Angew Chem Int Edit Engl* 1982;21:354–355.
17. Meier T, Mews R. N-Fluorosulfonyl-Sulfurtetrafluoridimide, FSO_2NSF_4. *J Fluorine Chem* 1989;42:81–85.
18. Mews R, Keller K, Glemser O, Seppelt K, Thrasher JS. Acyclic sulfur nitrogen fluorine compounds, thiazyl trifluoride. *Inorg Synth* 1986;24:12-14(a) Bis(imidosulfurous Difluoridato-N)-mercury(II), 14–16; (b) Thiazyl fluoride, 16–17; (c) N-Haloimidosulfurous difluorides $XNSF_2$, 18–21.
19. Clifford AF, Kobayashi CS. The preparation and properties of N-fluoroformyliminosulfur difluoride. *Inorg Chem* 1965;4:571–574.
20. Thrasher JS. (Fluorocarbonyl)imidosulfurous difluoride. *Inorg Synth* 1986;24:10–12.
21. Krebs B, Meyer-Hussein E, Glemser O, Mews R. Crystal and molecular structure of mercury di-iminosulphurdifluoride: bond properties of the NSF_2 Group. *Chem Commun* 1968;1578–1579.
22. Glemser O, Mews R, Roesky HW. N-Fluorosulphur Difluoride Imide, $F\text{-}N{=}SF_2$. *Chem Commun* 1969;914.
23. Varwig J, Steinbeisser H, Mews R, Glemser O. Imidosulfenyl-halogenide$(CF_3)_2$ C=N-S-X aus Thiazylhalogeniden und Bis(trifluormethyl)diazomethan. *Z Naturforsch* 1974;29b:813–814.
24. Stahl I, Mews R, Glemser O. Schwefel-schwefeldifluoridimide. *Chem Ber* 1980;113: 2430–2433.

CHAPTER 17

Laboratory-Scale Synthesis of Gold Trifluoride and Uranium Hexafluoride

THOMAS M. KLAPÖTKE

Binary d- and f-block metal fluorides are versatile reagents for the preparation of various coordination compounds and can also be used as fluorinating or oxidizing agents. However, some of these metal fluorides are either commercially not available or can only be purchased in bulk quantities. As examples for such metal fluorides, here we describe the small-scale laboratory synthesis of gold trifluoride, AuF_3, and uranium hexafluoride, UF_6.

17.1 PREPARATIONS OF AuF_3 AND UF_3 USING ELEMENTAL FLUORINE AS A FLUORINATING AGENT

Preparation of AuF_3

$$Au(s) + 3/2\,F_2(g) \rightarrow AuF_3(s)$$

Apparatus A 100-mL, thick-walled quartz reaction vessel connected via a Swagelok glass-steel (Whitey) valve to a metal (Monel or steel) fluorine vacuum line, safety glasses, helmet with face shield, laboratory leather coat, protective leather gloves.

Chemicals Au wire with 0.25 mm diameter, dry fluorine gas.

Attention! Safety glasses, a helmet with a polycarbonate face shield, a thick leather coat, and protective gloves must be used at all times.

Caution! Fluorine and chlorine trifluoride are powerful oxidizers; all these materials (including AuF_3) are toxic. Suitable shielding is required, and protective clothing and face masks should be worn all times. Extensive care must be taken to avoid contact between fluorine, fluorides, and oxidizable materials.

Efficient Preparations of Fluorine Compounds, First Edition. Edited by Herbert W. Roesky.

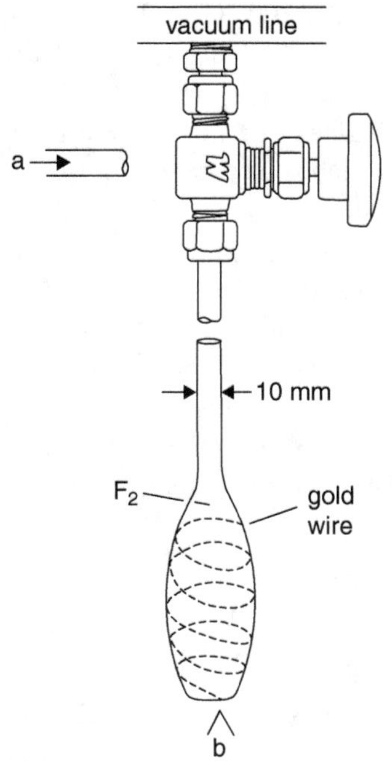

FIGURE 17.1 A 100-mL, thick-walled quartz reaction vessel for the preparation of AuF_3. a, one of the nozzles cooling the Whitey valve with compressed air. b, to initiate the reaction, the bottom of the quartz reaction vessel is rapidly heated for 10 s with a Bunsen burner.

All apparatus (including the quartz reaction vessel) were rinsed several times with $CFCl_3$ (Freon-11), dried and deactivated with ClF_3 at 100 °C and 2 bar for 8 h.

Experimental Procedure A gold wire (Aldrich, >99.9%, 0.25 mm diameter; 0.50 g, 2.54 mmol) was transferred into the passivated reactor (100-mL, thick-walled quartz reaction vessel closed by a Monel needle valve (Whitey) (Figure 17.1). Fluorine (Solvay, 0.42 g, 11.0 mmol, dried over KF) was condensed at −196 °C on the Monel vacuum line, and the mixture was warmed to room temperature. The bottom of the quartz reaction vessel was rapidly heated for 10 s to 800 °C with a Bunsen burner to ignite the gold wire in the fluorine atmosphere. After the reaction was complete, the vessel was re-cooled to room temperature. During the heating and the exothermic reaction, the needle valve was cooled with compressed air by a system of three nozzles. All material volatile at 20 °C was pumped off (through a soda lime absorber). The quartz vessel was taken into a dry box (N_2, 99.999%), opened, and the light orange-yellow AuF_3 was immediately transferred into flamed-out ampoules of Pyrex that were then flame-sealed. Because AuF_3 is very sensitive to atmospheric moisture and water, all handling and storage must be done under the utmost anhydrous conditions. Isolated yield: 0.42 g (62%) [1].

Characterization Data T_{subl} 300 °C. AuF_3 [253.96] F 22.8 (Calcd. 22.4) %. MS (EI, 70 eV, 150 °C) *m/z*: 508 (Au_2F_6), 254 (AuF_3), 235 (AuF_2), 216 (AuF), 197 (Au).

Waste Disposal The residues of AuF_3 have to be hydrolyzed with water and have to be collected in a labeled container for toxic heavy metal waste that has to be properly deposited.

Application Gold(III) fluoride is commercially not available but a useful starting material, for example, for the preparation of gold azides that themselves are suitable binary Au–N compounds for the preparation of ultrapure gold catalysts (thermal decomposition of gold azides) [2–6]:

$$AuF_3 + 3\,Me_3SiN_3 \rightarrow Au(N_3)_3 + 3\,Me_3SiF$$
$$Au(N_3)_3 \rightarrow Au + 4.5\,N_2$$

Preparation of UF_6

$$\text{Method I}: \quad U(s) + 3\,F_2(g) \rightarrow UF_6(s)$$
$$\text{Method II} :\rightarrow U(s) + 3\,ClF_3 \rightarrow UF_6(s) + 3\,ClF(g)$$

Apparatus A 120-mL Monel high-pressure bomb (Parr, series 4740) equipped with a Monel gage block assembly and rupture disk, connected via a Swagelok steel-steel (Whitey) valve to a metal (Monel or steel) fluorine vacuum line, safety glasses, helmet with face shield, laboratory leather coat, protective leather gloves.

Chemicals U turnings, dry fluorine gas.

Attention! Safety glasses and a helmet with a polycarbonate face shield, a thick leather coat and protective gloves must be used at all times.

Caution! Fluorine gas and uranium hexafluoride are powerful oxidizers; all these materials are toxic. Suitable shielding is required, and protective clothing and face masks should be worn all times. Extensive care must be taken to avoid contact between fluorine, uranium hexafluoride, and oxidizable materials.

All apparatus (including the Monel high-pressure bomb) were rinsed several times with $CFCl_3$ (Freon-11), dried and passivated with ClF_3 at 100 °C and 2 bar for 24 h.

Experimental Procedure *Method I*: In the dry box (N_2 99.999%), the Parr high-pressure Monel bomb is loaded with 7.5 g degreased and activated (rinsed with HNO_3) thick uranium chunks (Cerac, >99.7%). Fluorine (Solvay, 4.8 g, 126 mmol, dried over KF) was condensed at −196 °C on the Monel vacuum line, and the mixture was warmed to room temperature. The reactor was then heated to 280 °C for 12 h. After this time, the high-pressure bomb was re-cooled to room temperature. All material volatile at 20 °C was pumped off (through a soda-lime absorber). The UF_6 was purified by sublimation and

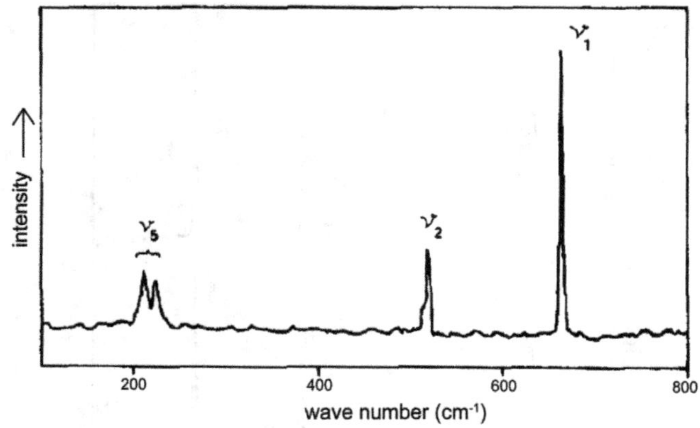

FIGURE 17.2 Raman spectrum of UF_6 (647.09 nm, 20 mW, 20 °C, 2 s/p).

stored in a passivated steel gas-mouse (40 mL). Because UF_6 is very sensitive to atmospheric moisture and water, all handling and storage must be done under the utmost anhydrous conditions. Isolated yield: 7.8 g (70%) [7, 8].

Characterization data Raman (647.09 nm, 20 mW, 20 °C, 2 s/p, $\Delta\nu$ in cm^{-1}): 211 (1) and 223 (1) (δ-FUF, ν_5), 517 (3) (ν-UF, ν_2), 663 (10) (ν-UF, ν_1) (Figure 17.2).

Method II: In the dry box (N_2 99.999%), the Parr high-pressure Monel bomb is loaded with 8.8 g degreased and activated (rinsed with HNO_3) thick uranium chunks (Cerac, >99.7%). Chlorine trifluoride, ClF_3 (Air Products, 12.0 g, 130 mmol), was condensed at −196 °C on the Monel vacuum line, and the mixture was warmed to room temperature. From outside, the reactor was cooled with compressed air by a system of three nozzles. *Caution*: The generated heat warms the outside surface of the bomb substantially. After 1 h all material volatile at −78 °C (generated ClF: mp −155.6 °C; bp −100.1 °C) was pumped off. The bomb was then warmed to −50 °C and traces of unreacted ClF_3 (ClF_3: mp −76.3 °C; bp +11.8 °C) were pumped off. The reactor was taken into the dry box, opened, and the white volatile UF_6 (UF_6: mp 64.05 °C/1139.6 Torr; subl. 56.5 °C) was immediately transferred into a flamed-out Pyrex vacuum sublimation system. After sublimation, white crystalline UF_6 was recovered. Isolated yield: 12.0 g (85%) [7, 8].

Characterization data Raman (neat solid, 647.09 nm, 30 mW, 20 °C, 1 s/p, $\Delta\nu$ in cm^{-1}): 211 (4) and 223 (3) (δ-FUF, ν_5), 519 (4) (ν-UF, ν_2), 664 (10) (ν-UF, ν_1).

Raman (1.5 mol L^{-1} solution in ClCN, 647.09 nm, 40 mW, 20 °C, 1 s/p, $\Delta\nu$ in cm^{-1}): 205 (1) (δ-FUF, ν_5), 395 (1) (δ-ClCN), 527 (1) (ν-UF, ν_2), 665 (10) (ν-UF, ν_1), 726 (3) (ν-Cl-CN) (Figure 17.3).

Waste Disposal The residues of UF_6 have to be hydrolyzed with water and have to be collected in a labeled container for toxic radioactive uranium waste that has to be properly deposited.

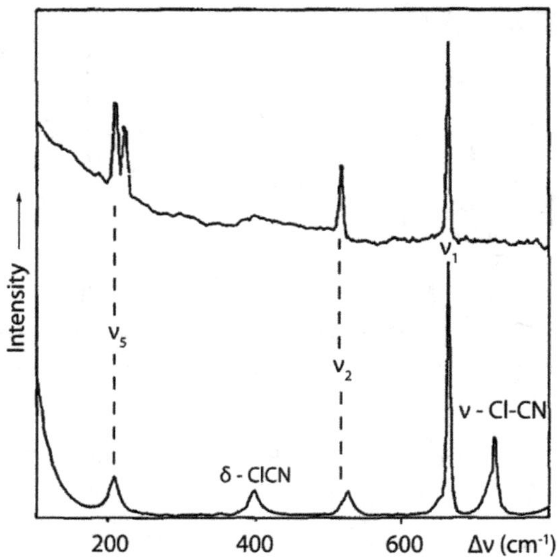

FIGURE 17.3 Raman spectra of neat UF_6 and UF_6 in ClCN solution.

Application Uranium hexafluoride is a good and mild oxidizing agent as shown from the following equations [7, 9]:

$$Au + 2\,CO + UF_6 \rightarrow [Au(CO)_2][UF_6]$$
$$Cp_2NbCl_2 + UF_6 \rightarrow [Cp_2NbCl_2][UF_6]$$
$$Ag + 2\,CH_3CN + UF_6 \rightarrow [Ag(CH_3CN)_2][UF_6]$$

UF_6 is also a good Lewis acid for the coordination of nitrile Lewis bases R–CN (R = H, Cl, CN, CH_2CN) [10–17]:

$$UF_6 + R - CN \rightarrow [UF_6(NC - R)] \quad (R = H,\ Cl,\ CN,\ CH_2CN)$$

REFERENCES

1. Tornieporth-Oetting IC, Klapötke TM. Laboratory scale direct synthesis of pure AuF_3. *Chem Ber* 1995;128:957–958.
2. Schröder D, Hrusak J, Tornieporth-Oetting IC, Klapötke TM, Schwarz H. Definitive experimental evidence for the existence of neutral gold(I) fluoride. *Angew Chem, Int Ed Engl* 1994;33:212–214.
3. Rèffy B, Kolonits M, Schulz A, Klapötke TM, Hargittai M. The intriguing gold trifluoride - molecular structure of monomers and dimers: an electron diffraction and quantum chemical study. *J Am Chem Soc* 2000;122:3127–3134.
4. Beck W, Klapötke TM, Klüfers P, Kramer G, Rienäcker CM. X-ray crystal structures and ab initio calculations of tetraphenylarsonium tetraazido-aurate(III) and azido(triphenylphosphine)gold(I). *Z Anorg Allg Chem* 2001;627:1669–1674.

5. Lear T, Marshall R, Gibson EK, Schütt T, Klapötke TM, Rupprechter G, Freund H-J, Winfield JM, Lennon D. Preparation of a highly structured, residue-free high surface area alumina-supported palladium catalyst. *Phys Chem Chem Phys* 2005;7:565–567.

6. Lear T, Marshall R, Lopez-Sanchez JA, Jackson SD, Klapötke TM, Bäumer M, Rupprechter G, Freund H-J, Lennon D. The application of IR spectroscopy to probe the surface morphology of alumina-supported palladium catalysts. *J Chem Phys* 2005;123:174706-1–174706-13.

7. Schulz A, Klapötke TM. Darstellung und Charakterisierung der kationischen Metallocen-Komplexe [Cp2Zr(CH3CN)3][AsF6]2 und [Cp2NbCl2][UF6] - Laborsynthese von UF6. *J Organomet Chem* 1993;461:91–94.

8. Tornieporth-Oetting IC. Uranium hexafluoride: laboratory scale synthesis, coordination behavior towards nitriles and Raman spectra. *Inorg Chim Acta* 1994;215:5–7.

9. Adelhelm M, Bacher W, Höhn EG, Jacob E. Dicarbonylgold(I) Hexafluorouranate(VI), $Au(CO)_2UF_6$. *Chem Ber* 1991;124:1559.

10. McGhee L, Winfield J. Oxidation of tellurium by molybdenum and uranium hexafluoride in acetonitrile and reactions between uranium hexafluoride and dichlorine or hydrogen chloride in acetonitrile. *J Fluorine Chem* 1992;57:147–154.

11. Iqbal J, Sharp DWA, Winfield J. Oxidation of silver metal and silver(I) cations by molybdenum, tungsten, and uranium hexafluorides and by the nitrosonium cation in acetonitrile. Preparation and properties of silver(III) hexafluorometalates(V). *J Chem Soc, Dalton Trans* 1989;461–464.

12. McGhee L, Rycroft DS, Winfield J. Oxidation of molecular bromine by uranium hexafluoride in acetonitrile. Preparation and properties of hexafluorouranates(V) containing positive bromine. *J Fluorine Chem* 1987;36:351–359.

13. Anderson GM, Iqbal J, Sharp DWA, Winfield J, Cameron JH, McLeod AG. Redox reactions involving molybdenum, tungsten and uranium hexafluorides in acetonitrile. *J Fluorine Chem* 1984;24:303–317.

14. Anderson GM, Winfield JM. Preparation and properties of bis(acetonitrile)iodine(I) hexafluoromolybdate(V) and hexafluorouranate(V). *J Chem Soc, Dalton Trans* 1986;337–341.

15. Sanyal DK, Sharp DWA, Winfield JM. Redox reactions of uranium hexafluoride. Comparison with phosphorus pentafluoride and preparation of copper(I) hexafluorouranate(V) [1]. *J Fluorine Chem* 1981;19:55–60.

16. Crawford M-J, Ellern A, Karaghiosoff K, Mayer P. Nitrate and perchlorate complexes of uranium(IV). *Inorg Chem* 2009;48:10877–10879.

17. Crawford M-J, Mayer P. Structurally characterized ternary U–O–N compound, UN_4O_{12}: $UO_2(NO_3)_2{\cdot}N_2O_4$ or $NO^+\,UO_2(NO_3)_3^-$? *Inorg Chem* 2005;44:8481–8485.

CHAPTER 18

Preparation of Transition Metal Fluorides using ClF_3

RALF HAIGES

General methods for the preparation of high oxidation state transition metal fluorides and oxyfluorides involve reactions with elemental fluorine at elevated temperature and pressure [1–4]. Handling of elemental fluorine can be difficult and special equipment and techniques are required. In addition, reactions with pure fluorine, especially at higher temperatures or pressures, require special safety precautions. Chlorine trifluoride, ClF_3, is a strong oxidizer and powerful fluorinating agent. At ambient temperature, ClF_3 is even more reactive than elemental fluorine. It is a liquid and can easily be handled in stainless steel or Monel equipment. It can be transferred through the vapor phase by condensation in a metal vacuum line.

Many metals react with ClF_3 already at ambient temperature under formation of the corresponding metal fluorides. With metal oxides, ClF_3 reacts as a deoxyfluorinating reagent, resulting in the formation of binary metal fluorides or metal oxyfluorides.

Rhenium oxypentafluoride, $ReOF_5$, is conveniently obtained in a single step from Re_2O_7 and ClF_3. When ClF_3 is used as solvent, no other rhenium-containing by-products are formed and the reaction is quantitative.

$$2Re_2O_7 + 10\,ClF_3 \rightarrow 4\,ReOF_5 + 5\,FClO_2 + 5\,ClF$$

$ReOF_5$ is volatile and can easily be separated from the only by-products ClF and $FClO_2$, and excessive ClF_3 by fractional condensation on a stainless steel vacuum line [5]. $ReOF_5$ is an off-white, relatively volatile solid with a melting point of 41 °C and boiling point of 73 °C and can be sublimed at room temperature [6]. It is an oxidizer and hydrolyzes readily when exposed to air. The compound is soluble in anhydrous HF.

$$Re + 3\,ClF_3 \rightarrow ReF_6 + 3\,ClF$$

Efficient Preparations of Fluorine Compounds, First Edition. Edited by Herbert W. Roesky.

Rhenium hexafluoride, ReF_6, is obtained in a similar way from rhenium metal and ClF_3. A slight excess of rhenium metal should be used in this reaction. Because of the formation of an adduct, ReF_6 and ClF_3 cannot easily be separated by fractional condensation. To simplify the isolation of pure ReF_6, the use of an excess of ClF_3 should be avoided. ReF_6 is a light yellow compound with a melting point of 18.5 °C and a boiling point of 33.7 °C [3]. It has a vapor pressure of 458 Torr (611 mbar) at 20.3 °C [3]. The compound is soluble in HF. ReF_6 is mainly used for the deposition of rhenium metal films in chemical vapor deposition processes for electronics and semiconductors.

The reaction of ClF_3 with molybdenum, tungsten, and osmium in anhydrous HF results in the quantitative formation of MoF_6, WF_6, and OsF_6, respectively. All three compounds can be isolated from the reaction mixture by fractional condensation.

$$M + 3\,ClF_3 \xrightarrow{HF} MF_6 + 3\,ClF \quad (M = Mo,\ W,\ Os)$$

When the less reactive metals ruthenium and iridium are treated with ClF_3 in anhydrous HF, they are only oxidized to the oxidation state +5 and hexafluorometalate(V) salts of the ClF_2^+ cation are formed.

$$2M + 7\,ClF_3 \rightarrow 2\,ClF_2MF_6 + 5\,ClF \quad (M = Ru,\ Ir)$$

After removing all volatiles from the reaction mixture in a vacuum, the salts ClF_2RuF_6 and ClF_2IrF_6 are isolated as yellow solids.

18.1 PREPARATION OF $ReOF_5$ FROM ReO_3 OR Re_2O_7

Apparatus A 175-mL stainless steel cylinder with stainless steel valve (e.g., Hoke 3232M4S) (Figure 18.1), stainless steel vacuum line equipped with Teflon/FEP U-tube vacuum traps (Figure 18.2) [5], glove box with an atmosphere of dry nitrogen, safety glasses, laboratory coat, protective gloves.

Chemicals ReO_3 or Re_2O_7, ClF_3, liquid nitrogen.

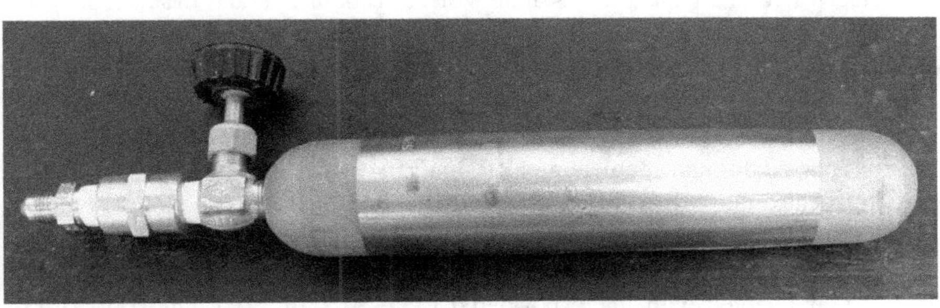

FIGURE 18.1 Stainless steel cylinder ($V = 175$ mL) with stainless steel valve. (For a color version of the figure, please see color plates.)

FIGURE 18.2 A stainless steel vacuum line. (For a color version of the figure, please see color plates.)

Attention! Safety glasses and protective gloves must be used at all times. As a safety precaution for the treatment of HF burns, calcium gluconate gel should be kept on hand.

Caution! Chlorine trifluoride is a toxic, corrosive, and highly reactive liquefied compressed gas packaged in cylinders as a liquid under its own vapor pressure of 1138 Torr (1517 mbar) at 21 °C. It is a very strong oxidizer and will react explosively with water and organic materials. Because of its high reactivity and toxicity, care should be taken while handling it. Improper storage, handling, or use of chlorine trifluoride can result in serious injury and/or property damage. $ReOF_5$ is a poisonous and corrosive strong oxidizer that can ignite organic materials. When exposed to air, ClF_3 and $ReOF_5$ produce fumes of hydrogen fluoride, a corrosive and poisonous gas that etches glass. Prolonged exposure to fumes of ClF_3, $ReOF_5$, and HF may cause pulmonary edema. All reactions should be carried out in a well-ventilated hood and behind safety shields.

Experimental Procedure A 175-mL stainless steel cylinder with a stainless steel valve is attached to a stainless steel vacuum line and evacuated. After about 5 min of pumping, the cylinder is filled with about 100 Torr of ClF_3. After 2–3 min, the cylinder is evacuated again and the gas passed through a Teflon/FEP trap cooled with liquid nitrogen. The cylinder is then filled with 200 Torr of ClF_3. After 2–3 more minutes, the cylinder is re-evacuated and the gas passed through another empty Teflon/FEP trap at −196 °C. This procedure is repeated until the collected solid in the Teflon/FEP trap is no longer orange. This indicates that no chlorine oxides are formed anymore and the inside of the steel cylinder is passivated by a coating with metal fluoride. The cylinder is now charged with finely powdered ReO_3 (9.37 g, 40.0 mmol) or Re_2O_7 (9.68 g, 20.0 mmol) in the nitrogen atmosphere of a dry box. The cylinder is then reattached to the stainless steel vacuum line, evacuated, and cooled with

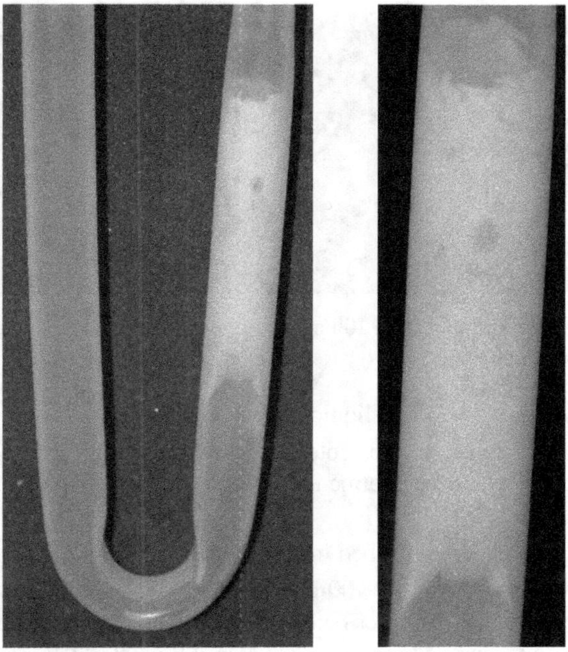

FIGURE 18.3 Teflon/FEP U-trap with crystals of $ReOF_5$. (For a color version of the figure, please see color plates.)

liquid nitrogen. ClF_3 (18.5 g, 200 mmol) is slowly and carefully condensed into the cylinder. The amount of ClF_3 can easily be metered by pressure and temperature measurement using a calibrated volume of the vacuum line and a pressure gauge. The cylinder valve is closed and the cylinder allowed warming to ambient temperature. After 10 h, the content of the steel cylinder is pumped out and fractionated through a series of cold traps at −10 °C, −64 °C, and −196 °C. The trap at −10 °C remains empty and the one at −196 °C contains a yellow-orange mixture of ClF, $FClO_2$, and excess ClF_3. The −64 °C trap yields 11.5 g (97%) of colorless $ReOF_5$ (Figure 18.3). The compound should be stored in passivated stainless steel or Monel cylinders.

Characterization Data ^{19}F NMR (in HF at 293 K): δ −36.0 (F_a), 183.4 (F_e) ppm [7, 8]. IR (5 cm cell with AgCl windows, 10 Torr, cm^{-1}): $\bar{\upsilon}$ 1973 (vw), 1451 (w), 1378 (w), 1279 (vw), 991 (s), 740 (m), 713 (vs), 641 (s) [9].

18.2 PREPARATION OF ReF_6 FROM RHENIUM

Apparatus A 100-mL Monel cylinder with a Monel valve (e.g., Hoke 3732M4M) (Figure 18.4), stainless steel vacuum line equipped with Teflon/FEP U-tube vacuum traps (Figure 18.2) [5], glove box with an atmosphere of dry nitrogen, safety glasses, laboratory coat, protective gloves.

FIGURE 18.4 Monel cylinder ($V = 100$ mL) with Monel valve. (For a color version of the figure, please see color plates.)

Chemicals Re powder, ClF_3, liquid nitrogen.

Attention! Safety glasses and protective gloves must be used at all times. As a safety precaution for the treatment of HF burns, calcium gluconate gel should be kept on hand.

Caution! Same as that described in the previous procedure for $ReOF_5$.

Experimental Procedure A 100-mL Monel cylinder with a Monel valve is passivated with ClF_3 as described in the previous procedure for $ReOF_5$. The cylinder is now charged with finely powdered Re metal (9.31 g, 50.0 mmol) in the nitrogen atmosphere of a dry box. At the stainless steel vacuum line, the cylinder is evacuated and cooled with liquid nitrogen. ClF_3 (12.9 g, 140 mmol) is condensed into the cylinder very slowly. The cylinder valve is closed and the cylinder warmed to ambient temperature. After 10 h, the content of the Monel cylinder is fractionated through a series of cold traps at −31 °C, −64 °C, −121 °C, and −196 °C. The trap at −31 °C remains empty and the one at −196 °C contains a colorless solid of frozen ClF. The trap at −64 °C contains yellow crystals of ReF_6 (Figure 18.5). The content of the −121 °C is fractioned

FIGURE 18.5 Teflon/FEP U-trap with solid ReF_6. (For a color version of the figure, please see color plates.)

two more times through traps at −64 °C and −196 °C. Combining the contents of all traps at −64 °C results in a yield of 13.87 g (99 %) of yellow ReF_6. The compound should be stored in passivated stainless steel or Monel cylinders.

Characterization Data IR (5-cm cell with AgCl windows, 10 Torr, cm^{-1}): $\bar{\upsilon}$ 1282 (vw), 717 (vs) [10].

18.3 PREPARATION OF MF_6 (M = Mo, W, Os) FROM MOLYBDENUM, TUNGSTEN, OR OSMIUM

Apparatus A 100-mL Monel cylinder with a Monel valve (e.g., Hoke 3732M4M), stainless steel vacuum line equipped with Teflon/FEP U-tube vacuum traps [5], glove box with an atmosphere of dry nitrogen, safety glasses, laboratory coat, protective gloves.

Chemicals Mo, W, or Os powder, ClF_3, anhydrous HF, liquid nitrogen.

Attention! Safety glasses and protective gloves must be used at all times. As a safety precaution for the treatment of HF burns, calcium gluconate gel should be kept on hand.

Caution! Chlorine trifluoride is a toxic, corrosive, and highly reactive liquefied compressed gas packaged in cylinders as a liquid under its own vapor pressure of 1138 Torr (1517 mbar) at 21 °C. It is a very strong oxidizer and will react explosively with water and organic materials. Because of its high reactivity and toxicity, care should be taken while handling it. Improper storage, handling, or use of chlorine trifluoride can result in serious injury and/or property damage. MoF_6, WF_6, and OsF_6 are poisonous and corrosive. When exposed to air, ClF_3, MoF_6, WF_6, and OsF_6 produce fumes of hydrogen fluoride. Anhydrous hydrogen fluoride is a corrosive and poisonous gas that etches glass. Burns with HF are usually very serious, with the potential for significant complications due to fluoride toxicity. Areas of the skin that have been exposed to HF have to be treated immediately with calcium gluconate gel. Concentrated HF liquid or vapor may cause severe burns, metabolic imbalances, pulmonary edema, and life-threatening cardiac arrhythmias. Even moderate exposures to concentrated HF may rapidly progress to fatality if left untreated. All reactions should be carried out in a well-ventilated hood and behind safety shields.

Experimental Procedure In the dry nitrogen atmosphere of a glove box, a 100-mL Monel cylinder with a Monel valve is loaded with Mo, W, or Os powder (13.0 mmol), respectively. The cylinder is attached to a stainless steel vacuum line, evacuated, and cooled with liquid nitrogen. Anhydrous HF (5.0 g, 250 mmol) and ClF_3 (3.79 g, 41.0 mmol) are slowly condensed into the cylinder. The cylinder valve is closed and the cylinder warmed to ambient temperature. After 48 h, the volatile material from the cylinder is fractioned through traps at −84 °C and −196 °C. The trap at −84 °C yields 12.5 mmol (96 %) of colorless metal hexafluoride.

Characterization Data MoF_6: IR (5-cm cell with AgCl windows, 5 Torr, cm^{-1}): $\bar{\upsilon}$ 913 (w), 745 (vs) [10]; WF_6: IR (5-cm cell with AgCl windows, 10 Torr, cm^{-1}): $\bar{\upsilon}$ 806 (w), 716 (vs) [11]; OsF_6: IR (5 cm cell with AgCl windows, 10 Torr, cm^{-1}): $\bar{\upsilon}$ 960 (w), 732 (vs) [11].

18.4 PREPARATION OF ClF_2MF_6 (M = Ru, Ir) FROM METALLIC RUTHENIUM AND IRIDIUM

Apparatus A 30-mL stainless steel cylinder with stainless steel valve (e.g., Hoke 3232M4S), stainless steel vacuum line equipped with Teflon/FEP U-tube vacuum traps (Figure 18.2) [5], glove box with an atmosphere of dry nitrogen, safety glasses, laboratory coat, protective gloves.

Chemicals Ru or Ir powder, ClF_3, anhydrous HF, liquid nitrogen.

Attention! Safety glasses and protective gloves must be used at all times. As a safety precaution, calcium gluconate gel should be kept on hand.

Caution! Chlorine trifluoride is a toxic, corrosive, and highly reactive liquefied compressed gas packaged in cylinders as a liquid under its own vapor pressure of 1138 Torr (1517 mbar) at 21 °C. It is a very strong oxidizer and will react explosively with water and organic materials. Because of its high reactivity and toxicity, care should be taken while handling it. Improper storage, handling, or use of chlorine trifluoride can result in serious injury and/or property damage. ClF_2MF_6 is poisonous and corrosive. It is a strong oxidizer that can ignite organic materials. When exposed to air, ClF_3 and ClF_2MF_6 produce fumes of hydrogen fluoride. Anhydrous hydrogen fluoride is a corrosive and poisonous gas that etches glass. Burns with HF are usually very serious, with the potential for significant complications due to fluoride toxicity. Areas of the skin that have been exposed to HF have to be treated immediately with calcium gluconate gel. Concentrated HF liquid or vapor may cause severe burns, metabolic imbalances, pulmonary edema, and life-threatening cardiac arrhythmias. Even moderate exposures to concentrated HF may rapidly progress to fatality if left untreated. All reactions should be carried out in a well-ventilated hood and behind safety shields.

Experimental Procedure In the dry nitrogen atmosphere of a glove box, a 30-mL stainless steel cylinder with a stainless steel valve is loaded with Ir or Ru powder (3.0 mmol), respectively. The cylinder is attached to a stainless steel vacuum line, evacuated, and cooled with liquid nitrogen. Anhydrous HF (1.0 g, 50 mmol) and ClF_3 (0.9 g, 10 mmol) are slowly condensed into the cylinder. The cylinder valve is closed and the cylinder warmed to ambient temperature. After 24 h, all volatile material is removed from the steel cylinder by pumping at room temperature for 4 h. Recrystallization of the pale yellow solid residue from HF at −20 °C yields 0.82 g (94%) of yellow ClF_2RuF_6 or 1.12 g (95% yield) of pale yellow ClF_2IrF_6, respectively.

Characterization Data ClF_2RuF_6 Raman (1064 nm, 200 mW, cm^{-1} [rel. intensity]): $\bar{\upsilon}$ 806 [2.9] ($\nu_3(B_1)$ $ClF_2{}^+$), 790 [7.5] ($\nu_1(A_1)$ $ClF_2{}^+$), 683 [10.0] (ν_1 $RuF_6{}^-$), 628 [3.5] (ν_3 $RuF_6{}^-$), 556 [2.3]/540 [1.1] (ν_2 $RuF_6{}^-$), 378 [0.5] (ν_2 (A_1) $ClF_2{}^+$), 271 [4.0] (ν_5 $IrF_6{}^-$), 234 [0.3] [12].

ClF_2IrF_6 Raman (1064 nm, 200 mW, cm^{-1} [rel. intensity]): $\bar{\upsilon}$ 800 [3.0] ($\nu_3(B_1)$ $ClF_2{}^+$), 791 [6.0]/787 [4.8] ($\nu_1(A_1)$ $ClF_2{}^+$), 674 [10.0] (ν_1 $IrF_6{}^-$), 653 [0.1]/645 [0.1] (ν_3 $IrF_6{}^-$), 568 [1.8]/558 [3.5] (ν_2 $IrF_6{}^-$), 383 [0.2] (ν_2 (A_1) $ClF_2{}^+$), 259 [2.6]/253 [2.4]/241 [1.8] (ν_5 $IrF_6{}^-$), 230 [0.5], 173 [0.4].

REFERENCES

1. Nakajima T, Zemva B, Tressaud A. *Advanced Inorganic Fluorides. Synthesis, Characterization and Application.* Amsterdam: Elsevier; 2000 and references therein.
2. (a) Ruff O, Kwasnik W, Ascher E. Die Fluorierung des Rheniums. *Z Anorg Allg Chem* 1932;209:113–122. (b) Ruff O, Kwasnik W. Rheniumfluoride. *Angew Chem* 1934;47:480.
3. Aynsley EE, Peacock RD, Robinson PL. The reaction of fluorine with the oxides and oxy-compounds of manganese and rhenium. *J Chem Soc* 1950;1622–1624.
4. Meshri DT. Fluorine compounds, inorganic, rhenium. *Kirk-Othmer Encyclopedia of Chemical Technology*. Chichester: Wiley-Interscience; 2000.
5. Christe KO, Wilson RD, Schack CJ. Chloryl fluoride. *Inorg Synth* 1986;24:3–6.
6. Sunder WA, Stevie FA. The synthesis and characterization of some oxide fluorides of rhenium and osmium. *J Fluorine Chem* 1975;6:449–463.
7. Bartlett N, Beaton S, Reeves LW, Wells EJ. ^{19}F N.M.R. spectra of heptavalent fluorides and oxide pentafluorides. *Can J Chem* 1964;42:2531–2540.
8. LeBlond N, Mercier HPA, Dixon DA, Schrobilgen GJ. Syntheses and structures of $TcOF_5$ and the $Tc_2O_2F_9{}^+$ cation and formation of the $TcOF_4{}^+$ cation in solution. *Inorg Chem* 2000;39:4494–4509.
9. Holloway JH, Selig H, Claassen HH. Vibrational spectra of IOF_5, $OsOF_5$, and $ReOF_5$. *J Chem Phys* 1971;54:4305–4311.
10. Claassen HH, Goodman GL, Holloway JH, Selig H. Raman spectra of MoF_6, TcF_6, ReF_6, UF_6, SF_6, and TeF_6 in the vapor state. *J Chem Phys* 1970;53:341–348.
11. Bosworth YM, Clark RJH, Rippon DM. The vapor phase raman spectra, raman band contour analyses, and coriolis constants of the spherical top molecules MF_6 (M = S, Se, Te, Mo, W, or U), $M(CH_3)_4$ (M = C, Si, Ce, Sn, or Pb), P_4, As_4 and OsO_4. *J Mol Spectrosc* 1973;46:240–255.
12. Bougon R, Cicha WV, Lance M, Meublat L, Nierlich M, Vigner J. Preparation, characterization, and crystal structure of $[ClO_2]^+[RuF_6]^-$. Crystal structure of $[ClF_2]^+[RuF_6]^-$. *Inorg Chem* 1991;30:102–109.

CHAPTER 19

Preparation of Fluorine-Containing Molecular Halides and Heteropolar Salts with Elements of Group 15 and Niobium and Tantalum Halides

LOTHAR KOLDITZ

One of the most efficient methods for preparing fluorine compounds is halogen metathesis, in most of the cases exchange of chlorine with fluorine is the preferred procedure. Proper fluorinating reagents have to be selected that do not result in difficult product separations.

A convenient fluorinating reagent for phosphorus, arsenic, as well as niobium and tantalum halides is arsenic trifluoride, AsF_3. It can be handled in dry glass apparatus, if moisture is strictly excluded. Herein, the preparation of well-defined halides of group 15 as well as of Nb and Ta is described.

Pure HF is a powerful fluorinating agent and has the advantage of eliminating volatile HCl but has the disadvantage of not allowing the use of glass apparatus.

Substances used and prepared herein are highly hygroscopic, and therefore experiments have to be carried out in Schlenk flasks under inert dry gas atmosphere (N_2 or Ar) in airtight systems. Flasks have to be charged in a dry box, and glass connections must be lubricated with fluorocarbon grease or polytetrafluoroethylene (PTFE) coatings.

Caution: AsF_3 is a dangerous and extremely toxic compound because of its skin-damaging property. Skin contact and inhalation of AsF_3 and $AsCl_3$ have to be strictly avoided. All experiments have to be carried out in a well-ventilated hood, and safety glasses and protective gloves must be used at all times. Contact of $AsCl_3$ and AsF_3 with moist air has to be avoided due to the formation of HCl and HF. Chemical residues after the preparation must be hydrolyzed, alkalized, and collected in labeled containers for toxic waste.

Efficient Preparations of Fluorine Compounds, First Edition. Edited by Herbert W. Roesky.

19.1 PREPARATION OF FLUORINE-CONTAINING COMPOUNDS WITH GROUP 15 ELEMENTS USING AsF_3 OR HF AS FLUORINATING REAGENTS

Preparation of $[PCl_4][PF_6]$

Apparatus Airtight glass equipment with three-necked, 500-mL glass flask, N_2 inlet, dropping funnel, magnetic stirrer, glass filter, safety glasses, protective gloves, laboratory coat.

Chemicals PCl_5, $AsCl_3$, AsF_3.

Attention and Caution! Safety glasses and protective gloves must be used at all times. Laboratory coat is necessary for protecting clothes. PCl_5, $AsCl_3$, and AsF_3 have to be handled in a well-ventilated hood. $AsCl_3$ and AsF_3 are extremely toxic.

Experimental Procedure An oven-dried, three-necked, 500-mL glass flask is charged in a dry box with PCl_5 (46 g, 0.11 M $[PCl_4][PCl_6]$), and connected with the apparatus. PCl_5 is dissolved in $AsCl_3$ (300 mL), then liquid AsF_3 (29 g, 0.22 mol) is added drop by drop. $[PCl_4][PF_6]$ immediately begins to deposit as a white crystalline salt. The reaction mixture must be stirred and slightly cooled to remain at room temperature. The deposit is filtered off by a glass frit within the apparatus and washed with $AsCl_3$. A stream of dry nitrogen is used to remove excess of $AsCl_3$. Yield: 35 g (100%; 0.11 M) [1].

Characterization Data $[PCl_4][PF_6]$ is a very hygroscopic white crystalline solid, subliming at 135 °C, soluble in CH_3CN, and very slightly soluble in $AsCl_3$, PCl_3, and $POCl_3$. Anion $[PF_6]^-$ is stable against hydrolysis even in alkaline solution. Dissolving $[PCl_4][PF_6]$ in 1 M KOH (0.78 g in 20 mL), evaporating in vacuo at 45 °C to about 3 mL, and washing crystals with ethanol results in pure $K[PF_6]$. Compound $[PCl_4][PF_6]$ reacts with excess AsF_3 to give PF_5.

Preparation of PF_5

Apparatus Airtight glass equipment with three-necked, 500-mL glass flask, magnetic stirrer, dropping funnel, glass traps.

Chemicals PCl_5, $AsCl_3$, AsF_3.

Attention and Caution! Safety glasses and protective gloves must be used at all times. Laboratory coat is necessary for protecting clothes. PCl_5, $AsCl_3$, and AsF_3 have to be handled in a well-ventilated hood. $AsCl_3$ and AsF_3 are extremely toxic.

Experimental Procedure $[PCl_4][PF_6]$ is produced as described in the previous experiment. After slightly warming to 25 °C, AsF_3 is added drop by drop in stoichiometric amounts to the flask containing the $[PCl_4][PF_6]$ [1].

$$3[PCl_4][PF_6] + 4AsF_3 \rightarrow 6\,PF_5 + 4\,AsCl_3$$

This reaction is quantitative. The resulting PF_5 is passed through a trap at –20 °C to remove $AsCl_3$ and AsF_3 and collected in a second trap cooled with liquid nitrogen.

Characterization Data PF_5 is a moisture-sensitive gas, condensing at –85 °C, with a mp at –92 °C.

Preparation of $[PBr_4][PF_6]$

Apparatus Airtight glass equipment with three-necked, 500-mL glass flask, N_2 inlet, dropping funnel, magnetic stirrer, glass filter, safety glasses, protective gloves, laboratory coat.

Chemicals PBr_5, AsF_3, dry CCl_4.

Attention and Caution! Safety glasses and protective gloves must be used at all times. Laboratory coat is necessary for protecting clothes. PBr_5, AsF_3, and CCl_4 have to be handled in a well-ventilated hood. $AsCl_3$ and PBr_5 are extremely toxic.

Experimental Procedure The suspension of PBr_5 (64.6 g, 0.15 mol) in 150 mL of dry CCl_4 is rapidly stirred and treated drop by drop with AsF_3 (22.2 g, 0.17 mol). The temperature of the suspension is kept between 30 °C and 50 °C. After adding the significant amount of AsF_3, the solution turned colorless; the brown PBr_5 has changed into a white crystalline compound. A yellow solution may indicate that moisture was not completely excluded or the precursor PBr_5 contains an excess of bromine. White reaction product is filtered off by a glass frit within the apparatus, and washed with CCl_4 to remove $AsBr_3$. Vacuum is used to remove excess of CCl_4. Yield: 31.2 g (84%) [2].

Characterization Data $[PBr_4][PF_6]$ is a white hygroscopic solid, soluble in AsF_3 at 0 °C, and less soluble in nonpolar solvents. Warming AsF_3 solutions of $[PBr_4][PF_6]$ causes Br–F exchange of $[PBr_4]^+$ to give PF_5. However, this exchange is much slower when compared with the reaction of AsF_3 with $[PCl_4][PF_6]$.

Preparation of $SbCl_4F$

Apparatus Airtight glass equipment with two-necked, 250-mL glass flask, dropping funnel, condenser, bulb tube.

Chemicals $SbCl_5$, AsF_3.

Attention and Caution! Safety glasses and protective gloves must be used at all times. Laboratory coat is necessary for protecting clothes. $SbCl_5$ and AsF_3 have to be handled in a well-ventilated hood. AsF_3 and $SbCl_5$ are very toxic chemicals.

Experimental Procedure When $SbCl_5$ (29.1 g, 0.1 mol) is mixed with AsF_3 (13.9 g, 0.1 mol), at once a slight raise of temperature occurs. The reaction is completed by using an oil bath at a temperature of 110 °C for 1 h. Excess

AsF_3 and $AsCl_3$ are removed in vacuum followed by distillation of the residue in high vacuum (10^{-5} mm Hg). The resulting product is condensed into a bulb tube. The bulb with the compound can be sealed off [3].

Characterization Data Colorless transparent square crystals, mp 83 °C.

$SbCl_4F$ crystallizes in the tetragonal system with lattice constants of $a =$ 12.87 Å and $c =$ 7.84 Å. The structure consists of cyclic $(SbCl_4F)_4$ units with F bridges between $SbCl_4$ groups [4]. The Sb–F bond lengths are longer when compared with single bonds, suggesting a strong heteropolar character in agreement with the Raman spectrum [5]. Cryoscopic molecular mass determinations in AsF_3 resulted in 138 (M/2 $SbCl_4F$ = 142) with the result that Sb–F–Sb bridges are no longer existent in AsF_3 solution giving solvated $SbCl_4^+$ and F^- ions.

$SbCl_4F$ can be used for converting C–Cl into C–F bonds [6].

Preparation of SbF_5

Apparatus PTFE, Monel equipment, reflux condenser (PTFE or Monel)

Chemicals $SbCl_5$, HF (water-free).

Attention and Caution! Safety glasses and protective gloves must be used at all times. Laboratory coat is necessary for protecting clothes. $SbCl_5$ and HF have to be handled in a well-ventilated hood. They are extremely toxic. Boiling point of HF is 19 °C.

Experimental Procedure Higher fluorination of $SbCl_5$ demands application of hydrogen fluoride. $SbCl_5$ is dissolved in HF at –75 °C to form $H_2F^+ SbCl_5F^-$. The chlorine–fluorine exchange starts at –40 °C by metathesis of two chlorine atoms in the temperature range between –40 °C and –10 °C. Further substitution occurs at about 0 °C, and at the boiling temperature of HF (19 °C), $H_2F^+ SbClF_5^-$ is formed. The latter is only slowly converted to $H_2F^+ SbF_6^-$ under refluxing condition with excess of HF. Before the excess of HF is removed, the chlorine must be completely exchanged to yield SbF_5, otherwise Cl_2 is formed that results in the formation of Sb(III)Sb(V) fluorides [7, 8].

An alternative preparation of SbF_5 is the reaction of SbF_3 with elemental fluorine.

Characterization Data Hygroscopic colorless liquid, bp 50 °C, mp 7 °C.

Preparation of $NbCl_4F$

Apparatus Airtight glass equipment with two-necked, 250-mL Schlenk flask, graduated glass tube, glass frit, N_2 inlet.

Chemicals $PCl_5 \cdot NbCl_5$, $AsCl_3$, AsF_3.

Attention and Caution! Safety glasses and protective gloves must be used at all times. Laboratory coat is necessary for protecting clothes. PCl_5, $NbCl_5$, $AsCl_3$, and AsF_3 have to be handled in a well-ventilated hood. $AsCl_3$ and AsF_3 are extremely toxic.

Experimental Procedure $PCl_5 \cdot NbCl_5$ is prepared according to the method described by Gut and Schwarzenbach [10]. It is heated to 110 °C with a small amount of $AsCl_3$, then $AsCl_3$ is added until the $PCl_5 \cdot NbCl_5$ is completely dissolved. The resulting solution is passed through a glass frit. Cooling of the solution resulted in the formation of crystals of composition $PCl_5 \cdot NbCl_5 \cdot AsCl_3$.

AsF_3 is added drop by drop to a solution of $PCl_5 \cdot NbCl_5$ in $AsCl_3$ saturated at room temperature (15 g/100 mL $AsCl_3$). During the addition of AsF_3, the cooling bath is kept at –20 °C. The formed PF_5 is immediately removed by bubbling nitrogen through the reaction mixture. After the addition of 0.204 mL AsF_3 (4.18 mmol) per g $PCl_5 \cdot NbCl_5$ (2.09 mmol), the initially deposited $NbCl_5$ is converted into $NbCl_4F$. The latter is filtered off by means of a glass frit and the adhering $AsCl_3$ is removed in vacuum.

For this reaction, it is important to use $PCl_5 \cdot NbCl_5$ *in statu nascendi*. Pure $NbCl_4F$ could not be produced when crystalline $NbCl_5$ is used as a precursor [9].

Characterization Data Slight yellow hygroscopic powder, mp 201 °C, conductivity of melt $8.08 \times 10^{-6}\ \Omega^{-1}\text{cm}^{-1}$. $NbCl_4F$ crystallizes in the triclinic system and forms cyclic tetrameric units of composition $(NbCl_4F)_4$.

Lattice constants are: $a = 12.01$ Å, $\alpha = 95.0°$, $b = 13.36$ Å, $\beta = 92.7°$, $c = 8.13$ Å, $\gamma = 93.2°$ [11].

Preparation of NbF_5

Apparatus Same equipment as that for the preparation of $NbCl_4F$.

Chemicals $PCl_5 \cdot NbCl_5$, AsF_3.

The preparation of $PCl_5 \cdot NbCl_5$ as described in the previous experiment.

Attention and Caution! Safety glasses and protective gloves must be used at all times. Laboratory coat is necessary for protecting clothes. $PCl_5 \cdot NbCl_5$ and AsF_3 have to be handled in a well-ventilated hood. They are extremely toxic.

Experimental Procedure $PCl_5 \cdot NbCl_5$ is directly treated with AsF_3 without any additional solvent to give NbF_5. The formed $AsCl_3$ is removed in vacuo (1 mm Hg) at 25 °C. This method is more convenient than the reaction of elemental fluorine with $NbCl_5$ [9].

Characterization Data Colorless needles, mp 78 °C.

Preparation of $TaCl_4F$

Apparatus Same equipment as that for the preparation of $NbCl_4F$. Airtight glass equipment with two-necked, 250-mL Schlenk flask, graduated glass tube, glass frit, N_2 inlet.

Chemicals $PCl_5 \cdot TaCl_5$, $AsCl_3$, AsF_3.

Attention and Caution! Safety glasses and protective gloves must be used at all times. Laboratory coat is necessary for protecting clothes. PCl_5, $TaCl_5$, $AsCl_3$, and AsF_3 have to be handled in a well-ventilated hood. $AsCl_3$ and AsF_3 are extremely toxic.

Experimental Procedure $PCl_5 \cdot TaCl_5$ is prepared according to the method described by Gut and Schwarzenbach [10]. It is heated and is dissolved in $AsCl_3$ with slightly warming. Colorless needles are deposited when the solution is slowly cooled. The crystals are recovered by filtration. Washing the crystals with $AsCl_3$ and removing the solvent by dry nitrogen or argon gas result in the pure compound of composition $PCl_5 \cdot TaCl_5 \cdot AsCl_3$.

A 100-mL Schlenk flask is charged in a dry box with a weighed amount of $PCl_5 \cdot TaCl_5 \cdot AsCl_3$. Solvent $AsCl_3$ is added (4 mL $AsCl_3$ per g of compound $PCl_5 \cdot TaCl_5 \cdot AsCl_3$). A clear solution is received by slowly warming the flask. Cooling the flask with wet ice produces a fresh suspension with reactive particles. AsF_3 is then added drop by drop (2 mol AsF_3 per mol $PCl_5 \cdot TaCl_5$ = 0.18 mL AsF_3 per g $PCl_5 \cdot TaCl_5$). After the addition, the reaction mixture remains for 12–15 h at room temperature. Adding a small amount of AsF_3 is favorable for completing the reaction. Cooling the solution to -10 °C results in the formation of colorless crystals. The crystals are recovered by filtration, washed with $AsCl_3$, and dried in vacuum (1 mm Hg) at 25 °C, which results in a 70% yield of $TaCl_4F$.

$TaCl_4F$ can also be prepared using liquid SO_2 as a solvent instead of $AsCl_3$. Under these conditions, PF_5 is removed together with SO_2 at room temperature [12].

Characterization Data Hygroscopic colorless needles, mp 214 °C. $TaCl_4F$ is isomorphic to $SbCl_4F$ and crystallizes in tetragonal system with constants $a = 12.71 \pm 0.08$ Å and $c = 7.84 \pm 0.05$ Å.

It forms tetrameric cyclic units of composition $(TaCl_4F)_4$ in the solid state with fluorine bridges between the Ta atoms [13].

Preparation of TaF_5

Apparatus Airtight glass equipment with two-necked, 250-mL Schlenk flask, graduated glass tube, glass frit, N_2 inlet.

Chemicals $PCl_5 \cdot TaCl_5$, $AsCl_3$, AsF_3.

Attention and Caution! Safety glasses and protective gloves must be used at all times. Laboratory coat is necessary for protecting clothes. $PCl_5 \cdot TaCl_5$, $AsCl_3$, and AsF_3 have to be handled in a well-ventilated hood. They are extremely toxic.

Experimental Procedure Reaction of $PCl_5 \cdot TaCl_5$ or $TaCl_5$ with excess of AsF_3 at room temperature and slightly warming results in the formation of TaF_5. This reaction is more convenient than the preparation from $TaCl_5$ and elemental fluorine [12].

Characterization Data Colorless crystals, mp 97 °C.

19.2 FLUORINE-CONTAINING MIXED HALIDES AS MOLECULAR AND HETEROPOLAR ISOMERS

Molecular and heteropolar isomers are known of phosphorus and arsenic halides, especially those containing fluorine. On the one hand, the tendency to form heteropolar salt structure increases with the formation of stable anions $[PF_6]^-$ or $[AsF_6]^-$. On the other hand, the stability of cations increases with chlorine containing cations of coordination number 4.

Commonly, in solid-state structure, the heteropolar isomer is preferred. Solid phosphorus(V) chloride exhibits the $[PCl_4][PCl_6]$ structure. Solutions of $[PCl_4][PCl_6]$ in polar solvents such as CH_3CN show ionic conductivity, whereas in nonpolar solvents such as CCl_4 the molecular form (PCl_5) is dissolved. The latter solutions show no ionic conductivity, if moisture is strictly excluded.

Preparation of PCl_4F (molecular)

Apparatus Airtight glass equipment with two-necked, 100-mL Schlenk flask and three connecting traps.

Chemicals $[PCl_4][PF_6]$

Attention and Caution! Safety glasses and protective gloves must be used at all times. Laboratory coat is necessary for protecting clothes. $[PCl_4][PF_6]$ has to be handled in a well-ventilated hood. It is extremely toxic.

Experimental Procedure An oven-dried, 100-mL Schlenk flask is charged with compound $[PCl_4][PF_6]$ in a dry box and connected with three in-line arranged traps. Vacuum (1 mm Hg) is kept in the equipment while warming the solid to the sublimation temperature of 135 °C. The first trap is cooled to –60 °C, the second one to –80 °C, and the third to –100 °C. In the first trap, molecular PCl_4F is deposited, in the second one PCl_2F_3 as a main product besides PF_5, and in the third trap PF_5 [14].

Characterization Data Molecular PCl_4F has a boiling point of ca. 67 °C and melts at –62 °C. It is stable at –60 °C. On warming to room temperature, a slow deposition of a white solid takes place, showing the same composition of PCl_4F.

Preparation of heteropolar $[PCl_4]F$

Apparatus A 100-mL, three-necked Schlenk flask, N_2 inlet, reflux condenser, two cooling traps.

Chemicals $[PCl_4][PF_6]$, $AsCl_3$.

Attention and Caution! Safety glasses and protective gloves must be used at all times. Laboratory coat is necessary for protecting clothes. $[PCl_4][PCl_6]$ and $AsCl_3$ have to be handled in a well-ventilated hood. They are extremely toxic.

Experimental Procedure A three-necked Schlenk flask is charged in a dry box with $[PCl_4][PF_6]$ (23 g). After adding 38 mL of $AsCl_3$, the equipment is arranged with a reflux condenser and three cooling traps. The suspension is warmed to 70–85 °C for 3 h. Gaseous PCl_2F_3 and PF_5 volatilize as gases. PCl_2F_3 (condensing at 8 °C) and PF_5 (bp –85 °C, mp –94 °C) are condensed in the traps cooled to –20 °C and –100 °C.

In the vertical reflux condenser, molecular PCl_4F is condensing. However, the main amount is dropping back to the Schlenk flask. Some of the liquid PCl_4F is converted in the condenser to the solid compound $[PCl_4]F$ that is contaminated with $[PCl_4][PF_6]$ from the reverse reaction of PCl_4F with PF_5.

The hot reaction solution is filtered off from white flakes by means of a glass frit and cooled in an ice bath for 12 h. The contamination of the crystals with $AsCl_3$ is removed in vacuum (1 mm Hg). Warming the crystals to 80 °C in vacuo removes the contamination of $[PCl_4][PF_6]$. Yield of $[PCl_4]F$ is 20% [14]. Its constitution may be $[PCl_4][PCl_4F_2]$.

Characterization Data Sublimation at about 175 °C, mp under pressure 177 °C, soluble in $AsCl_3$ and CH_3CN.

REFERENCES

1. Kolditz L. Tetrachlorphosphonium-hexafluorophosphat, $[PCl_4]^+[PF_6]^-$. *Z Anorg Allg Chem* 1956;284:144–152.
2. Kolditz L, Feltz A. Tetrabromophosphor(V)-hexafluorophosphat, $[PBr_4][PF_6]$. *Z Anorg Allg Chem* 1957;293:155–167.
3. Kolditz L. Tetrachloroantimon(V)-fluorid, $[SbCl_4]^+F^-$. *Z Anorg Allg Chem* 1957; 289:128–140.
4. Preiss H. Die Kristallstruktur der Verbindung $SbCl_4F$. *Z Chem* 1966;6:350–351.
5. Dehnicke K, Weidlein J. Ramanspektrum von $SbCl_4F$. *Chem Ber* 1965;98:1087–1092.
6. Kolditz L, Schultz S. Chlor-Fluor-Austauschreaktionen mit Antimonhalogeniden. *J Fluorine Chem* 1975;5:141–151.
7. Kolditz L, Daunicht H. Über die Reaktion des Antimon(V)-chlorides mit Fluorwasserstoff. *Z Anorg Allg Chem* 1959;302:230–236.
8. Ruff O, Plato W. Reaktion von $SbCl_5$ mit HF. *Ber Dtsch Chem Ges* 1904;37:673–683.
9. Kolditz L, Furcht G. Tetrachloroniob(V)-fluorid. *Z Anorg Allg Chem* 1961;312:11–17.
10. Gut R, Schwarzenbach G. Die Verbindungen $PCl_5 \cdot NbCl_5$ und $PCl_5 \cdot TaCl_5$. *Helv Chim Acta* 1959;42:2156–2163.
11. Preiss H. Die Kristallstruktur von $NbCl_4F$. *Z Anorg Allg Chem* 1968;362:13–18.
12. Kolditz L, Kürschner C, Calov U. Tetrachlorotantal(V)-fluorid. *Z Anorg Allg Chem* 1964;329:172–178.
13. Preiss H. Die Kristallstruktur von $TaCl_4F$. *Z Anorg Allg Chem* 1966;346:272–278.
14. Kolditz L. Tetrachlorphosphorfluorid, PCl_4F. *Z Anorg Allg Chem* 1956;286:307–316.
15. Kolditz L. Tetrachloroantimon(V)-fluorid, $[SbCl_4]^+F^-$. *Z Anorg Allg Chem* 1957;289:128–140, 135. (Preparation of $[AsCl_4][AsF_6]$).
16. Kolditz L, Weiss D, Calov U. Tetrachloroarsen(V)-hexafluoroantimonat, $[AsCl_4][SbF_6]$, und über Chlorofluoroantimonate. *Z Anorg Allg Chem* 1962;316:261–269.

17. Kolditz L, Schäfer W. Über die Verbindungen [SCl_3][AsF_6], [$SeCl_3$][AsF_6] und [$TeCl_3$][AsF_6]. *Z Anorg Allg Chem* 1962;315:35–45.
18. Ruff O. Reaktion von SCl_4 mit AsF_3. *Ber Dtsch Chem Ges* 1904;37:3513.
19. Kolditz L, Calov U. Molybdän(V)-dichloridtrifluorid, $MoCl_2F_3$. *Z Chem* 1966;6:431.
20. Kolditz L, Neumann V, Kilch G. Über Vanadinchloridfluoride. *Z Anorg Allg Chem* 1963;325:275–280.

CHAPTER 20

Fluoro and Fluorohydroxy Complexes of As, Sb, and Sn

LOTHAR KOLDITZ

The ions $[PF_6]^-$ and $[AsF_6]^-$ show remarkable stability; therefore, their tendency of forming these ions in mixed fluorides is high with the consequence that heteropolar structures are obtained. Generally, the coordination number of 6 is the preferred one of halogen compounds of higher periods.

The reaction of tetracoordinate arsenates with hydrofluoric acid (40%) gives fluorohydroxy arsenates having coordination number 6 at the arsenic atom. This type of reaction is also possible with the corresponding compounds of Sb and Sn.

Replacing OH^- in $[AsF_5OH]^-$ by F^- is not observed with hydrofluoric acid (40%). Higher concentrations or at least water-free HF is necessary for substituting the OH^- group by a fluoride anion. This step for exchanging OH^- group to give the AsF_6^- complex is slightly hindered. The ions $[SbF_6]^-$, $[GeF_6]^{2-}$, and $[SnF_6]^{2-}$ in comparison to $[PF_6]^-$ and $[AsF_6]^-$ are not highly stable against hydrolysis. Therefore, fluorohydroxy anions of these species can also be prepared by hydrolysis of the fluoride complexes.

20.1 PREPARATION USING HYDROFLUORIC ACID AND HYDROGEN FLUORIDE

Glass equipments are not useful for reactions in the presence of fluoride and HF. Therefore, polytetrafluoroethylene (PTFE) is highly recommended.

Caution! Hydrofluoric acid and also water-free hydrogen fluoride should not come into contact with the skin because this will produce harmful wounds. In case of exposure to HF, plenty of water should be used immediately to wash off the HF. Additionally, after treatment of the skin with water, a solution

Efficient Preparations of Fluorine Compounds, First Edition. Edited by Herbert W. Roesky.

of calcium gluconate should be used. Safety glasses, protecting gloves, and laboratory coat have to be used at all times in the laboratory. A well-ventilated hood is necessary for avoiding inhalation. Residues of the preparations must be hydrolyzed, alkalized, and collected in labeled containers for proper deposition.

Preparation of K[AsF_5OH]

Apparatus Pt dish.

Chemicals KH_2AsO_4, hydrofluoric acid (40%).

Attention and Caution! Glass equipments may not be used for reactions in the presence of fluoride and acid, but PTFE or Pt apparatus are necessary.

Hydrofluoric acid and also water-free hydrogen fluoride should not come into contact with the skin because this will produce harmful wounds. Safety glasses, protecting gloves, and laboratory coat have to be used at all times during preparation. A well-ventilated hood is necessary for avoiding inhalation. Residues of preparation must be hydrolyzed, alkalized, and collected in labeled containers for proper deposition.

Experimental Procedure KH_2AsO_4 (45 g) is dissolved in hydrofluoric acid (40%, 125 g) using a Pt dish. *Caution*: The reaction is exothermic. After cooling to room temperature, water and HF is removed until few crystals are formed. Further cooling results in the deposition of more crystalline material. The crystals are separated by filtration and washed with a small amount of ice water, then three times with ethanol, and finally twice with ethyl ether. Yield: 35 g, 62% [1].

Characterization Data Orthorhombic crystals, hydrolyzing in KOH.

Preparation of [$N(C_2H_5)_4$][SbF_6]

Apparatus Pt dish.

Chemicals Sb_2O_5, [$N(C_2H_5)_4$]OH, hydrofluoric acid (40%).

Attention and Caution! Glass equipments may not be used for reactions in the presence of fluoride and acid, but PTFE or Pt apparatus are necessary.

Caution! Hydrofluoric acid and also water-free hydrogen fluoride should not come in contact with the skin because this will produce harmful wounds. Safety glasses, protecting gloves, and laboratory coat have to be used at all times during preparation. A well-ventilated hood is necessary for avoiding inhalation. Residues of preparation must be hydrolyzed, alkalized, and collected in labeled containers for proper deposition.

Experimental Procedure Preparation of $[SbF_6]^-$ succeeds with hydrofluoric acid (40%). Higher concentrations of HF are not necessary. A saturated solution of [$N(C_2H_5)_4$]OH is added drop by drop to a solution of Sb_2O_5 (16 g, prepared by hydrolysis of $SbCl_5$) with 25 g hydrofluoric acid (40%) until formation of

$[N(C_2H_5)_4][SbF_6]$ is complete. The solid precipitate is recovered by filtration and followed by repeated washing with water, ethanol, and finally ethyl ether. The crude product is dried in vacuum and recrystallized from a water–ethanol mixture [7].

Characterizing Data White crystals, mp 328 °C.

Preparation of $[N(C_2H_5)_4][SbF_5OH]$

Apparatus Pt dish.

Chemicals $[N(C_2H_5)_4][SbF_6]$.

Attention and Caution! Glass equipments may not be used for reactions in the presence of fluoride and acid, but PTFE or Pt apparatus are necessary.

Hydrofluoric acid and also water-free hydrogen fluoride should not come into contact with the skin because this will produce harmful wounds. Safety glasses, protecting gloves, and laboratory coat have to be used at all times during preparation. A well-ventilated hood is necessary for avoiding inhalation of HF. Residues of preparation must be hydrolyzed, alkalized, and collected in labeled containers for proper deposition.

Experimental Procedure $[N(C_2H_5)_4][SbF_6]$ (1 g) is dissolved by heating in 50 mL of H_2O. After evaporation of the solution to half of the volume, undissolved material is filtered off, and the solution further evaporated until crystals of $[N(C_2H_5)_4][SbF_5OH]$ are formed [7].

Characterization Data Thin white crystal plates.

Preparation of $Na_2[SnF_5OH]$

Apparatus Pt dish.

Chemicals $Na_2[Sn(OH)_6]$, hydrofluoric acid (20%).

Attention and Caution! Glass equipments may not be used for reactions in the presence of fluoride and acid, but PTFE or Pt apparatus are necessary.

Caution! Hydrofluoric acid and also water-free hydrogen fluoride should not come into contact with the skin because this will produce harmful wounds. Safety glasses, protecting gloves, and laboratory coat have to be used at all times during preparation. A well-ventilated hood is necessary for avoiding inhalation. Residues of preparation must be hydrolyzed, alkalized, and collected in labeled containers for proper deposition.

Experimental Procedure $Na_2[Sn(OH)_6]$ (5.35 g) is dissolved with 50 mL H_2O, then 10 g hydrofluoric acid (20%) is slowly added under stirring. Deposit at the beginning is going in solution. After slow evaporation to half the previous volume and cooling, $Na_2[SnF_5OH]$ crystallizes from the solution. Yield: 4 g, 75% [12].

Characterization Data White crystals.

REFERENCES

1. Kolditz L, Röhnsch W. Über Fluoroarsenate. *Z Anorg Allg Chem* 1957;293:168–194.
2. Haase W. Beiträge zur Kristallchemie der Hydroxofluorokomplexe von Arsen und Antimon. *Chem Zvesti* 1965;19:167–171.
3. Dunken H, Haase W, Schönherr F. Die Kristallstruktur von Kaliumdihydroxotetrafluoroarsenat. *Z Chem* 1967;7:399–400.
4. Kolditz L, Sarrach D. Über Hydroxofluoroantimonate(V). *Z Anorg Allg Chem* 1957;293:132–146.
5. Kolditz L, Nussbücker B. Über Trimetatetrafluoroarsenat. *Z Anorg Allg Chem* 1961;312:299–306.
6. Kolditz L, Nussbücker B, Schönherr M. Über kondensierte Tetrafluoroarsenate. *Z Anorg Allg Chem* 1965;335:189–194.
7. Kolditz L, Nussbücker B. Über kondensierte Fluorooxoantimonate. *Z Anorg Allg Chem* 1965;337:191–199.
8. Kolditz L, Gitter M. Dekafluorooxodiarsenat(V), $[F_5AsOAsF_5]^{2-}$, und das analoge Antimonat. *Z Anorg Allg Chem* 1967;354:15–22.
9. Dunken H, Haase W. Die Kristallstruktur des Kalium-, Rubidium- und Ammoniumtetrafluoroarsenats. *Z Chem* 1963;3:433–434.
10. Kolditz L, Rehak W. Die Reaktion von SbF_5 mit Carbonat. *Z Anorg Allg Chem* 1965;341:100–102.
11. Kolditz L, Preiss H. Fluorooxocomplexe des Zinns und Germaniums. *Z Anorg Allg Chem* 1963;325:263–274.
12. Kolditz L, Preiss H. Fluorohydroxocomplexe des Germaniums und Zinns. *Z Anorg Allg Chem* 1963;325:252–262.

CHAPTER 21

Trifluoromethyl Compounds via Electrophilic and Nucleophilic Reactions

JEAN'NE M. SHREEVE

The availability of effective electrophilic [1–3] and nucleophilic [4] trifluoromethylating reagents has enhanced the opportunity for the introduction of the trifluoromethyl group(s) into inorganic or organic substrates via simple reactions to form a large variety of useful inorganic and organic compounds. A trifluoromethyl group has unique features, such as high electronegativity, stability, and lipophilicity. Thus, trifluoromethylated inorganic and organic compounds are increasingly important for developing new or more effective medicines and agricultural chemicals and new materials such as liquid crystals. The versatility of S-(trifluoromethyl)dibenzothiophenium trifluorosulfonate (**1**) [1,2] and (trifluoromethyl)trimethylsilane (**2**) [4] as excellent examples of electrophilic and nucleophilic reagents is shown in reactions with inorganic or organic nucleophiles and inorganic or organic electrophiles, respectively.

21.1 PREPARATION OF TRIFLUOROMETHYL-CONTAINING COMPOUNDS USING S-(TRIFLUOROMETHYL) DIBENZOTHIOPHENIUM TRIFLUOROSULFONATE (1) AS A REACTIVE ELECTROPHILE

Preparation of trifluoronitromethane (CF_3NO_2) [5–9]

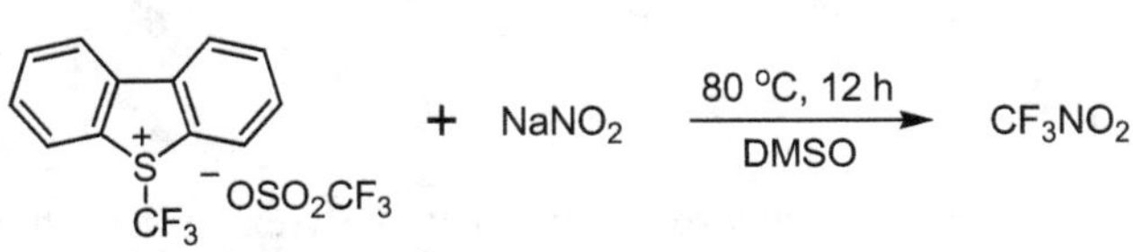

1

Efficient Preparations of Fluorine Compounds, First Edition. Edited by Herbert W. Roesky.

Apparatus A 50-mL Pyrex Schlenk flask with standard taper joint and vacuum stopcock, stir bar, magnetic stirrer, syringe (10 mL), vacuum transfer line, eye protection, lab coat, gloves, low vapor pressure stopcock grease.

Chemicals $NaNO_2$, S-(trifluoromethyl)dibenzothiophenium trifluorosulfonate, dry dimethyl sulfoxide (DMSO).

Attention! Eye protection and gloves must be used at all times.

Caution! All reactions should be carried out in a well-ventilated hood. All chemicals should be handled with extreme care assuming that they may be toxic and volatile.

Experimental Procedure Into a dry Pyrex Schlenk flask equipped with a standard taper joint and a magnetic stir bar, $NaNO_2$ (0.08 g, 1.2 mmol) and S-(trifluoromethyl)dibenzothiophenium trifluorosulfonate, **1** (0.33 g, 1 mmol) are placed. DMSO (10 mL) is added. The flask is closed with a vacuum stopcock attached to a standard taper outer joint. The reaction mixture is heated at 80 °C with stirring for 12 h. After cooling to ambient, the pure volatile product (bp ~ 30 °C) is transferred under vacuum to give (89.4%) colorless CF_3NO_2 (89.4%).

Characterization Data bp ~ 30 °C. ^{19}F NMR: δ −72.3 (t, $J_{F\text{-}N}$ = 14.54 Hz) ppm. IR (gas, KBr, cm^{-1}): 1627 (vs), 1612 (vs), 1309 (vs), 1280 (vs), 1272 (vs), 1167 (s), 1153 (s), 870 (w), 862 (w), 854 (w), 758 (m), 751 (m), 744 (m).

Application Trifluoronitromethane, CF_3NO_2, especially in combination with suitable small molecules gives compositions with chemical stability, low toxicity, low or nonflammability, and efficiency in use while reducing or eliminating the deleterious ozone depletion potential. Particularly attractive compositions or as a single molecule are suggested for use in the generation of information recording media, therapeutic delivery systems, gas and gaseous precursor-filled microspheres, in heat transfer fluids applications, and as gaseous dielectrics.

Waste Disposal Standard procedures for disposal of organic solvents and organic and inorganic materials should be used.

Preparation of 2-trifluoromethyl pyrrole

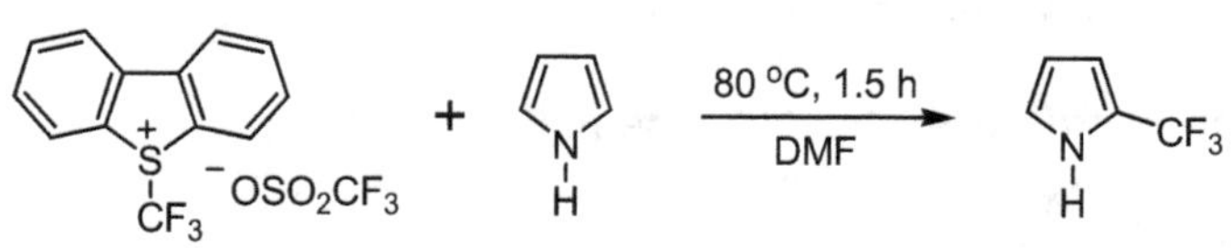

1

Apparatus A 50-mL, two-necked, round-bottom Pyrex flask, septum, T-shaped nitrogen inlet, stir bar, magnetic stirrer, syringe (10 mL), eye protection, lab coat, gloves.

Chemicals Pyrrole, S-(trifluoromethyl)dibenzothiophenium trifluorosulfonate (**1**), dry dimethylformamide (DMF).

Attention! Eye protection and gloves must be used at all times.

Caution! All reactions should be carried out in a well-ventilated hood. All chemicals should be handled with extreme care assuming that they may be toxic.

Experimental Procedure Into a dry, two-necked, 50-mL, round-bottom Pyrex flask equipped with a magnetic stir bar, a septum, and a T-shaped N_2 inlet containing 2.5 mmol of pyrrole in 10 mL dry DMF, S-(trifluoromethyl)-dibenzothiophenium trifluorosulfonate (**1**; 0.33 g, 1 mmol) is added. The resulting mixture is stirred at 80 °C for 1.5 h. After cooling to ambient, the mixture is extracted with diethyl ether to give a mixture of pyrrole and 2-trifluoromethyl pyrrole. Trifluoromethyl pyrrole is separated from the higher boiling pyrrole (bp 129–131 °C) via fractional distillation to give 0.11 g (90%) of colorless 2-trifluoromethyl pyrrole

Characterization Data bp 102–104 °C. ^{1}H NMR: δ 6.05 (1 H, dd, $J = 3.5$ and 2.5 Hz), 6.41 (1 H, dd, $J = 3.5$ and 2.4 Hz), 6.63 (1 H, dd, $J = 2.5$ and 2.4 Hz), 6.99 (1 H, s) ppm. ^{19}F NMR ($CFCl_3$): δ −59.0 (s) ppm. IR (liquid, NaCl): 3300, 1580, 1260, 740 cm^{-1}.

Application Various derivatives of 2-trifluoromethyl pyrrole have applications in pharmacology, as light and bright colored antifouling agents, and as insecticidal, acaricidal, and nematicidal agents.

Waste Disposal Standard procedures for disposal of organic solvents and organic and inorganic materials should be used.

21.2 PREPARATION OF TRIFLUOROMETHYL-CONTAINING COMPOUNDS USING (TRIFLUOROMETHYL)TRIMETHYLSILANE (2) AS A REACTIVE NUCLEOPHILE

Trifluoromethyl)trimethylsilane was prepared in 85% yield via condensation of CF_3Br and Me_3SiCl with $(Et_2N)_3P$ in benzonitrile at −78 °C [10].

Preparation of α-keto amide (4-[2-hydroxy-2-phenyl-2-(trifluoromethyl)acetyl]morpholine) [4, 10, 11]

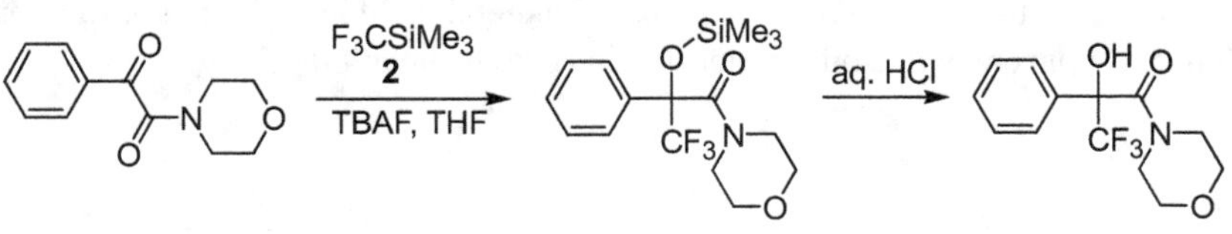

Apparatus A 50-mL, round-bottom Pyrex flask, T-shaped inlet, stir bar, magnetic stirrer, syringe (10 mL), eye protection, lab coat, gloves.

Chemicals 2-Morpholin-4-yl-2-oxo-1-phenylethanone (**3**), (trifluoromethyl)trimethylsilane, dry tetrahydrofuran (THF), ethyl ether, anhydrous tetrabutylammonium fluoride, anhydrous $MgSO_4$, 4 N HCl.

Attention! Eye protection and gloves must be used at all times.

Caution! All reactions should be carried out in a well-ventilated hood. All chemicals should be handled with extreme care assuming that they may be toxic.

Experimental Procedure Into a dry, two-necked, 50-mL, round-bottom flask equipped with a magnetic stir bar, and a T-shaped N_2 inlet containing 5 mL of dry THF, 2-morpholin-4-yl-2-oxo-1-phenylethanone (**3**; 1.095 g, 5 mmol) and (trifluoromethyl)trimethylsilane (0.852 g, 6 mmol) are added. Tetrabutylammonium fluoride (0.1 mL solution in THF, 0.1 mmol) is added drop by drop at room temperature. The reaction is exothermic, and the solution changes from colorless to yellowish brown. The mixture is stirred at 25 °C for 3 h and then hydrolyzed with 4 N HCl (6 mL) with stirring for 2 h.

The reaction mixture is diluted with water (20 mL), and the products extracted with ethyl ether (25 mL). The extract is dried over anhydrous $MgSO_4$ and filtered. Removal of ether at reduced pressure gives a colorless solid 4-[2-hydroxy-2-phenyl-2-(trifluoromethyl)acetyl]morpholine (95% yield).

Characterization Data mp 150–151 °C. ^1H NMR ($CDCl_3$): δ 3.45 (m, 8), 5.57 (1H, s, br), 7.38 (5H, m) ppm. ^{13}C NMR ($CDCl_3$): δ 66.1, 78.9 (q, $J_{C\text{-}C\text{-}F}$ = 28 Hz), 123.6 (q, $J_{C\text{-}F}$ = 282 Hz), 126.2, 128.9, 129.3, 134.5, 166.2 ppm. ^{19}F NMR ($CDCl_3$): δ −74.4 (s) ppm. MS (EI) *m/z* (species, relative intensity) (cm^{-1}): 289 (M^+, 38), 272 (M^+ − OH, 27), 114 ($CONC_4H_8O^+$, 100). IR (solid, KBr) 3295 (b), 1644 (s).

Application α-Hydroxy α-trifluoromethylated amides are attractive because of their structural analogy to some of the antiandrogens, which have found application in the treatment of prostate cancer in humans.

Waste Disposal Standard procedures for disposal of organic solvents and organic and inorganic materials should be used.

Preparation of hexa(trifluoromethyl)cyclotriphosphazene $(CF_3)_6P_3N_3$ [12, 13]

Hexafluorocyclotriphosphazene was prepared in good yield by refluxing hexachlorocyclotriphosphazene with sodium fluoride in acetonitrile at 80 °C.

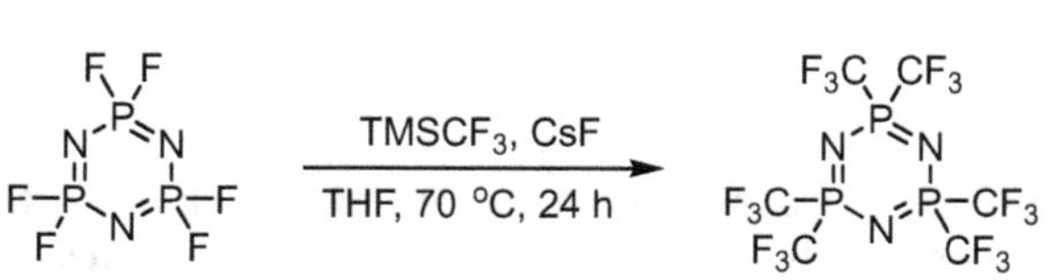

Apparatus A 100-mL, two-necked, round-bottom Pyrex flask, condenser, nitrogen inlet, stir bar, magnetic stirrer, syringe (25 mL), eye protection, lab coat, gloves.

Chemicals Hexafluorocyclotriphosphazene, (trifluoromethyl)trimethylsilane (**2**), dry THF, finely powdered anhydrous CsF (dried at 200 °C), dry trichloromethane.

Attention! Eye protection and gloves must be used at all times.

Caution! All reactions should be carried out in a well-ventilated hood. All chemicals should be handled with extreme care assuming that they may be toxic.

Experimental Procedure Hexafluorocyclotriphosphazene (4.98 g, 20 mmol) and (trifluoromethyl)trimethylsilane (**2**; 19.88 g, 140 mmol) are dissolved in 25 mL of dry THF in a two-necked, 100-mL flask equipped with a condenser. Anhydrous CsF (0.075, 0.5 mmol) is added directly from the oven to the stirred mixture. The mixture is heated at 70 °C under dry nitrogen with stirring for 24 h. The desired product begins to precipitate at the end of the reaction. The solvent is decanted at low temperature, and 25 mL of dry $CHCl_3$ is added. The flask is stoppered and warmed to room temperature with agitation. (*Caution!* If the stopper is released for any appreciable time at room temperature, loss of the product results because of its volatility.) The mixture is cooled to −50 °C to precipitate the product. The residual THF and $CHCl_3$ are decanted. The process is repeated once more with an additional 25 mL of trichloromethane. Purification is completed by a brief trap-to-trap distillation at −45 °C. Yield: 9.88 g (90%).

Characterization Data mp 64 °C (in sealed capillary). ^{31}P NMR ($CDCl_3$): δ 3.12 (s, $J_{P\text{-}F}$ = 130 Hz) ppm. ^{19}F NMR ($CDCl_3$): δ −73.65 (d, $J_{P\text{-}F}$ = 130 Hz) ppm. ^{13}C NMR ($CDCl_3$): δ 118.89 (dq, $J_{C\text{-}P}$ = 150 Hz, $J_{C\text{-}F}$ = 290 Hz) ppm. MS (EI) *m/z* (%): 549 (1, M^+), 480 (100, $M^+ - CF_3$), 69 (20, CF_3^+). IR (gas phase, KBr) (cm^{-1}): ν 1299 (s), 1212 (vs), 1164 (m), 1139 (s), 771 (m).

Application Fluoroalkyl-substituted phosphazenes possess useful properties as fire retardants and as fire-resistant lubricants and hydraulic fluids. The fluoroalkyl group should decrease the flammability of phosphazene compared to non-fluorinated aryl and alkyl derivatives.

Waste Disposal Standard procedures for disposal of organic solvents and organic and inorganic materials should be used.

REFERENCES

1. Umemoto T, Ishihara S. Power-variable electrophilic trifluoromethylating agents. S-, Se-, and Te-(trifluoromethyl)dibenzothio-, -seleno-, and -tellurophenium salt system. *J Am Chem Soc* 1993;115:2156–2164.
2. Umemoto T. Electrophilic perfluoroalkylating agents. *Chem Rev* 1996;96:1757–1777.

3. Yang J-J, Kirchmeier RL, Shreeve JM. New electrophilic trifluoromethylating agents. *J Org Chem* 1998;63:2656–2660.
4. Singh RP, Shreeve JM. Perfluoroalkylation of simple inorganic molecules: a one step route to novel perfluoroalkylated compounds. *Chem Commun* 2002;1818–1819.
5. Jander J, Haszeldine RN. Reactions of fluorocarbon radicals. XIII. Alternative syntheses, properties, and spectra of trifluoronitroso- and trifluoronitromethane. *J Chem Soc* 1954;912–919.
6. Haszeldine RN. Polyhalonitro compounds. Br Patent 770619. 1957 March 20; *Chem Abstr* 1953;51:81625.
7. Lu N, Thrasher JS. The direct synthesis of trifluoronitromethane, CF_3NO_2. *J Fluorine Chem* 2002;117:181–184; and references therein.
8. Muralidharan K, Chakraborty R, Shreeve JM. Electrophilic trifluoromethylation of simple inorganic salts: a one step route to trifluoronitromethane, CF_3NO_2. *J Fluorine Chem* 2004;125:1965–1966.
9. Singh RR, Van der Puy M, Poss AJ, Shankland IR, inventors; Honeywell International, assignee. Compositions and methods comprising trifluoronitromethane. WO Patent 2010/148003. 2010 Dec 23.
10. Nelson DW, O'Reilly NJ, Speier J, Gassman PG. Synthesis and solvolysis of 7-(perfluoroalkyl)-7-bicyclo[2.2.1]heptyl derivatives. *J Org Chem* 1994;59:8157–8171.
11. Singh RP, Kirchmeier RL, Shreeve JM. TBAF catalyzed direct nucleophilic trifluoromethylation of α-keto amides with (trifluoromethyl)trimethylsilane. *J Org Chem* 1999;64:2579–2581.
12. Schmutzler R. Trimeric phosphonitrile fluoride. *Inorg Synth* 1967;9:76–78.
13. Singh RP, Vij A, Kirchmeier RL, Shreeve JM. A novel synthesis of hexakis(trifluoromethyl)cyclotriphosphazene: single crystal X-ray structures of $N_3P_3(CF_3)_6$ and $N_3P_3F_6$. *Inorg Chem* 2000;39:375–377.

CHAPTER 22

Introduction of Fluorine into Compounds via Electrophilic and Nucleophilic Reactions

JEAN'NE M. SHREEVE

A large number of reagents may be used to introduce fluorine into inorganic and organic substrates. While elemental fluorine and some of its more reactive derivatives are useful in fluorinating a varied range of molecular types, it is often possible to prepare fluorinated derivatives using milder and more readily handled reagents by the uninitiated by using off-the-shelf compounds. Fluorine itself is a marvelous electrophilic reagent toward nucleophiles; however, it is becoming more and more difficult to obtain even in small quantities and then to store to the satisfaction of safety officers. Therefore, here we have selected only two reagents from a cast of thousands that can be readily obtained, have reasonable shelf life, and which are easily handled even by inexperienced undergraduates. 1-Chloromethyl-4-fluoro-1,4-diazoniabicyclo[2.2.2]octane bis(tetrafluoroborate) (Selectfluor®, Sigma-Aldrich) (**1**) [1] as a reactive electrophile and bis(2-methoxyethyl)aminosulfur trifluoride (Deoxo-Fluor®, Sigma-Aldrich) (**2**) [2] as a reactive nucleophile.

22.1 PREPARATION OF FLUORINE-CONTAINING COMPOUNDS USING 1-CHLOROMETHYL-4-FLUORO-1,4-DIAZONIABICYCLO-[2.2.2]OCTANE BIS(TETRAFLUOROBORATE (SELECTFLUOR) (1) AS A REACTIVE ELECTROPHILE

Preparation of *tert*-Butyldifluoroamine [$(CH_3)_3CNF_2$] [3]

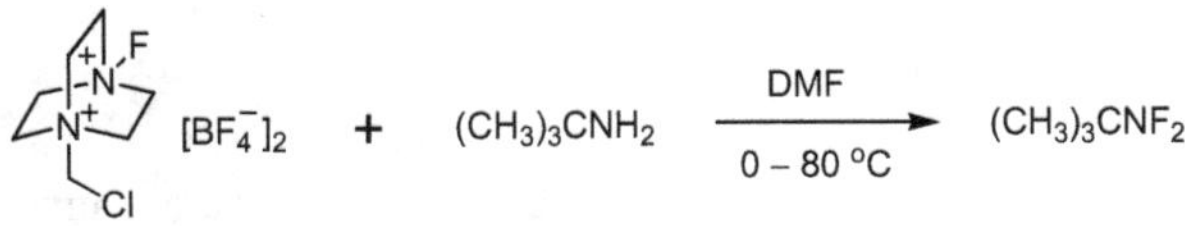

1

Efficient Preparations of Fluorine Compounds, First Edition. Edited by Herbert W. Roesky.

Apparatus A 50-mL Pyrex flask, stir bar, magnetic stirrer, nitrogen T-tube, vacuum apparatus to perform trap-to-trap separation, eye protection, lab coat, gloves.

Chemicals *tert*-Butylamine, Selectfluor, anhydrous dimethylformamide (DMF) (or dimethylacetamide (DMA)).

Attention! Eye protection and gloves must be used at all times.

Caution! All reactions should be carried out in a well-ventilated hood. All chemicals should be handled with extreme care assuming that they may be toxic and volatile.

Experimental Procedure To a solution of Selectfluor (**1**; 1.06 g, 3 mmol) in 3.5 mL of DMF or DMA in a dry, single-necked, 50-mL, round-bottom flask equipped with a magnetic stir bar, *tert*-butylamine (0.985 g, 1.35 mmol) in 1.5 mL of DMF (or DMA) is added drop by drop at 0 °C. After stirring for 0.5 h, the reaction mixture is stirred for an additional 1.5 h at 25 °C. Low temperature trap-to-trap distillation under vacuum allowed separation of the solvent (trap at −20 °C) from the product, which may be stopped in a trap at −78 °C or −100 °C (80% yield). *tert*-Butyldifluoroamine is stable in a sealed Pyrex glass container under refrigeration for ~3 months.

Characterization Data bp 56–57 °C. ^{1}H NMR (200 MHz, CD_3CN): δ 1.24 (t, 9H, J_{HF} = 1.6 Hz) ppm. ^{19}F (188 MHz, CD_3CN): δ 28.7 (br s) ppm. ^{13}C NMR (50 MHz, CD_3CN): 22.8, 70.6 (t, J = 6.8 Hz) ppm; IR (gas phase, cm^{-1}): 2991 (vs), 2944 (s), 1485 (s), 1375 (vs), 1253 (m), 1222 (m), 980 (vs), 880 (vs), 587 (m).

Application Alkyl difluoroamines are valuable as high-energy oxidizing agents in propellants for use in rockets and missiles. Their boiling points are considerably higher than that of liquid oxygen, a commonly used oxidizer, and other oxidizers for liquid fuels, which enhance storage in missiles to enable ready firing as required.

Waste Disposal Standard procedures for disposal of organic solvents and organic and inorganic materials should be used.

Preparation of 3-Fluoro-3-methyloxindole [4]

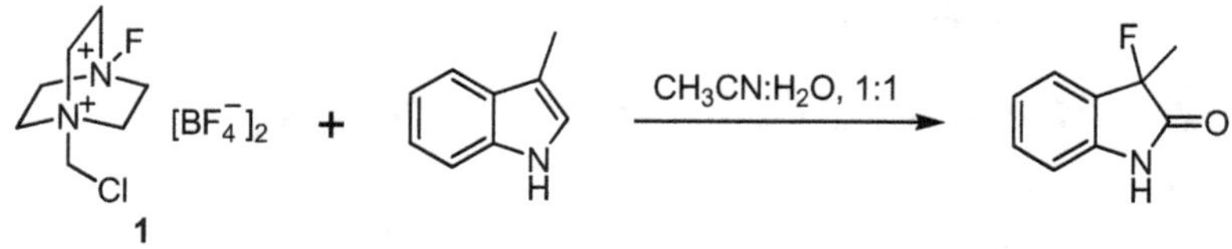

Apparatus A 50-mL, Pyrex flask, nitrogen inlet, stir bar, magnetic stirrer, syringe (5 mL), eye protection, lab coat, gloves.

Chemicals 3-Methylindole, Selectfluor (**1**), acetonitrile, silica gel column, Na_2SO_4, 4% HCl, $NaHCO_3$, ethyl acetate.

Attention! Eye protection and gloves must be used at all times.

Caution! All reactions should be carried out in a well-ventilated hood. All chemicals should be handled with extreme care assuming that they may be toxic and volatile.

Experimental Procedure To a dry, single-necked, 50-mL, round-bottom flask equipped with a magnetic stir bar and T-shaped inlet that contains a stirring solution of 3-methylindole (0.13 g, 1.0 mmol) in acetonitrile/water (1:1) (5 mL), Selectfluor (**1**; 1.05 g, 3.0 mmol) is added. The reaction is stirred under nitrogen at 25 °C for 12 h. The reaction mixture is diluted with ethyl acetate (50 mL), washed with water (10 mL), 4% HCl (10 mL), a saturated solution of sodium bicarbonate (10 mL), and brine (10 mL), and then dried over Na_2SO_4. The solvent is removed under reduced pressure to give a residue that is purified by column chromatography on silica gel eluting with ethyl acetate/hexane to give the product in pure state (71% yield).

Characterization Data mp 87–89 °C. ^{19}F NMR (254 MHz, CD_3Cl): δ −153.7 (q, J_{HF} = 22.2 Hz) ppm. ^{1}H NMR (270 MHz, $CDCl_3$): δ 8.33 (br s, 1H, NH), 7.42 (d, 1H, J = 7.6 Hz, ArH), 7.30–7.36 (m, 1H, ArH), 7.10 (t, J = 7.6 Hz, 1H, ArH), 6.92 (d, 1H, J = 7.5 Hz, ArH), 1.78 (d, 3H, J = 22.1, 13.5 Hz, CH_3) ppm. IR (KBr): 3208, 1734 cm^{-1}.

Application Indoles are convenient precursors in the synthesis of many biologically active molecules, and their derivatives are also useful probes for the study of enzymatic mechanisms and metabolic pathways. Fluorinated analogs containing indole moieties can be useful tools for the elucidation of various biological processes. 3-Fluorooxindoles are potential mimics of the corresponding oxindoles, 2- and 3-hydroxyoxindoles, that are often found in natural products, biologically active compounds, and metabolites of indoles.

Waste Disposal Standard procedures for disposal of organic solvents and organic and inorganic materials should be used.

Preparation of Ethyl 2,2-difluoroacetoacetate (Microwave Assisted) [5,6]

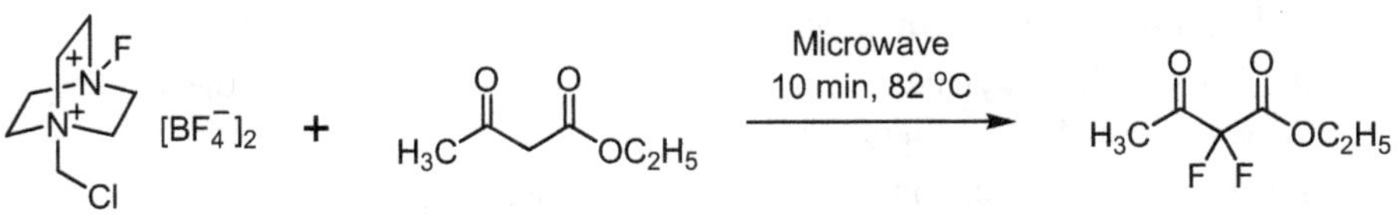

1

Apparatus Microwave tube with sealable septum, microwave apparatus (100 W), eye protection, lab coat, gloves.

Chemicals Ethyl acetoacetate, Selectfluor, acetonitrile, tetrabutylammonium hydroxide (TBAH) in methanol.

Attention! Eye protection and gloves must be used at all times.

Caution! All reactions should be carried out in a well-ventilated hood. All chemicals should be handled with extreme care assuming that they may be toxic and volatile.

Experimental Procedure Into a septum-sealed microwave tube with stir bar, ethyl acetoacetate (0.130 g, 1 mmol), Selectfluor (**1**; 1.05 g, 3 mmol), acetonitrile (1 mL), and TBAH (1 M solution in CH_3OH, 2 mL) are placed. The reaction mixture is irradiated in a monomode microwave cavity at 82 °C for 10 min. The solution is cooled rapidly to room temperature by passing compressed air through the microwave cavity. Any insoluble materials are removed by filtration. The solvent is removed from the filtrate in vacuo, and the crude products are purified directly by flash chromatography over silica gel (78%).

Characterization Data bp 75 °C (41 mm). ^1H NMR (300 MHz, $CDCl_3$): δ 4.35 (q, J = 7.1 Hz, 2H), 2.39 (J = 1.5 Hz, 3H), 1.33 (J = 7.1 Hz, 3H) ppm. ^{19}F (282 MHz, $CDCl_3$): δ −113.7 (s, 2F) ppm. IR (cm^{-1}): 1763 ($CO_2C_2H_5$), 1738 (C=O).

Application This family of esters finds use as solvents or additives in lithium battery electrolytes.

Waste Disposal Standard procedures for disposal of organic solvents and organic and inorganic materials should be used.

22.2 PREPARATIONS OF FLUORINE-CONTAINING COMPOUNDS USING BIS(2-METHOXYETHYL)AMINOSULFUR TRIFLUORIDE (DEOXO-FLUOR) (2) AS A REACTIVE NUCLEOPHILE

Preparation of Bis[2-(2-fluoroethoxy)ethyl] ether [7]

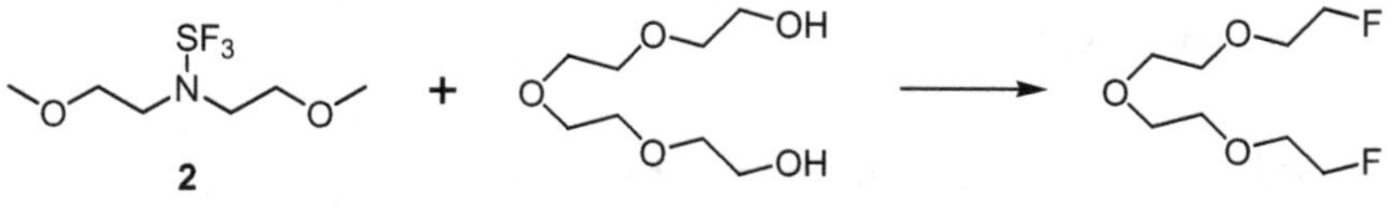

Apparatus A 50-mL Pyrex flask, stir bar, magnetic stirrer, 10-mL dropper, silica gel column, eye protection, lab coat, gloves.

Chemicals Tetraethylene glycol (bis[2-(2-hydroxyethoxy)ethyl] ether), methylene chloride, Deoxo-Fluor, $NaHCO_3$.

Attention! Eye protection and gloves must be used at all times.

Caution! All reactions should be carried out in a well-ventilated hood. All chemicals should be handled with extreme care assuming that they may be toxic and volatile.

Experimental Procedure To a stirred solution of tetraethylene glycol (0.777 g, 4 mmol) in 4 mL of CH_2Cl_2 at 0 °C in a dry, single-necked, 50-mL, round-bottom flask equipped with a magnetic stir bar, Deoxo-Fluor (1.550 g, 8.2 mmol) is added drop by drop as a neat liquid. After 1 min, the cold bath is removed and the reaction mixture is stirred at room temperature for 15 h. The progress of the reaction was monitored by thin layer chromatography (TLC). On completion, the reaction is worked up by slowly adding saturated sodium bicarbonate solution followed by washing with water. The pure product is isolated by flash chromatography on silica gel using hexane/dichloromethane mixture (91% yield).

Characterization Data Dense liquid. ^{1}H NMR (200 MHz, CD_3Cl): δ 3.66–3.70 (m, 10H), 3.78 (t, 2H, $J = 4.2$ Hz), 4.55 (dt, 4H, $J = 4.2$ Hz, $J = 47.7$ Hz) ppm. ^{19}F (188 MHz, CD_3Cl): δ −223.21 (tt, 2F, $J = 47.7$ Hz, $J = 28.0$ Hz) ppm. IR (KBr film): 2952, 1458, 1351, 1296, 1247, 1118, 1047, 943, 873 cm^{-1}.

Application The replacement of hydroxyl with fluorine is an extensively used strategy for enhancement of biological activity in the design of biologically important molecules.

Waste Disposal Standard procedures for disposal of organic solvents and organic and inorganic materials should be used.

Preparation of 1,1,2,2-Tetrafluoro-1,2-diphenylethane [8]

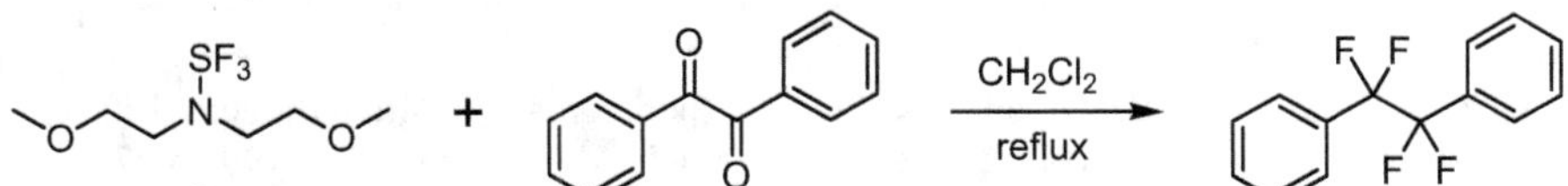

2

Apparatus A 50-mL Pyrex flask, stir bar, magnetic stirrer, 5-mL dropper, separatory funnel, eye protection, lab coat, gloves.

Chemicals 1,2-Diphenylethane-1,2-dione (benzil), dichloromethane, Deoxo-Fluor, $NaHCO_3$, $MgSO_4$, C_2H_5OH.

Attention! Eye protection and gloves must be used at all times.

Caution! All reactions should be carried out in a well-ventilated hood. All chemicals should be handled with extreme care assuming that they may be toxic and volatile.

Experimental Procedure 1,2-Diphenylethane-1,2-dione (0.420 g, 2 mmol) is dissolved in dichloromethane (5 mL). Deoxo-Fluor (0.113 g, 6 mmol) is added at room temperature and followed by the addition of two drops of ethanol (to generate a catalytic amount of HF). The reaction mixture is heated at 60 °C for 24 h. The reaction is quenched by the slow addition of aqueous saturated $NaHCO_3$ solution until effervescence is complete. The dichloromethane layer is separated and dried over anhydrous $MgSO_4$. It is filtered and removal of the solvent leaves the product (75% yield).

Characterization Data mp 115 °C. ^{1}H NMR (200 MHz, CD_3Cl): δ 7.3–7.5 (m, 10H) ppm. ^{19}F (188 MHz, CD_3Cl): δ −112.26 (s, 4F) ppm. IR (KBr film): 1449, 1255, 1145, 1078, 933, 878, 751, 696, 657 cm^{-1}.

Application The methodology required to convert ketones or diketones to difluoro or tetrafluoro species is widely needed and often used to prepare compounds with modified properties that are useful in a variety of systems.

Waste Disposal Standard procedures for disposal of organic solvents and organic and inorganic materials should be used.

REFERENCES

1. Singh RP, Shreeve JM. Recent highlights in electrophilic fluorination with 1-chloromethyl-4-fluoro-1,4-diazoniabicyclo[2.2.2]octane bis(tetrafluoroborate). *Acc Chem Res* 2004;37:31–44.
2. Singh RP, Shreeve JM. Recent advances in nucleophilic fluorination reactions of organic compounds using Deoxofluor and DAST. *Synthesis* 2002, 2561–2578.
3. Singh RP, Shreeve JM. The first application of Selectfluor® in electrophilic fluorination of amines: a new route to -NF_2, -NHF, and NF Compounds. *Chem Commun* 2001;1196–1197.
4. Takeuchi Y, Tarui T, Shibata N. A novel and efficient synthesis of 3-fluorooxindoles from indole mediated by Selectfluor. *Org Lett* 2000;2:639–642.
5. Xiao J-C, Shreeve JM. Microwave-assisted rapid electrophilic fluorination of 1, 3-dicarbonyl derivatives with Selectfluor®. *J Fluorine Chem* 2005;126:475–478.
6. Xu Z-Q, DesMarteau DD, Gotoh Y. N-fluorobis[(perfluoroalkyl)sulfonyl]imides. Efficient reagents for the fluorination of 1,3-dicarbonyl derivatives. *J Fluorine Chem* 1992;58:71–79.
7. Ye C, Shreeve JM. Rearrangements accompanying fluorination of 1, 2-amino alcohols and 1, 2-diols with Deoxofluor. *J Fluorine Chem* 2004;125:1869–1872.
8. Singh RP, Majumder U, Shreeve JM. Difluoro- and tetrafluorination of dicarbonyl compounds. *J Org Chem* 2001;66:6263–6267.

CHAPTER 23

The "Real" Iodine and Bromine Monofluorides

SHLOMO ROZEN

23.1 ADDING XF (X = Br, I) ACROSS VARIOUS DOUBLE BONDS

Adding the elements of BrF or IF is an old procedure in organic chemistry. It has been usually achieved by a two-step reaction performed by a mixture of two reagents, one a source of an electrophilic bromine or iodine such as *N*-bromosuccinimide (NBS) or *N*-iodosuccinimide (NIS), and the other a source of nucleophilic fluorine such as HF or some HF/amine [1]. This combination of reagents is usually depicted by the symbol [XF]. There are several limitations to this general method. One of them is the prolonged reaction times and the other its potency. The reaction proceeds well with isolated double bonds, but has problems with deactivated or acetylenic π centers. These drawbacks are eliminated when the real XF made directly from the elements is used.

When nitrogen-diluted fluorine was passed through a cold (−75 °C) suspension of I_2 or Br_2 in $CFCl_3$, the original color of the reaction mixture changed to a brown suspension in the case of I_2 and to a pale yellow one in the case of Br_2. The main reactive species resulting from these reactions are the known IF [2], and the never fully described before BrF [3]. These two reagents are much more reactive than either I_2 or Br_2 and practically react with any π systems in dark at −75 °C in a matter of seconds [4].

The addition of IF across a double bond is a fully regiospecific process due to the electrophilicity of the bromine or the iodine. Aliphatic olefins such as 1-octene react with IF to give 1-iodo-2-fluoro derivatives in around 70% yield. Despite its high reactivity, an aldehyde function may be present, because IF is not a strong oxidizer. Thus, citronellal forms only one regiospecific adduct, 3,7-dimethyl-7-fluoro-6-iodo-1-octanal, in good yield. The reaction is also stereospecific, and cyclohexene and

Efficient Preparations of Fluorine Compounds, First Edition. Edited by Herbert W. Roesky.

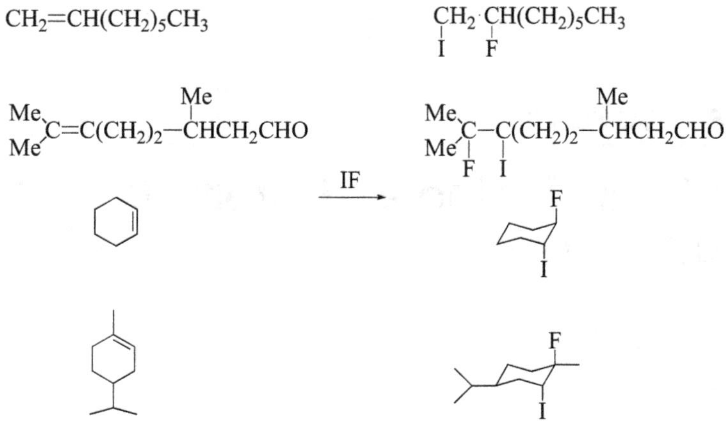

SCHEME 23.1 Adding IF to olefins.

the more hindered 1-menthenyl-4-*i*-propyl resulted exclusively in *trans*-1-fluoro-2-iodocyclohexane and 1β-fluoro-α-methyl-2α-iodo-4β-isopropylcyclohexane respectively (Scheme 23.1).

The iodine could serve as an entry for obtaining other fluorine derivatives. Thus, for example, the 1-fluoro-2-iodocycloheptane obtained from cycloheptene could be easily hydrolyzed on a silica gel to produce the corresponding fluorohydrin. Using *t*-BuOK, which is a stronger base than water and not as nucleophilic, causes an elimination of HI, and fluoroolefins were formed in excellent yield (Scheme 23.2).

With solvents consisting only of $CFCl_3$, $CFCl_3/CHCl_3$ (EtOH free), or $CFCl_3$/dry CH_3CN, the reaction of BrF with olefins proved to be too uncontrollable and unselective even at −75 °C, and only bromine- and fluorine-containing tars were obtained. When, however, a proton donor such as EtOH or *i*-PrOH was added to the BrF suspension before the reaction with an olefin, a surprising taming effect on the BrF was observed and the corresponding expected adducts were formed in 70–80% yield. Being more reactive than IF, bromine monofluoride was also readily added across enones, where the adduct could eliminate HF and give bromo olefins (Scheme 23.3).

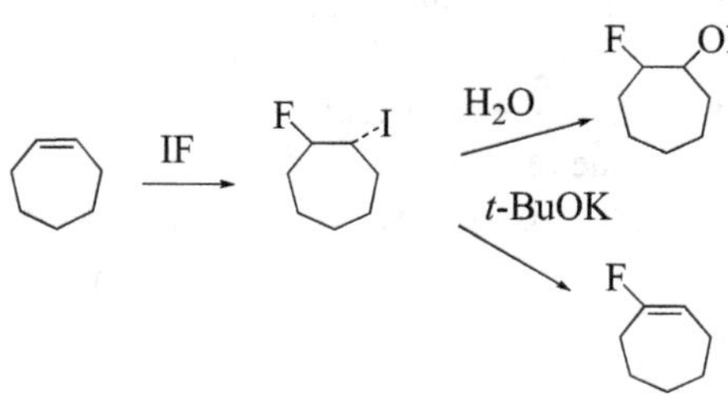

SCHEME 23.2 Reactions on the iodine of the fluoroiodo adducts.

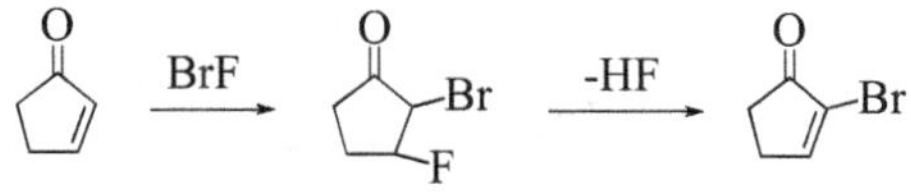

SCHEME 23.3 Adding bromine monofluoride across π centers.

Experimental Procedure A suspension of 25 g of I_2 (100 mmol) in 450 mL of $CFCl_3$ was cooled to −75 °C. Nitrogen-diluted (10%) fluorine (total of 50 mmol F_2) is bubbled into the mixture during a period of about 10 h. A brown suspension of mixture of IF and I resulted in a ^{19}F NMR resonance of −161 ppm. A cold, 20-mL $CHCl_3$ solution of 10–20 mmol of the olefin is added in one portion to the IF mixture at −75 °C. After a few minutes at that temperature, the reaction was stopped by pouring it into 500 mL of dilute thiosulfate solution, washing with water until neutral, drying over $MgSO_4$, and evaporating. The crude product was usually purified by chromatography on a short silica gel column and, if needed, also by high-performance liquid chromatography (HPLC). In a similar way, BrF was also prepared possessing a ^{19}F NMR resonance at −158 ppm. About 200 mL of precooled $CHCl_3$ containing a few milliliters of EtOH was added and stirred for 15 min resulting in a clear solution. Subsequently, a cold (−78 °C) solution of 10 mmol of an olefin in $CHCl_3$ was added, allowed to react for a few minutes, and worked up as described above.

Characterization Data for *trans*-1-fluoro-2-iodomenthan A cold $CHCl_3$ solution of 1.38 g of 1-menthenyl-4-*i*-propyl was allowed to react with IF for 5 min and then worked up as described above. Purification by chromatography using petroleum ether as the eluent gave 1.28 g of the product (45% yield) as an oil. ^{1}H NMR: δ 4.47 (1H, dt, $J_{HH} = 2.5$ Hz, $J_{HF} = 9$ Hz), 1.59 (3H, d, $J = 22$ Hz), 0.89 (6H, d) ppm. ^{19}F NMR: δ −145 (m) ppm. MS *m/z*: 284 M^+, 264 $(M - HF)^+$.

While the triple bond is usually resistant to various sources of [XF], it is not a match for the real IF and BrF [5]. Reacting slightly more than twofold excess of XF with 1-hexyne at −75 °C, 1,1-diiodo-2,2-difluorohexane was obtained in good yield and in less than 5 min. The electrophilic iodine atom in IF attacks initially the more electron-rich terminal carbon of 1-hexyne followed by attack of the fluoride on the resulting secondary carbonium ion. The fluoroiodo olefin thus obtained reacts immediately with a second molecule of IF, which adds in such a way as to produce the more stable carbocation on the carbon attached to the fluorine atom. As with olefins, BrF benefited from a small amount of a proton donor such as EtOH. It enabled it to react with 1-hexyne forming 1,1-dibromo-2,2-difluorohexane in 60% yield. When only 1 equiv of BrF was used, 1-bromo-2-fluoro-1-hexene was formed in 50% yield.

The addition of XF is not confined to terminal acetylenes. Thus, 2-butyne reacts cleanly with IF to give 2,2-difluoro-3,3-diiodo butane in better than 85% yield. While many other acetylene derivatives reacted according to the

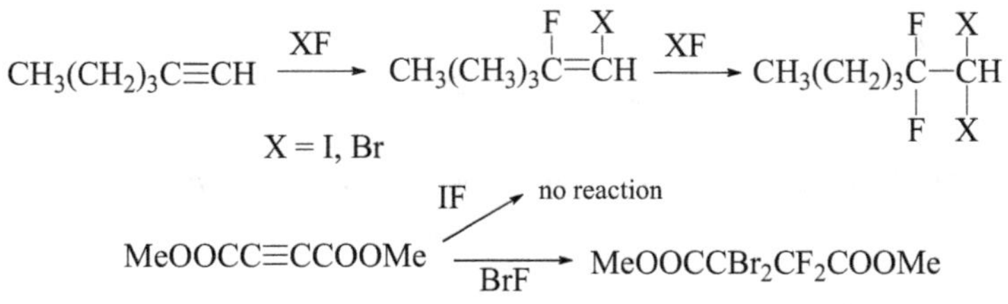

SCHEME 23.4 Adding XF across acetylenic bonds.

pattern outlined above, dimethyl acetylenedicarboxylate gave some slightly different results. This compound contains one of the most unreactive triple bonds toward electrophilic attack. When it was treated with NBS/HF/Py, no reaction was observed. The same was true when molecular IF was brought in contact with it for 3 days at −75 °C. However, even this acetylenic derivative was not resistant toward BrF and, after 12 h at −75 °C, dimethyl 2,2-dibromo-3,3-difluorosuccinate was obtained in 70% yield (Scheme 23.4).

Experimental Procedure For the reaction of dimethyl acetylenedicarboxylate with BrF; 2.8 g (20 mmol) of dimethyl acetylenedicarboxylate was dissolved in 50-mL cold $CHCl_3$ and then added to more than 50 mmol of BrF dissolved in a cold mixture of $CFCl_3$ and $CHCl_3$ (1:1). The reaction was monitored by gas chromatography (GC) and after 5 min practically no reaction was observed. The reaction mixture was stirred overnight at −75 °C, and then worked up as described in the previous experimental procedure. Chromatography of the crude product with 30% EtOAc in PE purified the dimethyl 2,2-dibromo-3,3-difluorosuccinate that was obtained in 70% yield.

Characterization Data for Dimethyl 2,2-dibromo-3,3-difluorosuccinate IR: 1730 cm^{-1}. ^1H NMR: δ 3.938 (s), 3.941 (s) ppm. ^{19}F NMR: δ −101.0 (s) ppm. ^{13}C NMR: δ 110.39 (t, $^1J_{CF}$ = 265 Hz), 55.0, 54.1, 53.64 (CBr_2, t, $^2J_{CF}$ = 29 Hz) ppm. MS *m/z*: 296, 294, 292 $[M-(CH_3O)-(CH_3)]^+$, 218, 216, 214 $(CBr_2COO)^+$.

Chemicals F_2, $CFCl_3$, $CHCl_3$ I_2, Br_2, EtOH.

Attention! Safety glasses and well-ventilated area must be used at all times.

Caution! Fluorine is a strong oxidizer and corrosive material. An appropriate vacuum line made from copper or Monel in a well-ventilated area should be constructed for working with this element. If elementary precautions are taken, then despite the mythical fears associated with this element, working with diluted F_2 is relatively simple and safe.

Waste Disposal During preparation of IF and BrF, unconsumed fluorine, some HF which is always present, and traces of Br_2 should be trapped with soda lime producing harmless CaF_2 and $CaBr_2$. At the end, the reaction should be poured into dilute thiosulfate solution (destroying any excess of BrF, IF, I_2, Br_2, and HF).

23.2 FROM THE CARBONYL TO THE CF_2 GROUP

Iodine monofluoride was found to be useful as a mild reagent for converting carbonyls into the important CF_2 group [6]. The transformation was accomplished via various hydrazones and oxime derivatives. The main idea of this subject was to use the fact that the electrophilic iodine in IF should polarize the imine bond enabling the not hydrated, and hence relatively strong nucleophilic fluoride, to form the desired carbon fluorine bond. The procedure does not usually require a tedious purification of the hydrazones, and the crude products were good enough substrates for the reaction. Thus reacting acetophenone hydrazone with IF at −78 °C for an hour resulted in 80% yield of 1,1-difluoro-1-phenylethane. Other benzylic hydrazones also provided very satisfactory results. It should be mentioned though that in some cases small amounts of the difluoroiodo derivatives were also formed. Those could be converted to the desired difluoro compounds by treating the reaction mixture with reagents such as $LiAlH_4$ or Bu_3SnH.

Aromatic aldehydes were also good candidates for this reaction as evident from 4-nitro- and 4-cyanobenzaldehyde, which were converted to the corresponding α,α-difluorotoluenes.

The reaction with IF is not confined only to benzylic carbonyls. The cyclic derivative 4-*t*-butylcyclohexanone hydrazone, for example, is converted to the 4,4-difluoro-*t*-butylcyclohexane in 65% yield. Steroidal ketones were also suitable substrates. It should be mentioned that these reactions are slower than with the benzylic derivatives, but raising the temperature to −10 °C or even to room temperature accelerates them considerably. *N,N*-Dimethylhydrazones, dinitrophenylhydrazones (DNP), semicarbazones, and oximes could also react satisfactory (Scheme 23.5).

Experimental Procedure Between 5 and 20 mmol of the substrate in 30-mL $CHCl_3$ solution was added to IF suspension either cooled to −78 °C for benzylic derivatives or to room temperature for all other hydrazones. The reaction was

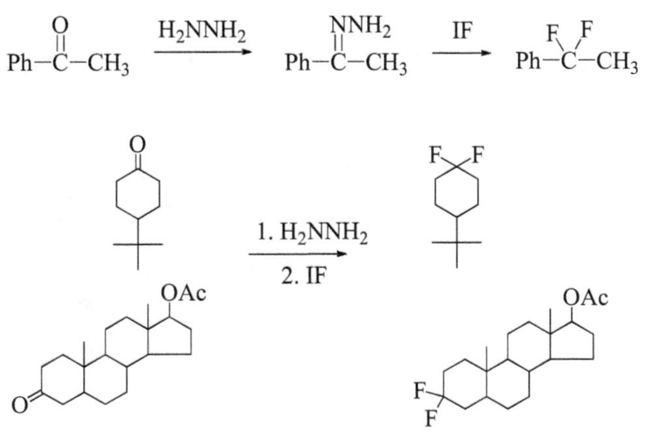

SCHEME 23.5 CO to CF_2 transformation using IF.

carried under vigorous stirring using a vibromixer and usually with an excess of about 4 mol/equiv of IF. The advancing reaction was monitored by either GC or thin layer chromatography (TLC) after treating aliquots with thiosulfate. On completion, the reaction was poured into 400 mL of 10% $Na_2S_2O_3$ solution, and stirred until colorless. The organic layer was washed with sodium bicarbonate till neutral, dried over anhydrous $MgSO_4$, and the solvent evaporated. The crude product was usually purified by vacuum flash chromatography using silica gel 60 H (Merck) and mixtures of petrol ether/EtOAc as eluent.

Characterization Data for 4,4-difluoro-1-*t*-butylcyclohexane oil ^{1}H NMR: δ 1.0–2.2 (9H, m), 0.88 (9H, s) ppm. ^{19}F NMR: δ −92 (1F, d, J = 234 Hz), −103.6 (1F, ddt, $^1J = 234$, $^2J = 40$, $^3J = 11$ Hz) ppm.

REFERENCES

1. Olah GA, Nojima M, Kerekes I. *Synthesis* 1973;780–783.
2. Hauptschein M, Braid M. *J Am Chem Soc* 1961;83:2383–2386.
3. Naumann D, Lehmann E. *J Fluorine Chem* 1975;5:307–321.
4. Rozen S, Brand M. *J Org Chem* 1985;50:3342–3348.
5. Rozen S, Brand M. *J Org Chem* 1986;51:222–225.
6. Rozen S, Zamir D. *J Org Chem* 1991;56:4695–4700.

CHAPTER 24

The Versatile Chemistry of Acetyl Hypofluorite: The First Hypofluorite Not Bonded to a Polyhaloalkyl Group

SHLOMO ROZEN

24.1 SYNTHESIS OF ACETYL HYPOFLUORITE

The OF group has been known for a long time. F–O–F could be made, for example, by passing F_2 over ice surface [1], although its synthetic use was quite limited [2]. Prager and Thompson prepared several, quite useful at the time, perfluoroalkylhypofluorites such as $CF_2(OF)_2$ and $CF_3CF(OF)_2$ [3]. Another family of hypofluorites consists of perfluoroacyl hypofluorites [4], but the most important hypofluorite was the relatively stable CF_3OF, which found quite a few applications in organic chemistry mainly as an electrophilic fluorinating agent [5]. The problem with all these hypofluorites was that they were quite expensive, not easy to make, and not commercially available, except CF_3OF that could be purchased for a short time in the late 1960s. By examining all the above reagents, it could be noticed that they have one thing in common. Their fluoroxy group is attached to a perfluorinated alkyl because otherwise, it was believed, HF elimination will take place instantaneously and the reagent will decompose. As with many legends, this was also discredited, and it was found that passing dilute F_2 through sodium acetate solvated with water, or better with acetic acid in either $CFCl_3$ or CH_3CN, produced an oxidizing solution of acetyl hypofluorite AcOF (Scheme 24.1) in very good yields, which could be titrated to determine its concentration [6]. All physical data on AcOF were measured and Appelman et al. even isolated it in its pure form [7], but for any practical use its solution was used without any further purification.

Experimental Procedure About 10% fluorine in nitrogen was bubbled through a suspension of 8 g of solvated (with AcOH) sodium acetate in 400 mL solvent ($CFCl_3$/AcOH, 10:1) at −75 °C using an efficient vibromixer. The progress

Efficient Preparations of Fluorine Compounds, First Edition. Edited by Herbert W. Roesky.

$$CH_3COONa\text{-}H_2O\text{-}AcOH \xrightarrow[\substack{CHCl_3/CFCl_3 \\ \text{or } CH_3CN}]{F_2;\ \text{low temp}} \boxed{CH_3COOF}$$

typical concentrations of 0.2–0.3 molar.

SCHEME 24.1 The formation of acetyl hypofluorite.

of the reaction was monitored by treating aliquots with KI and titrating the liberated iodine.

Characterization Data for AcOF Although for synthetic purposes AcOF is always used in a solution of $CFCl_3/CHCl_3$ or CH_3CN, its spectral and physical properties have been established. ^{1}H NMR: δ 2.12 (d, $^4J_{HF} = 3.6$ Hz) ppm. ^{19}F NMR: δ + 168 ($W_{h/2} = 15$ Hz) ppm. It fully decomposes thermally (in the injector of a gas chromatography (GC)) to form CH_3F and CO_2 (positively identified) in 1:1 ratio [8]. AcOF has a bp of −53 °C and mp of −96 °C. Its IR, UV, and MS were also measured [7].

Chemicals F_2, $CFCl_3$, $CHCl_3$ or CH_3CN and AcOH, AcONa, or AcOK.

Attention! Safety glasses and well-ventilated area must be used at all times.

Caution! Fluorine is a strong oxidizer and a very corrosive material. An appropriate vacuum line made from copper or Monel in a well-ventilated area should be constructed for working with this element. If elementary precautions are taken, then despite the mythical fears associated with this element, working with diluted F_2 is relatively simple and safe.

In regard to AcOF Unless absolutely necessary, do not attempt to isolate and purify it because it may be highly explosive when not in solution.

Waste Disposal During preparation of AcOF, unconsumed fluorine and some HF, which is always present, should be trapped with soda lime producing harmless CaF_2. After any use of AcOF, the reaction mixture should be poured into dilute bicarbonate solution (destroying any excess of AcOF and HF) producing AcONa and NaF.

24.2 ADDITION TO COMPOUNDS WITH DOUBLE BONDS

Most fluorinating agents possessing the OF moiety do not add satisfactorily across the double bonds of olefins, but the milder acetyl with its electrophilic fluorine does. For example, it adds itself across isolated double bonds such as cyclohexene in 60% yield, across benzylic ones such as dibenzosuberenone (55% yield), and even across certain α,β-enones represented here by *trans*-ethyl cinnamate in 57% yield (Scheme 24.2) [9].

Two main features should be noted. The regiochemistry clearly points to the fact that the oxygen-bound fluorine is indeed the electrophilic part of AcOF, and that the stereochemistry results from the dominantly *syn* addition. This somewhat surprising fact stems from the ionic nature of the reaction where the intermediate is a cage

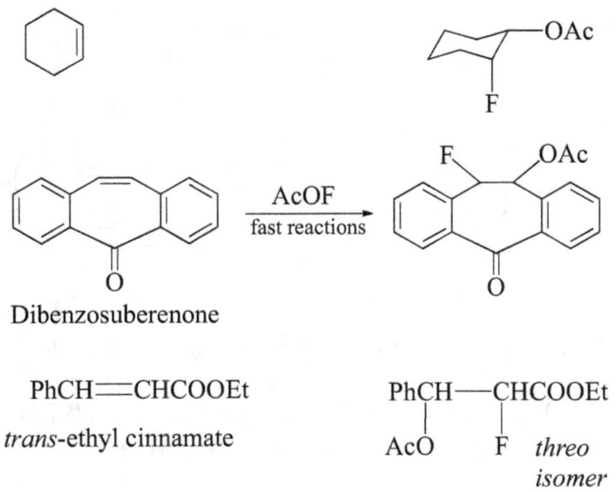

SCHEME 24.2 Adding acetyl hypofluorite to olefins.

of pair of ions, one of them the very short living α-fluorocarbonium ion. Thus, when AcOF reacted with *trans*-4-acetoxystilbene, practically only the *threo*-isomer was formed in 50% yield. If, however, the α-fluorocarbocation is somewhat more stabilized as in the case of *trans*-4-methoxystilbene, it had enough time to rotate around the central C–C bond, and consecutively, the stereoselectivity was reduced to only 4:1 *threo*/*erythro* ratio. These results also helped to discredit any 4-center addition mechanism (Scheme 24.3).

Experimental Procedure Fluorination of dibenzosuberenone was carried out with 6.2 g (30 mmol) dissolved in cold (−78 °C) $CHCl_3$. Cold AcOF solution (60 mmol) was added drop by drop. The reaction was completed after 5 min and the crude product chromatographed using 20% EtOAc in petrol ether (PE) as eluent. About 4.68 g (55% yield) of the adduct was isolated.

Characterization Data for Dibenzosuberenone mp 92 °C (from Et_2O). IR: 1740, 1650 cm^{-1}. MS *m*/*z*: 284 M^+ 264 $(M - HF)^+$, 225 $(M - OAc)^+$. ^1H

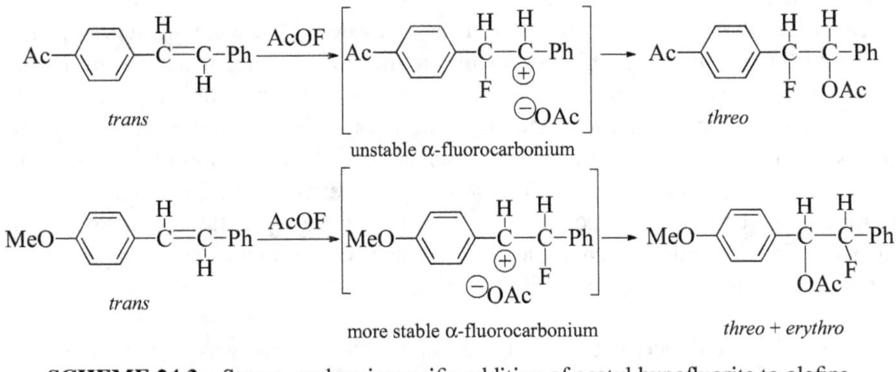

SCHEME 24.3 Stereo- and regiospecific addition of acetyl hypofluorite to olefins.

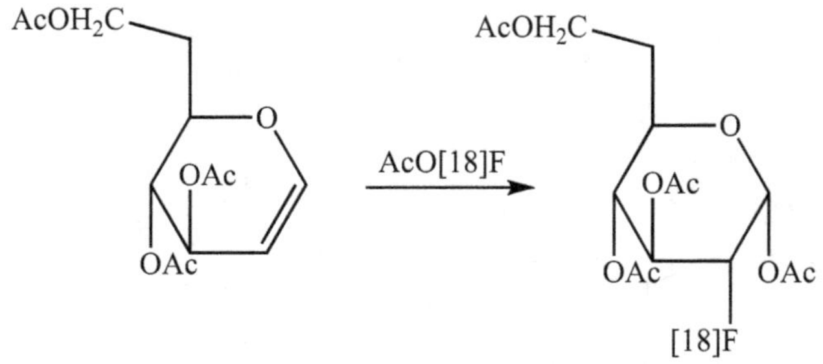

SCHEME 24.4 The first meaningful synthesis of [18]FDG.

NMR: δ 7.1–8.0 (8H, m), 6.27 (1H, dd, $^1J = 17.8$, $^2J = 0.7$ Hz), 5.75 (1H, dd, $^1J = 46.5$, $^2J = 0.7$ Hz), 1.90 (3H, s) ppm. ^{19}F NMR: δ −174.5 (dd, $^1J = 46.5$, $^2J = 17.8$ Hz) ppm.

Application Probably the most important application of this reagent was its immediate mobilization in favor of developing the area of positron emitting tomography (PET). [18]Fluorodeoxyglucose ([18]FDG) with its positron emitting isotope [18]F (half-life of 110 min) was badly needed because in many cases the living body, and especially the brain, will not distinguish between it and glucose itself. At the time, the problem was that the radiochemical yield of the [18]FDG was not greater than 0.05%. Because [18]F[19]F could be generated in good yield from neon and F_2, the immediate conclusion was that AcO[18]F could be prepared in almost 50% radiochemical yield. This reagent was then added to deoxyglucose, mainly in a fast *syn* mode, forming the [18]FDG in higher than 40% radiochemical yield, elevating the initial yield by three orders of magnitude and helping to spread the PET all over the world in becoming a standard tool for medical diagnosis (Scheme 24.4) [10].

24.3 REACTIONS WITH ENOL DERIVATIVES FORMING THE α-FLUOROCARBONYL MOIETY

There are many sources for electrophilic fluorine. A partial list includes $FClO_3$, XeF_2, R_fOF, R_fCOOF, and various NF derivatives all made with elemental fluorine. The electrophilic fluorine in AcOF is probably the most affordable, as apart from F_2 its synthesis involves only common salts of acetic acid, and the technical procedure is very simple. One of the basic reactions is the fluorination of enols forming the α-fluorocarbonyl derivatives. Thus, for example, one could make α-fluorotetralone [11], fluorovalium [12], and 2-fluoro-1,3-dicarbonyl derivatives [13]. Of special interest was the construction of α-fluorocarboxylic acids, including α-fluoroibuprofen and other fluorinated acids, which could not be made by any other method such as α-fluoro-β,β,β-trimethylpropionic acid (Scheme 24.5) [14].

Experimental Procedure Methyl 3,3-dimethylbutyrate was converted to its trimethylsilyl ketene acetal, oil, $bp_{(12\,mm)}$: 60 °C. ^{1}H NMR: δ 3.72 (1H, s),

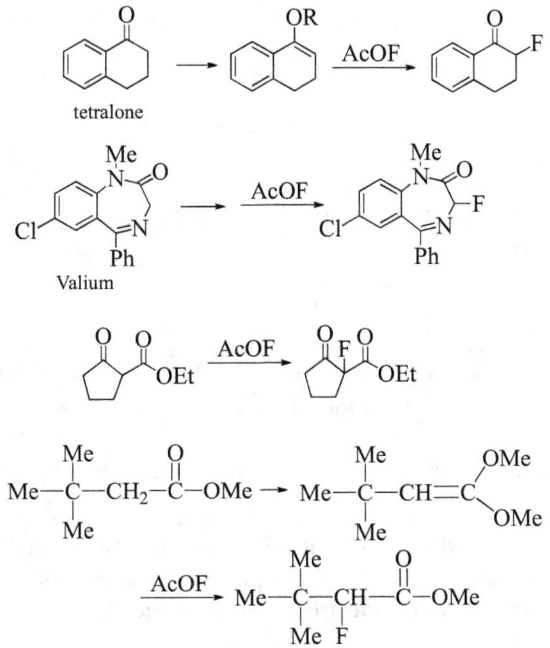

SCHEME 24.5 Reacting acetyl hypofluorite with various enols.

3.49 (3H, s), 1.05 (9H, s), 0.21 (9H, s) ppm. ^{13}C NMR: δ 152.41, 96.67, 54.09, 31.01, 29.67, −0.4 ppm. That compound was then reacted with AcOF to give methyl 3,3-dimethyl-2-fluorobutyrate in 83% yield as an oil.

Characterization Data for Methyl 3,3-dimethyl-2-fluorobutyrate IR: 1755 cm^{-1}. ^{1}H NMR: δ 4.55 (1H, d, J = 49 Hz), 3.79 (3H, s), 1.03 (9H, s) ppm. ^{19}F NMR: δ −194.18 (d, J = 49 Hz) ppm. ^{13}C NMR: δ 168.87 (d, J = 25 Hz), 95.46 (d, J = 187 Hz), 51.24, 34.32 (d, J = 19 Hz), 24.96 ppm. MS, with supersonic GC–MS using electron ionization of the sample vibrationally cooled in a supersonic molecular beam, reveals the molecular ion peak at m/z 148 M^{+}.

24.4 FLUORINATION OF AROMATIC COMPOUNDS

Possessing electrophilic fluorine, AcOF can fluorinate activated aromatic rings such as phenol and aniline derivatives. The fluorination takes place mainly on the *ortho* position. This considerable regioselectivity is a result of addition–elimination process along the *ipso-ortho* position. In certain cases, it was possible to isolate the adduct [15]. The reactions are fast and are used quite extensively for introducing the positron emitting ^{18}F isotope in aromatic rings such as tyrosine and dopamine [16]. A different route for achieving fluoroaromatics was through substitution of a metal such as Sn, Ge, or Si already attached to an aromatic ring (Scheme 24.6) [17].

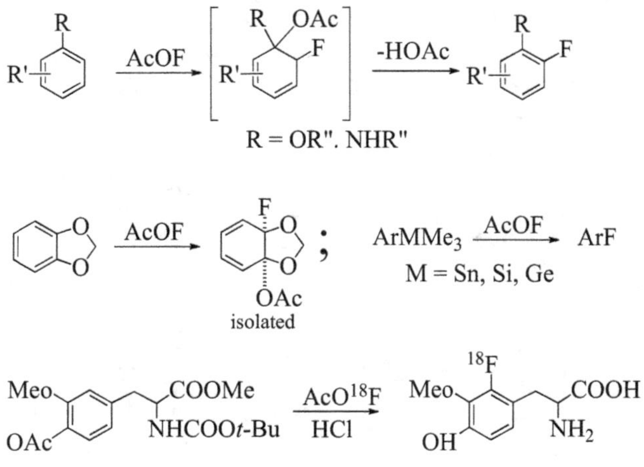

SCHEME 24.6 Synthesis of fluoroaromatics using acetyl hypofluorite.

The reaction of substituting metals with fluorine could also be used on nonaromatic compounds. Because the reaction is electrophilic in nature, it proceeds with a full retention of configuration, as the electrophile attacks the electrons of the C–M bond (Scheme 24.7).

Experimental Procedure A cold solution of anisole in CH_2Cl_2 was added in one portion to AcOF solution (twofold excess). The *ortho*-fluoro derivative obtained in 77% yield was separated from the *para* one (8% yield) by chromatography. The physical data of both isomers matched those in the literature.

The fluorination of the β-methoxy-α-mercury derivative of benzalacetophenone was carried out on a 10-mmol scale using twofold excess of AcOF. No reaction took place at −78 °C, but the starting material was fully consumed when the reaction mixture was allowed to warm up to 0 °C. After flash chromatography using 10% EtOAc in PE as eluent, the corresponding fluoroether was isolated as an oil in 80% yield.

Characterization Data for *erythro*-1,3-diphenyl-1-fluoro-2-methoxy-3-propanone IR 1685 cm^{-1}. ^1H NMR: δ 6.5–7.20 (10H, m), 5.40 (1H, dd,

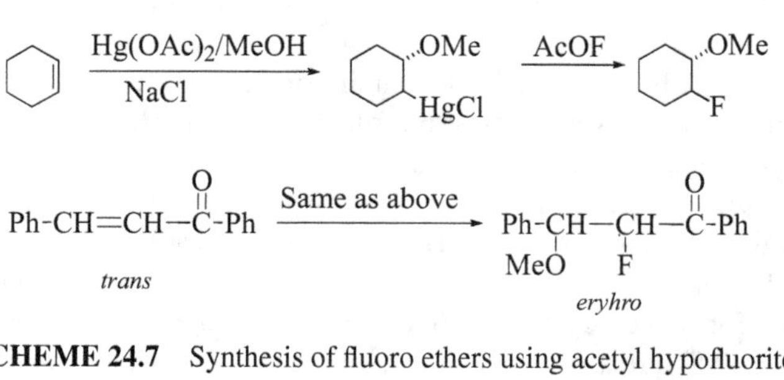

SCHEME 24.7 Synthesis of fluoro ethers using acetyl hypofluorite.

$^1J = 50$ Hz, $^2J = 3.5$ Hz), 4.72 (1H, dd, $^1J = 24$ Hz, $^2J = 3.5$ Hz), 3.23 (3H, s) ppm. ^{19}F NMR: δ −198.73 (dd, $^1J = 50$ Hz, $^2J = 24$ Hz) ppm.

REFERENCES

1. Appelman EH, Jache AW. *J Am Chem Soc* 1987;109:1754–1757.
2. Bolte G, Haas A. *Chem Ber* 1984;117:1982–1986.
3. Prager JH, Thompson PG. *J Am Chem Soc* 1967;89:2263–2267.
4. (a) Rozen S, Lerman O. *J Org Chem* 1980;45:672–678. (b) Barnette WE, Wheland RC, Middleton WJ, Rozen S. *J Org Chem* 1985;50:3698–3701.
5. Hesse RH. *Isr J Chem* 1978;17:60–70.
6. Rozen S, Lerman O, Kol M. *J Chem Soc, Chem Commun* 1981;443–444.
7. Appelman EH, Mendelsohn MH, Kim H. *J Am Chem Soc* 1985;107:6515–6518.
8. Hebel D, Lerman O, Rozen S. *J Fluorine Chem* 1985;30:141–146.
9. Rozen S, Lerman O, Kol M, Hebel D. *J Org Chem* 1985;50:4753–4758.
10. (a) Shiue CY, Salvadori PA, Wolf AP, Fowler JS, MacGregor RR. *J Nucl Med* 1982;23:899. (b) Jewett DM, Potocki JF, Ehrenkaufer RE. *J Fluorine Chem* 1984;24:477–484.
11. Rozen S, Brand M. *Synthesis* 1985;665–667.
12. Luxen A, Barrio JR, Satyamurthy N, Bida GT, Phelps ME. *J Fluorine Chem* 1987;36: 83–92.
13. Lerman O, Rozen S. *J Org Chem* 1983;48:724–727.
14. Rozen S, Hagooly A, Harduf R. *J Org Chem* 2001;66:7464–7468.
15. Lerman O, Tor Y, Hebel D, Rozen S. *J Org Chem* 1984;49:806–813.
16. (a) Chirakal R, Firnau G, Couse J, Garnett ES. *Int J Appl Radiat Isot* 1984;35:651–656. (b) Mislankar S, Gildersleeve DL, Wieland DM, Massin CC, Mulholland GK, Toorongian SA. *J Med Chem* 1988;31:362–366. (c) Adam MJ, Jivan S. *J Labelled Compd Radiopharm* 1992;31:39–43.
17. Coenen HH, Moerlein SM. *J Fluorine Chem* 1987;36:63–75.

CHAPTER 25

Direct Fluorination of Organic Compounds with Elemental Fluorine

SHLOMO ROZEN

25.1 FLUORINATION AT TERTIARY INACTIVATED CARBONS

Since Moissan's first experiments with F_2 and organic materials, which ended with the decomposition of the reactants, no one in organic chemistry was interested in working with this element. It was labeled as too reactive and hence very destructive, so it practically disappeared from the mind of chemists dealing with selective reactions. It was a wrong attitude. Some years ago, it was shown that under suitable conditions, F_2 can indeed perform very specific electrophilic front attack reactions on relatively electron-rich tertiary C–H bonds. Thus, when *trans*-4-methylcyclohexyl *p*-nitrobenzoate (**1**) was treated with 3–4% fluorine in N_2, a major product, identified as *trans*-4-fluoro-4-methylcyclohexyl nitrobenzoate (**2**), was obtained in 60% yield. When the *cis* isomer (**3**) was similarly treated, only the *cis*-4-fluoro-4-methylcyclohexyl *p*-nitrobenzoate (**4**) was formed in 65% yield. The lack of any isomerization in these reactions clearly indicated a nonradical mechanism. Similar results were obtained with *trans*- and *cis*-4-*t*-butylcyclohexyl *p*-nitrobenzoate (**5** and **6**, respectively). The tertiary hydrogen at C-4 was stereoselectively substituted by fluorine, forming the corresponding fluoro derivatives **7** and **8** in 60% and 83% yield, respectively.

The ionic nature of the substitution could be further shown by monitoring the fluorination of an equimolar mixture of **1** and **5**, or **3** and **6**. Despite considerable steric hindrance of the bulky *t*-butyl group, the more electron-rich C_4–H bonds geminal to the *t*-Bu were substituted faster than the corresponding hydrogens geminal to the methyl group in **1** and **3**, forming first the fluorinated derivatives **7** and **8** (Scheme 25.1).

The mechanism through which this unusual substitution proceeds is the nonclassical, two-electron, three-center carbonium ion, which inevitably leads to products with full retention of configuration. The chloroform is important in this reaction.

Efficient Preparations of Fluorine Compounds, First Edition. Edited by Herbert W. Roesky.

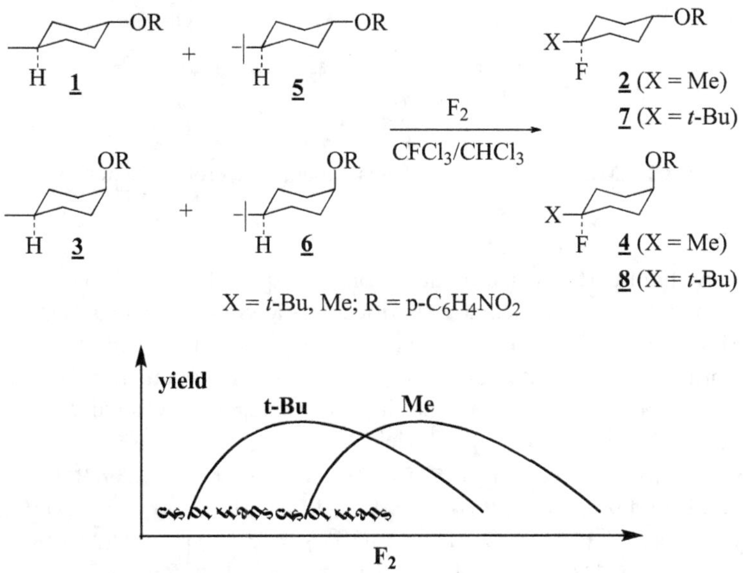

SCHEME 25.1 Fluorination of 4-alkylcyclohexanols.

It provides the necessary polar medium, acts as a radical scavenger, and supplies a somewhat acidic hydrogen that serves as an acceptor through hydrogen bonding for the emerging fluoride anion resulting in lower activation energy of the transition state (Scheme 25.2) [1].

Experimental Procedure Fluorination of *trans*-4-methylcyclohexyl *p*-nitrobenzoate (**1**) was performed with 0.5 g (1.9 mmol) using 5% F_2 in N_2. After the reaction reached its best yield (monitored by gas chromatography (GC)), it was poured into 500 mL of water; the organic layer was washed with $NaHCO_3$ solution followed by water until neutral. The organic layer was dried over $MgSO_4$, and the solvent evaporated. The crude product was usually purified by vacuum flash chromatography, using Silica gel 60-H (Merck) and if needed

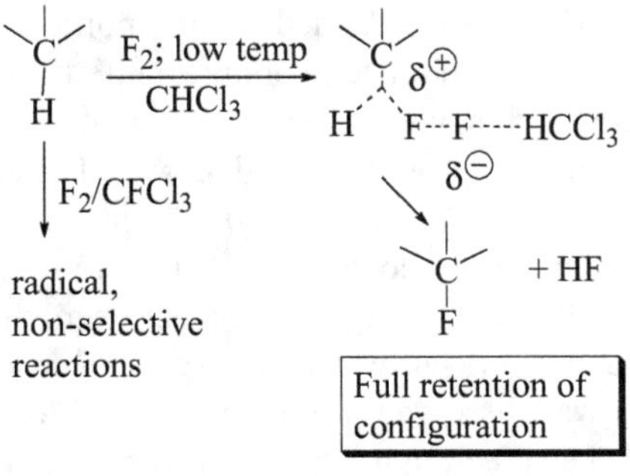

SCHEME 25.2 The two-electron, three-center mechanism of tertiary fluorination.

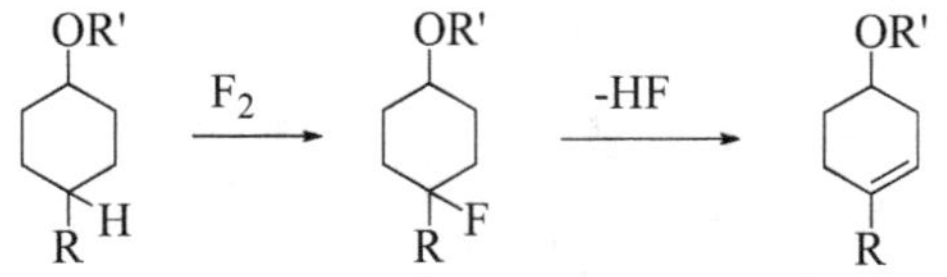

SCHEME 25.3 Forming olefins through dehydrofluorination.

also by high-performance liquid chromatography (HPLC). If the crude reaction mixture is not immediately purified, it is advisable to add a drop or two of pyridine or hexamethyldisilazane to capture the small amount of HF that may be formed with time. Without these bases, the small amount of HF will autocatalyze additional elimination, and the compounds will eventually decompose. Pure **2** was thus isolated in 60% yield.

Characterization Data mp 115 °C (from MeOH). ^{1}H NMR: δ 8.24 (4H, AB, $J = 8$ Hz), 4.95 (1H, m, $W_{h/2} = 24$ Hz), 1.26 (3H, d, $J_{HF} = 21.4$ Hz), 2.4-1.0 (8H, m) ppm. ^{19}F NMR: δ −151.3 (m, $W_{h/2} = 93$ Hz) ppm. ^{13}C NMR: δ 92.54 (d, $J = 167.6$ Hz), 26.60 (d, $J = 19$ Hz), 34.75 (d, $J = 22$ Hz), 27.04 (s), 73.5 (s), 123.54, 130.76, 136.09, 150.58, 164.23 ppm. MS: *m*/*z* 281 (M^+), 261 [$(M - HF)^+$].

Apparatus A standard vacuum line (brass or Monel) for diluting the pure commercial fluorine (95%) with N_2 should be constructed as described [2]. An efficient mixing (preferably with a vibromixer for breaking the gas bubbles) is essential. The reactors are standard glass vessels.

Application One of the important applications of this reaction is the easy dehydrofluorination that could be performed. This consists of a chemical activation of a site that can hardly be approached because it is located far from any chemical anchor, such as π center, heteroatom, and alike. Scheme 25.3 shows the formation of a double bond at such center. Once one has the appropriate olefin at hand, there are numerous ways allowing further chemical applications.

Dehydrofluorination Procedures Three methods could be used for the HF elimination. *Method A*: The corresponding fluoro derivative (1.0 mmol) was dissolved under nitrogen in dry benzene and cooled to about 10 °C. Freshly distilled $BF_3{\cdot}OEt_2$ (8 mL) was added in one portion, the reaction mixture was allowed to warm to room temperature and stirred for additional 4 h. Cold diluted HCl solution was added, the organic layer was washed with bicarbonate and worked-up as described above. The resulting olefins were usually chromatographed by HPLC. *Method B*: The fluoro compound (1 mmol) was dissolved in dry ether and cooled to 0 °C under nitrogen. About 3 mmol of MeMgI solution in ether was added, and the reaction mixture was stirred at room temperature for 12 h. Work-up was as above. *Method C*: The corresponding fluoride (1 mmol) was dissolved in ethylene glycol to which 2 g of aqueous NaOH was added. The reaction mixture was stirred for 12 h at 105 °C, poured into water, extracted with several portions of CH_2Cl_2, and worked-up as usual.

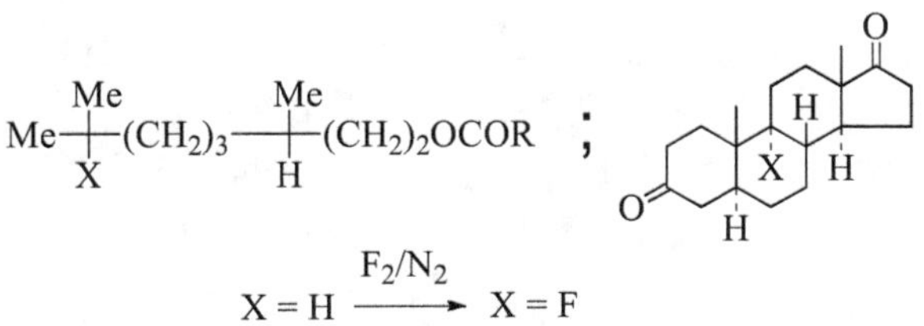

SCHEME 25.4 Fluorination of aliphatics and steroids.

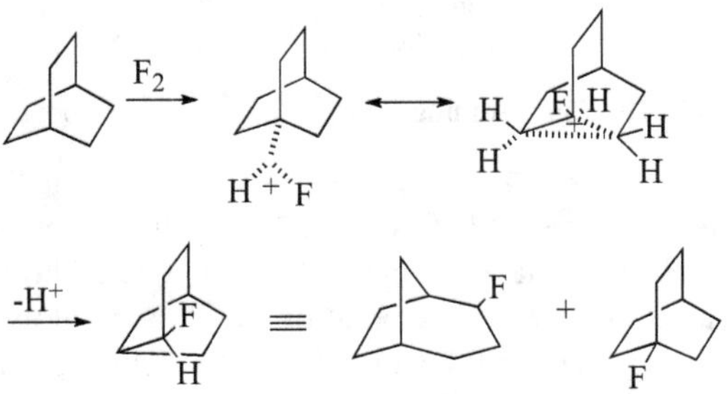

SCHEME 25.5 Fluorination of bicyclic compounds.

Preparation of Other Types of Tertiary Monofluoro Compounds The fluorination procedure described above could be carried out with a wide array of molecules such as straight chains [3], steroids [4], and bicyclic molecules [5] (Schemes 25.4 and 25.5). In all cases, the stereospecificity is retained while the regiospecificity, in the case of more then one tertiary hydrogen, is governed by the electron density of the tertiary C–H bond. In other words, the higher the hybridization on *p*, the easier the substitution. This hybridization is mainly governed by the distance of the relevant C–H bond from electronegative groups and by geometrical consideration. The two-electron, three-center mechanism (non-classical carbocation) will not result in eliminations and rearrangements with the exception of [2.2.2]bicyclooctane, which is another proof for the unusual front side attack.

25.2 ADDING FLUORINE TO HYDROCARBON π CENTERS

Addition of F_2 to Olefins

Addition of halogens (Cl_2, Br_2, and in some cases I_2) to double bonds has been a standard procedure for almost two centuries. They all add in an *anti* mode. It is quite surprising to find out that F_2 does not share the same generality, the reason being the high enthalpy release during the reaction. Low reaction temperature, high dilution of the fluorine gas, and especially polar solvents (such as a few percent of EtOH) can efficiently suppress radical reactions and are the key factors for successful addition

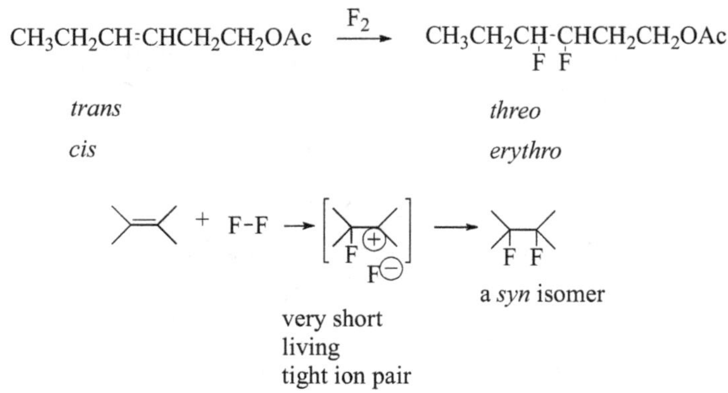

SCHEME 25.6 Adding fluorine to C–C double bonds.

of F_2 across various p centers. What is more unlike the other halogens, fluorine adds itself almost exclusively via a *syn* mode. Such unique addition originates from initial nucleophilic attack of the double bond on the fluorine atom of the somewhat polarized F_2 molecule (see Section 24.2 and 25.1). The resulting α-fluorocarbocation is of course highly unstable, and the tight ion pair collapses before any rotation around the C–C bond takes place. This is also supported by the fact that no nucleophilic attack by the EtOH was detected, an attack that should have taken place if a loose ion pair would have resulted (Scheme 25.6). In a typical reaction, *trans*-3-hexene-1-ol acetate was fluorinated with 1–2% F_2 in N_2 until all starting material was exhausted. The *threo*-difluoro derivative was obtained in 55% yield. When the reaction was repeated with *cis*-3-hexene-1-ol acetate, only the *erythro*-difluoro derivative was formed in 50% yield [6].

Characterization Data for the *erythro*-Difluoro Derivative Oil; IR: 1720 cm^{-1}. ^{1}H NMR: δ 4.59 (1H, dm, $^2J_{HF} = 48$ Hz, $^2J_{HH} = 4$ Hz), 4.43–4.17 (3H, m), 2.07 (3H, s), 1.05 (3H, t, $J_{HH} = 7.5$ Hz), 2.0–1.2 (4H, m) ppm. ^{19}F NMR: δ −196.1 (1F, m, $W_{h/2} = 105$ Hz), −194.7 (1F, m, $W_{h/2} = 103$ Hz) ppm. ^{13}C NMR: δ 170.8 (CO), 94.67 (C-3, dd, $^1J_{CF} = 206$ Hz, $^2J_{CF} = 25$ Hz), 90.54 (C-4, dd, $^1J_{CF} = 186$ Hz, $^2J_{CF} = 27$ Hz), 60.29 (C1), 30.07 (C-2, dd, $^2J_{CF} = 25$ Hz, $^3J_{CF} = 4$ Hz), 23.82 (C-5, dd, $^2J_{CF} = 25$ Hz, $^3J_{CF} = 4$ Hz), 20.7, 9.19 ppm. MS *m/z*: 121 $(M - OAc)^+$.

Characterization Data for the *threo*-Difluoro Derivative Oil; IR: 1720 cm^{-1}. ^{1}H NMR: δ 4.60 (1H, 4 dt, $^2J_{HF} = 48$ Hz, $^3J_{HF} = 23$ Hz, $J_{HH} = 3$ and 9.5 Hz), 4.44–4.16 (3H, m), 2.07 (3H, s), 1.04 (3H, t, $J_{HH} = 7.5$ Hz), 1.95–1.25 (4H, m) ppm. ^{19}F NMR: δ −201.3 (1F, m, $W_{h/2} = 99$ Hz), −194.7 (1F, m, $W_{h/2} = 100$ Hz) ppm. ^{13}C NMR: δ 170.8 (CO), 94.42 (C-3, dd, $^1J_{CF} = 190$ Hz, $^2J_{CF} = 20$ Hz), 89.93 (C-4, dd, $^1J_{CF} = 190$ Hz, $^2J_{CF} = 22$ Hz), 60.31 (C1), 30.01 (C-2, dd, $^2J_{CF} = 23$ Hz, $^3J_{CF} = 5$ Hz), 23.58 (C-5, dd, $^2J_{CF} = 22$ Hz, $^3J_{CF} = 4$ Hz), 20.79, 9.27 ppm. MS *m/z*: 160 $(M - HF)^+$, 121 $(M - OAc)^+$.

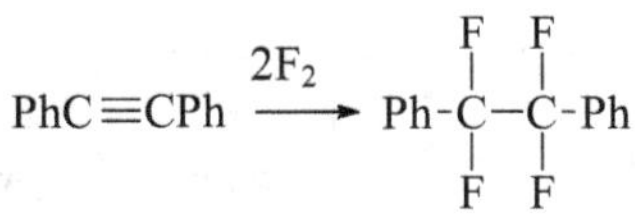

SCHEME 25.7 Adding fluorine to aryl acetylenes.

Addition of F_2 to Aryl Acetylenes

Compounds with the general formula of $ArCF_2CF_2Ar'$ were obtained by several indirect methods as direct addition of this halogen to triple bonds was described to give "complicated mixtures" [7]. Other indirect approaches were the reaction of Deoxo-Fluor® with some benziles [8] or the use of combination of reagents such as nitrosonium tetrafluoroborate ($NO^+BF_4^-$) and pyridinium polyhydrogen fluoride (PPHF) [9]. Adding F_2 across aromatic triple bonds in a direct one-step reaction is, however, possible. Again, as with olefins, a suitable solvent and efficient stirring are of utmost importance. Thus, for example, when diluted fluorine (3–10% F_2 in N_2) was bubbled slowly through a cold solution (−78 °C) of diphenylacetylene in a mixture of $CHCl_3/CFCl_3/EtOH$ as a solvent, 1,2-diphenyltetrafluoroethane was formed in 75% yield (Scheme 25.7). Monoarylacetylenes are also suitable for this reaction, but acetylenes with nonaromatic groups attached to the sp carbon did not give satisfactory results [10].

Characterization Data for Diphenyltetrafluoroethane White crystals 75% yield; mp: 120.3–121.0 °C. ^{1}H NMR: 7.40–7.48 (10H, m) ppm. ^{13}C NMR: 116.8 (tt, $^1J_{C\text{-}F} = 253$ Hz, $^2J_{C\text{-}F} = 36$ Hz), 127.1 (t, $J_{C\text{-}F} = 4$ Hz) 128.2, 131.1 ppm. ^{19}F NMR: −112.2 (4F, s) ppm. MS (EI) (*m/z*) $(M)^+$: 254.

25.3 RADICAL FLUORINATION OF POLYFLUORO ETHERS

Perfluorovinyl ethers ($R_fOCF{=}CF_2$) constitute of an important class of commercial monomers that are used to modify crystallinity in fluoroplastics or are principal components in fluoroelastomers. The first step toward the preparation of these compounds was reacting tetrafluoroethylene (TFE) with polyfluoroalkoxides. The Achilles' heel of these monomers are the few remaining hydrogens in the molecules, because the polymers that were to be prepared eventually were earmarked for some demanding applications. The best way to eliminate this hurdle is to substitute these hydrogens with fluorine atoms. Generally, such kind of transformation could be achieved by using fluorine radicals generated from F_2. It was clear, however, that before any fluorination procedure could be applied, the double bonds would have to be protected and later easily regenerated. The protection issue was solved by chlorinating the vinyl with Cl_2 at 5–15 °C in the absence of solvent to form the dichloro adducts in high yields. These derivatives, which have only two hydrogens on the etheric methylene group, were treated with fluorine under irradiation. The first hydrogen was easily substituted by fluorine, but the second one required a longer time and an excess of F_2.

$$CF_3CF_2CH_2ONa + CF_2{=}CF_2 \longrightarrow CF_3CF_2CH_2OCF{=}CF_2 \xrightarrow{Cl_2}$$

$$CF_3CF_2CH_2OCFClCF_2Cl \xrightarrow[hv]{F_2} CF_3CF_2CF_2OCFClCF_2Cl$$

$$\xrightarrow{Zn} CF_3CF_2CF_2OCF{=}CF_2$$

SCHEME 25.8 The synthesis of perfluorovinyl ethers.

Roughly speaking, 3–4 mol/equiv was enough for replacing the first hydrogen while additional 5–6 mol/equiv was required for completion of the reaction. The yields of the perfluoro derivatives were around 70%. The dechlorination step was easily accomplished with zinc dust in DMF. The whole reaction scheme (for heptafluoropropyl trifluorovinyl ether (serving as an example)) is outlined in Scheme 25.8 [11].

Experimental Procedure NaH (60% oil suspension, 48 g, 1.2 mol) was suspended in anhydrous 1,4-dioxane (600 mL). Pentafluoropropanol (135 g, 0.9 mol) was added slowly while the reaction was kept at 15–20 °C with external cooling. After the addition was completed, the mixture was stirred at ambient temperature for 1–2 h. The pentafluoropropanol salt solution was then transferred into a 1-L autoclave, sealed, and heated for 20 h at 100 °C. The autoclave was cooled, evacuated, and pressured with TFE to 300 psi. The reaction was allowed to proceed for 40 h at 30–35 °C while the TFE pressure was maintained at 300 psi. The neat 1-dihydropentafluoropropyl trifluorovinyl ether was chlorinated with chlorine gas (bubbling into the substrate through a gas inlet tube) at 10–15 °C. The reaction was stopped when the conversion of the starting material was complete. The product was purified by distillation and afforded a clear, colorless liquid of the dichloro derivative in 88% yield. For the fluorination step, mixtures of 25–30% F_2 diluted with nitrogen were used. The appropriate substrate was dissolved either in perfluoro 2-butyl-THF (FC-75) or in Krytox GPL 100 (a fluorinated oil) that also contained about 5 g of pulverized NaF to absorb the released HF. The reaction mixture was cooled to −10 °C, stirred with the aid of a vibromixer, and irradiated with a 450-W, medium pressure mercury lamp. The reactions were monitored by GC/MS and were brought to completion; the mixture was poured into water, washed with bicarbonate till neutral, and the organic layer separated and dried over $MgSO_4$. The product, heptafluoropropyl trifluorovinyl ether (70% yield), was then distilled at 78–80 °C under atmospheric pressure.

Characterization Data for Heptafluoropropyl Trifluorovinyl Ether No protons were detectable in the ^{1}H NMR. ^{19}F NMR: δ −71.5 (bs, 2F, CF_2Cl), −77.5 (narrow m, 1F, CFCl), −82 (t, J = 7 Hz, 3F), −85.5 (m, 2F, CF_2O), −130.7 (narrow m, 2F, CF_3 $\mathit{CF_2}$ CF_2O) ppm; MS m/z: 300.9476 [$(M - Cl)^+$], (calcd 300.9477); 168.9773 $(CF_3CF_2CF_2)^+$ (calcd 168.9888); 150.9251, $(CFClCF_2Cl)^+$ (calcd 150.9329).

Chemicals F_2; $CFCl_3$, $CHCl_3$, and the corresponding organic substrate.

Attention! Safety glasses and protective gloves must be used at all times. All reactions should be carried out in a well-ventilated hood.

Caution! Fluorine is a strong oxidizer and a very corrosive material. An appropriate vacuum line made from copper or Monel in a well-ventilated area should be constructed for working with this element. If elementary precautions are taken, then despite the mythical fears associated with this element, working with diluted F_2 is relatively simple and safe.

Waste Disposal During the reaction, unconsumed fluorine and some HF are trapped with soda lime producing harmless CaF_2. At the end of the reaction, the remaining HF is treated with bicarbonate.

REFERENCES

1. Rozen S, Gal C. *J Org Chem* 1987;52:2769–2779.
2. Dayan S, Kol M, Rozen S. *Synthesis* 1999;1427–1430.
3. Rozen S, Gal C. *J Org Chem* 1987;52:4928–4933.
4. Rozen S, Ben-Shushan G. *J Org Chem* 1986;51:3522–3527.
5. Rozen S, Gal C. *J Org Chem* 1988;53:2803–2807.
6. Rozen S, Brand M. *J Org Chem* 1986;51:3607–3611.
7. McEwen WE, Guzikowski AP, Wolf AP. *J Fluorine Chem* 1984;25:169–193.
8. Chang Y, Tewari A, Adi A, Bae C. *Tetrahedron* 2008;64:9837–9842.
9. York C, Prakash GKS, Olah GA. *J Org Chem* 1994;59:6493–6494.
10. Singh RP, Majumder U, Shreeve JM. *J Org Chem* 2001;66:6263–6267.
11. Hung MH, Rozen S, Smart BE. *J Org Chem* 1994;59:4332–4335.

CHAPTER 26

The Surprising Chemistry of Bromine Trifluoride

SHLOMO ROZEN

26.1 INTRODUCTION

Bromine trifluoride (BrF_3) was, and still is, a somewhat problematic reagent mainly because of certain mythical fears and prejudice associated with it. Under the right conditions, however, it can perform very useful and unique reactions. These should be carried out with solvents such as $CHCl_3$, $CFCl_3$, CCl_4, CH_2Cl_2, or perfluoroethers. BrF_3 is commercially available, but can also be readily prepared in the laboratory by passing fluorine through bromine at about $+5$ °C. It can be stored indefinitely in Teflon® vessels.

BrF_3 is essentially a nucleophilic fluorinating agent. It possesses nonsolvated, and hence very reactive, fluorides along with a strong electrophilic bromine. Occasionally, this presents certain problems for reactions with compounds having activated aromatic rings because it can brominate them. The electrophilic bromine is also a soft acid, a pivotal feature that governs most of its ionic reactions. It complexes itself around soft bases such as sulfur and nitrogen, which brings the nucleophilic fluorides close to their target as shown in Scheme 26.1.

Experimental Procedure BrF_3 is a commercial reagent, but it could also be readily prepared by anybody having a fluorine line. A typical procedure uses 0.2 mol bromine set in a copper reactor and cooled to 0–$+10$ °C. Pure fluorine (0.6 mol) was then slowly bubbled through bromine. Very little fluorine escapes (and trapped by a soda-lime trap) indicating that the reaction is practically quantitative. Under these conditions, BrF_5 is not formed in any appreciable amount. BrF_3 thus obtained is used without further purification and can be stored in Teflon vessels indefinitely.

Efficient Preparations of Fluorine Compounds, First Edition. Edited by Herbert W. Roesky.

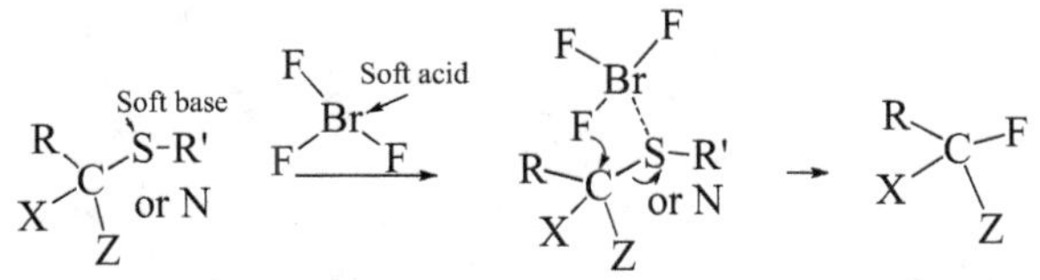

SCHEME 26.1 Reactions of BrF_3 with sulfur- and nitrogen-containing compounds.

Characterization Data for BrF_3 Mp 8.8 °C; bp 126 °C; d_{liq} 2.49; pale yellow when pure; commercially supplied in cylinders as a volatile liquid; orange red colored crystals at low temperatures; decomposes at high temperature to Br_2 and BrF_5. All reactions with BrF_3 are conducted in standard glass vessels.

Chemicals F_2, Br_2.

Attention! Safety glasses and well-ventilated area must be used at all times.

Caution! Fluorine is a strong oxidizer and a very corrosive material. An appropriate vacuum line made from copper or Monel in a well-ventilated area should be constructed for working with this element. If elementary precautions are taken, working with diluted F_2 is relatively simple and safe.

In Regard to BrF_3 It is a strong oxidizer and tends to react exothermically with water and oxygenated organic solvents, such as acetone or tetrahydrofuran (THF), as well as with paper, petroleum ether, and alike. Any work using BrF_3 should be conducted in a well-ventilated area, and caution and common sense should be exercised. Suitable solvents are $CFCl_3$, $CHCl_3$, CH_2Cl_2, or if solubility permits also perfluorinated ethers. Please note that BrF_3 will react with these solvents as well, but in a slow pace. Because its reactions with substrates are fast, this does not pose a major problem.

Waste Disposal Any excess of BrF_3 should be treated with care. Usually, the best way to destroy it is to add it carefully to thiosulfate solution until full decolorization.

26.2 FROM CARBONYLS, VIA IMINES, TO THE CF_2 MOIETY

Transformation of carbonyl to the CF_2 moiety could be achieved by IF (see Chapter 23), but it requires a large excess of IF that is not a commercial reagent and has to be prepared just before its use. What is more, azines, oxime methyl ethers, and some other imine derivatives are not reactive. BrF_3 compensates for these limitations [1].

Thus, adamantanyl azine easily prepared from the parent adamantanone and hydrazine reacted immediately with BrF_3 forming 2,2-difluoroadamantane in 95% yield. When the same reaction was performed on adamantane oxime methyl ether or adamantane dinitrophenylhydrazone (DNP), same product was obtained although in somewhat lower yield.

Similar results were obtained with other cyclic as well as aliphatic straight chain ketones such as 4-*t*-butylcyclohexanone, 2-decanone, and ketones with deactivated

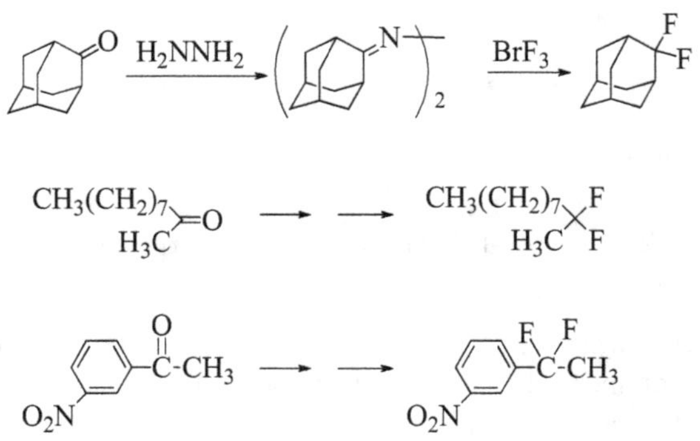

SCHEME 26.2 Azines to CF_2 moiety.

aromatic rings such as *m*-nitroacetophenone, producing 4,4-difluoro-*t*-butylcyclohexane, 2,2-difluorodecane, and 1,1-difluoro-*m*-nitroethylbenzene, respectively (Scheme 26.2).

This reaction could, in special cases, lead to very surprising results [2]. Thus, for example, when the oxime methyl ether of 1,1,1-trichloroethyl pyruvate was reacted with 1.5 mol/equiv of BrF_3, the *C* to *N* migration of the carboxylic acid took place (mechanism described in the original paper) forming *N*-(1,1-difluoroethyl)-*N*-methoxy-2,2,2-trichloroethyl carbamate (Scheme 26.3).

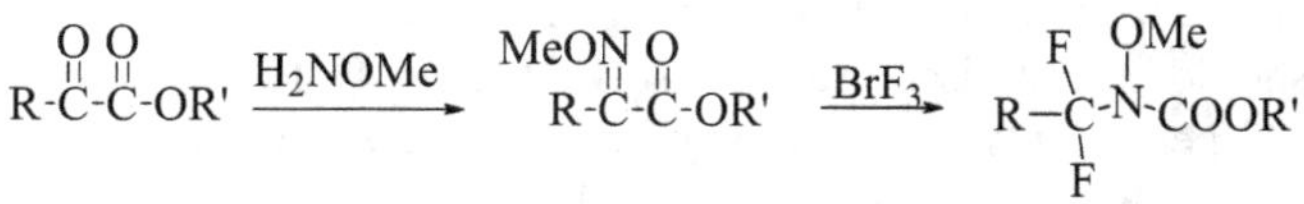

SCHEME 26.3 An unprecedented *C* to *N* ester group migration.

Experimental Procedure Cold (0 °C) solution of $CFCl_3$ (25 mL) containing 6 mmol (0.3 mL) of BrF_3 was slowly (10 min) added to a $CFCl_3$ solution of adamantanyl azine (1 equiv, 6 mmol), easily prepared from the parent adamantanone and hydrazine. An immediate reaction took place forming the known 2,2-difluoroadamantane in 95% yield [3].

26.3 FORMATION OF ACYL FLUORIDES

BrF_3 has found a use in converting carboxylic acids and derivatives into acyl fluorides [4]. Many free carboxylic acids do react directly with BrF_3 to form the corresponding acyl fluorides. Thus, for example, cyclohexane carboxylic acid forms cyclohexanoyl fluoride in 65% yield. Better results were obtained with acyl chloride serving as a substrate as shown by the acyl chloride of dehydrocholic acid, which when treated with BrF_3 formed immediately the corresponding acyl fluoride in good yield. It was found that *t*-butyl esters are also suitable and can serve as starting materials for making acyl fluorides, although the yields are somewhat lower. This is exemplified

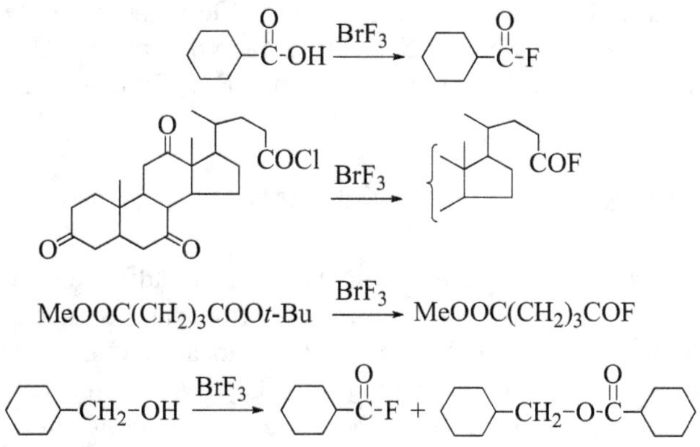

SCHEME 26.4 Formation of acyl fluorides.

with the *t*-butyl ester of 2-norbornyl acetic acid. When treated with BrF_3, it was converted to the respective acyl fluoride. *t*-Butyl esters are unique in this behavior as other aliphatic esters are not affected. Methyl *t*-butyl glutarate, for example, was converted to methyl glutaroyl fluoride in 55% yield, again in a fast reaction (Scheme 26.4).

Another interesting and related reaction takes place with primary alcohols. The reagent was able to oxidize the alcohol and convert it to the corresponding acyl fluoride, although this time the symmetrical ester is also formed as a by-product resulting from a secondary reaction of the acyl fluoride and unreacted alcohol (Scheme 26.4) [5].

Experimental Procedure The reactants that may be carboxylic acids, acyl chlorides, or *t*-butyl esters (1–10 mmol) were dissolved in 10–20 mL of $CFCl_3$ and cooled to 0 °C. About 1.1 mol/equiv of BrF_3 dissolved in the same solvent and cooled to the same temperature was added drop by drop. After the addition was completed, the reaction mixture was washed with $Na_2S_2O_3$ solution till colorless. The aqueous layer was extracted with CH_2Cl_2 and the organic layer was dried over $MgSO_4$.

For Alcohol Oxidation BrF_3 (1.6 mol/equiv, usually around 15 mmol) was dissolved in $CHCl_3$ (30 mL) and cooled to −5 °C. This solution was added drop by drop (10–15 min) to a cold (10 °C) solution of 10 mmol alcohol in 50 mL of either $CHCl_3$ or $CFCl_3$. After the addition was completed, the reaction mixture was added to water, the organic layer was separated as quickly as possible, dried over $MgSO_4$, and the organic solvent was evaporated. The acyl fluoride was then distilled under reduced pressure leaving the symmetrical esters as a distillation residue.

Characterization Data for Methyl Glutaroyl Fluoride It was obtained in 60% yield. IR: 1845 cm^{-1}. ^{1}H NMR: δ 2.59 (1H, t, $^1J = 7$ Hz) ppm. ^{19}F NMR: δ +45 (bs) ppm. MS *m*/*z* (CI): 149 (M + 1)$^+$.

Characterization Data for Cyclohexylacetyl Fluoride from the Oxidation Reaction It was obtained in 55% yield. IR: 1834 cm^{-1}. ^1H NMR: δ 2.51 ppm (1H, tt, $^1J = 10.5$ Hz, $^2J = 4$ Hz) ppm. ^{19}F NMR: δ +36.2 (bs) ppm.

26.4 REACTIONS OF BrF_3 WITH THIOCARBONYLS

As stated above, BrF_3 has a strong affinity to the soft basic sulfur atom. This helped in devising the synthesis of α,α-difluoroethers starting from various esters. The carbonyl moiety of an ester could be readily converted to thiocarbonyl using the Lawesson reagent. A consecutive reaction with BrF_3 at 0 °C provided after few seconds the corresponding α,α-difluoroethers [6]. When instead of thioester dithioesters were used, the whole carboxylic moiety could be replaced with the trifluoromethyl group. It should be mentioned that the reaction proceeds well with aromatic and tertiary acids, less effectively with secondary ones, and not at all with primary acids as they produce mainly tars because of their equilibrium with the enthiol form [7]. Such ideas found a very interesting use in the synthesis of the rare and most interesting trifluoromethyl ethers. Starting with a variety of alcohols, the corresponding xanthates could be readily made. Those xanthates (aliphatic and aromatic) in their turn were reacted with BrF_3 to produce very good yields of the trifluoromethyl ethers (Scheme 26.5) [8].

Experimental Procedure The xanthates (2–4 mmol) were dissolved in 30–60 mL dry $CFCl_3$ and cooled to 0 °C. A 3 mol/equiv of BrF_3 was dissolved in 30 mL of the same solvent, cooled to 0 °C, and added drop by drop to the xanthate solution over a period of 5–10 min. The reaction mixture was then washed with aqueous Na_2SO_3 until colorless, neutralized with aqueous $NaHCO_3$, extracted with CH_2Cl_2, dried over $MgSO_4$, and evaporated. The products were purified by fractional distillation.

Characterization Data for Octyl Trifluoromethyl Ether Octanoyl methyl xanthate (15 mmol) in 250 mL of $CFCl_3$ was reacted with 45 mmol (2.5 mL) of

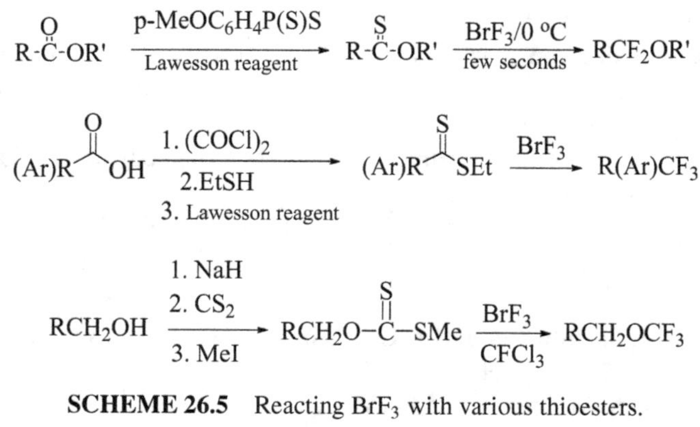

SCHEME 26.5 Reacting BrF_3 with various thioesters.

BrF_3 in 100 mL of $CFCl_3$ according to the general procedure described earlier. The product was isolated as a colorless oil, bp (171 mm) 95–97 °C in 82% yield. ^1H NMR: δ 3.94 (2H, t, J = 6.5 Hz), 1.68 (2H, quin, J = 6.6 Hz), 1.5–1.2 (10H, m), 0.89 (3H, t, J = 6.5 Hz) ppm. ^{19}F NMR: δ −61.13 (s) ppm. ^{13}C-NMR: δ 122 (q, J = 251 Hz), 67, 32, 29, 28, 25, 23, 14 ppm. MS (m/z): 198 M^+, 99 $(CH_2OCF_3)^+$, 69 $(CF_3)^+$.

Similar to the preparation of xanthates, alcohols in general can react with thiophosgene to form alkyl or aryl chlorothioformates. When these compounds were reacted with BrF_3, chlorodifluoromethyl ethers were formed in 60–85% yields. The reactions proceed under mild conditions and short reaction times. Incorporation of this moiety into organic molecules can often modify the biological and physiological activities and properties. In addition, the interest in preparing the OCF_2Cl moiety is also based on its ability to serve as a starting point for further chemical transformations [9]. With higher ratio of the aromatic and aliphatic alcohols to bromine trifluoride, thiocarbonate derivatives were obtained in good yields. The thiocarbonates could be reacted with BrF_3 forming linear symmetric and asymmetric difluoromethylenedioxy compounds in good yields under mild conditions [10]. The reaction with thiocarbonate was also extended to catechol derivatives and 1,2-dihydroxy heterocyclic compounds. The resulting difluoroaryldioxoles, many of them highly interesting for the pharmaceutical industry, were obtained in very good yields (Scheme 26.6) [11].

Experimental Procedure 2,3-Dihydroxypyridine (6.3 mmol) was dissolved in 40 mL dichloromethane and 12.6 mmol of 4-dimethylaminopyridine (DMAP).

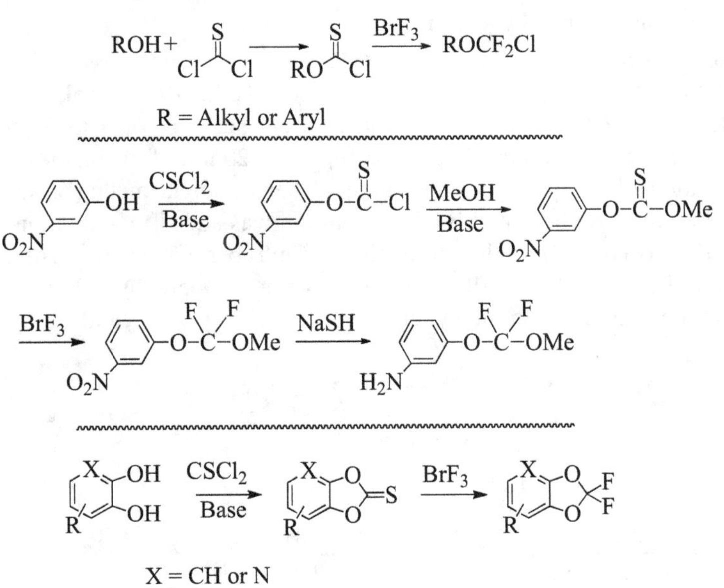

SCHEME 26.6 Reacting BrF_3 with various thiocarbonyls.

The suspension was cooled to 0 °C and 0.6 mL (7.8 mmol) thiophosgene was added drop by drop and then stirred for 2 h. The 1,3-dioxolo-4,5-pyridine-2-thione was isolated by filtration and purified by flash chromatography (90% yield). This thione was dissolved in 25 mL of $CHCl_3$ in a glass flask and cooled to 0 °C. BrF_3 (1 mol/equiv) was dissolved in 10 mL $CFCl_3$, cooled to 0 °C, and added drop by drop (about 1 min) at the same temperature using a glass dropping funnel. The reaction mixture was then washed with aqueous Na_2SO_3 till colorless, the aqueous layer was extracted three times with CH_2Cl_2, and the combined organic layers were dried over $MgSO_4$. Evaporation of the solvent followed by flash chromatography yielded the desired 2,2-difluoro-1,3-dioxolo-4,5-pyridine in 95% yield.

Characterization Data for 1,3-Dioxolo-4,5-pyridine-2-thione White crystals, mp 82.3–83.8 °C. ^{1}H NMR: δ 8.23 (1H, dd, ^1J = 6 Hz, ^2J = 1.5 Hz), 7.63 (1H, dd, ^1J = 8 Hz, ^2J = 1.5 Hz), 7.33 (1H, dd, ^1J = 8 Hz, ^2J = 6 Hz) ppm. ^{13}C NMR: δ 181.2, 155.6, 144.4, 138.9, 122.1, 118.5 ppm. MS (supersonic molecular beam) molecular ion at *m/z* 153 M^+.

Data for 2,2-Difluoro-1,3-dioxolo-4,5-pyridine Colorless oil. ^{1}H NMR: δ 7.89 (1H, dd, ^1J = 5 Hz, ^2J = 1.5 Hz), 7.33 (1H, dd, ^1J = 8 Hz, ^2J = 1.5 Hz), 7.06 (1H, dd, ^1J = 8 Hz, ^2J = 5 Hz) ppm. ^{13}C NMR: δ 153.1, 141.2, 136.7, 129.5 (t, J = 269 Hz), 119.7, 116.7 ppm. ^{19}F NMR: δ −50.5 (2F, s) ppm. This compound is very volatile and unstable but the supersonic GC/MS method revealed a strong molecular ion at *m/z* 159 $(M)^+$.

Application One of the important applications of the reaction leading to chlorodifluoromethyl ethers is opening perspectives for further chemical transformation based on the chlorine atom. One of these perspectives is the reduction of the chlorine atom to hydrogen (Scheme 26.7) [12].

Reduction Procedure Preparation of 3-chlorodifluoromethoxyaniline. To the solution of *m*-nitrophenyl difluorochloromethyl ether (0.025 mol), sodium sulfide $Na_2S{\cdot}9H_2O$ dissolved in 20 mL of water and 2 g $NaHCO_3$ were added. When the carbonate has dissolved completely, 20 mL of methanol was added. Filtration of the precipitated sodium carbonate left a methanol solution of sodium hydrosulfide (NaSH). This solution was added to a solution of 2.6 g (0.012 mol) of 1-chlorodifluoromethoxy-3-nitrobenzene in 20 mL of methanol and refluxed for 1 h. Purification by flash chromatography led to the desired 3-amino difluorochloromethyl ether, yellow oil, 2 g, 90% yield.

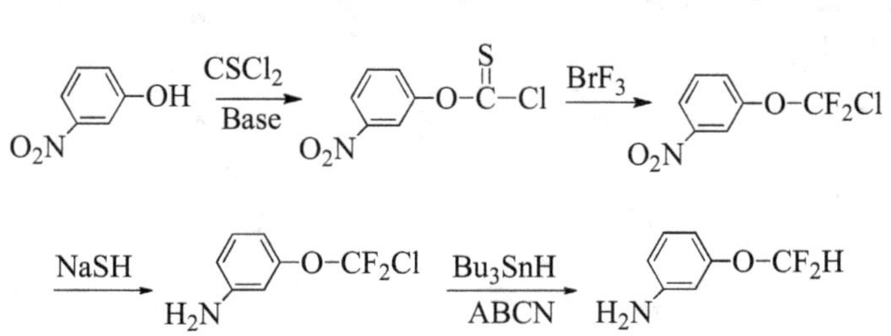

SCHEME 26.7 From $ROCF_2Cl$ to $ROCF_2H$.

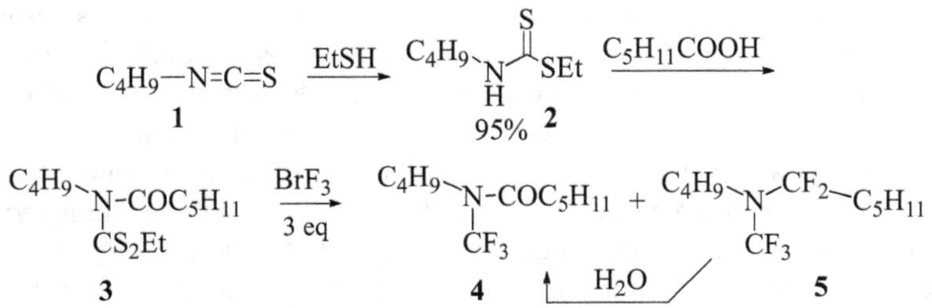

SCHEME 26.8 Synthesis of *N*-(trifluoromethyl)amides.

3-Chlorodifluoromethoxyaniline was then dissolved in 20 mL of anhydrous toluene, and 350 mg of 1,1′-azobis(cyclohexanecarbonitrile) (1.5 mmol) and 2.2 mL of tributyltin hydride (8.1 mmol) were added. The solution was refluxed for 8 h. After evaporation of the solvent, the product, 3-difluoromethoxyaniline, was purified by flash chromatography.

Characterization Data for 3-Difluoromethoxyaniline It is prepared from the difluorochloromethyl ether, 90% yield, yellow oil. ^{1}H NMR: δ 7.06 (1H, t, $J =$ 8 Hz), 6.46–6.35 (3H, m), 6.43 (1H, t, $J = 75$ Hz) ppm. ^{13}C NMR: δ 116.2 (t, $J = 258$ Hz), 152.5, 148.2, 130.4, 111.9, 108.4, 105.7 ppm. ^{19}F NMR: δ −80.6 (d, $J = 75$ Hz) ppm. MS (m/z): $[MH]^+$ calcd for $C_7H_8F_2NO$, 160; found, 160.

Another interesting reaction involving thiocarbonyl was developed with the ultimate goal for constructing the trifluoromethyl group attached to a nitrogen atom [13]. The family of isothiocyanates proved to be the best general method for *N*-(trifluoromethyl)amides. The isothiocyanate derivative (**1**) was reacted with ethanethiol to form ethyl alkyldithio carbamates (**2**) in almost quantitative yield (Scheme 26.8). Next, this compound was treated with carboxylic acid in the presence of *N*,*N*′-dicyclohexylcarbodiimide (DCC) and DMAP forming ethyl alkyl(alkanoyl)dithiocarbamate (**3**) in good yield. This derivative was then reacted with BrF_3 resulting in *N*-butyl-*N*-(trifluoromethyl)hexane amide (**4**) (Scheme 26.8). In many cases, the pentafluoro derivative of type **5** was also formed, but if needed it could be hydrolyzed back to the *N*-trifluorocarbamide **4** raising the overall yield to more than 80%.

Experimental Procedure A solution of the isothiocyanate (**1**) (26.0 mmol) was reacted with ethanethiol (3.3 mL, 39.0 mmol) in 40 mL of dry THF in the presence of Et_3N (0.3 mL, 26.0 mmol) and refluxed for 20 h. Evaporation of the solvent followed by flash chromatography yielded dithiocarbamate (**2**) in 95% yield. This product was reacted with 11 mmol of hexanoic acid in the presence of 3.5 g (16.9 mmol) of DCC and 134 mg (1.1 mmol) of DMAP in 50 mL of dichloromethane. The corresponding dithiocarbamate derivative (**3**) was thus formed in 70–95% yield. After evaporation of the solvent and short flash chromatography, **3** was used without further purification for the next fluorination step. Dithiocarbamate derivative (**3**) (3 mmol) was dissolved in

25 mL of $CHCl_3$ or $CFCl_3$ in a glass flask and cooled to 0 °C. The best result was achieved when BrF_3 (3 mol/equiv, 9 mmol) was dissolved in a few milliliters of $CFCl_3$, cooled to 0 °C, and added drop by drop to the reaction mixture using a glass dropping funnel at 0 °C. The reaction mixture was then washed with aqueous Na_2SO_3 till colorless, extracted with CH_2Cl_2, and dried over $MgSO_4$. Evaporation of the solvent followed by flash chromatography yielded fluorinated **4**. The smaller pentafluorinated fraction **5** was readily hydrolyzed back to **4** increasing its total yield to 85%. If there is no interest in compound **5**, the hydrolysis step could be performed on the crude mixture of **4** and **5**.

Characterization Data for *N*-Butyl-*N*-(trifluoromethyl)hexanamide (4) Oil. ^{1}H NMR: δ 3.54–3.49 (2H, m), 2.45 (2H, t, $J = 7.0$ Hz), 1.67–1.59 (2H, m), 1.51–1.48 (2H, m), 1.31–1.22 (6H, m), 0.90–0.88 (6H, m) ppm. ^{13}C NMR: δ 172.5, 122.1 (q, $J = 265$ Hz), 43.3, 34.9, 31.3, 31.1, 24.5, 22.4, 19.8, 13.8, 13.5 ppm. ^{19}F NMR: δ −51.8 (3F, s) ppm. IR: 1703.2 cm^{-1}. MS (CI) *m/z*: 240.2 $(MH)^+$.

26.5 REACTIONS OF BrF_3 WITH NONCYCLIC SULFUR ACTIVATORS LEADING TO THE IMPORTANT CF_2- AND CF_3-CONTAINING COMPOUNDS

Because sulfur is a good coordinating anchor for BrF_3, it was used frequently to form fluorine-containing compounds. One of the promising derivatives is tris(methylthio)methane, which possesses an acidic hydrogen. This compound readily reacts under basic condition with alkyl halides such as bromodecane to form tris(methylthio)undecane. It was then reacted with threefold excess of BrF_3 resulting in 1-methylthio-2-bromo-1,1-difluoroundecane in 70% yield. The reaction conditions are mild and consist of reaction temperature of 0 °C for a few minutes. The difluorinated products can be transformed to the trifluoro one by applying another 3 equiv of BrF_3. In all cases, the bromine atom could be easily removed by either replacing it with hydrogen or by elimination reactions (see later) [14]. The reaction between BrF_3 and the tris(methylthio) derivatives can be applied to a number of aliphatic chains with similar results (Scheme 26.9).

Experimental Procedure To a cold (−78 °C) THF solution of 1.8 mL (13.6 mmol) tris(methylthio)methane, 8.9 mL (14.24 mmol) of BuLi was added

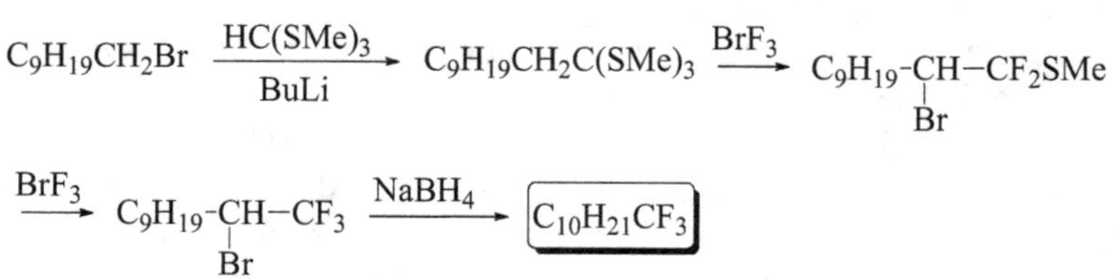

SCHEME 26.9 From RX to RCF_3.

under nitrogen. After stirring for 2 h at this temperature, 12 mmol of neat decyl bromide was added, the reaction mixture was left for additional 2 h at −78 °C, and stirred for another hour at room temperature. The usual work-up provides the corresponding tris(methylthio) derivative pure enough for the next step. This compound was dissolved in 10–15 mL of dry $CFCl_3$ and cooled to 0 °C. About 3 mmol (for obtaining the difluoromethylmethylsulfide products) or 10 mmol (for obtaining the trifluoromethyl compounds) of BrF_3 was dissolved in 10–15 mL of the same solvent, cooled to 0 °C, and added drop by drop to the tris(methylthio) solution, at the same temperature, over a period of up to 3 min. The reaction mixture was then washed with aqueous Na_2SO_3 till colorless and worked up as usual. Same procedure was applied when the bromo difluoromethylmethylsulfide derivative was reacted with 3 mmol of BrF_3 converting it to the corresponding trifluoromethyl one. To 1 mmol of the above bromo compound (di- or trifluorinated), 6 mmol of $NaBH_4$ dissolved in 15 mL of DMSO was added. The reaction mixture was stirred at 110 °C for 1–2 h. It was then acidified with aqueous HCl and worked up as usual.

Characterization Data for 1,1-Difluoro-1-methylthioundecane It was obtained in 70% yield (for the reduction step): oil. ^1H NMR: δ 2.28 (3H, s), 2.20–1.96 (2H, m) ppm. ^{13}C NMR: δ 130.8 (t, $J = 274$ Hz), 38.9 (t, $J = 23$ Hz), 31.9, 29.6, 29.5, 29.4, 29.3, 29.1, 23.2, 22.6, 14.0, 10.1 ppm. ^{19}F NMR: δ −76.74 (t, $J = 15$ Hz) ppm.

Application This reaction has an important potential for the field of positron emitting tomography (PET) because no good method exists for the preparation of ^{18}F CF_3 group. A difluoro sulfur compound (RCF_2SMe) could be oxidized to the sulfone derivative (RCF_2SO_2Me) and then reacted with the readily made Bu_4N[18]F, providing an efficient route for synthesizing RCF_2[18]F.

Moieties containing sulfur helped to construct and place the CF_3 group α to a primary carboxylate group of various carboxylic acids (**1**; Scheme 26.10) [15]. One of the best methods to place a sulfur atom near the α-position of an ester group is to react its corresponding enolate with carbon disulfide (CS_2) followed by methyl iodide (MeI) forming the disulfur compound **2** [16]. To substitute both sulfur atoms of the resulting 2-carbomethoxy-1,1-bis(methylsulfide)-1-alkene, 5 mol/equiv of BrF_3 had

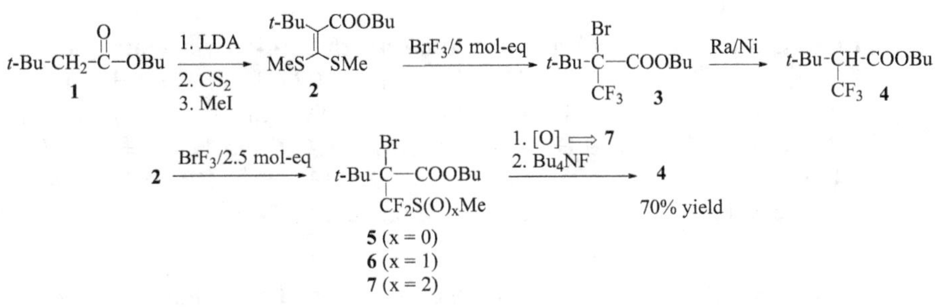

SCHEME 26.10 Constructing the CF_3 group α to a primary carboxylic acid.

to be used to form 2-bromo-2-trifluoromethyl carboxylate (**3**). The bromine atom could then be removed by Raney nickel and the desired α-trifluoromethyl ester (**4**) was obtained. However, the reaction was considerably improved, and the yields more than doubled, when a somewhat different route was used. When only 2.5 mol/equiv of BrF_3 was reacted for less than a minute with the disulfide **2**, a mixture of more than 85% of methyl 2-bromo-2-[difluoro(methylsulfide)methyl]alkanoate **5**, the respective sulfoxide **6**, and traces of the sulfone **7** were obtained. This mixture was not resolved but treated "as is" with an oxygen transfer agent ($HOF{\cdot}CH_3CN$ gave the best results) at room temperature, transferring the sulfur-containing compound to the corresponding sulfone **6** in a few minutes. The latter was reacted with Bu_4NF eliminating both the bromine and the sulfone group to give the target α-trifluoromethylalkanoate **4** in overall yields of up to 70% based on the starting ester. It should be mentioned here that this method is also very suitable for introducing the important isotope ^{18}F into the CF_3 group for PET purposes. Scheme 26.10 describes the above two routes using butyl neopentanoate **1** as an example.

Experimental Procedure The general procedure for the preparation of 2-carbomethoxy-1,1-bis(methylsulfide)-1-alkene derivative (compound **2** in Scheme 26.10) had been described in Reference 16. This compound (1 mmol) was dissolved in 10–15 mL of dry $CFCl_3$. About 2.5 mmol of BrF_3 was dissolved in 10 mL of the same solvent, cooled to 0 °C, and added drop by drop for 1–2 min to compound **2**. The reaction was quenched with saturated aqueous Na_2SO_3 and washed until colorless. The aqueous layer was extracted with CH_2Cl_2 and the organic layer was dried over $MgSO_4$. The oily crude product that contained mainly the difluoromethylsulfide (**5**) and the difluoromethylsulfoxide (**6**) derivatives was dissolved in 15 mL of $CHCl_3$ and oxidized with threefold excess of $HOF{\cdot}CH_3CN$ [17] at 0 °C for 20 min. The reaction was terminated with $NaHCO_3$ solution, extracted with CH_2Cl_2, and dried over $MgSO_4$. After removing the solvent, the reaction mixture contained mainly the difluoromethylsulfonyl derivative (**7**) which, if desired, could be isolated by flash chromatography. The crude derivative **7** (300 mg) was dissolved in 15 mL of THF and 1.05 mol/equiv of Bu_4NF (1-M solution in THF) was added at room temperature and stirred for 30 min. Evaporation of the solvent followed by flash chromatography gave the desired butyl 2-trifluoromethyl-2-*t*-butylacetate (**4**) in 70% overall yield.

Characterization Data for Butyl 2-trifluoromethyl-2-*t*-butylacetate (4) IR 1739 cm^{-1}. ^{1}H NMR: δ 4.17 (2H, t, $J = 7$ Hz), 3.02 (1H, q, $J = 9$ Hz), 1.69–1.62 (2H, m), 1.46–1.39 (2H, m), 1.14 (9H, s), 0.92 (3H, t, $J = 7$ Hz) ppm. ^{19}F NMR: δ −67 (d, $J = 9$ Hz) ppm. ^{13}C NMR: 167.2, 125.1 (q, $J = 282$ Hz), 65.1, 59.3 (q, $J = 25$ Hz), 32.8, 30.4, 28.5, 19.0, 13.5 ppm. HRMS (CI) (*m*/*z*): $(MH)^+$ calcd for $C_{11}H_{19}F_3O_2$, 241.1415; found, 241.1406.

For placing the trifluoromethyl group at the α-position of a secondary carboxylic acid, a somewhat different route had to be taken [18]. Ethyl cycloheptyl carboxylate

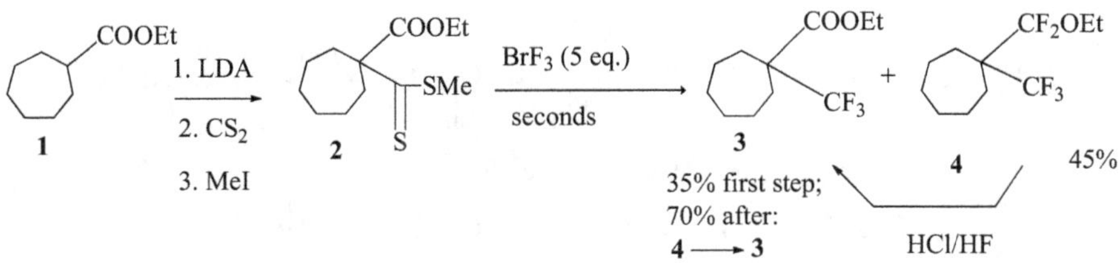

SCHEME 26.11 Constructing the CF_3 group α to a secondary carboxylic acid.

(**1**) could serve as an illustrative example. When this ester was reacted with lithium diisopropylamide (LDA), CS_2, and MeI, methyl 2-carboethoxydithiocycloheptanoate (**2**) (Scheme 26.11) was obtained in very good yield. Unlike primary carboxylic acids, the products here do not have an olefinic bond. Despite that, using a large excess (5 mol/equiv) of BrF_3 resulted, after a few seconds, directly into the desired ethyl 2-trifluoromethylcycloheptanoate (**3**), although in 35% yield only. It was accompanied by 1-(difluoromethylene ethyl ether)-1-trifluoromethylcycloheptane (**4**) (45% yield). This could be treated with an aqueous alcoholic mixture of HCl/HF for a few hours regenerating the ester moiety in higher than 80% yield, raising the overall yield of compound **3** to 70% (Scheme 26.11).

Experimental Procedure Ester **1** (10 mmol) was dissolved in 100 mL of dry THF and cooled to −78 °C. LDA (12.5 mmol) (1.5-M solution in cyclohexane) was added, the cooling bath was removed, and the mixture was stirred for 2 h. The reaction mixture was cooled again to −78 °C and CS_2 (about 40 mmol) was added. The brown solution was stirred for another hour, cooled again to −78 °C, and MeI (about 40 mmol) was added. After another 2 h, the reaction mixture was warmed to room temperature, poured into water, extracted with ether, dried over $MgSO_4$, and the solvent was removed. Compound **2** was isolated by flash chromatography as a brown-orange oil. Compound **2** (1 mmol) was dissolved in 10–15 mL of dry $CFCl_3$ and cooled to 0 °C. About 5 mmol of BrF_3 was dissolved in 10 mL of the same solvent, cooled to 0 °C, and added drop by drop, during 1–2 min, to the above solution. On completion, the reaction was quenched with saturated aqueous Na_2SO_3 and washed until colorless. The aqueous layer was extracted with CH_2Cl_2 and the organic layer was dried over $MgSO_4$. Evaporation of the solvent followed by purification by flash chromatography gave the desired product **3** in 35% yield, and product **4** in 45% yield. The pentafluoro derivative (**4**) was then transformed back to compound **3** by heating it to 100 °C for 6 h in 10 mL of the ethyl alcohol and 5 mL water containing 0.5 mL of HCl (32%) and 0.5 mL HF (40%). The regeneration of the ester was monitored by GC. The aqueous layer was extracted with ether, the organic solvent dried over $MgSO_4$, evaporated, and the residue was purified by flash chromatography. The desired product **3** was thus obtained in 70% yield.

Characterization Data for Ethyl 2-trifluoromethylcycloheptanoate (3) It was obtained in 70% overall yield: oil. IR 1731 cm^{-1}. ^1H NMR: δ 4.22 (2H, q,

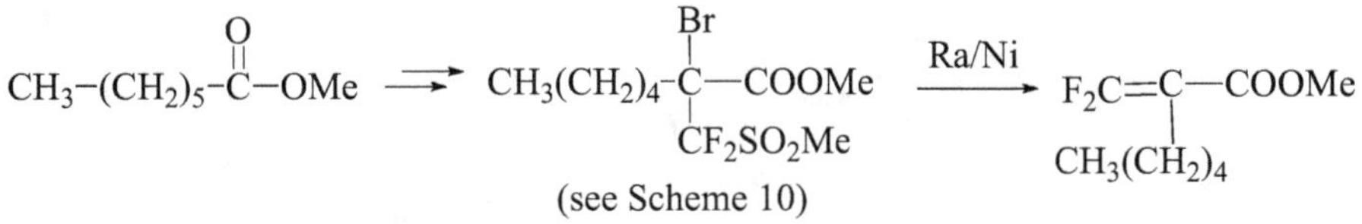

SCHEME 26.12 Constructing β,β-difluoroacrylates.

$J = 7$ Hz), 1.24 (3H, t, $J = 7$ Hz) ppm. ^{19}F NMR: δ −73.3 (s) ppm. ^{13}C NMR: δ 170.2, 127.7 (q, J = 283 Hz), 61.6, 55.8 (q, J = 23 Hz), 30.4, 30.0, 23.3, 13.8 ppm. HRMS (CI) (*m*/*z*): $(MH)^+$ calcd for $C_{11}H_{18}F_3O_2$ 239.1258; found, 239.1262.

The reaction leading to the construction of α-trifluoromethyl carboxylic acid (Scheme 26.10) could also lead to various α-alkyl-β,β-difluoroacrylates. This was achieved by treating compounds of type **7** with Raney nickel instead of Bu_4NF (Scheme 26.12) [19]. Methyl heptanoate can serve as an example. After the corresponding bromodifluoro sulfone was obtained and treated with Raney nickel, the desired methyl 2-pentyl-β,β-difluoroacrylate was formed in 55% overall yield.

Experimental Procedure Following the preparation of the α-bromo-α-difluoromethyl sulfone methyl heptanoate (as described in the section relevant to Scheme 26.10), it was dissolved in 15 mL of THF and 3 g of Raney nickel was added. The reaction mixture was stirred for 30 min at room temperature and filtered through silica. The methyl 2-pentyl-β,β-difluoroacrylate was isolated by flash chromatography in 60% yield.

Characterization Data for 2-Pentyl-β,β-difluoroacrylate Oil. IR 1718 cm^{-1}. ^{1}H NMR: δ 3.78 (3H, s), 2.31–2.05 (2H, m), 0.90 (3H, t, $J = 7$ Hz) ppm. ^{19}F NMR: δ −70.1 (1F, dm, $J = 3$ Hz), −75.0 (1F, dm, $J = 3$ Hz) ppm. ^{13}C NMR: δ 167.3 (dd, ^1J = 13 Hz, ^2J = 6 Hz), 161.7 (dd, ^1J = 310 Hz, ^2J = 295 Hz), 90.6 (dd, ^1J = 23 Hz, ^2J = 5 Hz), 53.9, 33.1, 30.1, 26.3, 24.2, 15.8 ppm. MS (CI) (*m*/*z*): 193 $(MH)^+$.

BrF_3 is also a very useful reagent for transforming either aldehydes or ketones into terminal difluoroolefins resembling the Wittig reaction, which leads to terminal olefins. Thus, a lithium salt of the commercial bis(methylthio)methane was reacted with trimethylsilyl chloride and then with 5-nonanone (**1**) (Scheme 26.13) to produce 2-butyl-1,1-bis(methylthio)hexene (**2**). Reacting this derivative with 2 mol/equiv of BrF_3 gave an oily mixture that contained 2-bromo-2-butyl-1,1-difluoro-1-methylthiohexane (**3**, $x = 0$) as the main product along with some of the corresponding sulfoxide and sulfone derivatives (**3**, $x = 1$ and 2 correspondingly). Because sulfone is a better leaving group than sulfide or sulfoxide, the crude mixture **3** ($x = 0$, 1, 2) was treated with the HOF·CH_3CN complex, producing the sulfone derivative 1-bromo-1-butyl-1-[difluoro-1-(methylsulfonyl)methyl]pentane (**3**, $x = 2$) in a few minutes. This

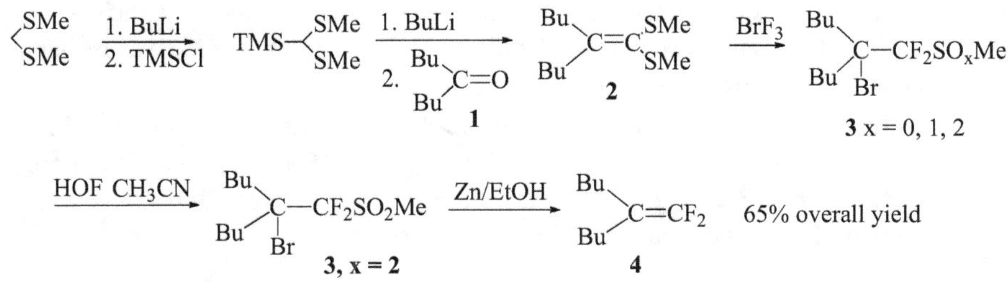

SCHEME 26.13 The construction of terminal difluoroolefins.

derivative was then brought in contact with activated Zn eliminating both the sulfone group and the bromine atom forming 1,1-difluoro-2-butylhex-1-ene (**4**) in 65% yield.

Experimental Procedure The preparation of 2-butyl-1,1-bis(methylthio)hex-1-ene starts with 8 g of bis(methylthio)methane (74 mmol) dissolved in 100 mL of dry THF and cooled to −78 °C under inert atmosphere. BuLi (53 mL, 1.6 M in hexane) was added and the mixture was allowed to warm slowly (4 h) to 0 °C. The reaction mixture was cooled again to −78 °C and 17 g (160 mmol) of trimethylsilyl chloride was added. The reaction mixture was allowed to warm to room temperature and stirred for another 15 h at that temperature. Water was added, the organic layer was separated, and the aqueous layer was extracted with pentane. The combined organic layers were washed with water, dried over K_2CO_3, filtered, and evaporated. The product bis(methylthio)trimethylsilylmethane can be isolated by distillation (75 °C/10 mbar) in 90% yield. Four grams of it (27 mmol) in 100 mL of dry THF was cooled to −78 °C under inert atmosphere and 16.9 mL of BuLi (1.6 M in hexane) was added and the mixture allowed reaching 0 °C in 4 h. The reaction mixture was cooled again to −78 °C and 27 mmol of 5-nonanone (**1**) was added. The reaction mixture was allowed to reach room temperature and then stirred for another 12 h. Water was added, the organic layer was separated, and the aqueous one was extracted with pentane. The combined organic layers were washed with water, dried over K_2CO_3, filtered, and evaporated. The intermediate **2** was isolated by flash chromatography in about 90% yield. Compound **2** was dissolved in 15 mL dry $CFCl_3$ and cooled to 0 °C. About 10 mmol of BrF_3 was dissolved in 15 mL of the same solvent, cooled to 0 °C, and added drop by drop during 1–2 min. The reaction was quenched with saturated aqueous Na_2SO_3 and washed until colorless. The aqueous layer was extracted with CH_2Cl_2 and the organic layer was dried over $MgSO_4$. Evaporation of the solvent left the oily crude 2-bromo-1,1-difluoro-1-methylthiononane (**3**, $x = 0 + 1 + 2$). It was dissolved in 40 mL of CH_2Cl_2, oxidized with threefold excess of $HOF{\cdot}CH_3CN$ at 0 °C for 10 min, and quenched with $NaHCO_3$. The product was extracted with CH_2Cl_2, dried over $MgSO_4$, and evaporated giving an oily mixture containing the 2-bromo-1-[difluoro-1-(methylsulfonyl)methyl]nonane derivative (**3**, $x = 2$), which could be isolated if needed by flash chromatography. This product (1 g) was dissolved in 8 mL of ethanol and 1.2 g of activated zinc powder was added.

The mixture was refluxed for 15–30 min. After filtration of the zinc, the solvent was evaporated and the desired 1,1-difluoro-2,2-dibutylethane (**4**) was isolated by flash chromatography in 65% yield.

Characterization Data for 1,1-Difluoro-2,2-dibutylethane (4) Oil. ^{1}H NMR: δ 1.97–1.90 (4H, m), 0.93 (6H, t, $J = 7$ Hz) ppm. ^{19}F NMR: δ −97.2 (s) ppm. ^{13}C NMR: δ 153.2 (t, $J = 282$ Hz), 88.8 (t, $J = 18$ Hz), 29.6, 25.5, 22.5, 13.8 ppm. MS (supersonic GC/MS) (*m*/*z*): 176 $(M)^+$.

26.6 REACTIONS OF BrF_3 WITH DITHIANES

Dithianes proved to be very effective complexation moiety for BrF_3 bringing its nucleophilic fluorine close to their target (the carbon attached to the two sulfur atoms). The products are the corresponding 1,1-difluoromethyl alkanes ($RCHF_2$) generally obtained in 60–75% yield [20]. There are several ways of constructing dithianes and one of them is the reaction of an alkyl bromide with the lithium derivative of dithiane itself. The reaction proceeds well with primary alkyl halides. The limiting step for secondary alkyl halides is the relatively low yield of the dithiane preparation, although circumventing routes do exist and are described below (Scheme 26.14).

Experimental Procedure The procedure for reaction of the 2-alkyl-1,3-dithiane [21] with BrF_3 is as following: a 2-decane-1,3-dithiane (1 mmol) was dissolved in 10–15 mL of dry $CFCl_3$. About 3 mmol of BrF_3 was dissolved in the same solvent, cooled to 0 °C, and added drop by drop to the dithiane solution. The reaction mixture was then washed with saturated aqueous $Na_2S_2O_3$ till colorless. The aqueous layer was extracted with CH_2Cl_2 and the organic layer was dried over $MgSO_4$. Evaporation of the solvent followed by purification by flash chromatography gave the target 1,1-difluoroundecane.

Characterization Data for 1,1-Difluoroundecane It was obtained in 75% yield. ^{1}H NMR: δ 5.79 (1H, tt, $J = 57$ Hz, $J = 5$ Hz), 1.77–1.84 (2H, m), 1.28 (16H, br s), 0.9 (3H, t, $J = 7$ Hz) ppm. ^{13}C NMR: δ 117.4 (t, $J = 236$ Hz), 34.0 (t, $J = 20$ Hz), 31.8, 29.5, 29.3, 29.0, 22.6, 22.0, 14.0 ppm. ^{19}F NMR: δ −116 (dt, $J = 57$ Hz, $J = 17$ Hz) ppm.

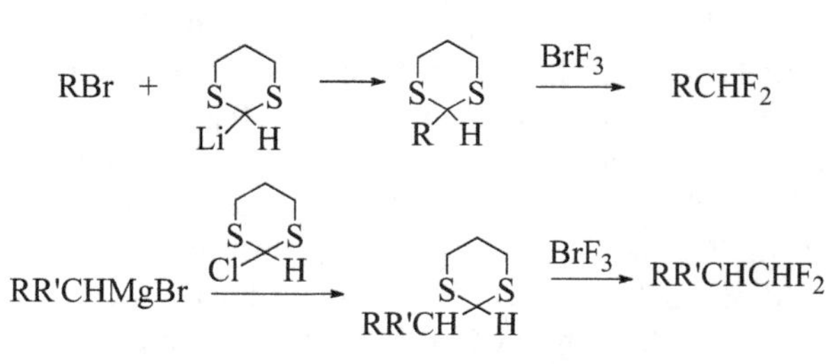

SCHEME 26.14 From RX to RCF_2H via dithianes.

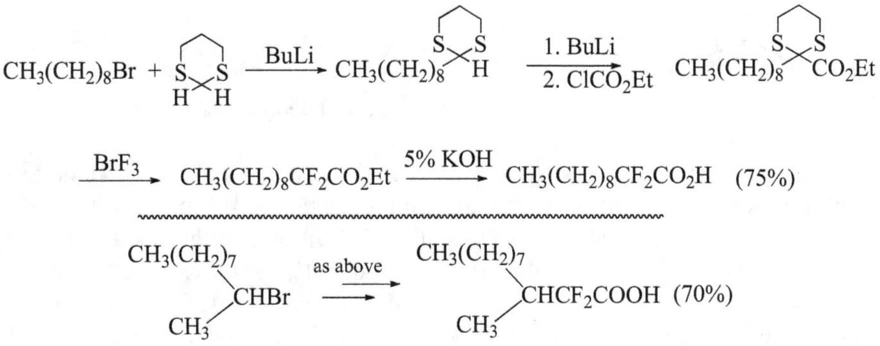

SCHEME 26.15 From RX to RCF_2COOH via dithianes.

This method could also be used for forming α,α-difluoroesters both primary and secondary from the respective alkyl halides [22]. Thus, decylbromide was first reacted with the lithium salt of the dithiane itself and then with ethyl chloroformate to produce 2-decyl-2-ethoxycarbonyl-1,3-dithiane. This compound was reacted for 1–2 min at 0 °C with threefold excess of BrF_3 and the expected ethyl 2,2-difluorododecanoate was formed in 75% yield. This method could be used for the formation of relatively hindered difluoroesters as well. 2-Bromodecane, for example, was transformed to its dithiane derivative and then reacted with BrF_3 to produce ethyl 2,2-difluoro-3-methylundecanoate in 70% yield. If needed the difluoroesters could be hydrolyzed to the corresponding acids by refluxing with 5% KOH in $EtOH/H_2O$ for 1 h in nearly quantitative yield (Scheme 26.15).

Experimental Procedure General procedure for preparing the 2-alkyl-1,3-dithiane derivatives has been long developed [23]. 2-(2-Methyl)nonyl-2-ethoxycarbonyl-1,3-dithiane (1 mmol) was dissolved in 10–15 mL of dry $CFCl_3$. About 3 mmol of BrF_3 was dissolved in the same solvent, cooled to 0 °C, and added drop by drop to the dithiane solution. The reaction mixture was then washed with saturated aqueous $Na_2S_2O_3$ till colorless. The aqueous layer was extracted with CH_2Cl_2 and the organic layer was dried over $MgSO_4$. Evaporation of the solvent followed by purification by flash chromatography gave the target ethyl 2,2-difluorododecanoate.

Characterization Data for Ethyl 2,2-difluoro-3-methylundecanoate It was obtained in 70% yield. oil; IR: 1759 cm^{-1}. ^1H NMR: δ 4.32 (2H, q, $J = 7$ Hz), 1.35 (3H, t, $J = 7$ Hz), 1.02 (3H, t, $J = 7$ Hz), 0.88 (3H, t, $J = 7$ Hz) ppm. ^{13}C NMR: δ 164.4 (t, $J = 33$ Hz), 117.8 (t, $J = 250$ Hz), 62.5, 37.7 (t, $J = 22$ Hz), 31.7, 29.5, 29.4, 29.3, 28.7, 22.5, 13.9, 13.2, 11.8 ppm. ^{19}F NMR: δ −113.3 (dd, $J_1 = 40$, $J_2 = 15$ Hz) ppm. HRMS (CI) (m/z): $(MH)^+$ calcd for $C_{14}H_{27}F_2O_2$, 265.1979; found, 265.1983.

The α-position of a carboxylic acid is not the only position that could be difluorinated. As a matter of fact, any carboxylic acid possessing a keto group along its chain

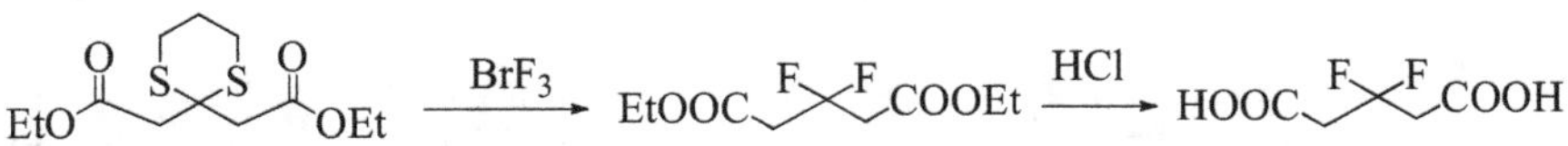

SCHEME 26.16 Formation of β,β-difluoroacids.

can serve as a substrate. Among these keto acids, the easier to make are β-ketoesters via the well-known Claisen type of condensation reaction [24]. The 1,3-dithiane of ethyl 3-oxohexadecanoate was prepared similarly to all other dithianes. It was then reacted for 1–2 min at 0 °C with a threefold excess of BrF_3 and the expected ethyl 3,3-difluorohexadecanoate was formed in 75% yield. In the same way, the 1,3-dithiane of diethyl 3-oxoglutarate was converted to diethyl β,β-difluoroglutarate in 60% yield. Hydrolysis with concentrated HCl for 4 h gave the known 3,3-difluoro glutaric acid in higher than 95% yield (Scheme 26.16).

Experimental Procedure The dithiane (5 mmol) was dissolved in 10–20 mL of $CFCl_3$ and cooled to 0 °C. About 3 mol/equiv of BrF_3 was dissolved in the same solvent, cooled to 0 °C, and added drop by drop to the reaction mixture during 1–2 min. After the addition was completed, the reaction mixture was washed with $Na_2S_2O_3$ solution till colorless. The aqueous layer was extracted with CH_2Cl_2 and the organic layer was dried over $MgSO_4$. Evaporation of the solvent followed by flash chromatography (using 5% ethyl acetate in petroleum ether as eluent) gave the desired ethyl 3,3-difluoro glutarate.

Characterization Data for Diethyl β,β-difluoroglutarate It was obtained in 60% yield. IR: 1745 cm^{-1}. ^1H NMR: δ 4.19 (4H, q, $J = 7$ Hz), 3.25 (4H, t, $J = 15$ Hz), 1.28 (6H, t, $J = 7$ Hz) ppm. ^{13}C NMR: δ 167.3 (t, $J = 9$ Hz), 119.8 (t, $J = 242$ Hz), 61.4, 40.8 (t, $J = 27.5$ Hz), 14.2 ppm. ^{19}F NMR: δ −89.5 (quin, $J = 14.7$ Hz) ppm. HRMS (ESI–QqTOF) (*m*/*z*): $(M + Na)^+$ calcd for $C_9H_{14}F_2O_4$, 247.0752; found, 247.0668.

Another unique transformation, made possible by the dithiane moiety, was the replacement of a carbonyl's oxygen with the pharmaceutically important difluoromethyl group. The key step of the reaction is treating the carbonyl compounds with 2-(trimethylsilyl)-1,3-dithiane then reducing the resulting ketene dithioacetal and reacting the formed dithiane with BrF_3 (Scheme 26.17) [25]. It should be noted that 2-(trimethylsilyl)-1,3-dithiane is readily available either commercially or through a simple preparation using 1,3-dithiane and trimethylsilyl chloride. Its lithium salt reacts with aldehydes and ketones to form ketene dithioacetals, which can be reduced

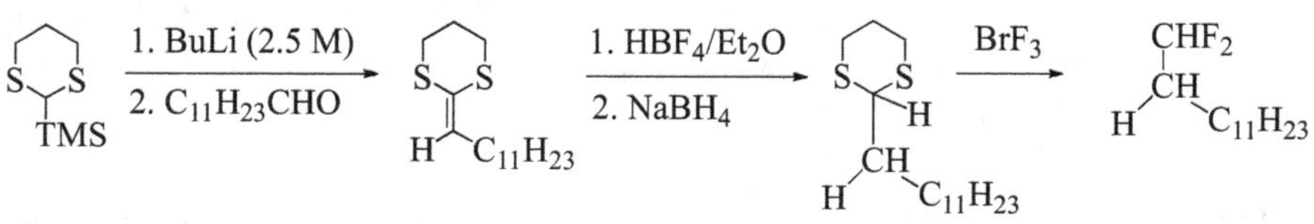

SCHEME 26.17 Replacing the carbonyl's oxygen with the difluoromethyl group.

with tetrafluoroboric acid and sodium borohydride to the corresponding 2-alkyl-1,3-dithianes. Dodecanal can serve as an example.

Experimental Procedure 2-Trimethylsilyl-1,3-dithiane (4.81 g, 25 mmol) was dissolved in dry THF (50 mL) under nitrogen. The solution was cooled to −78 °C, and BuLi (12 mL, 30 mmol) was added. The reaction mixture was allowed to warm to 0 °C over 5 h. It was then cooled again to −78 °C, and dodecanal (30 mmol) was added. The solution was stirred overnight, allowed to warm up to room temperature, poured into water (50 mL), and extracted with CH_2Cl_2 (50 mL). The combined organic phases were dried over $MgSO_4$, the solvent was evaporated, and the crude residue was subjected to flash chromatography (using petroleum ether as eluent). There is no need for analytical purification at this stage as the formed materials were clean enough for the following reductive step that used tetrafluoroboric acid (3 mL, 54% in ether, 22 mmol). This acid was added at room temperature to the stirred solution of the ketene dithioacetal (10 mmol) in dry acetonitrile (25 mL). A brown solution was obtained and after 30 min it was cooled to 0 °C and powdered sodium borohydride (0.7 g, 18.5 mmol) was gradually added. The resulting suspension of 2-dodecyl-1,3-dithiane was stirred overnight at room temperature, poured into an aqueous ammonium chloride solution (10%, 100 mL), and extracted with ether (100 mL). The combined organic phases were washed with water (50 mL), dried over Na_2SO_4, and filtered. Evaporation of the solvent followed by flash chromatography (using petroleum ether as eluent) gave the desired dithiane pure enough for the final fluorination step. The resulting 2-dodecyl-1,3-dithiane (3 mmol) was dissolved in 10–20 mL of $CFCl_3$ and cooled to 0 °C. BrF_3 (2.5 mol/equiv) was dissolved in 20 mL of the same solvent, cooled to 0 °C, and added drop by drop to the reaction mixture during 1–2 min. After the addition was completed, the reaction mixture was washed with $Na_2S_2O_3$ solution till colorless. The aqueous layer was extracted with CH_2Cl_2 (50 mL) and the organic phase was dried over $MgSO_4$. Evaporation of the solvent followed by flash chromatography (using petroleum ether as eluent) gave the desired 1,1-difluorotridecane.

Characterization Data for 1,1-Difluorotridecane It was obtained in 75% yield; colorless oil. ^{1}H NMR: δ 5.79 (1H, tt, J = 57 Hz, J = 5 Hz), 1.97–1.66 (2H, m), 1.26 (20H, br s), 0.88 (3H, t, J = 7 Hz) ppm. ^{13}C NMR: δ 117.4 (t, J = 239 Hz), 34.0 (t, J = 20 Hz), 31.8, 29.5, 29.4, 29.3, 29.1, 29.0, 28.7, 28.6, 22.6, 22.0 (t, J = 5 Hz), 14.0 ppm. ^{19}F NMR: δ −116.2 (dt, J = 57 Hz, J = 17 Hz) ppm. MS (EI) (m/z): 220 M^+.

REFERENCES

1. Rozen S, Mishani E, Bar-Haim A. *J Org Chem* 1994;59:2918.
2. Rozen S, Ben-David I. *J Org Chem* 2001;66:496–500.

3. Prakash GKS, Reddy VP, Li XY, Olah GA. *Synlett* 1990;594–596.
4. Cohen O, Sasson R, Rozen S. *J Fluorine Chem* 2006;127:433–436.
5. Rozen S, Ben-David I. *J Fluorine Chem* 1996;76:145–147.
6. Rozen S, Mishani E. *J Chem Soc, Chem Commun* 1993;1761–1762.
7. Cohen O, Mishani E, Rozen S. *Tetrahedron* 2010;66:3579–3582.
8. Ben David I, Rechavi D, Mishani E, Rozen S. *J Fluorine Chem* 1999;97:75–78.
9. Hagooly Y, Sasson R, Welch MJ, Rozen S. *Eur J Org Chem* 2008;2875–2880.
10. Hagooly Y, Rozen S. *J Org Chem* 2008;73:6780–6783.
11. Hagooly Y, Welch MJ, Rozen S. *Org Biomol Chem* 2011;9:902–905.
12. Hagooly Y, Cohen O, Rozen S. *Tetrahedron Lett* 2009;50:392–394.
13. Hagooly Y, Gatenyo J, Hagooly A, Rozen S. *J Org Chem* 2009;74:8578–8582.
14. Hagooly A, Ben-David I, Rozen S. *J Org Chem* 2002;67:8430–8434.
15. Hagooly A, Rozen S. Chem Commun 2004;594–595.
16. Konen DA, Pfeffer PE, Silbert LS. *Tetrahedron* 1976;32:2507–2512.
17. Rozen S, Bareket Y. *J Org Chem* 1997;62:1457–1462.
18. Hagooly A, Rozen S. *J Org Chem* 2004;69:7241–7245.
19. Hagooly A, Rozen S. *J Org Chem* 2004;69:8786–8788.
20. Sasson R, Hagooly A, Rozen S. *Org Lett* 2003;5:769–771.
21. Seebach D, Corey EJ. *J Org Chem* 1975;40:231–237.
22. Hagooly A, Sasson R, Rozen S. *J Org Chem* 2003;68:8287–8289.
23. Corey EJ, Erickson BW. *J Org Chem* 1971;36:3553–3560.
24. Cohen O, Rozen S. *Tetrahedron* 2008;64:5362–5364.
25. Cohen O, Hagooly Y, Rozen S. *Tetrahedron* 2009;65:1361–1365.

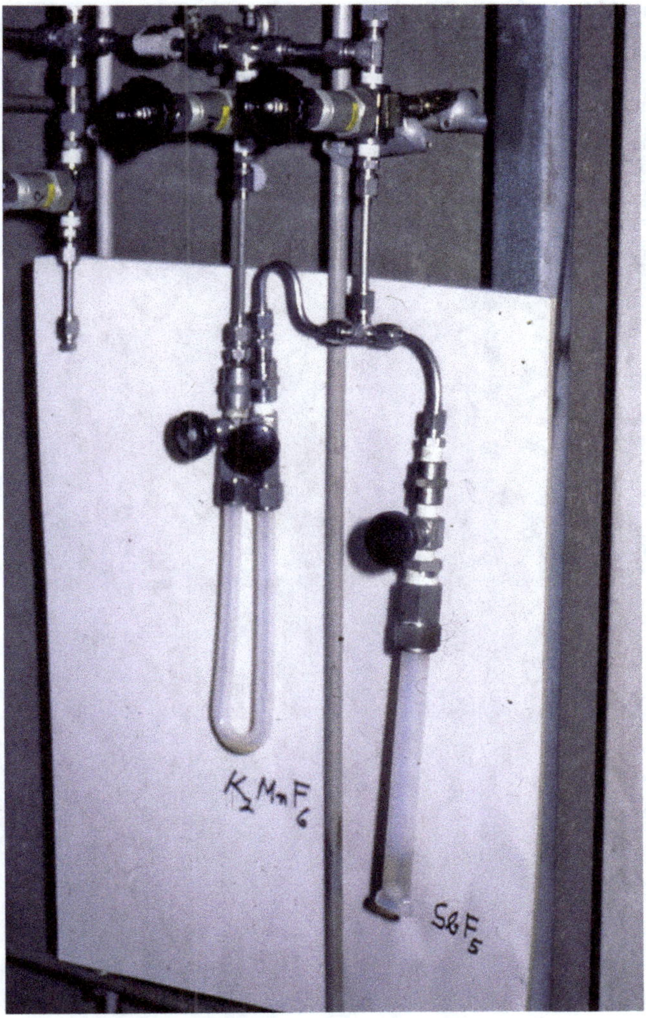

FIGURE 1.1 The original apparatus used in 1986 by Christe for the first chemical synthesis of F_2.

Efficient Preparations of Fluorine Compounds, First Edition. Edited by Herbert W. Roesky.

FIGURE 1.2 The original apparatus used in 1986 by Christe for the assay of F_2. On the left is the Teflon_FEP U-tube reactor containing the F_2 gas. It is connected to a glass bulb containing the mercury, and the whole system is connected to the Heise gauge for the pressure measurements.

FIGURE 9.1 Setup for Willard-Winter Distillation Method.

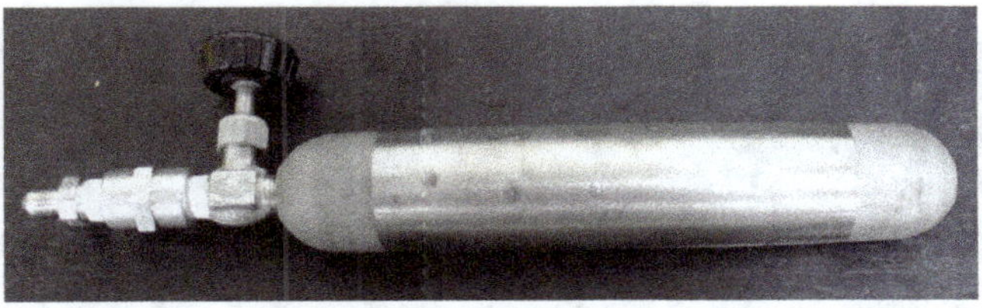

FIGURE 18.1 Stainless steel cylinder (V = 175 mL) with stainless steel valve.

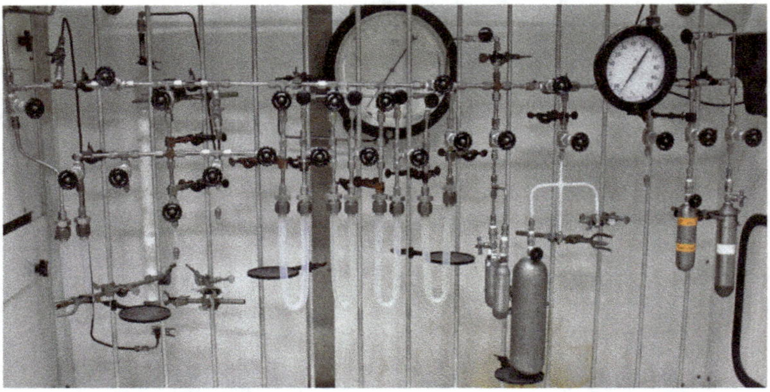

FIGURE 18.2 A stainless steel vacuum line.

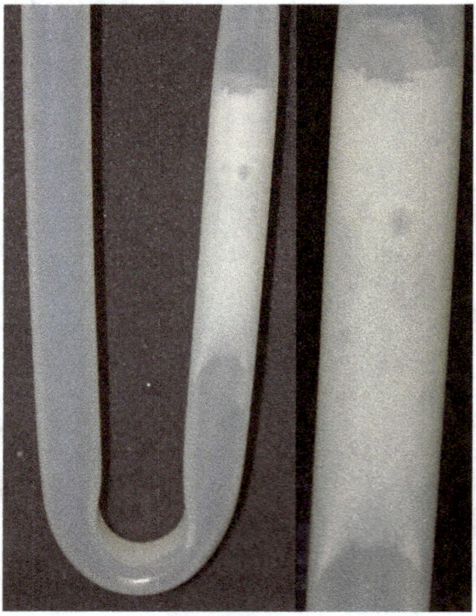

FIGURE 18.3 Teflon/FEP U-trap with crystals of $ReOF_5$.

FIGURE 18.4 Monel cylinder (V = 100 mL) with Monel valve.

FIGURE 18.5 Teflon/FEP U-trap with solid ReF_6.

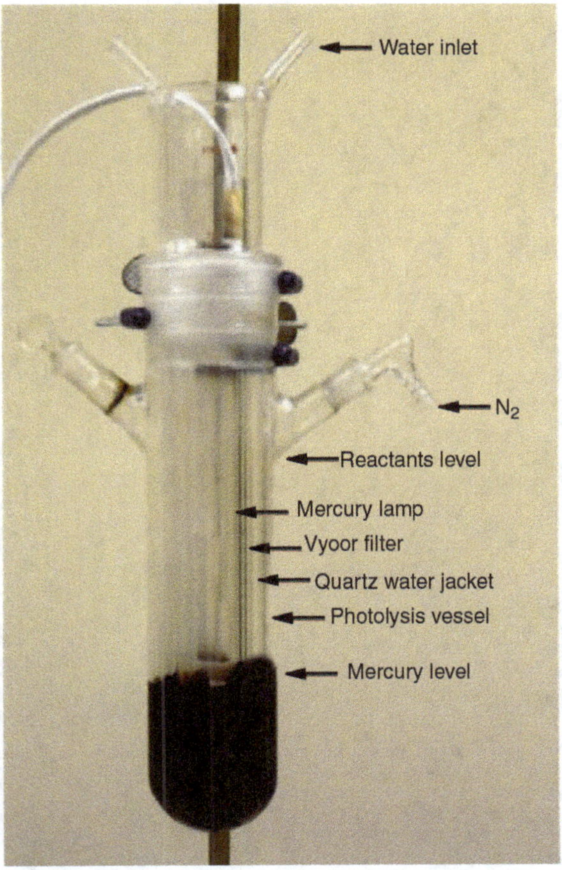

FIGURE 49.1 Apparatus for the photocycloaddition.

FIGURE 59.2 Original membranes from copolymers based on VDF and sulfonic acid containing fluorinated comonomers.

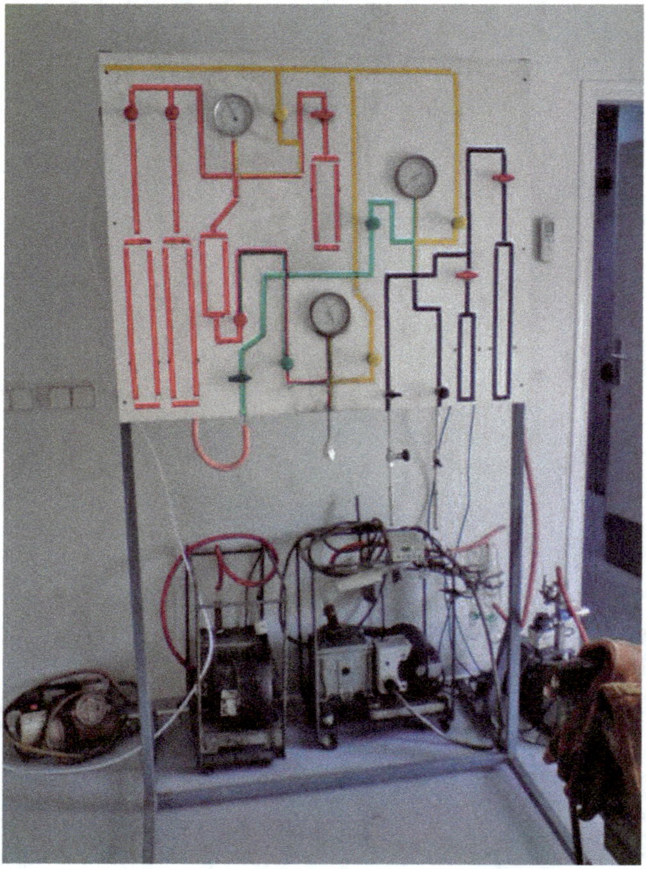

FIGURE 59.3 Picture of manifold used to prepare the sealed Carius tubes for the radical copolymerization of VDF with fluorofunctional comonomers.

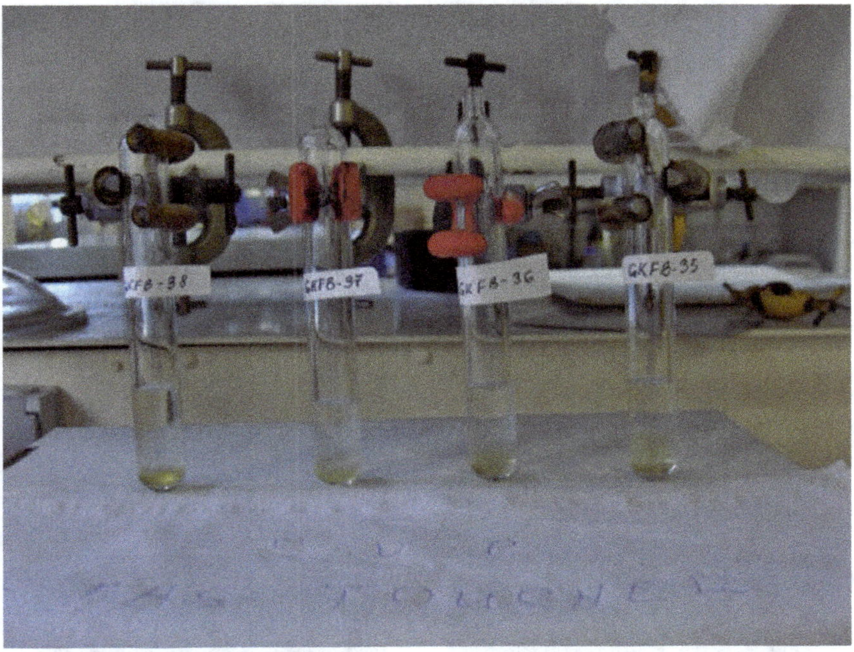

FIGURE 59.4 Picture of Carius tubes after the radical copolymerization of VDF with α,β-difluoroacrylic acid.

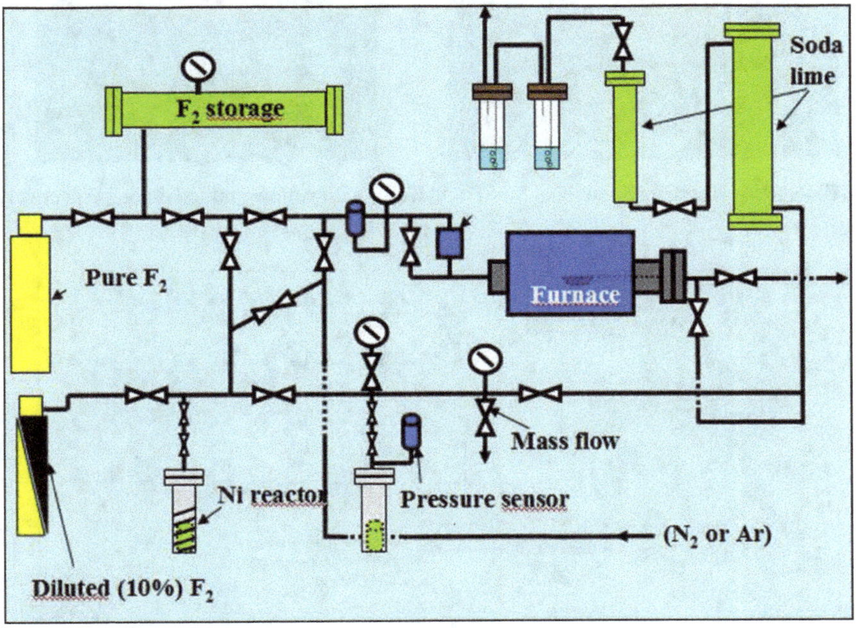

FIGURE 64.1 Scheme of a fluorination apparatus using F_2-gas.

FIGURE 64.2 Sulfurization apparatus for preparing lanthanum sulfofluoride LaSF.

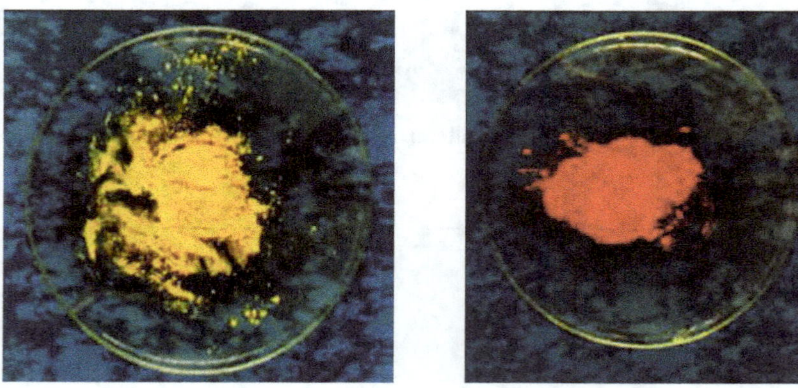

FIGURE 64.4 Inorganic pigments SmSF and CeSF with yellow and red color, respectively.

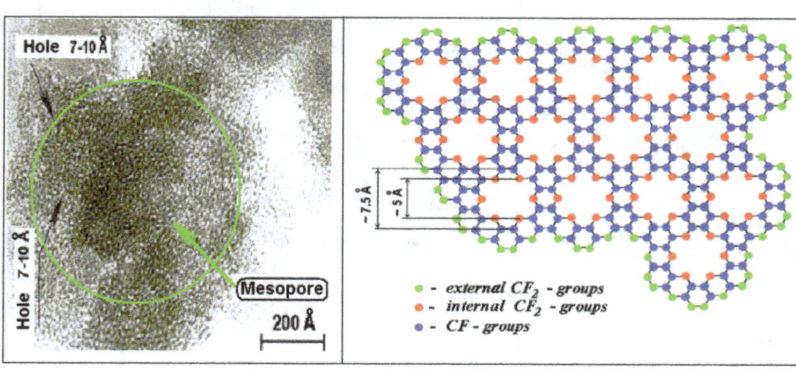

FIGURE 66.3 Image of microparticle surface and assumed structure for $CF_{1.33}$ [7, 8].

CHAPTER 27

Preparation of Silicon- and Sulfur-Based Fluorinated Methane Derivatives as Versatile Fluoromethylation Reagents

G.K. SURYA PRAKASH AND FANG WANG

Despite the fact that fluoromethylated compounds can be derived through C–F bond forming reactions with fluorine gas (F_2), SF_4 derivatives, fluorides, and/or electrophilic fluorinating reagents, the direct introduction of various fluoromethyl groups using fluoromethyl synthons prevails under many synthetic regimes because of their superior efficacy and functional group compatibility [1]. In principle, fluoromethylations can be achieved via nucleophilic, electrophilic, radical, as well as carbene pathways. Although seemingly feasible, the nucleophilic and the electrophilic fluoromethylating reactions, particularly the trifluoromethylations, were indeed quite challenging for many years. For example, the trifluoromethyl anion has been found to possess extreme lability due to the vicinal negative charge–lone pair repulsion, which leads to the rapid decomposition of the anion into fluoride and singlet difluorocarbene [2]. On the one hand, attempts to prepare trifluoromethyl analogs of the organometallic reagents, such as the Grignard or lithium reagents, commonly used in nucleophilic alkylations, have always failed. On the other hand, whereas the trifluoromethyl cation has been frequently observed in mass spectrometric studies [3], electrophilic trifluoromethylation was found to be unproductive using trifluoromethyl iodide and trifluoromethyl sulfonates [4]. This unusual inertness of the CF_3 moiety toward nucleophiles can be attributed to (a) the reverse polarization of the CF_3–halogen and CF_3–O bonds and (b) the steric inaccessibility of the CF_3 group for nucleophilic attack. To address these inherent synthetic challenges, extensive efforts have been made on the development of highly efficient nucleophilic and electrophilic fluoromethylation reagents. Over the past three decades, a series of silicon-based [5]

Efficient Preparations of Fluorine Compounds, First Edition. Edited by Herbert W. Roesky.

and sulfur-based [6] fluorinated methane derivatives have been prepared and applied in fluoromethylations of various organic frameworks as versatile reagents.

27.1 PREPARATION OF (TRIFLUOROMETHYL)TRIMETHYLSILANE (TMSCF$_3$, THE RUPPERT–PRAKASH REAGENT) AS A CF$_3^-$ ANION EQUIVALENT AND A DIFLUOROCARBENE PRECURSOR

(Trifluoromethyl)trimethylsilane (TMSCF$_3$) can be prepared through several synthetic routes. The original procedure involves the reaction of chlorotrimethylsilane (TMSCl) with a complex between trifluoromethyl bromide (CF$_3$Br) and hexaethylphosphorus triamide [(Et$_2$N)$_3$P] in benzonitrile [7, 8]. To avoid the use of ozone-depleting CF$_3$Br, an alternative protocol has been developed, which uses phenyl trifluoromethyl sulfoxide (PhSOCF$_3$) as a trifluoromethyl source to transfer the CF$_3$ group to TMSCl in the presence of Mg(0) [9]. Other less frequently used methods have also been shown [10].

Route 1. Preparation of TMSCF$_3$ using TMSCl, CF$_3$Br, and [(Et$_2$N)$_3$P] (Scheme 27.1) [7, 8]

Apparatus A 2-L, three-necked flask fitted with a sealed high-torque mechanical stirrer, a cold-finger condenser (30 cm in length and 8 cm in diameter), a 500-mL Ace dry ice gas condenser trap, a rubber septum, an oil bubbler, a 600-mL pressure-equalizing dropping funnel, a glass stopper, a joint adapter, cooling baths, a 50-mL separatory funnel, a distillation apparatus, a 15-cm column packed with glass helices, safety goggles, laboratory coat, and protective gloves.

Chemicals CF$_3$Br, TMSCl, anhydrous benzonitrile, hexaethylphosphorus triamide, magnesium sulfate (MgSO$_4$), acetone, potassium hydroxide, and calcium hydride.

Attention! Safety glasses and protective gloves must be used at all times.

Caution! Because of their toxicity and/or volatility, care should be taken to avoid inhalation of CF$_3$Br, TMSCl, benzonitrile, hexaethylphosphorus triamide, and acetone or contact of their solution with the skin. All reactions should be carried out in a well-ventilated hood. Due the extreme coldness, acetone–dry ice baths should be handled carefully.

Experimental Procedure A 2-L, three-necked flask is oven dried and equipped with an efficient, overhead, sealed mechanical stirrer, a cold-finger condenser, and a rubber septum. The top outlet of the condenser is attached to an oil bubbler. The flask is flushed with dry nitrogen and charged with TMSCl (118.8 g,

$$\text{TMSCl} + \text{CF}_3\text{Br} \xrightarrow[\text{PhCN}]{(\text{Et}_2\text{N})_3\text{P}} \text{TMS-CF}_3$$

SCHEME 27.1 Preparation of TMSCF$_3$ using TMSCl, CF$_3$Br, and (Et$_2$N)$_3$P.

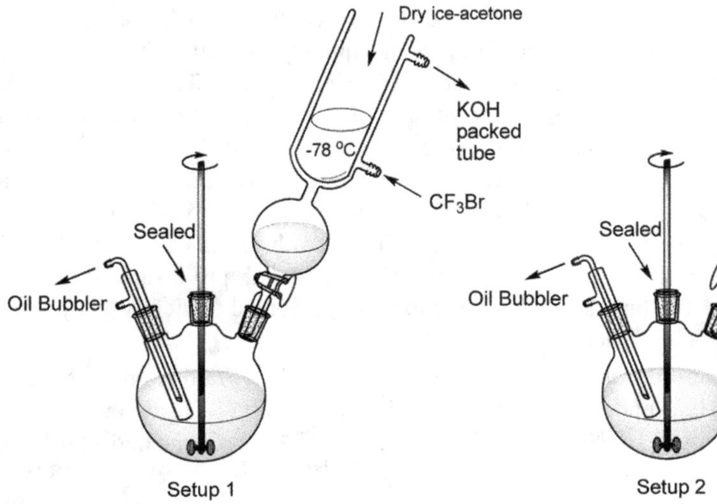

FIGURE 27.1 Reaction setups for the preparation of $TMSCF_3$ using CF_3Br, TMSCl, and hexaethylphosphorus triamide.

1.09 mol), which is distilled over calcium hydride just before use, in 100-mL anhydrous benzonitrile. The septum is quickly replaced with a 500-mL dry ice gas condenser trap under nitrogen protection. The outlet of the trap is connected with a tube filled with potassium hydroxide to protect from moisture, and the inlet is connected with a cylinder of CF_3Br through Tygon tubing. The 2-L flask is immersed in a dry ice–acetone bath (−30 °C), and the condensers are filled with dry ice–acetone mixture maintained at −78 °C (Figure 27.1, Setup 1). CF_3Br (250 mL as liquid, 485 g, 3.25 mol) is condensed into the reservoir, and is gradually added to the reaction pot at −30 °C with rapid mechanical stirring, meanwhile the reservoir is slowly warmed to −45–50 °C. The resulting white slurry is further cooled to −60 °C (the reaction mixture solidifies if the temperature decreases below −60 °C). The dry ice gas condenser trap is then disconnected under nitrogen and quickly replaced with a 600-mL pressure-equalizing dropping funnel containing a solution of hexaethylphosphorus triamide (325.0 g, 1.31 mol, used as received) in 250-mL anhydrous benzonitrile (Figure 27.1, Setup 2). This solution is added to the white slurry mixture with stirring over a period of 2.5 h, and the reaction mixture is maintained at −60 °C. On completion of the addition, the reaction mixture is further stirred at −60 °C for 1 h. The reaction mixture is allowed to warm to room temperature (25 °C) over a period of 14 h, during which time a yellowish mixture can be observed. The condenser and dropping funnel are replaced with a glass stopper and a joint adapter with its glass tube connected to two 250-mL dry ice-acetone-cooled traps. Aspirator vacuum (~20 mm Hg) is then applied and the reaction mixture is gradually warmed to 50 °C to remove the volatile materials over a period of 3 h. The cooling baths are removed and the material in the traps is brought to 0 °C. The colorless liquid is rapidly transferred to a 250-mL separatory funnel, and quickly washed with ice-cold water (100 mL × 3). The organic layer is separated (on top). The product is dried over anhydrous $MgSO_4$ (5 g), and the liquid is decanted into a 250-mL Erlenmeyer flask. The product is fractionally distilled through a 15-cm column packed with glass helices. Three fractions are collected. The first minor fraction (bp 45–54 °C) and the second major fraction (bp 54–55 °C) are found to contain the main quantity of $TMSCF_3$. The third minor fraction (bp 55–65 °C) mainly consists of hexamethyldisiloxane with just a small quantity of $TMSCF_3$. The first and second fractions are combined to yield 116.9 g (75%) of clear liquid product, bp 54–55 °C.

Characterization Data ^1H NMR (200 MHz, $CDCl_3$, TMS reference): δ 0.25 (s, 9H, $Si(CH_3)_3$) ppm. ^{13}C NMR (50.0 MHz, $CDCl_3$): δ 131.7 (q, $J = 321.9$), −5.2 (CH_3–Si) ppm. ^{19}F NMR (188.0 MHz, $CDCl_3$, $CFCl_3$ reference): δ −66.1 ppm. ^{29}Si NMR (39.7 MHz, $CDCl_3$): δ +4.7 (q, $^2J(^{29}Si–^{19}F) = 37.9$) ppm. MS ($m/z$): 123 ($M^+ - 19$).

Route 2. Preparation of $TMSCF_3$ via Mg(0)-Mediated Reductive Trifluoromethylation of TMSCl and $PhSOCF_3$ (Scheme 27.2)

Apparatus A 250-mL Schlenk flask, a rubber septum, a 10-mL syringe with a needle, a cooling trap (liquid N_2), a magnetic stir bar, an ice bath, a 100-mL separatory funnel, a fractional distillation apparatus, safety goggles, laboratory coat, and protective gloves.

Chemicals TMSCl, $PhSOCF_3$ (commercially available, otherwise can be prepared via a known procedure [11]), Mg turnings, anhydrous dimethylformamide (DMF), and activated 4-Å molecular sieves.

Attention! Safety glasses and protective gloves must be used at all times.

Caution! Because of their toxicity and/or volatility, care should be taken to avoid inhalation of $PhSOCF_3$, TMSCl, and DMF or contact of them. All reactions should be carried out in a well-ventilated hood. Due the extreme coldness, the liquid N_2 trap should be handled carefully.

Experimental Procedure Into a 250-mL, oven-dried, Schlenk flask, Mg turnings (1.14 g, 47.5 mmol) and TMSCl (11.8 g, 109 mmol) in 50-mL DMF were added at 0 °C under inert atmosphere. The reaction mixture was stirred for 2 min before the slow addition of $PhSOCF_3$ (4.62 g, 23.8 mmol) in 5-mL DMF via a syringe. The reaction mixture was stirred at room temperature at 0 °C for 30 min, thereafter at room temperature for another 1.5 h until all the starting material was converted into $TMSCF_3$ (monitored by ^{19}F NMR). The reaction

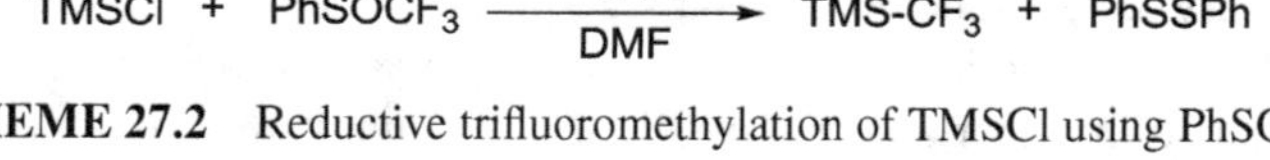

SCHEME 27.2 Reductive trifluoromethylation of TMSCl using $PhSOCF_3$.

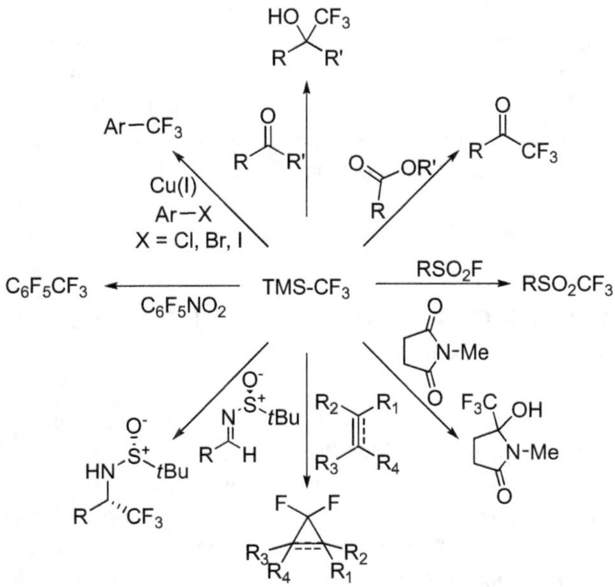

SCHEME 27.3 Typical trifluoromethylation using $TMSCF_3$.

flask was then connected to the liquid N_2 trap before the application of vacuum. The low-boiling fractions were collected into the trap followed by warming to room temperature. The volatile fractions were washed with ice water (50 mL × 3) and quickly dried over activated molecular sieves. The organic matter was fractionally distilled using a 30-cm-long column to afford $TMSCF_3$ (2.73 g, 81% yield). Under the similar reaction conditions, other trifluoromethylsilanes can also be prepared.

Applications Prepared by Ruppert in 1984 [8], the synthetic application of $TMSCF_3$ as a trifluoromethyl anion (CF_3^-) was first shown by Prakash and coworkers in the nucleophilic trifluoromethylation of carbonyl compounds [12]. Initiated by catalytic amounts of fluorides, such as cesium fluoride (CsF) and tetra-*n*-butylammonium fluoride (TBAF), $TMSCF_3$ can readily react with ketones and aldehydes to render the corresponding alcohols (from their silyl ethers) in good yields. Since then, a broad spectrum of electrophiles was found to react with $TMSCF_3$, including aldehydes, ketones, esters, imines [13], nitriles [14], nitrones [15], and alkyl halides [6,16] (Scheme 27.3). More importantly, the use of the Ruppert–Prakash reagent further allows the efficient stereoselective synthesis of various chiral trifluoromethylated organic compounds possessing unique biological and pharmaceutical properties [17]. Interestingly, $TMSCF_3$ has also been found to be a versatile difluorocarbene precursor [18]. In the presence of anhydrous fluoride sources, $TMSCF_3$ can release CF_3^- anion, which readily decomposes to fluoride and singlet difluorocarbene. In the presence of alkenes and alkynes, difluorocyclopropanes and difluorocyclopropenes, respectively, can be obtained.

Moreover, $TMSCF_3$ has been used as a trifluoromethyl source for the preparation of $CuCF_3$ and its ligated derivatives, which are capable of trifluoromethylation of aromatic halides (Scheme 27.4, Eqs. 27.2, 27.8 and 27.10) [19], aryl boronic acids (Scheme 27.4, Eq. 27.6) [20], terminal alkynes (Scheme 27.4, Eq. 27.5) [21], and indole derivatives (Scheme 27.4, Eq. 27.9) [22]. In particular, (trifluoromethyl)triethylsilane, an analog of $TMSCF_3$, has also shown significant viability in transition metal-catalyzed–mediated aromatic trifluoromethylation reactions (Scheme 27.4, Eqs. 27.1, 27.3 and 27.4) [23].

27.2 PREPARATION OF TRIFLUOROMETHYL PHENYL SULFONE ($PhSO_2CF_3$) AS A CF_3^- ANION EQUIVALENT

Trifluoromethyl phenyl sulfone ($PhSO_2CF_3$) is usually prepared through the oxidation of trifluoromethyl phenyl sulfide ($PhSCF_3$) using various oxidizing reagents. $PhSCF_3$ can be prepared via several synthetic protocols. The original procedure treated trichloromethyl phenyl sulfide ($PhSCCl_3$) with antimony trifluoride (SbF_3) to yield the titled compound in 70% yield [24]. An improved method used triethylamine trihydrofluoride as the fluoride source, which reacts with $PhSCCl_3$ under microwave irradiation [25]. Alternatively, $PhSCF_3$ can be obtained in 60% yield by direct trifluoromethylthiolation of iodobenzene using methyl fluorosulfonyldifluoroacetate and S_8 [26]. A more feasible preparative method was achieved by reacting CF_3^-, generated from the deprotonation of trifluoromethane (CF_3H) with potassium *tert*-butoxide (*t*BuOK) in DMF, with diphenyl disulfide (PhSSPh) [27]. $PhSO_2CF_3$ can also be directly prepared using $TMSCF_3$ (the Ruppert–Prakash reagent) as a trifluoromethyl source, which reacts with benzenesulfonyl fluoride ($PhSO_2F$) [28] or methyl benzenesulfonate ($PhSO_3Me$) [29] to render the trifluoromethylated product in high yields.

Route 1. Preparation of $PhSO_2CF_3$ using CF_3H and PhSSPh (Scheme 27.5) [27]

Apparatus A 1-L, three-necked, round-bottomed flask equipped with a dry ice condenser, a magnetic stirring bar, and two rubber septa, a dry ice/ethylene glycol/acetone bath, a long needle for bubbling CF_3H, a vacuum distillation apparatus, a 1-L separatory funnel, a distillation apparatus, a 30-cm-long distillation column, a 250-mL, round-bottomed flask, a reflux condenser, a 250-mL separatory funnel, a filter funnel, safety goggles, laboratory coat, and protective gloves.

Chemicals CF_3H, PhSSPh, *t*BuOK, anhydrous DMF, 30 wt% hydrogen peroxide, acetic acid, dichloromethane, ethyl acetate (EtOAc), acetone, $MgSO_4$, and sodium hydroxide (NaOH), brine.

Attention! Safety glasses and protective gloves must be used at all times.

Caution! Because of their toxicity, irritation, and/or volatility, care should be taken to avoid inhalation of CF_3H, PhSSPh, *t*BuOK, hydrogen peroxide, DMF, dichloromethane, EtOAc, acetic acid, acetone, and NaOH or contact of their

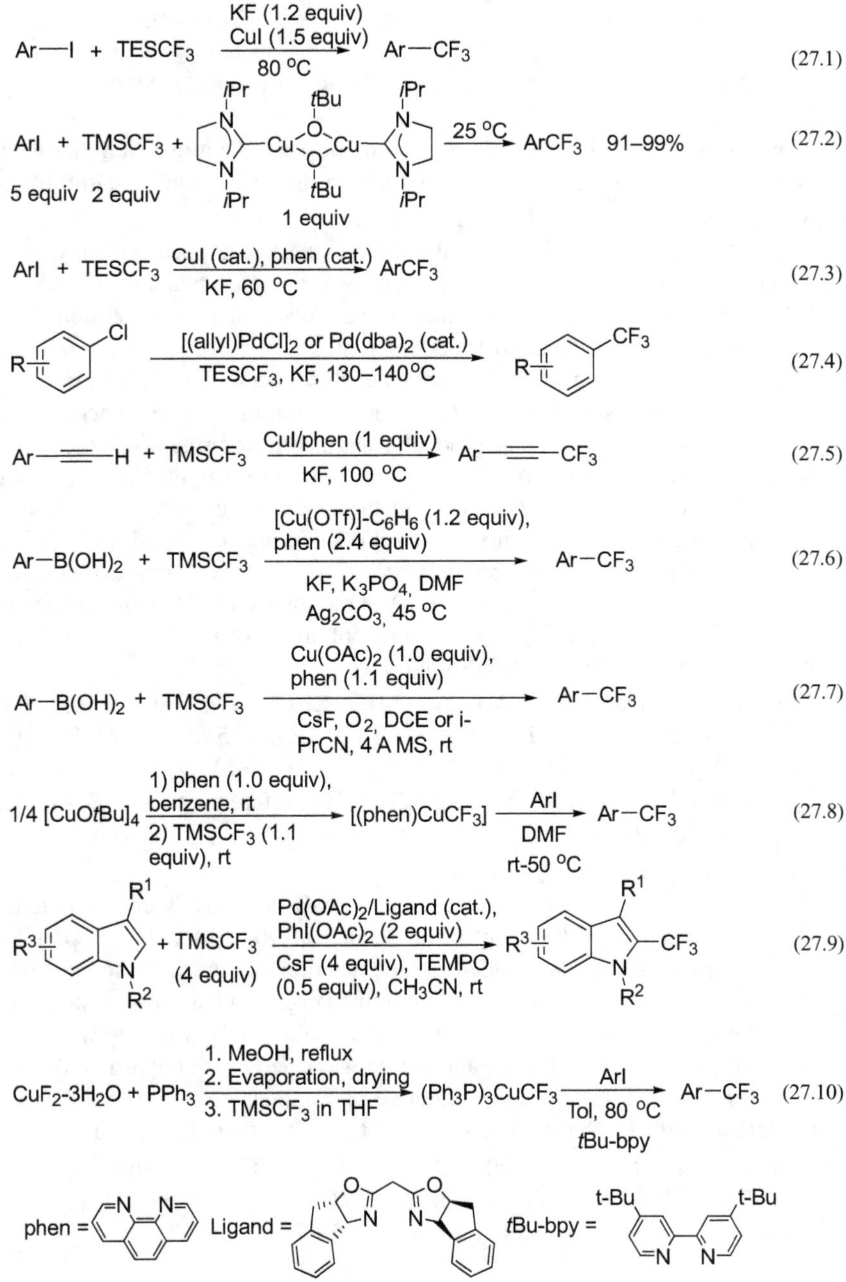

SCHEME 27.4 Transition metal-catalyzed–mediated trifluoromethylation of aromatics, alkenes, and alkynes using $TMSCF_3$ and $TESCF_3$.

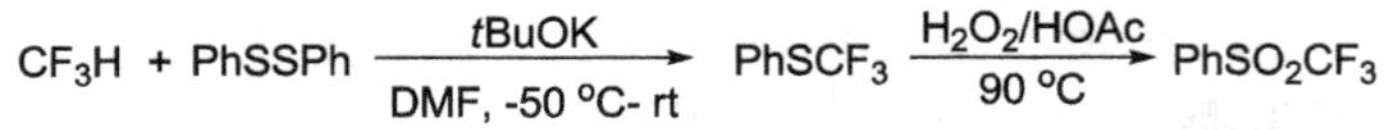

SCHEME 27.5 Preparation of $PhSO_2CF_3$ using CF_3H and PhSSPh.

solutions with the skin. All reactions should be carried out in a well-ventilated hood. Due the extreme coldness, the cold bath should be handled carefully.

Experimental Procedure *Step 1: Preparation of $PhSCF_3$*

Into a 1-L, three-necked, round-bottomed flask equipped with a dry ice condenser, a magnetic stirring bar, and two rubber septa, PhSSPh (85 g, 0.39 mol) and *t*BuOK (60 g, 0.53 mol) were added under N_2 protection. After the addition of anhydrous DMF (600 mL), the reaction mixture was cooled to $-40 \sim -50$ °C using a dry ice/ethylene glycol/acetone bath. CF_3H (70 g, 1.0 mol) was then slowly bubbled into the reaction mixture via a needle over a period of 4 h. The flask was gradually warmed to room temperature over a period of 5 h, and the reaction mixture was stirred overnight. Crude products $PhSCF_3$ and DMF were distilled from the reaction mixture under vacuum. The distillate was poured into water (600 mL), and extracted with EtOAc (200 mL × 2) in a 1-L separatory funnel. The combined organic phase was washed with water and dried over $MgSO_4$. Fractional distillation under vacuum (bp: 55 °C/30 mmHg) using a 30-cm-long column gave $PhSCF_3$ as a colorless liquid (54.3 g, 81% based on PhSSPh used).

Characterization Data ^1H NMR (500 MHz, $CDCl_3$, TMS reference): δ 7.40 (t, $J = 7.8$ Hz, 2H) ppm, 7.47 (t, $J = 7.4$ Hz, 1H), 7.65 (d, $J = 7.8$ Hz, 2H). ^{13}C NMR (125 MHz, $CDCl_3$): δ 124.4, 127.7 (q, $J = 309$ Hz), 129.47, 130.81, 136.37 ppm. ^{19}F NMR (470 MHz, $CDCl_3$, $CFCl_3$ reference): δ −43.3. MS (EI, 70 eV): $m/z = 178$.

Step 2: Preparation of $PhSO_2CF_3$

A mixture of $PhSCF_3$ (5 g, 28 mmol) and 30 wt% aqueous hydrogen peroxide (30 mL) in acetic acid (50 mL) was heated at 90 °C for 21 h. After the reaction, brine (40 mL) was added and the reaction mixture was extracted with dichloromethane (50 mL × 2). The combined organic phase was washed with cold NaOH aqueous solution (10 wt%) twice, followed by washing with brine and water successively. The organic phase was dried over anhydrous $MgSO_4$, and the solvent was removed via rotatory evaporation to give pure $PhSO_2CF_3$ as a colorless liquid (5.28 g, 90%), which can be used without further purification.

Characterization Data ^1H NMR (500 MHz, $CDCl_3$, TMS reference): δ 7.69 (t, $J = 7.7$ Hz, 2H), 7.86 (t, $J = 7.6$ Hz, 1H), 8.05 (d, $J = 7.8$ Hz, 2H) ppm. ^{19}F NMR (470 MHz, $CDCl_3$, $CFCl_3$ reference): δ −78.9 ppm.

Route 2. One-Step Synthesis of $PhSO_2CF_3$ using $TMSCF_3$ and $PhSO_2F$ (Scheme 27.6) [28]

Apparatus A 50-mL Schlenk flask equipped with a magnetic stirring bar and a rubber septum, a vacuum distillation apparatus, a 100-mL separatory funnel, a filter funnel, safety goggles, laboratory coat, and protective gloves.

$$TMSCF_3 + PhSO_2F \xrightarrow[\text{Petroleum ether, 25 °C}]{(Me_2N)_3S^+Me_3SiF_2^-} PhSO_2CF_3$$

SCHEME 27.6 One-step synthesis of $PhSO_2CF_3$ using $TMSCF_3$ and $PhSO_2F$.

Chemicals $TMSCF_3$, $PhSO_2F$, tris(dimethylamino)sulfonium difluorotrimethyl siliconate (TASF), petroleum ether (PE), and $MgSO_4$.

Attention! Safety glasses and protective gloves must be used at all times.

Caution! Because of their toxicity, irritation, and/or volatility, care should be taken to avoid inhalation of $TMSCF_3$, $PhSO_2F$, TASF, and PE or contact of their solutions with the skin. The reaction should be carried out in a well-ventilated hood.

Experimental Procedure To a 50-mL Schlenk flask, $TMSCF_3$ in PE (10 mL) was added to a suspension of $PhSO_2F$ (1.60 g, 10 mmol) and TASF (275 mg, 1 mmol) in PE at 25 °C over a period of 10–15 min under Ar atmosphere. The reaction mixture was further stirred for 0.5 h before quenching with water (30 mL). The reaction mixture was extracted with PE (30 mL), and organic phase was washed with water (50 × 4 mL), dried over $MgSO_4$, and concentrated. The pure product was obtained via vacuum distillation (bp 118–119 °C/1.5 mm Hg).

Applications Due to the extremely electron-deficient nature of the trifluoromethylsulfonyl group ($CF_3SO_2^-$), the sulfur atom in $PhSO_2CF_3$ can readily accept an electron or be attacked by a nucleophile, which leads to the release of the trifluoromethyl anion species. Prakash et al. have described the reductive trifluoromethylation of chlorosilanes using $PhSO_2CF_3$ (Scheme 27.7, Eq. 27.1) [9]. Mediated by Mg^0 in DMF, a series of trifluoromethylsilanes was obtained in moderate to high yields. It has been shown that nucleophilic trifluoromethylation of non-enolizable carbonyl compounds can be achieved through the reaction of carbonyl compounds and $PhSO_2CF_3$ in the presence of excess amounts of *t*BuOK in DMF (Scheme 27.7, Eq. 27.2) [9]. Under similar conditions, trifluoromethylations of iodobenzene and PhSSPh were also shown (Scheme 27.7, Eqs. 27.3 and 27.4) [30]. More recently, Zhao et al. have shown the Mg^0-mediated reductive trifluoromethylation of aldehydes (Scheme 27.7, Eq. 27.5) [31]. Avoiding the use of strong bases, such as *t*BuOK, the protocol was found to be applicable to both non-enolizable and enolizable substrates to render trifluoromethylated alcohols in low to high yields.

27.3 PREPARATION OF DIFLUOROMETHYL PHENYL SULFONE ($PhSO_2CF_2H$) AS A CF_2H^- ANION EQUIVALENT AND A CF_2^{2-} DIANION EQUIVALENT

Difluoromethyl phenyl sulfone ($PhSO_2CF_2H$) is usually prepared through the oxidation of difluoromethyl phenyl sulfide ($PhSCF_2H$), which can be obtained through the reaction between sodium thiophenoxide (PhSNa) and chlorodifluoromethane ($ClCF_2H$, a difluorocarbene precursor) [32,33]. An alternative procedure was shown

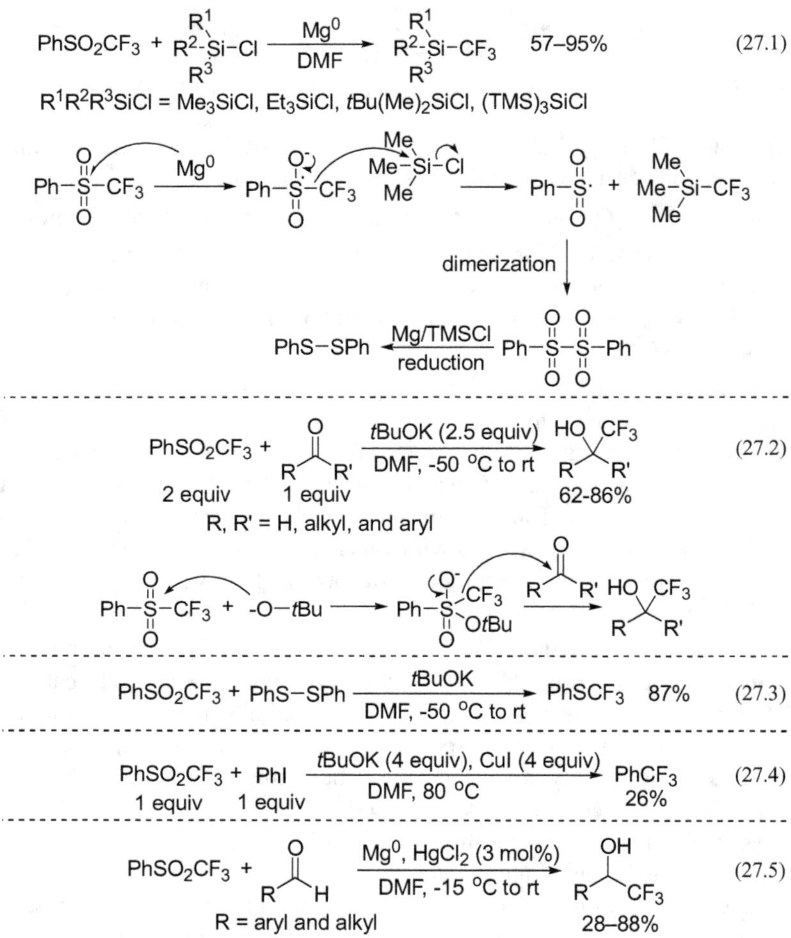

SCHEME 27.7 Typical trifluoromethylation using $TMSCF_3$.

by Chen et al., who prepared the title compound in one step by treating fluorosulfonyldifluoroacetic acid ($FSO_2CF_2CO_2H$, a difluorocarbene precursor) with sodium benzene sulfinate ($PhSO_2Na$) to yield $PhSO_2CF_2H$ in 65% yield [34]. In addition to the above-mentioned methods, another synthetic route has also been reported, however, with fewer efficacies [35].

Preparation of $PhSO_2CF_2H$ using $ClCF_2H$ and PhSNa (Scheme 27.8) [32,33,36]

Apparatus A 1-L, three-necked flask equipped with a dry ice condenser, a dropping funnel, a rubber septum and a magnetic stirring bar, a dry ice/ethylene glycol/acetone bath, an ice bath, a long needle for bubbling CF_2ClH, a fractional distillation apparatus, a 1-L separatory funnel, a 250-mL, round-bottomed flask,

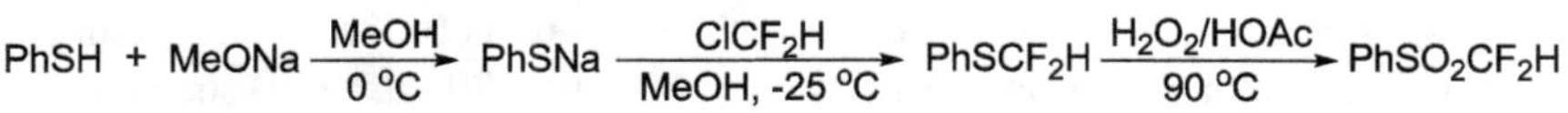

SCHEME 27.8 Preparation of $PhSO_2CF_2H$ using $ClCF_2H$ and PhSNa.

a reflux condenser, a 250-mL separatory funnel, a filter funnel, safety goggles, laboratory coat, and protective gloves.

Chemicals $ClCF_2H$, thiophenol (PhSH), sodium metal (Na), anhydrous methanol, 30 wt% hydrogen peroxide aqueous solution, acetic acid, dichloromethane, diethyl ether (Et_2O), 5 wt% NaOH aqueous solution, 10 wt% $NaHCO_3$ aqueous solution, saturated Na_2SO_3 aqueous solution, $MgSO_4$, and brine.

Attention! Safety glasses and protective gloves must be used at all times.

Caution! Because of their toxicity, irritation, and/or volatility, care should be taken to avoid inhalation of CF_2ClH, PhSH, Na, hydrogen peroxide, methanol, dichloromethane, Et_2O, acetic acid, NaOH, $NaHCO_3$, and Na_2SO_3 or contact of their solutions with the skin. All reactions should be carried out in a well-ventilated hood. Due the extreme coldness, the cold bath should be handled carefully.

Experimental Procedure *Step 1: Preparation of $PhSCF_2H$*

Under Ar atmosphere, into a 1-L, three-necked flask equipped with a dry ice condenser, a dropping funnel, a rubber septum and a magnetic stirring bar, sodium (65 g, 2.82 mol) was added. Methanol (600 mL) was carefully added into the flask at 0 °C under Ar atmosphere with caution (brisk hydrogen evolution occurred). The mixture was stirred for another 8 h until all the sodium was consumed, and PhSH (100 g, 0.91 mol) was added at 0 °C. The reaction mixture was stirred at room temperature for 3 h and subsequently cooled to −25 °C. $ClCF_2H$ (102 g, 1.18 mol) was slowly bubbled into the reaction mixture via a needle over a period of 7 h. The reaction mixture was gradually warmed to room temperature and stirred overnight before the addition of ice water (30 mL). Volatile materials (methanol and CH_3OCF_2H by-product) were removed through fractional distillation. The residue was washed with water (100 mL) and extracted with dichloromethane (50 mL × 3). The combined organic phase was washed with 5 wt% NaOH aqueous solution (30 mL × 3) and water (30 mL × 3). After being dried over $MgSO_4$, the organic mixture was fractionally distilled to afford $PhSCF_2H$ as a colorless liquid (61.2 g, 42 %).

Characterization Data ^1H NMR (500 MHz, $CDCl_3$, TMS reference): δ 7.59 (d, $J = 7.8$ Hz, 2H), 7.41 (m, 3H), 6.83 (t, $J = 56.8$ Hz, 1H) ppm. ^{13}C NMR (125 MHz, $CDCl_3$): δ 121.0 (t, $J = 276.6$ Hz), 126.1, 129.4, 129.8, 135.3 ppm. ^{19}F NMR (470 MHz, $CDCl_3$, $CFCl_3$ reference): δ −91.9 (d, $J = 57.2$ Hz) ppm.

Step 2: Preparation of $PhSO_2CF_2H$

A mixture of $PhSCF_2H$ (30 g, 0.19 mol), 30 wt% aqueous hydrogen peroxide (64 mL, 0.63 mol), and acetic acid (80 mL) in a 250-mL, round-bottomed flask was heated at 90 °C. After 20 h, brine (150 mL) was added, and the reaction mixture was extracted with ether (60 mL × 3). The combined organic phase was washed with 10 wt% $NaHCO_3$ aqueous solution (100 mL × 5), saturated Na_2SO_3 aqueous solution (20 mL × 3), and water (20 mL × 3) successively. The organic phase was dried over anhydrous $MgSO_4$, and the solvent was removed to give pure $PhSO_2CF_2H$ as a colorless liquid (36.2 g, 98%).

Characterization Data ^1H NMR (500 MHz, $CDCl_3$, TMS reference): δ 8.00 (d, $J = 7.9$ Hz, 2H), 7.81 (t, $J = 7.6$ Hz, 1H), 7.66 (t, $J = 7.8$ Hz, 2H), 6.20 (t, $J = 53.5$ Hz, 1H) ppm. ^{19}F NMR (470 MHz, $CDCl_3$, $CFCl_3$ reference): δ −122.2 (d, $J = 53.4$ Hz) ppm.

Applications Although $PhSO_2CF_2H$ was known as early as 1960 [33], its synthetic applications as CF_2H^- or CF_2^{2-} equivalents were not described until much later. Stahly showed the nucleophilic addition of $PhSO_2CF_2^-$ to various aldehydes rendering the corresponding carbinols, which can be reductively desulfonated into α-difluoromethylated alcohols (Scheme 27.9, Eq. 27.1) [36]. An enantioselective variant of this reaction was facilitated through the application of cinchona alkaloid-derived ammonium salts as chiral catalysts, which gave chiral carbinols with 4–64% enantiomeric excesses (Scheme 27.9, Eq. 27.2) [37]. $PhSO_2CF_2H$ was found to react with a series of esters to yield the corresponding α,α-difluorinated ketones (Scheme 27.9, Eq. 27.3) [38]. $PhSO_2CF_2H$ can also react with primary alkyl iodides to afford substituted products, which can be converted to 1,1-difluoro-1-alkenes and 1,1-difluoromethylalkanes under basic and reductive conditions, respectively (Scheme 27.9, Eq. 27.4) [39]. The reaction between $PhSO_2CF_2H$ and chiral *N*-(*tert*-butylsulfinyl)imines showed high diastereoselectivity (Scheme 27.9, Eq. 27.5) [40]. Similarly, $PhSO_2CF_2H$ can also react with chiral *N*-(*tert*-butylsulfinyl)ketimines to afford the corresponding optically active α-difluoromethyl amines in moderate to high yields (Scheme 27.9, Eq. 27.6) [41]. In addition, $PhSO_2CF_2H$ has been used in the synthesis of β-difluoromethylated or β-difluoromethylenated alcohols and amines through its reactions with 1,2-cyclic sulfates and sulfamidates, respectively (Scheme 27.9, Eqs. 27.7 and 27.8) [42]. Intriguingly, Prakash et al. reported an efficient one-pot synthesis of *anti*-2,2-difluoropropane-1,3-diols using $PhSO_2CF_2H$ as a difluoromethylene dianion (CF_2^{2-}) equivalent (Scheme 27.9, Eq. 27.9) [43].

27.4 PREPARATION OF [(PHENYLSULFONYL)DIFLUOROMETHYL] TRIMETHYLSILANE ($PhSO_2CF_2TMS$) AS A CF_2H^- ANION EQUIVALENT

[(Phenylsulfonyl)difluoromethyl]trimethylsilane ($PhSO_2CF_2TMS$) was originally prepared via the oxidation of phenyl (trimethylsilyl)difluoromethyl sulfide ($PhSCF_2TMS$), which can be obtained by Mg^0-mediated trimethylsilylation of bromodifluoromethyl phenyl sulfide ($PhSCF_2Br$) [9]. An improved method was later achieved by treating bromodifluoromethyl phenyl sulfone ($PhSO_2CF_2Br$) with *n*-butyl lithium (*n*BuLi) in the presence of TMSCl (Scheme 27.10) [44].

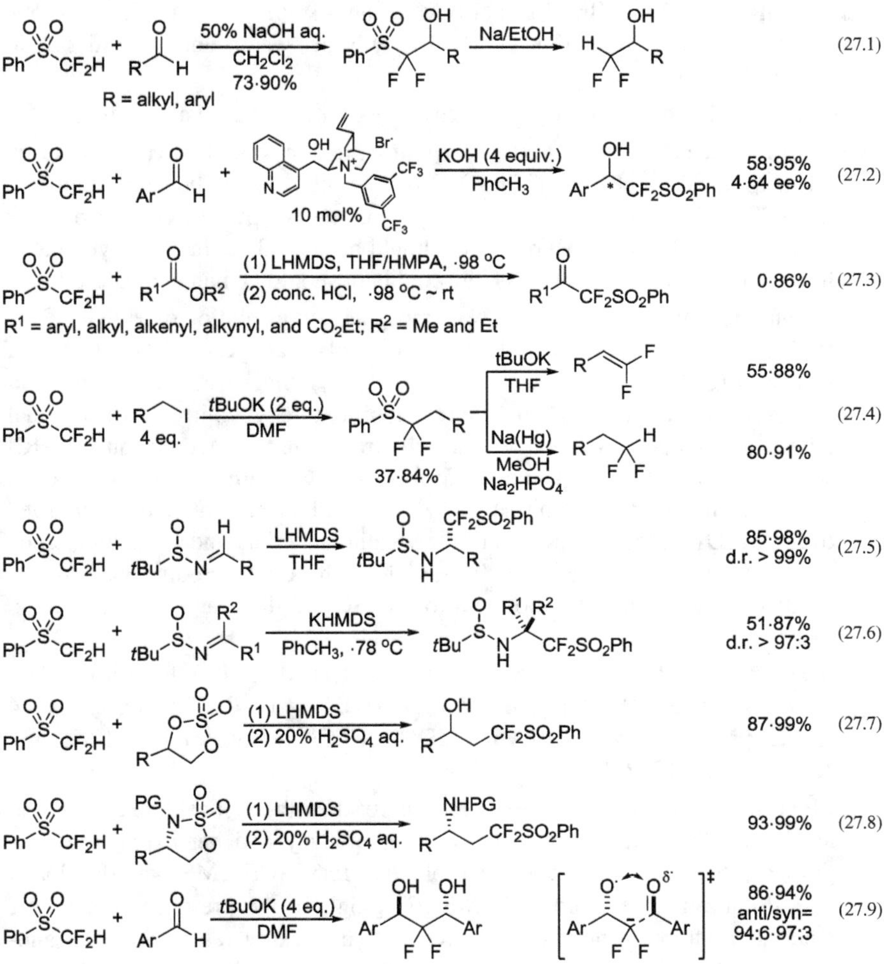

SCHEME 27.9 Synthetic applications of $PhSO_2CF_2H$.

Preparation of $PhSO_2CF_2TMS$ using $PhSO_2CF_2Br$ and TMSCl [45b]

Apparatus A 250-mL Schlenk flask equipped with a magnetic stirring bar and a rubber septum, an acetone–dry ice bath, a 500-mL separatory funnel, a vacuum distillation apparatus, a filter funnel, safety goggles, laboratory coat, and protective gloves.

$$PhSO_2CF_2Br + TMSCl \xrightarrow[\text{THF, -78 °C}]{n\text{BuLi}} PhSO_2CF_2TMS$$

SCHEME 27.10 Preparation of $PhSO_2CF_2TMS$ using $PhSO_2CF_2Br$ and TMSCl.

Chemicals $PhSO_2CF_2Br$, *n*BuLi (1.6-M hexanes solution), TMSCl, anhydrous THF, aqueous HCl solution (1 M), Et_2O, brine, water, and sodium sulfate (Na_2SO_4).

Attention! Safety glasses and protective gloves must be used at all times.

Caution! Because of their toxicity, irritation, and/or volatility, care should be taken to avoid inhalation of $PhSO_2CF_2Br$, *n*BuLi (1.6-M hexanes solution), TMSCl, THF, aqueous HCl solution, Et_2O, and Na_2SO_4 or contact of their solutions with the skin. All reactions should be carried out in a well-ventilated hood. Due the extreme coldness, the cold bath should be handled carefully.

Experimental Procedure Under N_2 atmosphere, to a solution of $PhSO_2CF_2Br$ (6.0 g, 22 mmol) and TMSCl (4.5 mL, 33 mmol) in anhydrous THF (105 mL), *n*BuLi (1.6 M hexanes solution, 15.8 mL, 25 mmol) was added slowly at −78 °C over a period of 1.5 h. The reaction mixture was stirred for additional 2 h at same temperature and then carefully transferred into a cold aqueous HCl solution (1 M, 100 mL). The mixture was extracted with Et_2O (70 mL × 3), and the combined organic phase was washed with brine, water, and then dried over Na_2SO_4. After the removal of the volatile materials under vacuum, crude product was obtained (5.73 g, 98% yield). The crude product was further purified via vacuum distillation to afford $PhSO_2CF_2TMS$ as a colorless liquid (5.10 g, 87%).

Characterization Data ^1H NMR (500 MHz, $CDCl_3$, TMS reference): δ 7.95 (d, $J = 8.0$ Hz, 2H), 7.74 (t, $J = 7.5$ Hz, 1H), 7.61 (t, $J = 7.5$ Hz, 2H), 0.44 (s, 9H) ppm. ^{19}F NMR (470 MHz, $CDCl_3$, $CFCl_3$ reference): δ −112.9 ppm. MS(EI) (*m*/*z*): 175 (M^+).

Applications Although $PhSO_2CF_2H$ can react with enolizable ketones and aldehydes, the reaction suffers from low efficacy and harsh reaction conditions [45]. To overcome these problems, $PhSO_2CF_2TMS$ was developed as a difluoromethyl analog of TMSCF3 [45a]. In the presence of catalytic amounts of fluoride initiators, the reagent can readily react with both enolizable and non-enolizable carbonyl compounds to yield difluoromethylated carbinols in good yields (Scheme 27.11, Eq. 27.1). Under similar reaction conditions, the stereoselective synthesis of α-difluoromethyl-β-amino alcohols was achieved exploiting $PhSO_2CF_2TMS$ as a nucleophilic difluoromethylating reagent (Scheme 27.11, Eq. 27.2) [45b]. An enantioselective difluoromethylation of carbonyl compounds with $PhSO_2CF_2TMS$ has been achieved using cinchonium fluoride catalysts to afford chiral α-difluoromethylated alcohols with low to moderate enantiomeric excesses [37]. Moreover, the nucleophilic substitution reaction between alkyl halides and $PhSO_2CF_2^-$, generated from $PhSO_2CF_2TMS$, was also reported recently (Scheme 27.11, Eq. 27.3) [46]. In addition to the above-mentioned applications, $PhSO_2CF_2TMS$ has also been used as a precursor for preparations of other versatile difluoromethylating reagents. A $PhSO_2CF_2$-iodine(III) reagent (Scheme 27.11, Eq. 27.4) [47] was developed as an electrophilic difluoromethylating reagent, which can be obtained by treating the corresponding acetate-iodine(III) compound with $PhSO_2CF_2TMS$ in the presence of a fluoride initiator [48]. Phenylsulfonyl

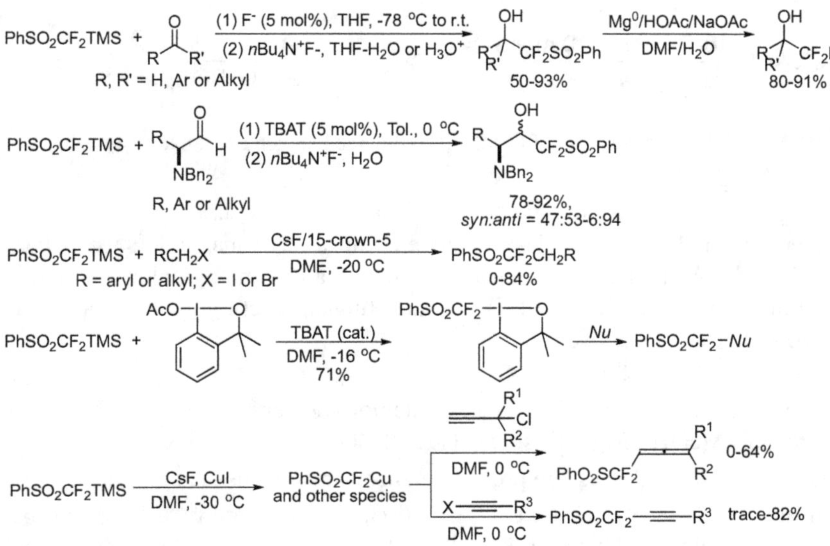

(27.1)

(27.2)

(27.3)

(27.4)

(27.5)

SCHEME 27.11 Synthetic applications of $PhSO_2CF_2TMS$.

difluoromethylcopper species can be prepared by treating $PhSO_2CF_2TMS$ with CsF and copper iodide (CuI) in DMF (Scheme 27.11, Eq. 27.5). Propargyl chlorides and alkynyl halides can undergo reactions with these species to give $PhSO_2$-containing allenes and alkynes [49].

27.5 PREPARATION OF FLUOROMETHYL PHENYL SULFONE ($PhSO_2CH_2F$) AS A CH_2F^- ANION EQUIVALENT

Fluoromethyl phenyl sulfone ($PhSO_2CH_2F$) is obtained via the oxidation of fluoromethyl phenyl sulfide ($PhSCH_2F$) [50]. The precursor $PhSCH_2F$ can be prepared through the halogen exchange reaction between chloromethyl phenyl sulfide ($PhSCH_2Cl$) and potassium fluoride (KF) [51]. Alternatively, $PhSCH_2F$ can be obtained through the treatment of methyl phenyl sulfoxide with deoxofluorinating reagents, such as diethylaminosulfur trifluoride (DAST) [52] and diethylaminodifluorosulfinium tetrafluoroborate (XtalFluor-E) [53]. Robins and Wnuk have also described the preparation of $PhSCH_2F$ using DAST and methyl phenyl sulfide in quantitative yield [54]. More recently, Zhang et al. reported an efficient synthetic approach toward $PhSCH_2F$ using chlorofluoromethane (CH_2FCl) and sodium thiolate (PhSNa) [55].

Route 1. Preparation of $PhSO_2CH_2F$ via the Halogen Exchange Approach (Scheme 27.12) [50, 51]

Apparatus A 500-mL, three-necked, round-bottomed flask equipped with a magnetic stirring bar and three rubber septa, a reflux condenser fitted with a

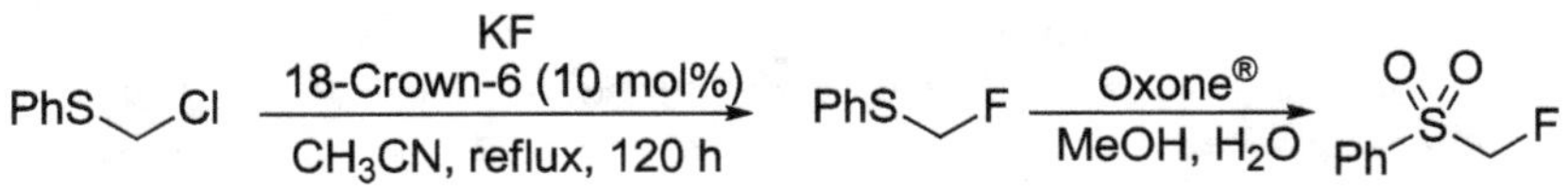

SCHEME 27.12 Preparation of $PhSO_2CH_2F$ via the halogen exchange approach.

nitrogen inlet adaptor, a syringe, an oil bath, an ice bath, a 1-L separatory funnel, a 2-L Erlenmeyer flask equipped with a large magnetic stirring bar, a 1-L addition funnel, a 2-L separatory funnel, a chromatographic column, a 500-mL round-bottomed flask, a Büchner funnel, safety goggles, laboratory coat, and protective gloves.

Chemicals Spray-dried KF, $PhSCH_2Cl$, 18-crown-6, Oxone®, anhydrous acetonitrile (CH_3CN), distilled water, dichloromethane (CH_2Cl_2), Na_2SO_4, methanol (MeOH), $MgSO_4$, silica gel (230–400 mesh), and hexanes.

Attention! Safety glasses and protective gloves must be used at all times.

Caution! Because of their toxicity, irritation, and/or volatility, care should be taken to avoid inhalation of KF, $PhSCH_2Cl$, 18-crown-6, Oxone, CH_3CN, CH_2Cl_2, Na_2SO_4, MeOH, $MgSO_4$, silica gel (230–400 mesh), and hexanes or contact of their solutions with the skin. All reactions should be carried out in a well-ventilated hood. Due the extreme heat, the oil bath should be handled carefully.

Experimental Procedure *Step 1: Preparation of $PhSCH_2F$*

In a glove box, a 500-mL, three-necked, round-bottomed flask equipped with a magnetic stirring bar was charged with spray-dried KF (44 g, 0.76 mol, activated under c. 1 mm Hg vacuum at 120 °C for 24 h) and 18-crown-6 (10 g, 37.5 mmol). The flask was sealed with three rubber septa and transferred to a fume hood. A reflux condenser fitted with a nitrogen inlet adaptor was quickly attached to one of the necks of the flask. Anhydrous CH_3CN (250 mL) and $PhSCH_2Cl$ (51 mL, 60 g, 0.38 mol) were successively added to the flask via a syringe under N_2 atmosphere. The reaction mixture was heated to reflux with stirring in an oil bath (100–105 °C) for 120 h, and then cooled in an ice bath. The reaction mixture was diluted with ice water (250 mL) and transferred into a 1-L separatory funnel. The mixture was extracted with CH_2Cl_2 (5 × 100 mL). The combined organic layer was washed with water (150 mL), dried over Na_2SO_4, and filtered. The solvent was removed to give a brownish oily residue (46.5 g, 86%), which was immediately subjected to oxidation.

Characterization Data ^{1}H NMR (500 MHz, $CDCl_3$, TMS reference): δ 7.47–7.50 (m, 2H), 7.28–7.26 (m, 3H), 5.70 (d, J = 53.1 Hz, 2H) ppm. ^{19}F NMR (470 MHz, $CDCl_3$, $CFCl_3$ reference): δ −182.0 (t, J = 52.4 Hz, 1F) ppm.

Step 2: Oxidation of $PhSCH_2F$ to $PhSO_2CH_2F$ To a 2-L Erlenmeyer flask equipped with a large magnetic stirring bar, Oxone (492 g, 1.6 mol $KHSO_5$) and distilled water (750 mL) were added. The mixture was placed in an ice bath. A solution of $PhSCH_2F$ (45.5 g, c. 0.32 mol) in methanol (750 mL) was added slowly from an addition funnel over a period of c. 1 h. The reaction

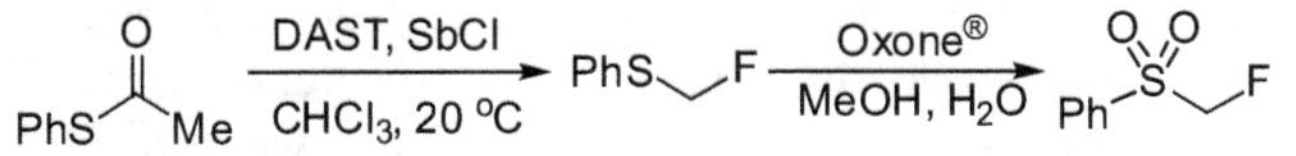

SCHEME 27.13 Preparation of $PhSO_2CH_2F$ using $PhSOCH_3$ and DAST.

mixture was then gradually warmed to room temperature and stirred for 12 h. Methanol was removed via rotary evaporation. The residue was extracted with CH_2Cl_2 (5 × 150 mL) in a 2-L separatory funnel. The combined organic layer was dried over $MgSO_4$, filtered, and concentrated to c. 200 mL. The solution was then filtered through a plug of silica gel (230–400 mesh, 500 mL), and washed with CH_2Cl_2 (c. 1 L). The filtrate was concentrated via rotary evaporation and further dried under vacuum to result in slightly yellowish oil, which slowly solidified at room temperature. The solid was stirred with hot hexanes (c. 250 mL, 60–65 °C) for 20 min, which formed two layers. On cooling to 0 °C, the bottom layer gradually crystallized to yield white crystals of $PhSO_2CH_2F$ (44.0 g, 79%), which were collected by filtration on a Büchner funnel and washed with 100 mL cold hexanes.

Characterization Data ^1H NMR (300 MHz, $CDCl_3$, TMS reference): δ 7.60–7.80 (m, 5H), 5.15 (d, $J = 47.1$ Hz, 2H) ppm. ^{19}F NMR (282 MHz, $CDCl_3$, $CFCl_3$ reference): δ −211.2 (t, $J = 47.4$ Hz, 1F) ppm. MS (EI) (*m/z*): 175 (M^+).

Route 2. Preparation of $PhSO_2CH_2F$ via the Reaction Between $PhSOCH_3$ and DAST (Scheme 27.13) [52]

Apparatus A 1-L, three-necked, round-bottomed flask equipped with a magnetic stirring bar, a thermometer, and an air condenser, a cooling bath, a 2-L separatory funnel, a filter funnel, safety goggles, laboratory coat, and protective gloves.

Chemicals Methyl phenyl sulfoxide (PhSOMe), DAST, antimony trichloride ($SbCl_3$), saturated sodium bicarbonate aqueous solution, NaOH, brine, potassium carbonate (K_2CO_3), and chloroform ($CHCl_3$).

Attention! Safety glasses and protective gloves must be used at all times. Gas evolution can occur during the workup of the reaction.

Caution! Because of their toxicity, irritation, and/or volatility, care should be taken to avoid inhalation of PhSOMe, DAST, $SbCl_3$, NaOH, K_2CO_3, and $CHCl_3$ or contact of their solutions with the skin. All reactions should be carried out in a well-ventilated hood. Due the extreme heat, the oil bath should be handled carefully.

Experimental Procedure To a 1-L, three-necked, round-bottomed flask, equipped with a magnetic stirring bar, a thermometer, and an air condenser, PhSOMe (25.2 g, 0.18 mol) and trichloromethane (150 mL) were added. The flask was placed in a cooling bath containing 3 L water and kept at 20 °C. DAST (38.5 g, 31.6 mL, 0.24 mol) was added to the flask, followed by $SbCl_3$

(0.50 g, 0.0022 mol), and an additional 50 mL trichloromethane. The light yellow reaction mixture was stirred under Ar atmosphere. An exothermic reaction was observed after 2–8 h, and a dark orange solution was formed. The reaction mixture was carefully poured, with stirring, into saturated sodium bicarbonate aqueous solution containing 10 g (0.25 mol) NaOH (600 mL) at 0 °C (gas evolution occurred). After 10 min, the trichloromethane layer was separated and the aqueous layer was extracted with trichloromethane (3 × 100 mL). The combined organic layers were washed with saturated sodium bicarbonate aqueous solution (250 mL), brine, and dried over potassium carbonate successively. Trichloromethane was removed with a rotary evaporator to result in yellow orange oil as crude $PhSCH_2F$ (c. 29 g). The crude product was used immediately in the oxidation reaction as mentioned in Step 2.

Applications $PhSO_2CH_2F$ can react with ketones and aldehydes to yield the corresponding β-fluoro-alcohols, which can be converted to terminal vinyl fluorides (Scheme 27.14, Eq. 27.1) [53b]. The in situ treatment of $PhSO_2CHF^-$ anion with diethyl chlorophosphate [$ClP(O)(OEt)_2$] led to the formation of diethyl 1-fluoro-1-(phenylsulfonyl)methanephosphonate anion, which readily undergoes reaction with carbonyl compounds to yield the corresponding vinyl fluorides (Scheme 27.14, Eq. 27.2) [56]. In addition, $PhSO_2CH_2F$ has been used in the synthesis of various optically active fluoromethylated amines [57]. In 2006, Li et al. disclosed the stereoselective

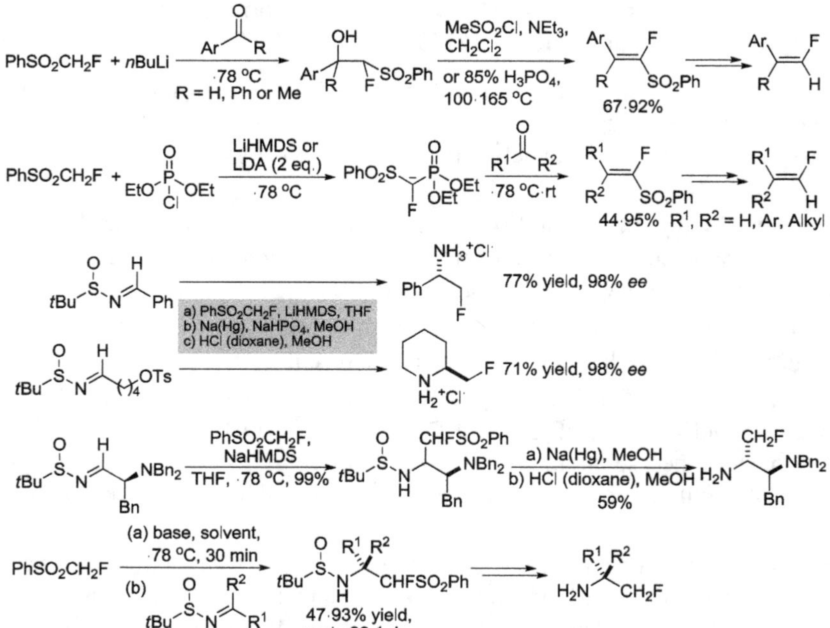

SCHEME 27.14 Synthetic applications of $PhSO_2CH_2F$.

monofluoromethylation of homochiral *N*-(*tert*-butylsulfinyl)imines, which could afford chiral β-fluoromethylated amines with high enantiomeric excesses (Scheme 27.14, Eqs. 27.3 and 27.4) [58]. Likewise, $PhSO_2CH_2F$ was also found to react with chiral α-amino *N-tert*-butanesulfinimines (Scheme 27.14, Eq. 27.5) [41b] and *N*-(*tert*-butylsulfinyl)ketimines (Scheme 27.13, Eq. 27.6) [59] to render various α-monofluoromethylated amines with high enantiomeric purities.

27.6 PREPARATION OF α-FLUOROBIS(PHENYLSULFONYL) METHANE AS A CH_2F^- ANION EQUIVALENT

Fluorobis(phenylsulfonyl)methane (FBSM) was originally prepared by Shibata and coworkers [60] and Hu and coworkers [61] independently in 2006 as a versatile CH_2F^- anion equivalent through the electrophilic fluorination of bis(phenylsulfonyl)methane using Selectfluor®. Hu and coworkers also described a superior synthetic route based on the sulfoxidation of $PhSO_2CH_2F$ followed by oxidation, which avoids the tedious purification step necessary in the original synthesis [38]. Prakash et al. have designed a practical one-step synthesis of FBSM with $PhSO_2CH_2F$ and less costly $PhSO_2F$, rendering FBSM with high efficacy and purity [62]. In addition to the above-mentioned synthetic protocols, FBSM can also be obtained via electrochemical fluorination approach, which is, however, not frequently used in synthetic organic chemistry laboratories [63].

Route 1. Preparation of FBSM via Electrophilic Fluorination of Bis(phenylsulfonyl)methane (Scheme 27.15) [64]

Apparatus A 50-mL Schlenk flask, a rubber septum, a magnetic stir bar, an ice bath, a 100-mL separatory funnel, a chromatography column, safety goggles, laboratory coat, and protective gloves.

Chemicals Bis(phenylsulfonyl)methane, Selectfluor, sodium hydride (NaH, 60% oil dispersion), anhydrous THF, anhydrous CH_3CN, $MgSO_4$, EtOAc, saturated aqueous ammonium chloride, brine, silica gel, hexane, and dichloromethane.

Attention! Safety glasses and protective gloves must be used at all times. NaH reacts violently with water and can ignite in air, and should be handled under inert atmosphere.

Caution! Because of their toxicity, irritation, and/or volatility, care should be taken to avoid inhalation or contact of the chemicals mentioned above. All procedures should be carried out in a well-ventilated hood.

Experimental Procedure To a nitrogen-protected, 50-mL, Schlenk flask containing bis(phenylsulfonyl)methane (2.70 g, 9.0 mmol) in THF (25 mL), NaH (60% oil dispersion that is rinsed with pentane, 240 mg, 6.0 mmol) is added slowly with stirring at 0 °C. The temperature is maintained at 0 °C for 30 min, and a mixture of finely ground Selectfluor powder (2.1 g, 6.0 mmol) and CH_3CN

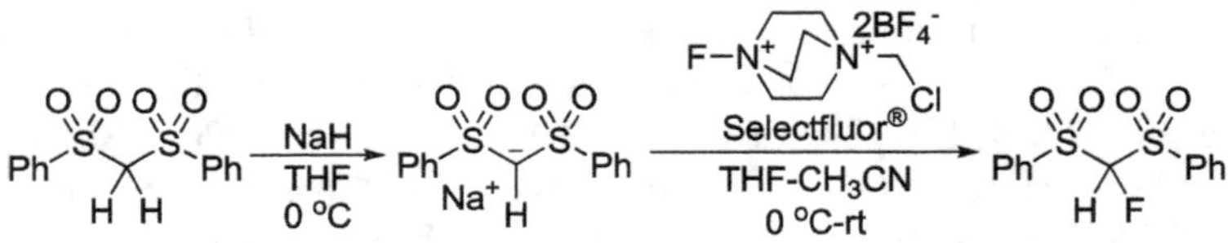

SCHEME 27.15 Electrophilic fluorination of bis(phenylsulfonyl)methane.

(5 mL) is added at 0 °C. The reaction mixture is warmed to room temperature and stirred for another 12 h. The reaction mixture is quenched by saturated aqueous ammonium chloride. The resulting mixture is then extracted by EtOAc (50 mL × 3). The combined organic layer is washed with brine and dried over anhydrous $MgSO_4$. The solvent is removed under vacuum and the residue is purified by silica gel column chromatography using hexane/dichloromethane as the eluent to afford FBSM (1.6 g, 75%) as a white solid.

Characterization Data Mp 114–114.5 °C (from hexane). ^{1}H NMR (200 MHz, $CDCl_3$, TMS reference): δ 5.70 (1H, d, J = 45.8 Hz, C*H*F), 7.55–7.65 (4H, m, Ar), 7.70–7.80 (2H, m, Ar), 7.95–8.05 (4H, d, J = 7.6 Hz, Ar) ppm. ^{13}C NMR (50 MHz, $CDCl_3$): δ 105.3 (d, J = 263.4), 129.2, 129.8, 134.9, 135.4 ppm. ^{19}F NMR (188 MHz, $CDCl_3$, $CFCl_3$ reference): δ −167.2 (d, J = 45.8 Hz) ppm. IR (KBr) ν: 1354, 1174 cm^{-1}. MS (ESI-TOF) *(m/z)*: 314 (M^+), 173 (M^+ $-SO_2Ph$), 141 (M^+ $-PhSO_2CHF$).

Route 2. Preparation of FBSM via Oxidation of Sulfoxidated Product of $PhSO_2CH_2F$ (Scheme 27.16) [38]

Apparatus A 20-mL Schlenk tube, a rubber septum, a 3-mL syringe, a magnetic stir bar, a cooling bath (dry ice–acetone), a 50-mL separatory funnel, a chromatography column, a filter funnel, safety goggles, laboratory coat, and protective gloves.

Chemicals Methyl sulfinate, $PhSO_2CH_2F$ (commercially available, and can be prepared via a known procedure [50]), lithium hexamethyldisilazide (LHMDS, 1 M in THF), *m*-chloroperoxybenzoic acid (mCPBA), anhydrous THF, $MgSO_4$, EtOAc, saturated aqueous HCl, saturated aqueous $NaHCO_3$, silica gel, hexane, PE, and dichloromethane.

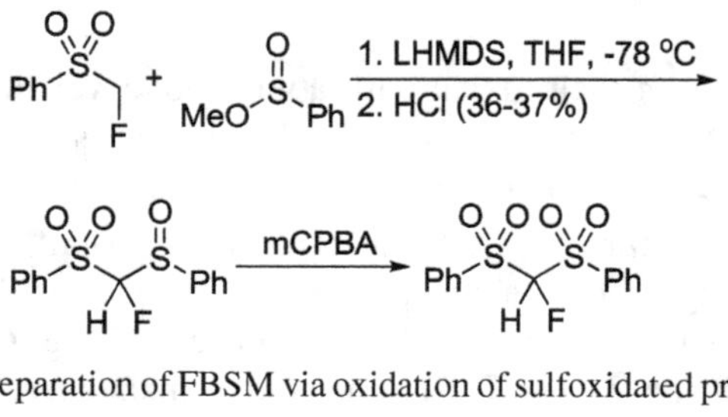

SCHEME 27.16 Preparation of FBSM via oxidation of sulfoxidated product of fluoromethyl phenyl sulfone.

Attention! Safety glasses and protective gloves must be used at all times. mCPBA is a flammable solid and contact with heat or oxidizable material should be avoided.

Caution! Because of their toxicity, irritation, and/or volatility, care should be taken to avoid inhalation or contact of the chemicals mentioned above. All procedures should be carried out in a well-ventilated hood.

Experimental Procedure Under N_2 atmosphere, methyl sulfinate (187 mg, 1.2 mmol), $PhSO_2CH_2F$ (174 mg, 1.0 mmol), and anhydrous THF (5.0 mL) are added into a Schlenk tube, which is cooled to −78 °C. LHMDS (1.0 M in THF, 2.0 mL, 2.0 mmol) is added drop by drop with vigorous stirring at the same temperature for 30 min. The reaction mixture is quenched with saturated aqueous HCl (36–37%, 2 mL) at this temperature, followed by extraction with EtOAc (20 mL × 3). The combined organic phase is dried over $MgSO_4$ before the removal of the solvents. The crude product is further purified by silica gel column chromatography (PE/EtOAc 3:1 as eluent) to afford the phenylsulfinyl sulfone (285 mg, 95%).

To a solution of the obtained phenylsulfinyl sulfone (298 mg, 1.0 mmol) in CH_2Cl_2 (5 mL), mCPBA (264 mg, 85% purity, 1.3 mmol) is added in one portion at 0 °C. The reaction mixture is warmed to room temperature and stirred for 6 h. The reaction mixture is then diluted with CH_2Cl_2 (50 mL) and washed with saturated aqueous $NaHCO_3$ (20 mL × 3). The organic layer is dried over $MgSO_4$, and the solvents are evaporated under vacuum. The crude product is purified by silica gel column chromatography (PE/EtOAc 1:1–1:2 as eluent) to afford FBSM (304 mg, 96%). Thus, the overall yield of this two-step synthetic protocol is 91%.

Characterization Data [Fluoro (phenylsulfinyl) methylsulfonyl] benzene ($PhSOCHFSO_2Ph$) A mixture of two diastereomers in a ratio of 2:1. ^1H NMR (300 MHz, $CDCl_3$, TMS reference): δ 5.56 (d, J = 46 Hz, 0.67H, C*H*F), 5.58 (d, J = 47 Hz, 0.33H, C*H*F), 7.52–7.68 (m, 5H, Ar), 7.70–7.82 (m, 3H, Ar), 8.00–8.07 (m, 2H, Ar) ppm. ^{19}F NMR (282 MHz, $CDCl_3$, $CFCl_3$ reference): δ −182.4 (d, J = 46 Hz, 0.67F), −167.6 (d, J = 47 Hz, 0.33F) ppm. IR (KBr) ν: 1477, 1447, 1336, 1312, 1158 cm^{-1}.

Route 3. Preparation of FBSM via Sulfonation of $PhSO_2CH_2F$ using $PhSO_2F$ (Scheme 27.17) [62]

Apparatus A 20-mL Schlenk tube, a rubber septum, a 10-mL syringe a magnetic stir bar, a cooling bath (dry ice–acetone), a 50-mL separatory funnel, a filter funnel, safety goggles, laboratory coat, and protective gloves.

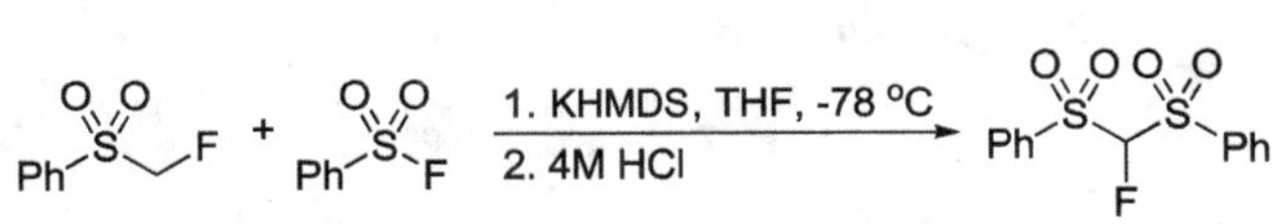

SCHEME 27.17 One-step preparation of FBSM using $PhSO_2CH_2F$ and $PhSO_2F$.

Chemicals $PhSO_2F$, $PhSO_2CH_2F$, potassium hexamethyldisilazide (KHMDS), anhydrous THF, $MgSO_4$, aqueous HCl (4 M), and dichloromethane.

Attention! Safety glasses and protective gloves must be used at all times.

Caution! Because of their toxicity, irritation, and/or volatility, care should be taken to avoid inhalation or contact of the chemicals mentioned above. All procedures should be carried out in a well-ventilated hood.

Experimental Procedure $PhSO_2CH_2F$ (348 mg, 2 mmol) and $PhSO_2F$ (320 mg, 2 mmol) are dissolved in anhydrous THF (10 mL) in a Schlenk tube under inert atmosphere. The solution is cooled to −78 °C. KHMDS (499 mg, 5 mmol, in 5 mL anhydrous THF) is added drop by drop to the Schlenk flask. The reaction mixture is stirred for 30 min at −78 °C before being poured into 4 M HCl aqueous solution (20 mL). The resulting mixture is washed with water and

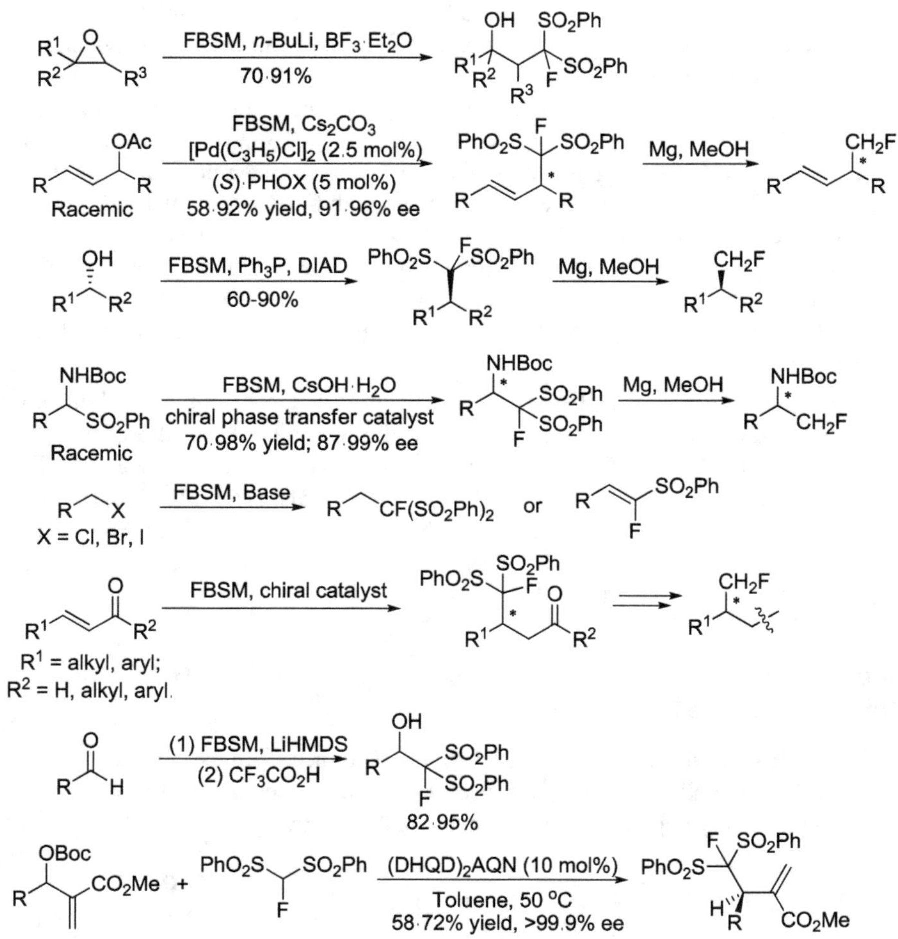

SCHEME 27.18 Monofluoromethylations using α-fluorobis(phenylsulfonyl)methane (FBSM).

extracted with CH_2Cl_2 (15 mL × 3). The combined organic layer is dried over $MgSO_4$, and the solvents are evaporated to afford an oily product, which slowly solidifies after standing over a period of time (598 mg, 95%). ^{1}H NMR and ^{19}F NMR spectroscopy show the product is highly pure (>95%).

Applications FBSM has been developed as a versatile nucleophilic monofluoromethylating reagent (a CH_2F^- anion equivalent). Owing to the presence of the two phenylsulfonyl groups, FBSM is more acidic than $PhSO_2CH_2F$, thereby undergoing feasible deprotonation to render rather stable α-fluorocarbanion [65]. Thus, a variety of nucleophilic monofluoromethylation reactions has been achieved using FBSM, such as the ring-opening reaction of epoxides and aziridines [61], the allylic monofluoromethylation reaction [66], the Mitsunobu reaction [67], conjugate addition reactions [68], the Mannich reaction [69], the aldol reaction [70], the Morita–Baylis–Hillman reaction [71], as well as many other reactions (Scheme 27.18) [57, 72]. In particular, the facile reductive removal of the sulfonyl groups allows the introduction of unfunctionalized CH_2F motif using FBSM, thereby prevailing over many other monofluoromethylating reagents. In addition, FBSM can be further converted to fluoroiodobis(phenylsulfonyl)methane, which has been used as a viable radical monofluoromethylating reagent [73].

27.7 PREPARATION OF *S*-(DIFLUOROMETHYL)-*S*-PHENYL-2,3,4,5-TETRAMETHYLPHENYLSULFONIUM TETRAFLUOROBORATE (DPTPT) AS A CF_2H^+ CATION EQUIVALENT

S-(Difluoromethyl)-*S*-phenyl-2,3,4,5-tetramethylphenylsulfonium tetrafluoroborate (DPTPT) [74] is an electrophilic difluoromethylating reagent, which is analogous to the electrophilic trifluoroemethylating reagents developed by Yagupolskii and Umemoto [75]. DPTPT was obtained through a two-step procedure. The triflate salt of DPTPT was prepared via the reaction of difluoromethyl phenyl sulfoxide ($PhSOCF_2H$) with 1,2,3,4-tetramethylbenzene in the presence of triflic anhydride (Tf_2O). The product was then subjected to an anion exchange reaction with sodium tetrafluoroborate ($NaBF_4$), rendering DPTPT in 51% overall yield.

Preparation of DPTPT using $PhSOCF_2H$ and 1,2,3,4-Tetramethylbenzene (Scheme 27.19) [74]

Apparatus A 150-mL Schlenk flask equipped with a magnetic stirring bar and a rubber septum, an ice bath, a syringe with a needle, a 250-mL separatory funnel, a filter funnel, safety goggles, laboratory coat, and protective gloves.

Chemicals $PhSOCF_2H$, 1,2,3,4-tetramethylbenzene, Tf_2O, anhydrous Et_2O, dichloromethane, $NaBF_4$ aqueous solution (1 M), and anhydrous $MgSO_4$.

Attention! Safety glasses and protective gloves must be used at all times.

Caution! Because of their toxicity, irritation, and/or volatility, care should be taken to avoid inhalation of $PhSOCF_2H$, 1,2,3,4-tetramethylbenzene, Tf_2O,

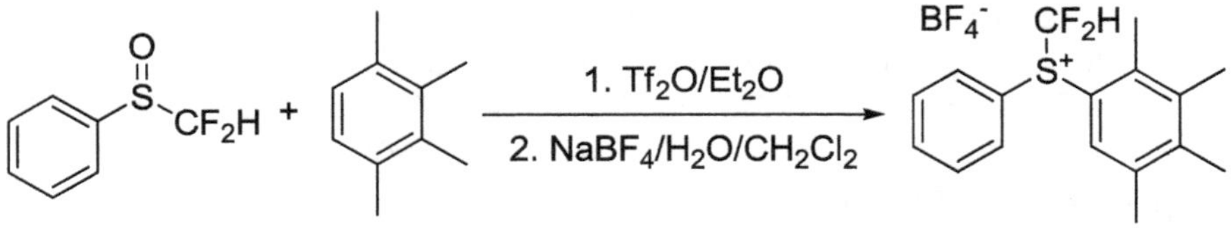

SCHEME 27.19 Preparation of DPTPT using $PhSOCF_2H$ and 1,2,3,4-tetramethylbenzene.

Et_2O, dichloromethane, $NaBF_4$ aqueous solution, and anhydrous $MgSO_4$ or contact of their solutions with the skin. All reactions should be carried out in a well-ventilated hood. Due the extreme coldness, the cold bath should be handled carefully.

Experimental Procedure To a stirred solution of $PhSOCF_2H$ (4.00 g, 25 mmol) and 1,2,3,4-tetramethylbenzene (3.35 g, 25 mmol) in anhydrous Et_2O (60 mL) at 0 °C under Ar, Tf_2O (7.0 g, 25 mmol) was added in small portions over a period of 2 h. The reaction mixture was stirred for 20 min at the same temperature, and the formed oil was separated from the Et_2O phase under nitrogen. Then, anhydrous Et_2O (30 mL) was added to the oil and the reaction mixture was stirred again. This procedure was repeated four times. The resulting oil was dissolved in dichloromethane (50 mL). The dichloromethane solution was extracted with $NaBF_4$ aqueous solution (1 M, 5 × 100 mL), and the organic layer was dried over anhydrous $MgSO_4$. The drying agent was filtered off and the dichloromethane was removed in vacuum. The product was obtained as a brown semisolid (5.9 g, 51%).

Characterization Data ^1H NMR (400 MHz, $CDCl_3$, TMS reference): δ 8.12 (t, $J = 47.4$ Hz, 1H), 7.65–7.95 (m, 5H), 7.49 (s, 1H), 2.57 (s, 3H), 2.40 (s, 3H), 2.32 (s, 3H), 2.31 (s, 3H) ppm. ^{13}C NMR (101 MHz, $CDCl_3$, TMS reference): 17.0, 17.1, 18.0, 21.1, 113.0, 118.90 (t, $J = 297.5$ Hz), 119.0, 129.0, 131.6, 131.9, 135.5, 138.9, 139.1, 140.0, 145.3 ppm. ^{19}F NMR (376 MHz, $CDCl_3$, $CFCl_3$ reference): δ −99.9 (d, $J = 53.4$ Hz, 1F), −100.6 (d, $J = 53.4$ Hz, 1F), −152.0 (s, 1F), −152.1 (s, 3F) ppm. HRMS (FAB) m/z: Calcd for $C_{17}H_{19}F_2S^+$, 293.1170; found, 293.1170.

Applications DPTPT was found to be a versatile reagent enabling electrophilic difluoromethylation of various nucleophiles, including CD_3OD, triflates, tertiary amines, phosphines, and imidazole derivatives (Scheme 27.20) [74]. Noticeably, an analogous solid-phase-bound electrophilic difluoromethylating reagent has also been synthesized, which facilitates the purification-free difluoromethylations of triflates and imidazoles [76].

27.8 PREPARATION OF *S*-(FLUOROMETHYL)-*S*-PHENYL-2,3,4,5-TETRAMETHYLPHENYLSULFONIUM TETRAFLUOROBORATE (FPTPT) AS A CH_2F^+ CATION EQUIVALENT

S-(Fluoromethyl)-*S*-phenyl-2,3,4,5-tetramethylphenylsulfonium tetrafluoroborate (FPTPT) was prepared by Prakash et al. as a novel electrophilic fluoromethylating

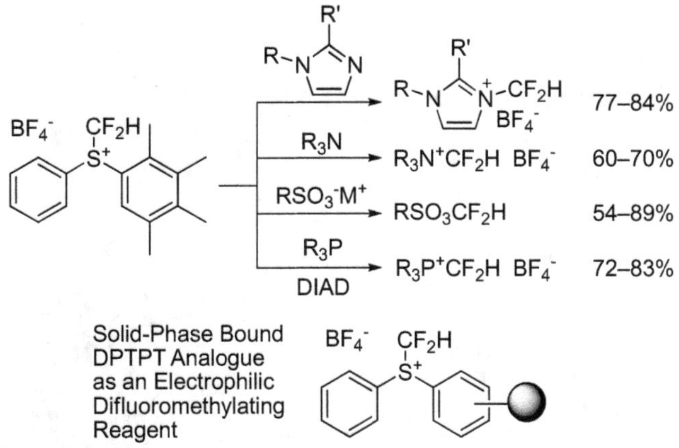

SCHEME 27.20 Synthetic applications of DPTPT in electrophilic difluoromethylations.

reagent [77]. Similar to its difluoromethyl counterpart, the reagent was synthesized through the treatment of fluoromethyl phenyl sulfoxide ($PhSOCH_2F$) with 1,2,3,4-tetramethylbenzene in the presence of triflic anhydride (Tf_2O) and the subsequent anion exchange.

Preparation of FPTPT using $PhSOCH_2F$ and 1,2,3,4-Tetramethylbenzene (Scheme 27.21) [77]

Apparatus A 150-mL Schlenk flask equipped with a magnetic stirring bar and a rubber septum, an ice bath, a syringe with a needle, a 250-mL separatory funnel, a filter funnel, safety goggles, laboratory coat, and protective gloves.

Chemicals $PhSOCH_2F$, 1,2,3,4-tetramethylbenzene, Tf_2O, anhydrous Et_2O, dichloromethane, $NaBF_4$ aqueous solution (1 M), and anhydrous $MgSO_4$.

Attention! Safety glasses and protective gloves must be used at all times.

Caution! Because of their toxicity, irritation, and/or volatility, care should be taken to avoid inhalation of $PhSOCH_2F$, 1,2,3,4-tetramethylbenzene, Tf_2O, Et_2O, dichloromethane, $NaBF_4$ aqueous solution, and anhydrous $MgSO_4$ or contact of their solutions with the skin. All reactions should be carried out in a well-ventilated hood. Due the extreme coldness, the cold bath should be handled carefully.

Experimental Procedure To a stirred solution of $PhSOCH_2F$ (3.00 g, 19 mmol) and 1,2,3,4-tetramethylbenzene (2.54 g, 19 mmol) in anhydrous Et_2O (45 mL)

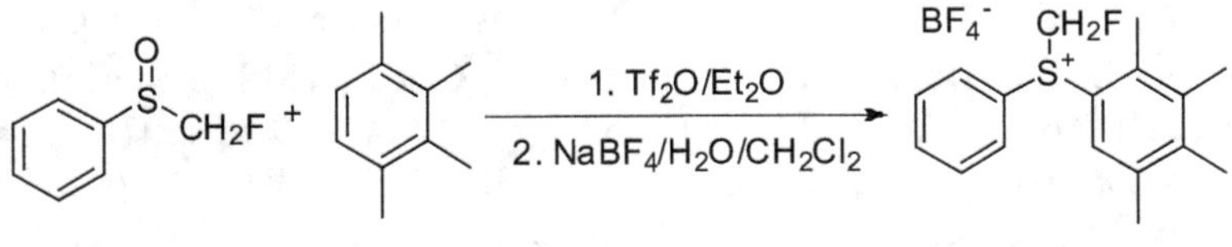

SCHEME 27.21 Preparation of FPTPT using $PhSOCH_2F$ and 1,2,3,4-tetramethylbenzene.

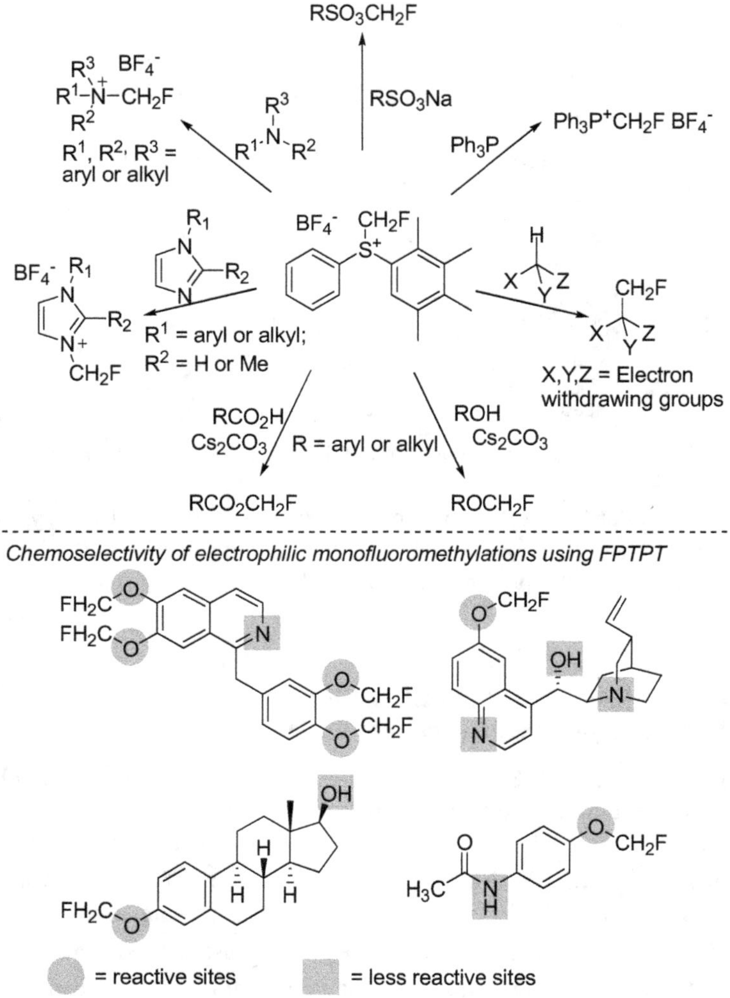

SCHEME 27.22 Synthetic applications of FPTPT in electrophilic fluoromethylations.

at 0–5 °C under Ar, Tf_2O (5.36 g, 19 mol) was added drop by drop over a period of 30 min. The reaction mixture was stirred at the same temperature range for 1 h. The precipitated triflate salt was filtered off and washed with Et_2O five times. The triflate salt was then dissolved in 60 mL dichloromethane and washed with $NaBF_4$ aqueous solution (1 M, 5 × 80 mL). The organic layer was dried over $MgSO_4$ and the solvent was evaporated under vacuum. The resulting solid was further dried under vacuum to give FPTPT as a white powder (6.88 g, 81%).

Characterization Data ^{1}H NMR (400 MHz, $CDCl_3$, TMS reference): δ 7.76 (m, 3H), 7.67 (m, 2H), 7.43 (s, 1H), 6.55 (dd, $J = 47.0, 9.7$ Hz, 1H), 6.46 (dd, $J = 47.0, 9.6$ Hz, 1H), 2.50 (s, 3H), 2.39 (s, 3H), 2.31 (s, 3H), 2.29 (s, 3H) ppm. ^{13}C NMR (101 MHz, $CDCl_3$, TMS reference): 17.1, 17.2, 17.9, 21.4, 89.9 (d,

$J = 242$ Hz), 116.4, 121.4, 121.4, 128.6, 128.6, 131.1, 131.6, 134.6, 137.7, 138,4, 139.6, 144.2 ppm. ^{19}F NMR (376 MHz, $CDCl_3$, $CFCl_3$ reference): δ 207.8 (t, $J = 47.1$ Hz, 1F), −152.1 (s, 3F), −152.0 (s, 1F). HRMS (FAB) m/z: Calcd for $C_{17}H_{20}FS^+$, 275.1264; found, 275.1257. Elemental analysis: Calcd: C, 56.37%; H, 5.57%. Found: C, 56.23%; H, 5.43%.

Applications FPTPT can react with a variety of nucleophiles to afford the corresponding monofluoromethylated products (Scheme 27.22) [77]. Compared with the substrate scope of electrophilic difluoromethylations using DPTPT, a broader spectrum of nucleophiles was found to readily react with FPTPT, including alkoxides, acetate, triflates, tertiary amines, phosphines, imidazole derivatives, and carbon nucleophiles. Noticeably, the protocol also showed remarkable chemoselectivity, which preferentially monofluoromethylates phenolic groups over many other nucleophiles (Scheme 27.22).

REFERENCES

1. (a) Uneyama K. *Organofluorine Chemistry*. Oxford: Blackwell; 2006, chapter 7. (b) Kirsch P. *Modern Fluoroorganic Chemistry: Synthesis, Reactivity, Applications*. Weinheim. Wiley-VCH; 2004, chapter 2.
2. Adolph HG, Kamlet MJ. Fluoronitroaliphatics. I. The effect of α fluorine on the acidities of substituted nitromethanes. *J Am Chem Soc* 1966;88:4761–4763.
3. Olah GA, Heiliger L, Prakash GKS. Stable carbocations. Part 276. Trihalomethyl cations. *J Am Chem Soc* 1989;111:8020–8021.
4. Umemoto T. Electrophilic perfluoroalkylating agents. *Chem Rev* 1996;96:1757–1777, and reference therein.
5. (a) Prakash GKS, Hu J. New nucleophilic fluoroalkylation chemistry. In: *Fluorinated Synthons*; Soloshonok VA, editor; ACS Symposium Series 911; Washington, DC: American Chemical Society, 2005. (b) Prakash GKS, Hu J. Selective fluoroalkylations with fluorinated sulfones, sulfoxides and sulfides. *Acc Chem Res* 2007;40:921–930. (c) Hu J. Nucleophilic, radical, and electrophilic (phenylsulfonyl) difluoromethylations. *J Fluorine Chem* 2009;130:1130–1139. (d) Hu J, Zhang W, Wang F. Selective difluoromethylation and monofluoromethylation reactions. *Chem Commun* 2009:7465.
6. (a) Prakash GKS, Yudin AK. Perfluoroalkylation with organosilicon reagents. *Chem Rev* 1997;97:757–786. (b) Singh RP, Shreeve JM. Nucleophilic trifluoromethylation reactions of organic compounds with (trifluoromethyl)trimethylsilane. *Tetrahedron* 2000;56:7613–7632. (c) Prakash GKS, Mandal M. Nucleophilic trifluoromethylation tamed. *J Fluorine Chem* 2001;112:123–131. (d) Shibata N, Mizuta S, Kawai H. Recent advances in enantioselective trifluoromethylation reactions. *Tetrahedron: Asymmetry* 2008;19:2633–2644.
7. Ramaiah P, Krishnamurti R, Prakash GKS. 1-trifluoromethyl-1-cyclohexanol. *Org Synth* 1995;72:232–236.
8. Ruppert I, Schlich K, Volbach W. Die ersten CF_3-substituierten organyl (chlor)silane. *Tetrahedron Lett* 1984;25:2195–2198.
9. Prakash GKS, Hu J, Olah GA. Preparation of tri- and difluoromethylsilanes via an unusual magnesium metal-mediated reductive tri- and difluoromethylation of chlorosilanes using tri- and difluoromethyl sulfides, sulfoxides, and sulfones. *J Org Chem* 2003;68:4457–4463.

10. (a) Pawelke GJ. Tetrakis (dimethylamino)ethylene/trifluoroiodomethane. A specific novel trifluoromethylating agent. *J Fluorine Chem* 1989;42:429–433. (b) Prakash GKS, Deffieux D, Yudin AK, Olah GA. Convenient ands afe electrochemical synthesis of (trifluoromethyl) trimethylsilane. *Synlett* 1994:1057–1058. (c) Aymard F, Nédélec J-Y, Périchon J. An efficient inexpensive electrochemical preparation of Ruppert's reagent. *Tetrahedron Lett* 1994;35:8623–8624. (d) Grobe J, Hegge J. Facile aluminum induced synthesis of (trifluoromethyl) trimethylsilane. *Synlett* 1995:641–642. (e) Prakash GKS, Yudin AK, Deffieux D, Olah GA. Facile preparation of (trifluoromethyl) tributyltin and transtrifluoromethylation of disilyl sulfides to the corresponding trifluoromethylsilanes. *Synlett* 1996: 151–153.

11. Russell J, Roques N. Effective nucleophilic trifluoromethylation with fluoroform and common base. *Tetrahedron* 1998;54:13771–13782.

12. (a) Prakash GKS, Krishnamurti R, Olah GA. Synthetic methods and reactions. 141. Fluoride-induced trifluoromethylation of carbonyl compounds with trifluoromethyltrimethylsilane (TMS-CF3). A trifluoromethide equivalent. *J Am Chem Soc* 1989;111:393–395. (b) Krishnamurti R, Bellew DR, Prakash GKS. Preparation of trifluoromethyl and other perfluoroalkyl compounds with (perfluoroalkyl)trimethylsilanes. *J Org Chem* 1991;56:984–989.

13. Blazejewski J-C, Anselmi E, Wilmshurst MP. Extending the scope of Ruppert's reagent: Trifluoromethylation of imines. *Tetrahedron Lett* 1999;40:5475–5478.

14. Huang A, Li H-Q, Massefski W, Saiah E. Direct trifluoromethylation of nitriles promoted by tetrabutylammonium bifluoride. *Synlett* 2009;2518–2520.

15. Nelson DW, Easley RA, Pintea BNV. Nucleophilic perfluoroalkylation of nitrones. *Tetrahedron Lett* 1999;40:25–28.

16. (a) Kim J, Shreeve JM. The first Cu(I)-mediated nucleophilic trifluoromethylation reactions using (trifluoromethyl) trimethylsilane in ionic liquids. *Org Biomol Chem* 2004;2: 2728–2734. (b) Tyrra W, Naumann D, Quadt S, Buslei S, Yagupolskii YL, Kremlev MM. Fluoride-mediated selective cross-coupling reactions of alkyl halides and trimethyl(perfluoroalkyl)silanes, Me3SiRf (Rf = CF_3, C_2F_5) in the absence of any catalysts. *J Fluorine Chem* 2007;128:813–817.

17. Ma J-A, Cahard D. Update 1 of: Asymmetric fluorination, trifluoromethylation, and perfluoroalkylation reactions. *Chem Rev* 2008;108:PR1–PR43.

18. (a) Prakash GKS. Nucleophilic perfluoroalkylation of organic compounds using perfluoroalkyltrialkylsilanes. In: Olah GA, Chambers RD, Prakash GKS, editors. *Synthetic Fluorine Chemistry*. New York: Wiley; 1992, chapter 10. (b) Wang F, Luo T, Hu J, Wang Y, Krishnan HS, Jog PV, Ganesh SK, Prakash GKS, Olah GA. Synthesis of gem-difluorinated cyclopropanes and cyclopropenes: Trifluoromethyltrimethylsilane as a difluorocarbene source. *Angew Chem, Int Ed* 2011;50:7153–7157.

19. (a) Dubinina GG, Furutachi H, Vicic DA. Active trifluoromethylatinga gents from well-defined copper(I)–CF_3 complexes. *J Am Chem Soc* 2008;130:8600–8601. (b) Tomashenko OA, Escudero-Adán EC, Belmonte MM, Grushin VV. Simple, stable, and easily accessible well-defined CuCF3 aromatic trifluoromethylating agents. *Angew Chem, Int Ed* 2011;50:7655–7659. (c) Morimoto H, Tsubogo T, Litvinas ND, Hartwig JF. A broadly applicable copper reagent for trifluoromethylations and perfluoroalkylations of aryl iodides and bromides. *Angew Chem, Int Ed* 2011;50:3793–3798.

20. (a) Chu L, Qing F-L. Copper-mediated oxidative trifluoromethylation of boronic acids. *Org Lett* 2010;12:5060–5063. (b) Senecal TD, Parsons AT, Buchwald SL. Room temperature

aryl trifluoromethylation via copper-mediated oxidative cross-coupling. *J Org Chem* 2011;76:1174–1176.

21. Chu L, Qing F-L. Copper-mediated aerobic oxidative trifluoromethylation of terminal alkynes with Me_3SiCF_3. *J Am Chem Soc* 2010;132:7262–7263.
22. Mu X, Chen S, Zhen X, Liu G. Palladium-catalyzed oxidative trifluoromethylation of indoles at room temperature. *Chem Eur J* 2011;17:6039–6042.
23. (a) Urata H, Fuchikami T. A novel and convenient method for trifluoromethylation of organic halides using $CF_3SiR'3$/KF/Cu(I) system. *Tetrahedron Lett* 1991;32:91–94. (b) Oishi M, Kondo H, Amii H. Aromatic trifluoromethylation catalytic in copper. *Chem Commun* 2009:1909–1911. (c) Cho DJ, Senecal TD, Kinzel T, Zhang Y, Watson DA, Buchwald SL. The palladium-catalyzed trifluoromethylation of aryl chlorides. *Science* 2010;328:1679–1681.
24. Yagupol'skii LM, Marenets MS. Phenyl trifluoromethyl sulfides and phenyl trifluoromethyl sulfones with substituents in the p-position. *Russ J Gen Chem* 1954;24:885–891.
25. Kremsner JM, Rack M, Pilger C, Kappe CO. Microwave-assisted aliphatic fluorinechlorine exchange using triethylamine trihydrofluoride (TREAT-HF). *Tetrahedron Lett* 2009;50:3665–3668.
26. Chen Q-Y, Duan J-X. Direct trifluoromethylthiolation of aryl halides using methyl fluorosulfonyldifluoroacetate and sulfur. *J Chem Soc, Chem Commun* 1993:918–919.
27. (a) Hu J. Ph.D. Thesis, Investigations in selective fluorinations: novel synthetic methodologies and material syntheses. University of Southern California, Los Angeles, USA, 2002.
28. Kolomeitsev AA, Movchun VN, Kondratenko NV, Yagupolski YL. A convenient route to aryl trifluoromethyl sulfones by fluoride-catalyzed cross-coupling of arenesulfonyl fluorides with (trifluoromethyl) trimethylsilane and (trifluoromethyl) trimethylstannane. *Synthesis* 1990:1151–1152.
29. Singh RP, Cao G, Kirchmeier RL, Shreeve JM. Cesium fluoride catalyzed trifluoromethylation of esters, aldehydes and ketones with (trifluoromethyl) trimethylsilane. *J Org Chem* 1999;64:2873–2876.
30. Prakash GKS, Hu J, Olah GA. Alkoxide- and hydroxide-induced nucleophilic trifluoromethylation using trifluoromethyl sulfone or sulfoxide. *Org Lett* 2003;5:3253–3256.
31. Zhao Y, Zhu J, Ni C, Hu J. Magnesium metal-mediated reductive trifluoromethylation of aldehydes with phenyl trifluoromethyl sulfone. *Synthesis* 2010:1899–1904.
32. Hine J, Porter JJ. Methylene derivatives as intermediates in polar reactions. VIII. Difluoromethylene in the reaction of chlorodifluoromethane with sodium methoxide. *J Am Chem Soc* 1957;79:5493–5496.
33. Hine J, Porter JJ. The formation of difluoromethylene from difluoromethyl phenyl sulfone and sodium methoxide. *J Am Chem Soc* 1960;82:6178–6181.
34. Chen Q-Y, Wu S-W. Perfluoro- and polyfluorosulfonic acids. 21. Synthesis of difluoromethyl esters using fluorosulfonyldifluoroacetic acid as a difluorocarbene precursor. *J Org Chem* 1989;54:3023–3027.
35. Walkowiak J, Campo T, Ameduri B, Gouverneur V. Syntheses of mono-, di-, and trifluorinated styrenic monomers. *Synthesis* 2010:1883–1890.
36. Stahly GP. Nucleophilic addition of difluoromethyl phenyl sulfone to aldehydes and various transformations of the resulting alcohols. *J Fluorine Chem* 1989;43:53–66.

37. Ni C, Wang F, Hu J. Enantioselective nucleophilic difluoromethylation of aromatic aldehydes with Me3SiCF2SO2Ph and $PhSO_2CF_2H$ reagents catalyzed by chiral quaternary ammonium salts. *Beilstein J Org Chem* 2008;4(21).

38. Ni C, Zhang L, Hu J. Synthesis of fluorinated β-ketosulfones and gem-disulfones by nucleophilic fluoroalkylation of esters and sulfinates with di- and monofluoromethyl sulfones. *J Org Chem* 2009;74:3767–3771.

39. (a) Prakash GKS, Hu J, Wang Y, Olah GA. Difluoromethyl phenyl sulfone, a difluoromethylidene equivalent: Use in the synthesis of 1,1-difluoro-1-alkenes. *Angew Chem, Int Ed* 2004;43:5203–5206. (b) Prakash GKS, Hu J, Wang Y, Olah GA. Nucleophilic difluoromethylation of primary alkyl halides using difluoromethyl phenyl sulfone as a difluoromethyl anion equivalent. *Org Lett* 2004;6:4315–4317.

40. (a) Li Y, Hu J. Facile synthesis of chiral alpha-difluoromethyl amines from N-(tert-butylsulfinyl) aldimines. *Angew Chem, Int Ed* 2005;44:5882–5886. (b) Liu J, Li Y, Hu J. Stereoselective synthesis of di- and monofluoromethylated vicinal ethylenediamines with di- and monofluoromethyl sulfones. *J Org Chem* 2007;72:3119–3121.

41. Liu J, Hu J. Highly diastereoselective synthesis of alpha-difluoromethyl amines from N-tert-butylsulfinyl ketimines and difluoromethyl phenyl sulfone. *Chem Eur J* 2010;16:11443–11454.

42. Ni C, Liu J, Zhang L, Hu J. A remarkably efficient fluoroalkylation of cyclic sulfates and sulfamidates with $PhSO_2CF_2H$: Facile entry into beta-difluoromethylated or beta-difluoromethylenated alcohols and amines. *Angew Chem, Int Ed* 2007;46:786–789.

43. Prakash GKS, Hu J, Mathew T, Olah GA. Difluoromethyl phenyl sulfone as a selective difluoromethylene dianion equivalent: One-pot stereoselective synthesis of anti-2,2-difluoropropane-1,3-diols. *Angew Chem, Int Ed* 2003;42:5216–5219.

44. (a) Ni C, Hu J. Nucleophilic difluoromethylation of carbonyl compounds using $TMSCF_2SO_2Ph$ and Mg-0-mediated desulfonylation. *Tetrahedron Lett* 2005;46:8273–8277. (b) Liu J, Ni C, Wang F, Hu J. Stereoselective synthesis of alpha-difluoromethyl-beta-amino alcohols via nucleophilic difluoromethylation with Me(3)SiCF(2)SO(2)Ph. *Tetrahedron Lett* 2008;49:1605–1608.

45. Prakash GKS, Hu J, Wang Y, Olah GA. Convenient synthesis of difluoromethyl alcohols from both enolizable and non-enolizable carbonyl compounds with difluoromethyl phenyl sulfone. *Eur J Org Chem* 2005:2218–2223.

46. Zhu L, Li Y, Zhao Y, Hu J. Nucleophilic (phenylsulfonyl)difluoromethylation of alkyl halides using $PhSO_2CF_2SiMe_3$: Preparation of gem-difluoroalkenes and trifluoromethyl compounds. *Tetrahedron Lett* 2010;51:6150–6152.

47. (a) Eisenberger P, Gischig S, Togni A. Novel 10-I-3 hypervalent iodine-based compounds for electrophilic trifluoromethylation. *Chem Eur J* 2006;12:2579–2586. (b) Kieltsch I, Eisenberger P, Togni A. Mild electrophilic trifluoromethy-lation of carbon- and sulfur-centered nucleophiles by a hypervalent iodine(III)-CF_3 reagent. *Angew Chem, Int Ed* 2007;46:754–757. (c) Eisenberger P, Kieltsch I, Armanino N, Togni A. Mild electrophilic trifluoromethylation of secondary and primary aryl- and alkylphosphines using hypervalent iodine(III)-CF_3 reagents. *Chem Commun* 2008;1575–1577.

48. Zhang W, Zhu J, Hu J. Electrophilic (phenylsulfonyl)difluoromethylation of thiols with a hypervalent iodine(III)-CF_2SO_2Ph reagent. *Tetrahedron Lett* 2008;49:5006–5008.

49. Zhu J, Wang F, Huang W, Zhao Y, Ye W, Hu J. Copper-mediated fluoroalkylation reactions with [(phenylsulfonyl)-difluoromethyl]trimethylsilane: Synthesis of $PhSO_2CF_2^-$ containing allenes and alkynes. *Synlett* 2011:899–902.

50. McCarthy JR, Matthews DP, Paolini JP. Reaction of sulfoxides with diethylaminosulfur trifluoride: fluoromethyl phenyl sulfone, a reagent for the synthesis of fluoroalkenes. *Org Synth* 1995;72:209–215.

51. More KM, Wemple J. The synthesis of aryl fluoromethyl sulfoxides. *Synthesis* 1977;791–792.

52. (a) McCarthy JR, Peet NP, LeTourneau ME, Inbasekaran M. (Diethylamino) sulfur trifluoride in organic synthesis. 2. The transformation of sulfoxides to .alpha.-fluoro thioethers. *J Am Chem Soc* 1985;107:735–737. (b) Inbasekaran M, Peet NP, McCarthy JR, LeTourneau ME. A novel and efficient synthesis of fluoromethyl phenyl sulphone and its use as a fluoromethyl Wittig equivalent. *Chem Commun* 1985;678–679.

53. L'Heureux A, Beaulieu F, Bennett C, Bill DR, Clayton S, LaFlamme F, Mirmehrabi M, Tadayon S, Tovell D, Couturier M. Aminodifluorosulfiniumsalts: Selective fluorination reagents with enhanced thermal stability and ease of handling. *J Org Chem* 2010;75:3401–3411.

54. Robins MJ, Wnuk SF. Nucleic acid related compounds. 79. Efficient conversions of thioethers to .alpha.-fluoro thioethers with DAST or DAST/antimony(III) chloride. *J Org Chem* 1993;58:3800–3801.

55. Zhang W, Zhu L, Hu J. Electrophilic monofluoromethylation of O-, S-, and N-nucleophiles with chlorofluoromethane. *Tetrahedron* 2007;63:10569–10575.

56. (a) McCarthy JR, Matthews DP, Edwards ML, Stemerick DM, Jarvi ET. A new route to vinyl fluorides. *Tetrahedron Lett* 1990;31:5449–5452. (b) McCarthy JR, Matthews DP, Stemerick DM, Huber EW, Bey P, Lippert BJ, Snyder RD, Sunkara PS. Stereospecific method to (E) and (Z) terminal fluoroolefins and its application to the synthesis of 2′-deoxy-2′-fluoromethylenenucleosides as potential inhibitors of ribonucleoside diphosphate reductase. *J Am Chem Soc* 1991;113:7439–7440.

57. (a) Hu J, Zhang W, Wang F. Selective difluoromethylation and monofluoromethylation reactions. *Chem Commun* 2009:7465–7478. (b) Ni C, Hu J. Selective nucleophilic fluoroalkylations facilitated by removable activation groups. *Synlett* 2011:770–782.

58. Li Y, Ni C, Liu J, Zhang L, Zheng J, Zhu L, Hu J. Stereoselective nucleophilic monofluoromethylation of N-(tert-butanesulfinyl)imines with fluoromethyl phenyl sulfone. *Org Lett* 2006;8:1693–1696.

59. Liu J, Zhang L, Hu J. Stereoselective monofluoromethylation of N-tert-butylsulfinyl ketimines using pregenerated fluoro(phenylsulfonyl) methyl anion. *Org Lett* 2008;10:5377–5380.

60. Fukuzumi T, Shibata N, Sugiura M, Yasui H, Nakamura S, Toru T. Fluoro-bis (phenylsulfonyl) methane: As a fluoromethide equivalent and palladium-catalyzed enantioselective allylic monofluoromethylation. *Angew Chem, Int Ed* 2006;45:4973–4977.

61. Ni C, Li Y, Hu J. Nucleophilic fluoroalkylation of epoxides with fluorinated sulfones. *J Org Chem* 2006;71:6829–6833.

62. Prakash GKS, Wang F, Ni C, Thomas TJ, Olah GA. Efficient synthesis of α-(fluoro/chloro/methoxy) disulfonylmethane derivatives as tunable substituted methyl synthons via a new CS bond forming strategy. *J Fluorine Chem* 2010;131:1007–1012.

63. Nagura H, Fuchigami T. Regioselective electrochemical monofluorination of α-sulfonyl sulfides. *Synlett* 2008:1714–1718.

64. Ni C, Zhang L, Hu J. Nucleophilic fluoroalkylation of alpha,beta-enones, arynes, and activated alkynes with fluorinated sulfones: Probing the hard/soft nature of fluorinated carbanions. *J Org Chem* 2008;73:5699–5713.

65. Prakash GKS, Wang F, Shao N, Mathew T, Rasul G, Haiges R, Stewart T, Olah GA. A persistent α-fluorocarbanion and its analogues: Preparation,characterization, and computational study. *Angew Chem, Int Ed* 2009;48:5358–5363.

66. (a) Liu W-B, Zheng S-C, He H, Zhao X-M, Dai L-X, You S-L. Iridium-catalyzed regio- and enantioselective allylic alkylation of fluorobis (phenylsulfonyl) methane. *Chem Commun* 2009:6604–6606. (b) Zhao X, Liu D, Zheng S, Gao N. Highly regioselective Pd-catalyzed allylic alkylation of fluorobis (phenylsulfonyl)methane. *Tetrahedron Lett* 2011;52:665–667.

67. Prakash GKS, Chacko S, Alconcel S, Stewart T, Mathew T, Olah GA. Stereoselective monofluoromethylation of primary and secondary alcohols using a fluorocarbon nucleophile in Mitsunobu reaction. *Angew Chem, Int Ed* 2007;46:4933–4936.

68. (a) Prakash GKS, Zhao X, Chacko S, Wang F, Vaghoo H, Olah GA. Efficient 1,4-addition of α-substituted fluoro(phenylsulfonyl)methane derivatives to α,β-unsaturated compounds. *Beilstein J Org Chem* 2008:4(17). (b) Moon HW, Cho MJ, Kim DY. Enantioselective conjugate addition of fluorobis(phenylsulfonyl)methane to α,β-unsaturated ketones catalyzed by chiral bifunctional organocatalysts. *Tetrahedron Lett* 2009;50:4896–4898. (c) Alba A-N, Companyó X, Moyano A, Rios R. Formal highly enantioselective organocatalytic addition of fluoromethyl anion to α,β-unsaturated aldehydes. *Chem Eur J* 2009;15:7035–7038. (d) Zhang S, Zhang Y, Ji Y, Li H, Wang W. Catalytic enantioselective conjugate addition of fluorobis (phenylsulfonyl) methane to enals: Synthesis of chiral monofluoromethyl compounds. *Chem Commun* 2009;4886–4888.

69. Mizuta S, Shibata N, Goto Y, Furukawa T, Nakamura S, Toru T. Cinchona alkaloid-catalyzed enantioselective monofluoromethylation reaction based on fluorobis (phenylsulfonyl) methane chemistry combined with a mannich-type reaction. *J Am Chem Soc* 2007;129:6394–6395.

70. Shen X, Zhang L, Zhao Y, Zhu L, Li G, Hu J. Nucleophilic fluoromethylation of aldehydes with fluorobis (phenylsulfonyl) methane: The importance of strong LiO coordination and fluorine substitution for C-C bond formation. *Angew Chem, Int Ed* 2011;50:2588–2592.

71. Yang W, Wei X, Pan Y, Lee R, Zhu B, Liu H, Yan L, Huang K-W, Jiang Z, Tan C-H. Highly enantio- and diastereoselective synthesis of β-methyl-γ-monofluoromethyl-substituted alcohols. *Chem Eur J* 2011;17:8066–8070.

72. (a) Prakash GKS, Chacko S. Novel nucleophilic and electrophilic fluoroalkylation methods. *Curr Opin Drug Discovery Dev* 2008;11:793–802. (b) Valero G, Companyo X, Rios R. Enantioselective organocatalytic synthesis of fluorinated molecules. *Chem Eur J* 2011;17:2018–2037.

73. Prakash GKS, Ledneczki I, Chacko S, Ravi S, Olah GA. Stereoselective synthesis of fluorobis (phenylsulfonyl) methyl-substituted alkenes using free radical fluoroalkylation. *J Fluorine Chem* 2008;129:1036–1040.

74. Prakash GKS, Weber C, Chacko S, Olah GA. New electrophilic difluoromethylating reagent. *Org Lett* 2007;9:1863–1866.

75. (a) Shibata N, Matsnev A, Cahard D. Shelf-stable electrophilic trifluoromethylating reagents: A brief historical perspective. *Beilstein J Org Chem* 2010;6(65).

76. Prakash GKS, Weber C, Chacko S, Olah GA. New solid-phase bound electrophilic difluoromethylating reagent. *J Comb Chem* 2007;9:920–923.

77. Prakash GKS, Ledneczki I, Chacko S, Olah GA. Direct electrophilic monofluoromethylation. *Org Lett* 2008;10:557–560.

CHAPTER 28

Pentafluoroethyl Lithium: Reactions with Carbonyl Compounds and Epoxides

OLESYA KAZAKOVA AND GERD-VOLKER RÖSCHENTHALER

The introduction of fluorine-containing moieties such as CF_3 and/or C_2F_5 into organic molecules has recently found much interest, because their physical, chemical, and more important biological properties can be changed remarkably [1]. In modern medicine, fluorine substituents have gained special attention because they may affect either the drug metabolic stability or the binding affinity of protein–drug complexes [2]. Among this, electronic withdrawal properties of fluorinated fragments have also been found to be applicable for inorganic chemists. Due to the chemical inertness as well as strong π-acceptor ability, useful properties have been developed in coordination chemistry for such C_2F_5-containing compounds [3]. This capability of fluorinated derivatives, for instance of secondary and tertiary alcohols, has recently been used toward the formation of novel electrolytes for lithium-ion batteries [4].

Despite the significance of these compounds (e.g., alcohols, ketones), their syntheses are mostly based on a nucleophilic (perfluoro)alkylation reaction (RMgX and/or RLi) with the corresponding carbonyl compounds [5]. To the best of our knowledge, the usefulness of perfluorinated organolithium nucleophiles has grown strongly. However, to obtain such a nucleophile as C_2F_5Li, C_2F_5I was predominantly treated with MeLi in dry THF [6]. We have recently discovered a straightforward formation of C_2F_5Li via a simple deprotonation reaction of pentafluoroethane using *n*-butyllithium (*n*-BuLi) in Et_2O at -78 °C [5a].

At this temperature, the respective carbanion can be stored for several hours without decomposition. We have found an efficient, convenient, improved, and environmentally friendly synthetic approach to afford novel alcohols and ketones bearing the pentafluoroethyl group.

Efficient Preparations of Fluorine Compounds, First Edition. Edited by Herbert W. Roesky.

28.1 SYNTHESIS OF POLYFLUORINATED COMPOUNDS VIA C_2F_5Li [7]

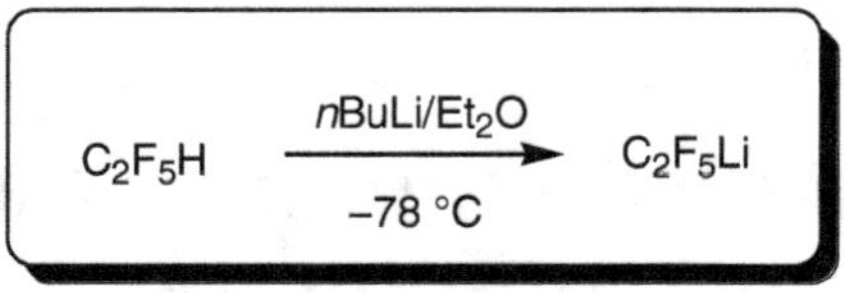

Synthesis of 2,2,3,3,3-Pentafluoro-1-phenylpropan-1-ol

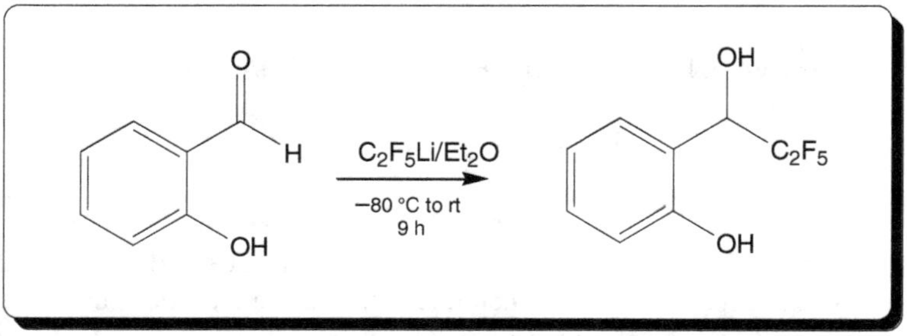

Apparatus Three-necked, 250-mL flask equipped with an internal thermometer, a dropping funnel, and a magnetic stirrer, N_2 inlet, Dewar flask (500 mL), safety glasses, laboratory coat, protective gloves.

Chemicals 2.5 M *n*-BuLi/hexane solution, C_2F_2H, dry ether, 2-hydroxybenzaldehyde, 6 N HCl, ether, $MgSO_4$, liquid N_2, ethanol for cooling bath.

Attention! Safety glasses and protective gloves must be used at all times.

Caution! Alkyllithium compounds are stored under inert gas to prevent loss of activity and for reasons of safety. *n*-BuLi reacts violently with water.

Experimental Procedure A round-bottomed flask equipped with an internal thermometer, a dropping funnel, and a magnetic stirrer was cooled to −85 °C (cooling bath) and a pentafluoroethane (1.2 g, 10 mmol) via a gas inlet tube was condensed into a cooled Et_2O (100 mL). 2.5 M *n*-BuLi/hexane solution (4.4 mL, 11 mmol) was then added drop by drop at −90 °C. The solution was kept at this temperature for an additional hour. 2-Hydroxybenzaldehyde (1.2 g, 10 mmol) was slowly added afterward and kept for 2–3 h. The solution was then left to reach ambient temperature and hydrolyzed with 6 N HCl. The aqueous layer was washed with Et_2O and dried over $MgSO_4$. The crude product was distilled under reduced pressure to give pure 2,2,3,3,3-pentafluoro-1-phenylpropan-1-ol: yellow liquid (92% yield).

Characterization Data bp 65 °C (0.1 mbar).

^{1}H NMR (200 MHz, $CDCl_3$): δ 4.35 (br s, 1H, OH), 5.62 (dd, $J = 12.8$ Hz, $J = 3.0$ Hz, 1H), 6.71–7.02 (m, 5H) ppm. ^{13}C NMR (50 MHz): δ 82.0 (t, $J_{C\text{-}F} = 29.8$ Hz), 122.2 (tq, $^1J_{C\text{-}F} = 286.8$ Hz, $J_{C\text{-}F} = 35.1$ Hz), 123.4, 124.9, 125.0

(qt, $J_{C\text{–}F} = 277.4$ Hz, $J_{C\text{–}F} = 34.9$ Hz), 128.1, 128.9, 129.1, 134.5 ppm. ^{19}F NMR (188 MHz, $CDCl_3$): δ −82.0 (s, 3F), −120.0 (d, $J_{F\text{–}F(A)} = 254.6$ Hz, 1F), −132.0 (d, $J_{F\text{–}F(B)} = 254.6$ Hz 1F) ppm. MS (EI, 70 eV, 200 °C) *m*/*z*: 107 (100%) $[M - C_2F_5]^+$, 225.1 $[M]^-$ ($C_9H_6F_5O$).

Synthesis of 5,5,6,6,6-Pentafluorohexane-3-ol [7]

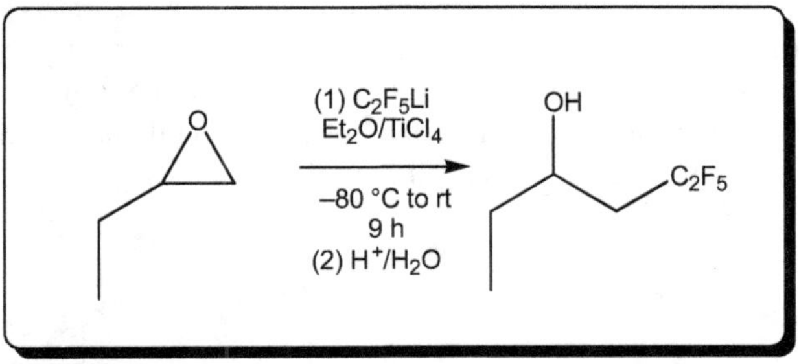

Apparatus Three-necked, 1-L flask equipped with an internal thermometer, a dropping funnel and a magnetic stirrer, N_2 inlet, Dewar flask (2 L), safety glasses, laboratory coat, protective gloves.

Chemicals *n*-BuLi/hexane solution (2.5 M), C_2F_2H, dry ether, 2-ethyloxirane, $TiCl_4$, 5% aqueous HCl, ether, aqueous $NaHCO_3$, $MgSO_4$, liquid N_2, ethanol for cooling bath.

Attention! Safety glasses and protective gloves must be used at all times.

Caution! Alkyllithium compounds are stored under inert gas to prevent loss of activity and for reasons of safety. *n*-BuLi reacts violently with water. Ethyloxirane is a probable human carcinogen. Hazards posed by titanium tetrachloride generally arise from the release of HCl. $TiCl_4$ is a strong Lewis acid, exothermically forming adducts with even weak bases such as THF and explosively with water, releasing HCl.

Experimental Procedure A round-bottomed flask equipped with an internal thermometer, a dropping funnel, and a magnetic stirrer was cooled to −85 °C (cooling bath) and pentafluoroethane (4.8 g, 40 mmol) via a gas inlet tube was condensed into a cooled Et_2O (400 mL). 2.5 M *n*-BuLi/hexane (17.6 mL, 44 mmol) was then added drop by drop at −90 °C. The solution was kept at this temperature for an additional hour.

Ethyloxirane (1.44 g, 20 mmol) was charged to a solution of pentafluoroethyl lithium (40 mmol) at −78 °C in dry diethyl ether. Then, titanium tetrachloride (3.8 g, 20 mmol) was slowly added under vigorous stirring. After 1 h, the reaction mixture was quenched with 5% HCl and extracted with diethyl ether and an aqueous solution of $NaHCO_3$. The combined organic phases were dried over $MgSO_4$. After the filtration of solids, the residue was concentrated and the crude product distilled under reduced pressure to give a pure 5,5,6,6,6-pentafluorohexane-3-ol colorless oil (73% yield).

Characterization Data ^{1}H NMR (200 MHz, $CDCl_3$): δ 0.99 (t, 3H, $J_{H\text{-}H} = 7.4$ Hz, CH_3), 1.60 (dq, 2H, $J_{H\text{-}H} = 7.4$ Hz, $J_{H\text{-}H} = 7.4$ Hz, CH_2), 1.90 (s, 1H, OH), 2.09–2.31 (m, 2H, CH_2CF_2), 4.02–4.1 (m, 1H, CH) ppm. ^{13}C NMR (50 MHz): δ 9.5, 30.5, 37.4, 66.6, 115.7 (tq, $J_{C\text{-}F} = 253$ Hz, $J_{C\text{-}F} = 37.8$ Hz),118.8 (qt, $J_{C\text{-}F} = 285$ Hz, $J_{C\text{-}F} = 35.7$ Hz) ppm. ^{19}F NMR (188 MHz): δ −87.4 (s, 3F, CF_3), −118.3 (m, 2F, CF_2) ppm. MS (CI positive, NH_3, 200 °C) *m/z* (%): 210 (34) $[M + NH_4]^+$, 192 (21) $[M]^+$ ($C_6H_9F_5O$), 163 (10) $[M - C_2H_5]^+$.

Application Due to chemical inertness as well as strong π-acceptor ability, useful properties have been developed in coordination chemistry for such C_2F_5-containing compounds. In other words, this capability of fluorinated derivatives, for instance secondary and tertiary alcohols, has recently been used toward the formation of novel electrolytes for lithium-ion batteries.

REFERENCES

1. (a) Welch TJ. *Tetrahedron* 1987;43:3123–3197. (b) Banks RE, Smart BE, Tatlow JC. *Organofluorine Chemistry: Principles and Commercial Applications*. New York: Plenum Press; 1994. (c) Hiyama T. *Organofluorine Compounds: Chemistry and Properties*. Berlin: Springer; 2000.
2. (a) Angelastro MR, Baugh LE, Bey P, Burkhart JP, Chen T-M, Durham SL, Hare CM, Huber EW, Janusz MJ, Koehl JR, Marquart AL, Mehdi S, Peet NP. Inhibition of human neutrophil elastase with peptidyl electrophilic ketones. 2. Orally active PG-Val-Pro-Val pentafluoroethyl ketones. *J Med Chem* 1994;37:4538–4553. (b) Leung D, Abbenante G, Fairlie DP. protease inhibitors: current status and future prospects. *J Med Chem* 2000;43:305–341.
3. (a) Ernst MF, Roddick DM. Synthesis and coordination properties of bis(bis(pentafluoroethyl)phosphino)ethane. *Inorg Chem* 1989;28:1624–1627. (b) Bejan D, Willner H, Ignatiev N, Lehmann WC. Synthesis and characterization of bis[bis(pentafluoroethyl) phosphinyl]imides, $M^+N[(C_2F_5)_2P(O)]_2^-$, M = H, Na, K, Cs, Ag, Me_4N. *Inorg Chem* 2008;47:9085–9089.
4. (a) Yoshimura S, Noma T, Yonezu I, inventors; Sanyo Electric Co. Ltd, assignee. Solute for nonaqueous electrolyte battery and nonaqueous electrolyte battery. Patent JP2000243437 A. 2000 Sept 8. (b) Nolan BG, Strauss SH. Nonaqueous lithium battery electrolytes based on bis(polyfluorodiolato)borates. *J Electrochem Soc* 2003;150:A1726–A1734. (c) Tsujioka S, Nolan BG, Takase H, Fauber BP, Strauss SH. Conductivities and electrochemical stabilities of lithium salts of polyfluoroalkoxyaluminate superweak anions. *J Electrochem Soc* 2004;151:A1418–A1423. (d) Barbarich TJ, Handy ST, Miller SM, Anderson OP, Grieco PA, Strauss SH. $LiAl(OC(Ph)(CF_3)_2)_4$: a hydrocarbon-soluble catalyst for carbon−carbon bond-forming reactions. *Organometallics* 1996;15:3776–3778. (e) Krossing I, Van Wüllen L. Superweak complexes of tetrahedral P4 molecules with the silver cation of weakly coordinating anions. *Chem Eur J* 2002;8:700–711. (f) Ivanova SM, Nolan BG, Kobayashi Y, Miller SM, Anderson OP, Strauss SH. Relative lewis basicities of six $Al(ORF)_4^-$ superweak anions and the structures of $LiAl\{OCH(CF_3)_2\}_4$ and [1-Et-3-Me-1,3-$C_3H_3N_2$][$Li\{Al\{OCH(CF_3)_2\}_4\}_2$]. *Chem Eur J* 2001;7:503–510.
5. (a) Marhold A, Henrich M, Kolomeitsev A, Barten J, Kadyrov A, Röschenthaler G-V, inventors; Bayer AG, assignee. Process for preparing polyfluoroalkyllithium compounds.

US patent 20030065135. 3 Apr 2003. (b) Königsmann HM, Lork E, Odinets IL, Shevchenko N, Nenajdenko V, Röschenthaler G-V. α-Pentafluoroethylated amines: a new synthetic approach. *Mendeleev Commun* 2006;16:141–143. (c) Shevchenko NE, Nenajdenko VG, Röschenthaler G-V. New method of preparation of C_2F_5Li and its reactions with cyclic imines and lactims: synthesis of α-pentafluoroethyl proline. *J Fluorine Chem* 2008;129:390–396. (d) Shevchenko NE, Balenkova SE, Röschentaler G-V. Practical synthesis of α-perfluoroalkyl cyclic imines and amines. *Synthesis* 2010;41:120–126. (e) Chernega AN, Kolomeitsev AA, Yagupolskij YL, Gentzsch A, Röschenthaler G-V. Molecular structures and reactivity of trifluoromethyltris(dialkylamino)phosphonium bromides. *J Fluorine Chem* 1994;70:271–275. (f) Sosnovskikh VYa, Sevenard DV, Usachev BI, Röschenthaler G-V. The first example of a preparative 1,4-perfluoroalkylation using (perfluoroalkyl)trimethylsilanes. *Tetrahedron Lett* 2003;44:2097–2099. (g) Sosnovskikh VYa, Usachev BI, Sevenard DV, Röschenthaler G-V. Regioselective nucleophilic 1,4-trifluoromethylation of 2-polyfluoroalkylchromones with (trifluoromethyl)trimethylsilane. Synthesis of fluorinated analogs of natural 2,2-dimethylchroman-4-ones and 2,2-dimethylchromenes. *J Org Chem* 2003;68:7747–7754.

6. (a) Gassman GP, O'Reilly JN. Nucleophilic addition of the pentafluoroethyl group to aldehydes, ketones, and esters. *J Org Chem* 1987;52:2481–2490 (b) Pooput C, Dolbier WR, Jr, Medebielle M. Nucleophilic perfluoroalkylation of aldehydes, ketones, imines, disulfides, and diselenides. *J Org Chem* 2006;71:3564–3568.
7. Duda B, Kalinovich N, Königsmann MH, Röschenthaler G-V. Pentafluoroethyllithium: reactions with carbonyl compounds and epoxides. *Tetrahedron* 2012, submitted.

CHAPTER 29

Synthesis of γ-Hydroxy-α,α-difluoromethylenephosphonates

OLESYA KAZAKOVA AND GERD-VOLKER RÖSCHENTHALER

Lewis acid-mediated cleavage of epoxides with various nucleophiles is an important transformation in organic synthesis [1]. The popularity of oxiranes as versatile intermediates in organic preparations generally emanates from their accessibility, reactivity, and the stereochemical predictability of their reactions. Despite, however, the numerous transformations already reviewed, little attention has been given to the opening of epoxides containing a difluoromethylenephosphonate moiety [2]. Moreover, very few examples are known in the literature, where a phosphorus- and fluorine-containing nucleophile has been used in these reactions [3].

In contrast to non-fluorinated hydroxyl-methylenephosphonates, the preparation of which has been studied largely by oxirane ring-opening reactions with phosphonomethyl organometallic reagents [4], this methodology has rarely been applied in the synthesis of fluorinated analogs [2]. Here are results on the Lewis acid-promoted ring-opening reactions of epoxides with lithium diethyl difluoromethylenephosphonate [5].

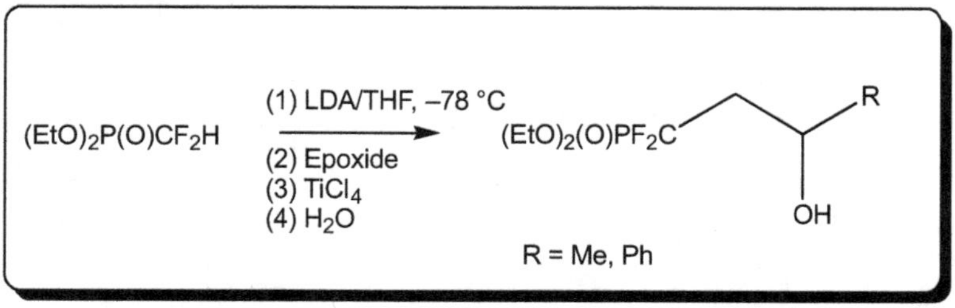

Efficient Preparations of Fluorine Compounds, First Edition. Edited by Herbert W. Roesky.

29.1 SYNTHESIS OF DIETHYL 1,1(DIFLUORO)-3-HYDROXYBUTYLPHOSPHONATE [5]

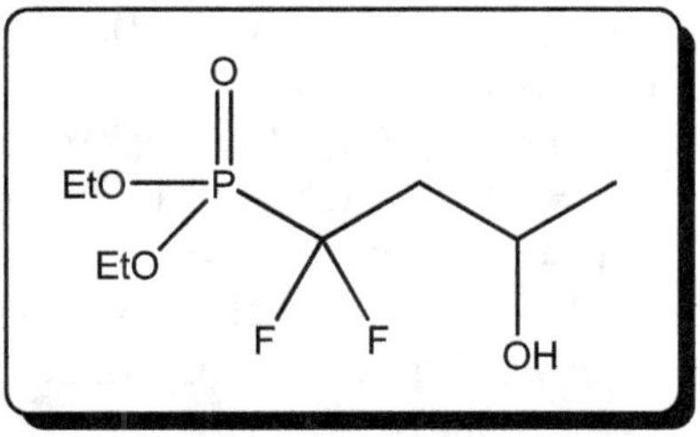

Apparatus Three-necked, 250-mL flask equipped with an internal thermometer, a dropping funnel, and a magnetic stirrer, N_2 inlet, Dewar flask (500 mL), safety glasses, laboratory coat, protective gloves.

Chemicals Lithium diisopropylamide/tetrahydrofuran (2.0 M; LDA/THF) solution, diethyl difluoromethylenephosphonate [6], dry THF, 2-methyloxirane, titanium tetrachloride ($TiCl_4$), NaCl, ether, $MgSO_4$, trichloromethane, hexane, silica gel, liquid N_2, ethanol for cooling bath.

Attention! Safety glasses and protective gloves must be used at all times.

Caution! LDA reacts violently with water. Methyloxirane is a probable human carcinogen. Hazards posed by $TiCl_4$ generally arise from the release of HCl. $TiCl_4$ is a strong Lewis acid, exothermically forming adducts with even weak bases such as THF and explosively with water, releasing HCl.

Experimental Procedure The reaction was carried out under an atmosphere of dry nitrogen. A round-bottomed flask contained dry THF equipped with an internal thermometer, a dropping funnel, and a magnetic stirrer, was cooled to −78 °C by means of a cooling bath.

Diethyl difluoromethylenephosphonate (2.0 g, 10.6 mmol) was added drop by drop to the solution of LDA (6.6 mL of 1.8 M solution, 11.8 mmol) and 60 mL of dry THF at −78 °C, and the reaction mixture was then stirred for further 1 h. After the formation of lithium difluoromethylenephosphonate was completed, appropriate amount of methyloxirane (0.63 g, 10.6 mmol) was added at −78 °C, followed by the careful addition of $TiCl_4$ (1.2 mL, 10.6 mmol). After stirring for further 2 h at −78 °C, the reaction mixture was quenched with water, and then NaCl was added until the saturation of water phase. The aqueous layer was then extracted with diethyl ether (2 × 20 mL), combined organic layers were dried with $MgSO_4$ and concentrated in vacuo. Flash column chromatography (trichloromethane/hexane, 2:1) afforded pure diethyl 1,1-difluoro-3-hydroxybutylphosphonate, slightly yellowish oil (62% yield).

Characterization Data ^{19}F NMR ($CDCl_3$): δ −107.5 (tdd, $^2J_{FbFa} = 303.7$ Hz, $^2J_{FbP} = 102.6$ Hz, $^3J_{FbH} = 15.9$ Hz, CF_b), −112.4 (tdd, $^2J_{FaFb} = 303.5$ Hz, $^2J_{FaP} = 110.3$ Hz, $^3J_{FaH} = 21.6$ Hz, CF_a) ppm. ^{31}P NMR ($CDCl_3$): δ 8.93 (dt, $^2J_{PF} = 106.2$ Hz) ppm. ^{1}H NMR ($CDCl_3$): δ 1.26 (d, 3H, CH_3), 1.39

(t, 6H, OCH_2CH_3), 2.22 (m, 2H, CF_2CH_2), 3.43 (br s, 1H, O*H*), 4.30 (m, 5H, OCH_2CH_3, C*H*OH) ppm. ^{13}C NMR ($CDCl_3$): δ 16.6 (d, OCH_2CH_3), 24.4 (s, CH_3), 45.3 (dt, $^2J_{CF} = 16.1$ Hz, $^2J_{CP} = 14.9$ Hz, CF_2CH_2), 62.6 (ddd, $^3J_{CP} =$ 7.4 Hz, $^3J_{CFa} = 3.4$ Hz, $^3J_{CFb} = 3.4$ Hz, *C*HOH), 65.8 (d, OCH_2CH_3), 121.1 (ddd, $^1J_{CFa} = 260.0$ Hz, $^1J_{CFb} = 260.0$ Hz, $^1J_{CP} = 213.1$ Hz, CF_2) ppm.

^{19}F NMR (D_2O): δ −110.9 (dddd, $^2J_{FbFa} = 302.8$ Hz, $^2J_{FbP} = 110.6$ Hz, $^3J_{FbH} = 21.3$ Hz, $^3J_{FbH} = 20.0$ Hz, CF_b), −113.3 (dddd, $^2J_{FaFb} = 303.0$ Hz, $^2J_{FaP} = 111.2$ Hz, $^3J_{FaH} = 22.0$ Hz, $^3J_{FaH} = 19.0$ Hz, CF_a) ppm. ^{31}P NMR (D_2O): δ 9.23 (t, $^2J_{PF} = 111.8$ Hz) ppm. ^{1}H NMR (D_2O): δ 1.25 (d, 3H, CH_3), 1.37 (t, 6H, OCH_2CH_3), 2.23 (m, 2H, CF_2CH_2), 3.40 (br s, 1H, O*H*), 4.32 (m, 5H, OCH_2CH_3, C*H*OH) ppm.

MS (CI^+) *m/z* (%): 510 $[2M + NH_4]^+$ (100), 247 $[M + H]^+$ (17). MS (CI^-) *m/z* (%): 245 $[M - H]^+$ (23). HRMS for $C_7H_{14}O_4PF_2$ $(M\text{-}CH_3)^+$: calcd 231.05954, found 231.05978.

Application Interest in the synthesis of γ-hydroxy-α,α-difluoromethylenephosphonates results mostly from their potential in the design of non-hydrolyzable analogs of biologically active phosphate esters. More recently, they have been described as substrates for NADH-linked *sn*-glycerol-3-phosphate dehydrogenase or as key intermediates in the synthesis of various phosphatase-resistant phosphonolipids [7].

REFERENCES

1. March J. *Advanced Organic Chemistry*. 3rd ed. New York: John Wiley; 1985.
2. Ozouf P, Binot G, Pommelet J-C, Lequeux TP. Syntheses of ω-hydroxy-α,α-difluoromethylphosphonates by oxacycle ring-opening reactions. *Org Lett* 2004;6:3747–3750. (a) Gillet JP, Sauvetre R, Normant JF. Di- and trifluorovinyllithium reagents. Synthesis 1986;5:355–360. (b) Normant JF. Synthesis of selectivity fluorinated substrates via organometallic reagents derived from $CF_2{=}CFCl$, $CF_2{=}CCl_2$, $CF_2{=}CH_2$. *J Organomet Chem* 1990;400:19–34.
3. (a) Smith JG. Synthetically useful reactions of epoxides. *Synthesis* 1984;8:629–656. (b) Tanaka H, Fukui M, Haraguchi K, Masaki M, Miyasaka T. Cleavage of a nucleosidic oxetane with carbanions: synthesis of a highly promising candidate for anti-HIV agents — a phosphonate isostere of AZT 5′-phosphate. *Tetrahedron Lett* 1989;30:2567–2570. (c) Li Z, Racha S, Dan L, El-Subbagh H, Abushanab E. A general and facile synthesis of β-and γ-hydroxy phosphonates from epoxides. *J Org Chem* 1993;58:5779–5783. (d) Azuhata T, Okamoto Y. Synthesis of diethyl 2-(trimethylsiloxy)-alkanephosphonates from epoxides and diethyl trimethylsilyl phosphite. *Synthesis* 1983;11:916–917. (e) Azuhata T, Okamoto Y. Synthesis of dialkyl 2-(dialkoxyphosphinyloxy)-alkanephosphonates. *Synthesis* 1984;5:417–424.
4. Montchamp JL, Migaud ME, Frost JW. Phosphonomethylation of cyclohexene oxides. *J Org Chem* 1993;58:7679–7684 and references cited therein.
5. Pajkert R, Kolomeitsev AA, Milewska M, Röschenthaler G-V, Koroniak H. $TiCl_4$ and Grignard reagent-promoted ring-opening reactions of various epoxides: synthesis of γ-hydroxy-α,α-difluoromethylenephosphonates. *Tetrahedron Lett* 2008;49:6046–6049.

6. Diethyl difluoromethylenephosphonate was prepared according to the procedure presented by Soborovskii and Baina (Soborovskii LZ, Baina NF. *J Gen Chem USSR (Engl Trans)* 1959;29:1115–1118): To the solution of diethyl phosphite (25.0 g, 0.18 mol) in dry THF (80 mL), Na metal (4.2 g, 0.18 mol) was added in portions. The formation of the salt $(EtO)_2P(O)Na$ was controlled by ^{31}P NMR, usually this process took ca. 24 h. Then, the flask was equipped with a cold-finger condenser (cooling mixture −80 °C). Freon 22 ($CHClF_2$) (19.0 g, 0.22 mol) was slowly bubbled through the reaction mixture while stirring and warming up to 30–35 °C (oil bath). The reaction mixture becomes more and more turbid. Afterward, all volatiles were condensed off in vacuo (0.04 Torr), in a cooled liquid nitrogen trap. From this mixture, THF was distilled off using Vigreux column (25 cm). The crude diethyl difluoromethylenephosphonate was distilled in vacuo (14 Torr) to afford pure product as colorless liquid (27.8 g, 82% yield). Bp 90 °C/14 Torr. ^{1}H NMR ($CDCl_3$): δ 1.38 (6H, m, OCH_2CH_3), 4.28 (4H, m, OCH_2CH_3), 5.88 (1H, dt, $^2J_{HP} = 26.9$ Hz, $^2J_{HF} = 49.0$ Hz, CF_2H) ppm. ^{19}F NMR ($CDCl_3$): δ −136.63 (dd, $^2J_{HF} = 48.3$ Hz, $^2J_{FP} = 91.4$ Hz) ppm. ^{31}P NMR (undecoupled, $CDCl_3$): δ 6.06 (dt, $^2J_{PH} = 27.2$, $^2J_{PF} = 91.4$ Hz) ppm.

7. (a) Nieschalk J, O'Hagan D. Monofluorophosphonates as phosphate mimics in bioorganic chemistry: a comparative study of CH_2-, CHF- and CF_2-phosphonate analogues of *sn*-glycerol-3-phosphate as substrates for *sn*-glycerol-3-phosphate dehydrogenase. *Chem Commun* 1995;7:719–720. (b) Nieschalk J, Batsanov AS, O'Hagan D, Howard JAK. Synthesis of monofluoro- and difluoro-methylenephosphonate analogues of *sn*-glycerol-3-phosphate as substrates for glycerol-3-phosphate dehydrogenase and the X-ray structure of the fluoromethylenephosphonate moiety. *Tetrahedron* 1996;52:165–176. (c) Chambers RD, Jaouhari R, O'Hagan D. Synthesis of a difluoromethylenephosphonate analogue of glycerol-3-phosphate. A substrate for NADH linked glycerol-3-phosphate dehydrogenase. *J Chem Soc, Chem Commun* 1988;17:1169–1170. (d) Chambers RD, Jaouhari R, O'Hagan D. Fluorine in enzyme chemistry. Part 2. The preparation of difluoromethylenephosphonate analogues of glycolytic phosphates. Approaching an isosteric and isoelectronic phosphate mimic. *Tetrahedron* 1989;45:5101–5108. (e) Moolenaar WH. Bioactive lysophospholipids and their G protein-coupled receptors. *Exp Cell Res* 1999;253:230–238. (f) Goetzl EJ, Dolezalova H, Kong Y, Hu YL, Jaffe RB, Kalli KR, Conover CA. Distinctive expression and functions of the Type 4 endothelial differentiation gene-encoded G protein-coupled receptor for lysophosphatidic acid in ovarian cancer. *Cancer Res* 1999;59:5370–5375. (g) Xu Y. Prestwich GD. Synthesis of chiral (α,α-difluoroalkyl)phosphonate analogues of (lyso)phosphatidic acid via hydrolytic kinetic resolution. *Org Lett* 2002;4:4021–4024.

CHAPTER 30

Synthesis of Pentafluoro-λ^6-sulfanyl-Substituted Acetylenes for Liquid Crystals

OLESYA KAZAKOVA, MAXIM PONOMARENKO, AND GERD-VOLKER RÖSCHENTHALER

It is a well-known fact that the introduction of a fluorine atom into organic compounds often influences the biological activity of pharmaceutical and agrochemical compounds to a considerable extent [1]. The pentafluoro-λ^6-sulfanyl (SF_5) group is one of the fluorine-containing substituents having attracted increasing interest in providing the respective compounds with low surface energy, high chemical resistance, thermal stability, high electronegativity, and lipophilicity, being important especially for agrochemical and pharmaceutical products [2]. Also a number of patents describe the properties of SF_5 derivatives as fungicides, herbicides, and insecticides [2–4]. The above-mentioned properties were often compared with those of the CF_3 group, and the advantages of the SF_5 group were hence referred to as a "super CF_3 function" [2,5]. The utility of pentafluoro-λ^6-sulfanyl-substituted compounds has been especially noted for the chemistry of polymeric products [6], energetic materials [2,7], liquid crystals (LCs) [2,3,8], and others.

The introduction of a SF_5 group into organic compounds has been known for the past 50 years, but the number of SF_5-containing compounds is limited by the availability of SF_5 sources or building blocks.

The most important and convenient method for the addition of the SF_5 group into organic compounds is based on radical addition of SF_5Hlg (Hlg = Cl, Br) to unsaturated substrates [9–14]. Such addition products are widely used for the synthesis of various SF_5-based unsaturated compounds ($SF_5CH{=}CH{-}$, $SF_5C{\equiv}CH{-}$), used in the synthesis of SF_5 aromatics [15], heterocycles [2,7,11,16], and carbocycles [9,17].

Efficient Preparations of Fluorine Compounds, First Edition. Edited by Herbert W. Roesky.

30.1 SYNTHESIS OF 4-[(*E*)-1-CHLORO-2-(PENTAFLUORO-λ6-SULFANYL)VINYL]-4′-PROPYL-1,1′-BI(CYCLOHEXYL) [17]

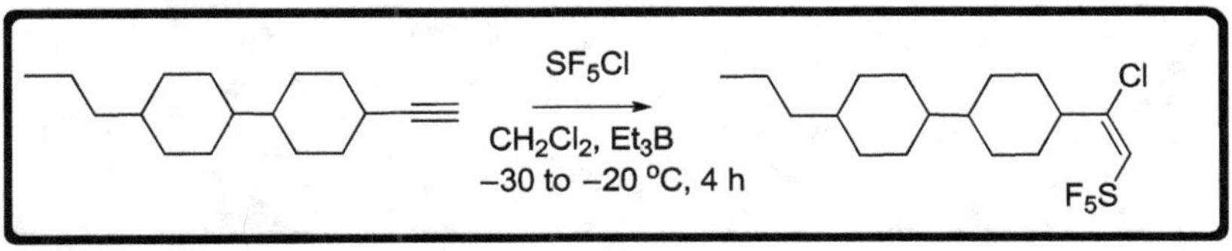

Apparatus A 250-mL, three-necked flask equipped with a dry ice reflux condenser, a nitrogen inlet, a septum and a magnetic stirrer, syringe, Dewar flask (500 mL), safety glasses, laboratory coat, protective gloves.

Chemicals SF_5Cl, dry *n*-hexane, 4-ethynyl-4′-propylbi(cyclohexane), triethylborane (1-M Et_3B/hexane solution), aqueous $NaHCO_3$ solution, $NaSO_4$, n-hexane, silica gel, ethanol for cooling bath, dry ice, liquid N_2.

Attention! Safety glasses and protective gloves must be used at all times.

Caution! SF_5Cl is toxic by inhalation. Et_3B catches fire spontaneously if exposed to air, and causes severe skin burns and eye damage. (Do not allow contact with air. Handle under inert gas.)

Experimental Procedure The reaction was carried out under an atmosphere of dry nitrogen. Into a three-necked flask equipped with a dry ice reflux condenser and a nitrogen inlet, alkyne 4-ethynyl-4′-propylbi(cyclohexane) (2.74 g, 11.79 mmol) in anhydrous *n*-hexane (40 mL) was added and cooled to −40 °C. Then SF_5Cl (4.8 g, 29.5 mmol, 2.5 equiv) was condensed into the solution. The solution was stirred at −40 °C for 5 min and Et_3B (0.1 g, 0.1 equiv, 1.2 mmol, 1 M in *n*-hexane) was added slowly using a syringe. The solution was vigorously stirred for 4 h at −30 °C to −20 °C, and then the mixture was warmed to room temperature. The mixture was hydrolyzed with aqueous $NaHCO_3$ solution and the organic layer dried over $NaSO_4$. The solvent was removed and the crude product was purified by filtration through a short column of silica gel (*n*-hexane). Removal of solvent provided crystals of 4-[(*E*)-1-chloro-2-(pentafluoro-λ6-sulfanyl)vinyl]-4′-propyl-1,1′-bi(cyclohexyl) (92% yield).

Characterization Data Colorless solid, mp 51–52 °C. ^{1}H NMR (200 MHz, $CDCl_3$): δ 0.70–1.45 (m, 17H), 1.55–1.80 (m, 9H), 3.06 (m, 1H), 6.53 (m, 1H, J_{HF} = 8.8 Hz, $HCSF_5$) ppm. ^{13}C NMR (50 MHz, $CDCl_3$): δ 14.8, 20.5, 29.3, 30.4, 30.5, 33.9, 38.0, 40.2, 42.8, 43.3, 43.7, 135.9 (quin, CSF_5, J = 20.6 Hz), 152.7 (quin, J = 7.0 Hz) ppm. ^{19}F NMR (188 MHz, $CDCl_3$): δ 67.1 (dm, 4F, J = 152.0 Hz, B_4-part), 83.1 (9 lines, 1F, A-part) ppm. MS (EI, 70 eV, 200 °C): *m/z* (%) 394 (19, M^+), 359 (3, M^+–Cl), 231 (11), 125 (81), 83 (61), 69 (100). HRMS for [M^+] ($C_{17}H_{28}ClF_5S$): calcd 394.1520, found 394.1534.

30.2 SYNTHESIS OF 4-[(PENTAFLUORO-λ^6-SULFANYL)ETHYNYL]-4′-PROPYL-1,1′-BI(CYCLOHEXYL) [17]

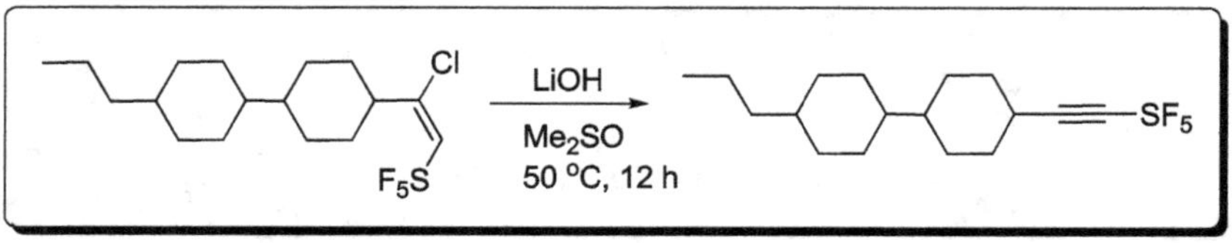

Apparatus A 100-mL flask equipped with a reflux condenser, an oil bath, and a magnetic stirrer, safety glasses, laboratory coat, protective gloves.

Chemicals 4-[(*E*)-1-chloro-2-(pentafluoro-λ^6-sulfanyl)vinyl]-4′-propyl-1,1′-bi(cyclohexyl), dimethyl sulfoxide (DMSO), LiOH, aqueous NaCl solution, *n*-hexane, $NaSO_4$, silica gel.

Attention! Safety glasses and protective gloves must be used at all times.

Caution! LiOH is toxic if swallowed, and causes severe skin burns and eye damage.

Experimental Procedure To a solution of 4-[(*E*)-1-chloro-2-(pentafluoro-λ^6-sulfanyl)vinyl]-4′-propyl-1,1′-bi(cyclohexyl) (1.5 g, 4.23 mmol) in Me_2SO (35 mL), LiOH (0.51 g, 21.15 mmol) was added. The mixture was stirred for 12 h at 50 °C, then poured into ice water (~100 mL), extracted with *n*-hexane (4 × 20 mL). The organic layers were combined, washed with brine, and dried over $NaSO_4$. The solvent was removed under reduced pressure, and then the residue was purified by chromatography (silica gel; *n*-hexane), giving alkyne 4-[(pentafluoro-λ^6-sulfanyl)ethynyl]-4′-propyl-1,1′-bi(cyclohexyl) (65% yield).

Characterization Data Colorless solid, mp 49–50 °C. ^1H NMR (200 MHz, $CDCl_3$): δ 0.70–2.15 (m, 26H), 2.29 (m, 1H) ppm. ^{13}C NMR (50 MHz, $CDCl_3$): δ 14.7, 20.4, 29.1, 29.5, 30.3, 32.1, 33.9, 38.0, 40.2, 42.6, 43.5, 82.2 (quin, CSF_5, $J = 40.8$ Hz, $J = 4.0$ Hz), 84.1 (m, $J = 7.5$ Hz) ppm. ^{19}F NMR (188 MHz, $CDCl_3$): δ 81.6–85.0 (9 lines, 1F, A-part), 88.1–89.2 (m, 4F, B_4-part) ppm. MS (EI, 70 eV, 200 °C): *m/z* (%) 358 (4, M^+), 315 (15, $M^+ - Pr$), 231 (9, $M^+ - SF_5$), 123 (33), 83 (63), 69 (100). HRMS for [M^+] ($C_{17}H_{27}F_5S$): calcd 358.1754, found 358.1752.

Application The pentafluoro-λ^6-sulfanyl (SF_5) group is one such fluorine-containing substituent that has attracted increasing interest, as it provides compounds with low surface energy, high chemical resistance, thermal stability, high electronegativity, and lipophilicity [2]. A number of patents and papers describing the synthesis and the properties of pentafluoro-λ^6-sulfanyl derivatives as new LCs with the suitable combination of physical properties have been published [3,8,18].

These unique properties, a tendency toward the formation of nematic mesophases, and the generation of large dielectric anisotropies in some terminal and substituted acetylenes have been reported in the literature. These materials have been compared with their hydrogenated analogs [19].

TABLE 30.1 The Physical Properties of the New SF_5-LCs in Comparison to Some of the Known SF_5 Analogues [17]

Structure	Phase sequence	$\Delta\varepsilon_{virt}$, Dielectric anisotropies	Δn_{virt} Birefringences
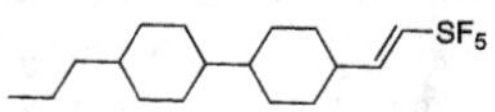	C 51 S_G? 65 I	10.5	0.0704
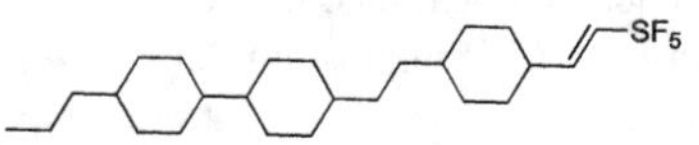	C 145 S_B 173 N 194.1 I	10.1	0.0641
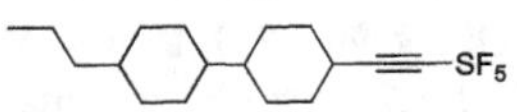	C 49 I	10.4	0.084
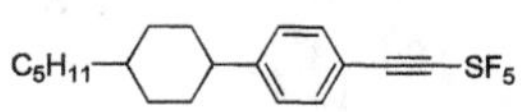	C 38 I	14.5	0.131
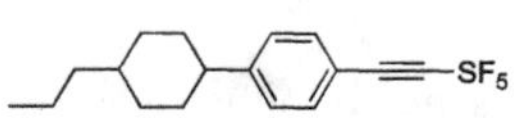	C 41 I	15.5	0.134

We have presented synthetic routes to SF_5-substituted acetylenes as novel LC molecules, based on radical addition of SF_5Cl to triple bonds of the corresponding LCs precursors. It was shown that interaction of SF_5Cl with alkynes initiated by Et_3B and followed by dehydrochlorination of the adducts is a convenient approach for preparation of such SF_5–C≡C– substituted LCs. The physical properties of the new LC were investigated. The latter showed very high dielectric anisotropies ($\Delta\varepsilon$) [17].

The physical properties [20] of the new SF_5 LCs (entries 3–5) in comparison to some of the known [21] (entries 1–2) SF_5 analogs are shown in (Table 30.1) [17].

The birefringences (Δn) of the SF_5 acetylenes, which were synthesized for the first time, are rather low and useful for reflective type LCDs. Most important are the observed high dielectric anisotropies ($\Delta\varepsilon$) of acetylenes (entries 4 and 5) compared to known LCs based on hypervalent sulfur fluorides and alkynes. Caused by the presence of the aromatic ring, the birefringence in acetylenes (entries 4 and 5) is increased when compared to acetylene (entry 3) but the polarity is also significantly higher [17].

REFERENCES

1. (a) Müller M, Faeh C, Diederich F. Fluorine in pharmaceuticals: looking beyond intuition. *Science* 2007;317:1881–1886. (b) Purser S, Moore PR, Swallow S, Gouverneur V. Fluorine in medicinal chemistry. *Chem Soc Rev* 2008;37:320–330. (c) Kirk KL. Fluorination in medicinal chemistry: methods, strategies, and recent developments. *Org Process Res Dev* 2008;12:305–321.

2. (a) Kirsch P, editor. *Modern Fluoroorganic Chemistry: Synthesis, Reactivity, Applications*. Weinheim, Germany: Wiley-VCH; 2004. (b) Synthesis of pentafluorosulfanylpyrazole and pentafluorosulfanyl-1,2,3-triazole and their derivatives as energetic materials by click chemistry Ye C, Gard GL, Winter RW, Syvret RG, Twamley B, Shreeve JM. *Org Lett* 2007;9:3841–3844. (c) Winter RW, Dodean RA, Gard GL. Fluorine-containing synthons. In: Soloshonok VA, editor. ACS Symposium Series 911. Washington, DC: ACS Press; 2005. p. 87–118. (d) Kirsch P, Röschenthaler G-V. Current fluoroorganic chemistry – new synthetic directions, technologies, materials, and biological applications. In: Soloshonok VA, Mikami K, Yamazaki T, Welch JT, Honek JF, editors. ACS Symposium Series 949. Washington: ACS Press; 2007. p. 221–243. (e) Welch JT, Lim DS. The synthesis and biological activity of pentafluorosulfanyl analogs of fluoxetine, fenfluramine, and norfenfluramine. *Bioorg Med Chem* 2007;15:6659–6666. (f) Lal GS, Syvret RG. Route to prepare 4-bromo-1-oxypentafluorosulfanylbenzene. *Chim Oggi* 2008;26:26–28. (g) Lentz D, Seppelt K. The $-SF_5$, $-SeF_5$, and $-TeF_5$ groups in organic chemistry. In: Akiba K, editor. *Chemistry of Hypervalent Compounds*. Wiley-VCH; 1999. p 295–325.
3. (a) Salmon R. Patent WO 9306089. 1993 Apr 1. (b) Howard MN, Jr, Sevenson TM. Patent WO 9516676. 1995 Dec 17.
4. Crowley PJ, Mitchell G, Salmon R, Worthington PA. Synthesis of some arylsulfur pentafluoride pesticides and their relative activities compared to the trifluoromethyl analogues. *Chimia* 2004;58:138–142.
5. Kirsch P, Binder JT, Lork E, Röschenthaler G-V. Liquid crystals based on hypervalent sulfur fluorides: Part 4. [1] Pentafluorosulfanyl alkanes and olefins. *J Fluorine Chem* 2006;127:610–619.
6. (a) Clair TL, Clair AK, Thrasher JS. US patent 5,220,070. 1993 June 15. (b) Winter RW, Nixon PG, Gard GL, Castner DG, Holcomb NR, Hu Y-H, Grainger DW. Photopolymerized acrylate copolymer films with surfaces enriched in sulfur pentafluoride ($-SF_5$) chemistry. *Chem Mater* 1999;11:3044–3049.
7. Abe T, Tao G-H, Joo Y-H, Winter RW, Gard GL, Shreeve JM. *Chem Eur J* 2009;15:9897–9904.
8. (a) Shigeyoshi N, Hidetaka S, inventors; UBE Industries Ltd, assignee. Method for producing pentafluorosulfanylbenzene compound and novel pentafluorosulfanylbenzene compound. WO2009028514. 2009 Mar 5. (b) Kirsch P, Tarumi K, Krause J. New sulfur pentafluoride derivatives. MERCK PATENT GMBHDE19748108. 1999 Oct 31
9. (a) Brel VK. Synthesis of new pentafluorosulfanylacrylates ($F_5SCH{=}CHCHO$, $F_5SCH{=}CHCN$, $F_5SCH{=}CHCOOCH_3$) and use of them as dienophiles in Diels-Alder reaction . *J Fluorine Chem* 2007;128:862–867. (b) Brel VK. Synthesis and epoxidation of 1,3-, 1,4-, and 1,5-alkadienes with pentafluoro-Λ^6-sulfanyl (SF_5) groups. *Synthesis* 2005;8:1245–1252. (c) Brel VK. Synthesis and Diels-Alder reactions of dienophiles with pentafluoro-λ^6-sulfanyl (SF_5) moiety. *Synthesis* 2006;2:339–343.
10. Case JR, Ray NH, Roberts HL. Sulphur chloride pentafluoride: reaction with unsaturated hydrocarbons. *J Chem Soc* 1961;2066–2070.
11. (a) Aït-Mohand S, Dolbier WR, Jr. New and convenient method for incorporation of pentafluorosulfanyl (SF_5) substituents into aliphatic organic compounds. *Org Lett* 2002;4:3013–3015. (b) Dolbier WR, Jr, Mitani A, Xu W, Ghiviriga I. Synthesis of novel pentafluorosulfanylfurans: two Retro-Diels–Alder approaches. *Org Lett* 2006;8:5573–5575. (c) Dolbier WR, Jr, Aït-Mohand S, Schertz TD, Sergeeva TA, Cradlebaugh JA,

Mitani A, Gard GL, Winter RW, Thrasher JS. A convenient and efficient method for incorporation of pentafluorosulfanyl (SF_5) substituents into aliphatic compounds. *J Fluorine Chem* 2006;127:1302–1310.

12. Winter RW, Gard GL. The reaction of SF_5Br with select 1,2-dihaloethylenes. *J Fluorine Chem* 2008;129:1041–1043.
13. Lim DS, Ngo SC, Lal SG, Minnich KE, Welch JT. Improved and facile addition reactions of pentafluorosulfanyl bromide. *Tetrahedron Lett* 2008;49:5662–5663.
14. Ponomarenko MV, Serguchev YA, Röschenthaler G-V. The reactions of 3,7-dimethylenebicyclo[3.3.1]nonane, norbornadiene and *cis,cis*-1,5-cyclooctadiene with pentafluoro-λ^6-sulfanyl chloride. *J Fluorine Chem* 2010;131:270–273.
15. (a) Hoover FW, Coffman DD. Synthesis and chemistry of ethynylsulfur pentafluoride. *J Org Chem* 1964;29:3567–3670. (b) Sergeeva TA, Dolbier WR, Jr. A new synthesis of pentafluorosulfanylbenzene. *Org Lett* 2004;6:2417–2419. (c) Winter RW, Gard GL. Synthesis of SF_5-benzene ($SF_5C_6H_5$) by the SF_5-halide method. *J Fluorine Chem* 2004;125:549–552.
16. Dolbier WR, Jr, Zheng Z. Preparation of pentafluorosulfanyl (SF_5) pyrrole carboxylic acid esters. *J Org Chem* 2009;74:5626–5628.
17. Ponomarenko MV, Kalinovich N, Serguchev YA, Bremer M, Röschenthaler G-V. Synthesis of pentafluoro-lambda-6–sulfanyl substituted acetylenes for novel liquid crystals. *J Fluorine Chem* 2012;135:68–74.
18. Goulding MJ, Duffy W, Adlem K, Kirsch P, Hahn A, Poetsch E, Binder W, Meyer V, Klasen-Memmer M, Heckmeier M, Lüssem G, inventors; Merck Patent, assignee. *Pyrans as liquid crystals*. EP 1482022 A1. 2004 Dec 1.
19. (a) Petrzilka M. Apolar acetylenic liquid crystals. *Mol Cryst Liq Cryst* 1984;111:347–358. (b) Buchecker R, Marck G, Schadt M. New nematics incorporating a 2,6-difluorophenyl acetylene group. *Mol Cryst Liq Cryst* 1995;260:93–105.
20. The application-oriented evaluation of liquid crystals for use in LCDs is centered around "virtual" clearing temperatures, electrooptical parameters, and viscosities. These data are obtained by extrapolation from a standardized nematic host mixture: $T_{NI,virt}$, $\Delta\varepsilon_{virt}$, and Δn_{virt} were determined by linear extrapolation from a 10% (w/w) solution in the commercially available Merck mixture ZLI-4792 (T_{NI} = 92.8 °C, $\Delta\varepsilon$ = 5.27, Δn = 0.0964). The values thus obtained are empirically corrected for changes in the order parameter. For the pure substances, the mesophases were identified by optical polarization microscopy, and the phase transition temperatures by differential scanning calorimetry (DSC). The helical twisting power (HTP) of 21 was measured at 20 °C in a 1% (w/w) solution of the analyte in the Merck liquid crystal mixture MLC-6260 (T_{NI} = 103.5 °C, $\Delta\varepsilon = 4.0$, $\Delta n = 0.088$).
21. Kirsch P, Binder JT, Lork E, Röschenthaler G-V. Liquid crystals based on hypervalent sulfur fluorides: Part 4. [1] Pentafluorosulfanyl alkanes and olefins. *J Fluorine Chem* 2006;127:610–619.

CHAPTER 31

Delocalized Lipophilic Cations as a Source of Naked Fluoride and Phase-Transfer Catalysts

NATALIYA KALINOVICH, OLESYA KAZAKOVA AND GERD-VOLKER RÖSCHENTHALER

There has been much interest in recent years in the generation of highly nucleophilic sources of soluble fluoride ion for application in organic and inorganic synthesis. One of the most useful approaches for naked fluoride has proved to be anhydrous tetramethylammonium fluoride (TMAF) or the salt $TDAE^{2+}2F^-$ (TDAE, tetrakis(dimethylamino)ethylene), recently synthesized from elemental fluorine, which could be quite easily prepared in hydrogen difluoride free form [1–5]. The main advantages of TMAF are the high thermal and chemical stability of the tetramethylammonium cation and the extreme high nucleophilicity of the naked fluoride ion.

Another approach includes the synthesis of hypervalent fluorosilicon or tin onium derivatives showing excellent fluorinating power and being a source of naked fluoride as well [6–10]. Among the fluoride ion sources of the last type, the most important is Tris(dimethylamino)sulfonium difluorotrimethylsilicate (TASF) [6]. The TAS counterion is remarkably stabilizing for many fluorinated anionic species, for example, perfluoroalkyl, perfluoroalkyloxy anions [11]. Hexamethylguanidinium (HMG) 1,1,1,3,3,3-hexafluoro-2-(trifluoromethyl)-2-propanide, $[C(NMe_2)_3]^+[C(CF_3)_3^-]$, a stable solid, was prepared from HMG^+F^-, HF_2^-, and perfluoroisobutene [12]. HMG^+ cation seems to be very attractive and useful for application in basic and applied organofluorine chemistry. We present simple preparative approach for HMG fluoride (HMGF) excluding or minimizing the content of hydrogen difluoride impurities [13].

Efficient Preparations of Fluorine Compounds, First Edition. Edited by Herbert W. Roesky.

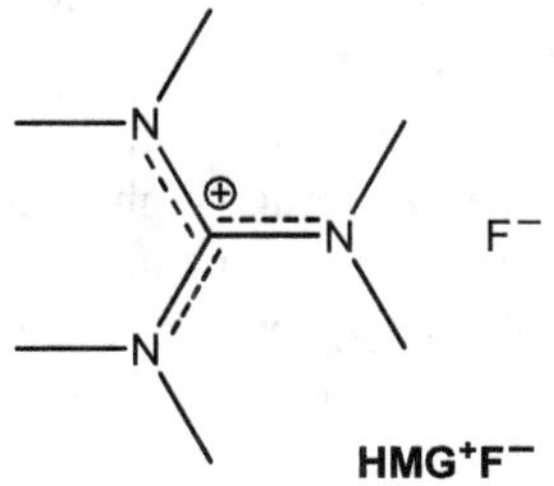

31.1 SYNTHESIS OF HMGF [13]

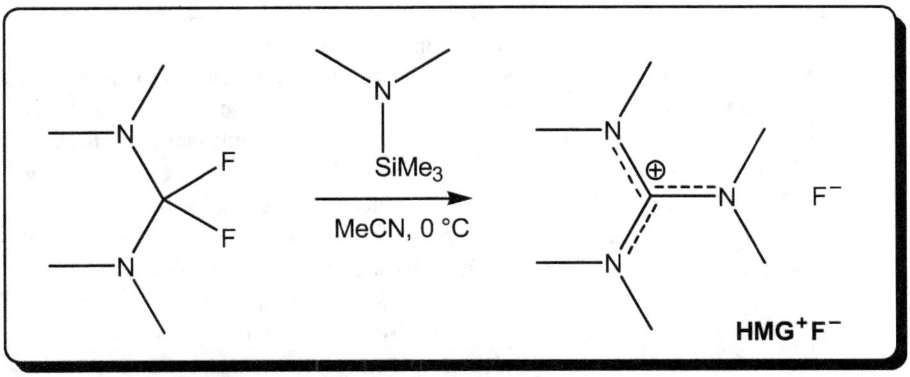

Apparatus Vacuum line, nitrogen inlet, sealed trap, a magnetic stirrer, Dewar flask, filter funnel, safety glasses, laboratory coat, and protective gloves.

Chemicals Bis(dimethylamino)difluoromethane [14], dimethylaminotrimethylsilane, acetonitrile (was distilled from phosphorus pentoxide and then stored over calcium hydride), Et_2O.

Attention! Safety glasses and protective gloves must be used at all times.

Caution! Dimethylaminotrimethylsilane is highly flammable liquid and its vapor causes severe skin burns and eye damage.

Experimental Procedure The reaction and all manipulations were conducted under an atmosphere of dry nitrogen. A mixture of 3.4 g (29 mmol) dimethylaminotrimethylsilane and 2.0 g (14.5 mmol) bis(dimethylamino) difluoromethane in acetonitrile (10 mL) was stirred for 2.5 h at 0 °C. After cooling to −30 °C, the reaction mixture was diluted with diethyl ether (30 mL), which was added over 5 min, and the precipitated product (salt) was filtered under nitrogen and washed with diethyl ether (10 mL) to yield HMGF (95% yield).

Characterization Data Mp 166–167 °C. ^{19}F NMR: δ −63.1 (s) ppm (CD_3CN, −30 °C). ^{1}H NMR: δ 2.95 (s) ppm [15, 16].

Application There is a growing interest in salts containing delocalized lipophilic cations (DLCs) due to their stability and high performance in numerous phase-transfer (PT) organic reactions proceeding under extreme conditions (e.g., high temperature, strongly basic media, powerful nucleophiles) [17, 18]. A new important aspect of their application is connected with the ability of some DLCs to selectively target mitochondria of carcinoma cells resulting in their selective killing. It is a novel and effective strategy intensely studied presently for the treatment of cancer [19]. Solid–liquid PT halogen exchange ("Halex" process) with alkaline fluorides is one of the two main techniques to produce selectively fluorinated aromatics on an industrial scale, complementary to diazotization of anilines in hydrogen fluoride or Schiemann reaction [20, 21]. Potassium fluoride presenting an optimal ratio between cost and reactivity is the most widely used reagent for the industrial synthesis of fluoroaromatics. In most of the Halex-type reactions, the rate-determining step is the fluoride anion/substrate reaction to generate the anionic Meisenheimer complex. The sparing solubility of potassium fluoride in aprotic solvents or aromatic substrates even at elevated temperatures results in low fluorination rates and formation of by-products. To overcome this problem, the range of PT catalysts, conventional tetraalkylammonium, phosphonium, and pyridinium salts, 18-crown-6, and polyethylene glycol were proposed [21, 22, 1]. Unfortunately, most of them show shortcomings, especially at elevated temperatures (c. 200 °C), that restrict their application on an industrial scale. Tetrakis(dialkylamino)phosphonium halides [23], however, with the robust and lipophilic phosphonium cation, were evaluated as powerful catalysts for some Cl/F exchange reactions [24]. There are no universal catalysts of halogen exchange reactions; so far, the choice depends strongly on the chloroaromatic substrate used and the degree of halogen activation. There is still a growing need to create robust, easily recyclable, cheap, and highly effective catalysts for this very important industrial process.

To improve the "Halex" fluorination process, we have synthesized a range of new salts with symmetrical and nonsymmetrical backbones (C–N–P^+, C–N–S^+, and S–N–P^+), all fully substituted with dialkylamino groups [25]. Among these new salts and already known aminodiphosphazenium [26] and 2-azaallenium [27] halides, novel catalysts of high potential for solid–liquid PT halogen exchange with KF in chloroaromatic compounds were evaluated [25]. Here is one of our invention about delocalized lipophilic halides for catalytic fluorination and furthermore hydrogen difluorides and difluorotrimethylsilicates as fluorinating reagents and Lewis bases easily soluble in aprotic solvents, which are of general interest in organic and organofluorine chemistry presents. The key point of our route to thermally stable salts stable toward the highly basic fluorides for the "Halex" process was the consideration that the guanidyl substituents being more efficient [28].

31.2 SYNTHESIS OF TRIS(TETRAMETHYLGUANIDO)SULFONIUM CHLORIDE [28]

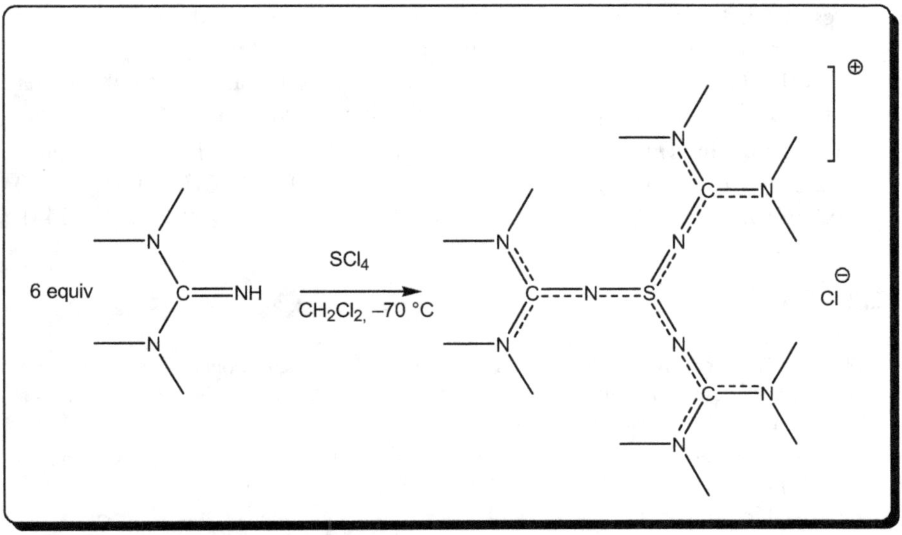

Apparatus Vacuum line, nitrogen inlet, sealed trap, a magnetic stirrer, Dewar flask, syringe, filter funnel, safety glasses, laboratory coat, and protective gloves.

Chemicals Sulfur dichloride in CH_2Cl_2 (1 M SCl_2/CH_2Cl_2), Cl_2, tetramethylguanidine, NaOMe, dry methanol, dry CH_2Cl_2, dry THF, dry Et_2O.

Attention! Safety glasses and protective gloves must be used at all times.

Caution! SCl_2 is toxic by inhalation, in contact with skin and if swallowed. SCl_2 hydrolyzes with release of HCl. It causes burns. There is limited evidence of a carcinogenic effect. Chlorine is a toxic gas that irritates the respiratory system. Because it is heavier than air, it tends to accumulate at the bottom of poorly ventilated spaces. Chlorine gas is a strong oxidizer, which may react with flammable materials. Tetramethylguanidine is highly flammable liquid and its vapor causes severe skin burns and eye damage. Sodium methoxide is highly caustic and the hydrolysis gives methanol, which is toxic and volatile. Methanol has a high toxicity in humans.

Experimental Procedure The reaction and all manipulations were conducted under an atmosphere of dry nitrogen.

To a stirred solution of sulfur dichloride (10.0 g, 97.1 mmol) in dichloromethane (100 mL), at −78 °C chlorine (6.9 g, 97.1 mmol) was condensed and the mixture was stirred for 5 min at this temperature. Then, the tetramethylguanidine (69.0 g, 600 mmol) was added in portions to keep the reaction temperature below −70 °C. The reaction mixture was allowed to warm to room temperature within 2 h. The solvent was pumped off in vacuo, the residue cooled to 0 °C, treated with sodium methoxide (15.7 g, 291.3 mmol) in methanol (120 mL), and the reaction mixture was allowed to warm to 22 °C. The solvent and tetramethylguanidine were removed in vacuo (0.05 Torr). Dissolving the residue in dichloromethane followed by the filtration from sodium chloride and evaporation of the solvent in vacuo gave tris(tetramethylguanido)sulfonium chloride (yield 38.6 g, 97%, purity 96.5%). Recrystallization from THF/ether at −30 °C led in 92% yield to the analytically pure compound.

Characterization Data Mp 115–116 °C. ^{1}H NMR (200.13 MHz, $CDCl_3$, ppm): δ 2.8 (s, 18H,) ppm. ^{13}C NMR (50.33 MHz, $CDCl_3$): δ 40.9 (CH_3, CH_3N), 165.0 (C=N) ppm. HRMS (*m/z*): 374,28257 (M^+, calcd for $C_{15}H_{36}N_9S$: 374,28144).

REFERENCES

1. Christe KO, Wilson WW, Wilson RD, Bau R, Feng J. Syntheses, properties, and structures of anhydrous tetramethylammonium fluoride and its 1:1 adduct with trans-3-amino-2-butenenitrile. *J Am Chem Soc* 1990;112:7619–7625.
2. Seppelt K. Gibt es das nackte F−-Ion? *Angew Chem* 1992;104:299–300; Does the naked fluoride ion exist? *Angew Chem, Int Ed* 1992;31:292–293.
3. Adams DJ, Clark JH. Nucleophilic routes to selectively fluorinated aromatics. *Chem Soc Rev* 1999;28:225–231.
4. Kolomeitsev AA, Seifert FU, Roeschenthaler G-V. Simple preparation of difluorophosphoranes using anhydrous zinc and tetramethylammonium fluorides. *J Fluorine Chem* 1995;71:47–49.
5. Chambers RD, Gray WK, Sandford G, Vaughan JFS. 'Naked fluoride ion' from elemental fluorine. *J Fluorine Chem* 1999;94:213–215.
6. Middleton WJ Homopolymers of polyfluorinated vinyl isocyanates. US19750624975,.Du Pont, 1976, 22 Jun.
7. Gingras M. Tetrabutylammonium difluorotripheylstannate as a new anhydrous synthetic equivalent to TBAF. *Tetrahedron Lett* 1991;32:7381–7384.
8. Hatanaka Y, Goda K, Okahara Y, Hiyama T. Highly selective cross-coupling reactions of aryl(halo)silanes with aryl halides: a general and practical route to functionalized biaryls. *Tetrahedron* 1994;50:8301–8316.
9. Pilcher PS, Ammon HL, DeShong P. Utilization of tetrabutylammonium triphenylsilyldifluoride as a fluoride source for nucleophilic fluorination. *J Am Chem Soc* 1995;117:5166–5167.
10. Brescia M-R, DeShong P.Stereoselective phenylation of allylic alcohol derivatives by palladium-catalyzed cross-coupling with hypervalent silicon complexes. *J Org Chem* 1998;63:3156–3157.
11. Smart BE, Middleton WJ, Farnham WB. Stable perfluoroalkyl carbanion salts. *J Am Chem Soc* 1986;108:4905–4907.
12. Delyagina NI, Igumnov SM, Snegirev VF, Knunyants IL. α-Fluoroalkylamides, a new source of nonhydrated fluoride Ion. 3.^{19}F NMR spectra of perfluoro carbanions generated by Means of N,N,N′,N′,-tetramethylformamidinium bifluoride *Bull Acad Sci USSR, Div Chem Sci* 1981;30:1836–1839.

13. Kolomeitsev AA, Bissky G, Kirsch P, Röschenthaler G-V. A simple route to hexamethylguanidinium fluoride. *J Fluorine Chem* 2000;103:159–161.

14. Bis(dimethylamino)difluoromethane was prepared according to the procedure presented by Kolomeitsev and Roeschenthaler [13]. To a stirred solution of 17.1 g (100 mmol) tetramethylchloroform amidinium chloride in CH_2Cl_2 (50 mL) held at 0 °C, 23.3 g (250 mmol) tetramethylammonium fluoride was added over 5 min. The cold mixture was stirred for additional 5 h, the precipitate filtered off, and washed twice with 25 mL pentane. The resulting solution was distilled using a 30-cm Vigreux column. Yield 13.1 g (95%). Purity 96–97% (^{1}H NMR); bp 99–100 °C. ^{19}F NMR: δ −97.0 (s) ppm [16].

15. Igumnov SM, Delyagina NI, Knunyants IL. α-Fluoroalkylamines, a new source of unhydrated fluoride ion. Communication 6. Hexamethylguanidinium fluoride. *Bull Acad Sci USSR, Div Chem Sci* 1986;35:1191–1193.

16. Fawcett FS, Tullock CW, Coffman DD. The chemistry of carbonyl fluoride. I. The fluorination of organic compounds. *J Am Chem Soc* 1962;84:4275–4285.

17. (a) Albanese D, Landini D, Maia A, Penso M. Key role of water for nucleophilic substitutions in phase-transfer-catalyzed processes: a mini-review Ind. *Eng Chem Res* 2001;40:2396–2401. (b) Albanese D, Landini D, Maia A, Penso M. Phase transfer catalysis: some recent applications in organic synthesis. *J Mol Catal A: Chem* 1999;150:113–131. (c) Makosza M, Bujok R. Cocatalysis by tetravalent tin compounds in phase-transfer catalyzed fluorination of alkyl halides and sulfonates. *Tetrahedron Lett* 2002;43:2761–2763.

18. Dehmlow EV, Knufinke V. *J Chem Res (S)* 1989;224–225.

19. (a) Fantin VR, Berardi MJ, Scorrano L, Korsmeyer SJ, Leder P. A novel mitochondriotoxic small molecule that selectively inhibits tumor cell growth. *Cancer Cell* 2002;2:29–42. (b) Modica-Napolitano JS, Aprille JR. Delocalized lipophilic cations selectively target the mitochondria of carcinoma cells. *Adv Drug Delivery Rev* 2001;49:63.

20. Langlois B, Gilbert L, Forat G. A new family of delocalized lipophilic cations . In: Desmurs J-R, Ratton S, editors. *Industrial Chemistry Library*. Amsterdam: Elsevier; 1996. Vol. 8, p 244.

21. (a) Clark HJ, Wails D, Bastock TW. *Aromatic Fluorination*. New York: CRC Press; 1996.

22. Sasson Y, Negussie S, Royz M, Mushkin N. Tetramethylammonium chloride as a selective and robust phase transfer catalyst in a solid–liquid halex reaction: the role of water. *J Chem Soc, Chem Commun* 1996;297–298.

23. (a) Issleib K, Lischewski M. Dimethylamino-Iminophosphorane. *Synth Inorg Met Org Chem* 1973;3:255–264. (b) Koidan GN, Marchenko P, Kudryavtsev AA, Pinchuk AM. *J Gen Chem USSR* 1982;52:1779–1787. (c) Marchenko P, Koidan GN, Povolotskii MI, Pinchuk AM. *J Gen Chem USSR* 1983;53:1364–1368.

24. (a) Pazenok SV, Kolomeitsev AA, inventors. Process for preparing fluorine-containing compounds. HOECHST RES & TECH GMBH & CO [DE]; KOLOMEITSEV ALEXANDER [UA]; PASENOK SERGEJ [DE] Patent WO 9805610. 1998 12 Feb; *Chem Abstr* 1998;128, 252250a. (b) Kolomeitsev AA, Yagupolskii YL, Yagupolskii LM, Pazenok SV, Appel WK, Pfirmann R, Wessel T. 12th European Symposium on Fluorine Chemistry; Berlin, Germany; 1998, Abstract PII-83. (c) Kolomeitsev AA, Kirij NV, Pazenok SV, Appel WK, Röschenthaler G-V. 14th Winter Fluorine Conference; St. Petersburg, Florida; 1999, Abstract 37, p 26.

25. (a) Henrich M, Marhold A, Kolomeitsev AA, Röschenthaler G-V, inventors; Bayer AG, assignee. Process for preparing nuclear-fluorinated aromatics. DE 10129057/EP 1266904,

2002 Dec 18. (b) Simultaneously to this patent application, the use of symmetrical and nonsymmetrical diphosphazenium salts as highly effective PT catalysts for halogen exchange aromatic fluorination has been claimed: Schanen V, Cristau H-J, Taillefer M, WO 02092226, 2002.

26. (a) Marchenko AP, Koidan GN, Pinchuk AM. *J Gen Chem USSR* 1983;53:583–589. (b) Schwesinger R, Link R, Thiele G, Rotter H, Honnert D, Limbach HH, Männle F. Stabile phosphazenium-ionen in der synthese - ein leicht zugängliches, extrem reaktives "nacktes" Fluoridsalz. *Angew Chem* 1991;103:1376–1378; Stable phosphazenium ions in synthesis—an easily accessible, extremely reactive "naked" fluoride salt. *Angew Chem, Int Ed* 1991;30:1372–1375.

27. Hamed A, Mueller E, Jochims JC. The reaction of 2,4,6-tris(trimethylsiloxy)-1,3,5-triazine with chlorocarbenium ions: a new access to 1-oxa-3-azabutatrienium and 2-azaallenium salts. *Tetrahedron* 1986;42:6645–6656.

28. Henrich M, Marhold A, Kolomeitsev AA, Kalinovich N, Röschenthaler G-V. A new family of delocalized lipophilic cations. *Tetrahedron Lett* 2003;44:5795–5798.

CHAPTER 32

Methyltrifluoropyruvate Imines Possessing *N*-oxalyl and *N*-Phosphonoformyl Groups: Precursors to a Variety of α-CF_3-α-Amino Acid Derivatives

OLESYA KAZAKOVA AND GERD-VOLKER RÖSCHENTHALER

Peptides modified by non-proteinogenic amino acids are useful building blocks for drug discovery. Therefore, the development of new synthetic pathways to unnatural amino acids containing different functionalities remains a constant challenge. In this connection, hydroxamic acid derivatives including *N*-oxalyl amino acids have attracted considerable attention due to their high activity in inhibiting medically important metalloenzymes [1]. For example, oxaloglycine and its derivatives are known to inhibit prolyl hydroxylase [2] involved in the biosynthesis of collagen. The hyperactivity of the enzyme leads to the accumulation of large amounts collagen in different organs and tissues that causes the life-threatening fibrotic diseases such as pulmonary fibroses, and liver and renal fibrosis. Furthermore, phosphonoformate (PFA, Foscarnet) is an effective antiviral compound, clinically used in the treatment of herpetic diseases and AIDS [3]. Incorporation of the PFA backbone into amides and amino acids is a promising strategy used currently for the development of new drug candidates based on selective inhibition of important enzymes, including matrix metalloproteinases (MMPs) [4]. On the one hand, the capacity of oxamic and phosphonoformic acid moieties for bidentate metal binding presumably plays a role in metalloenzyme inhibition. On the other hand, α-amino acids containing the trifluoromethyl (Tfm) group [5] are of particular interest due to the unique characteristics of the Tfm group, such as high electronegativity, electron density, steric hindrance, and hydrophobicity [6]. The advantages of peptides modified by Tfm amino acid include enhanced proteolytic stability, affinity for lipid bilayer membranes, as well

Efficient Preparations of Fluorine Compounds, First Edition. Edited by Herbert W. Roesky.

as stabilization of secondary supramolecular structures [7] owing to the ability of the fluorine atom to form hydrogen bonds [8, 9]. We have developed an effective strategy for the synthesis of novel imines of methyl trifluoropyruvate (MTFP) bearing oxalyl and phosphonoformyl groups at the nitrogen atom, and their subsequent transformations into the corresponding amino acid derivatives. The resulting α-Tfm-α-amino acids possessing *N*-oxalyl and *N*-phosphonoformyl groups are potentially useful candidates for the design of novel inhibitors of medically important enzymes.

MTFP [10] reacts readily with ethyl oxamate and diethyl carbamoylphosphonate at room temperature in the absence of any solvent to give, in high yields, the stable adducts (**1**), which could be dehydrated by standard procedure [11] to afford the corresponding imines (**2**) [12].

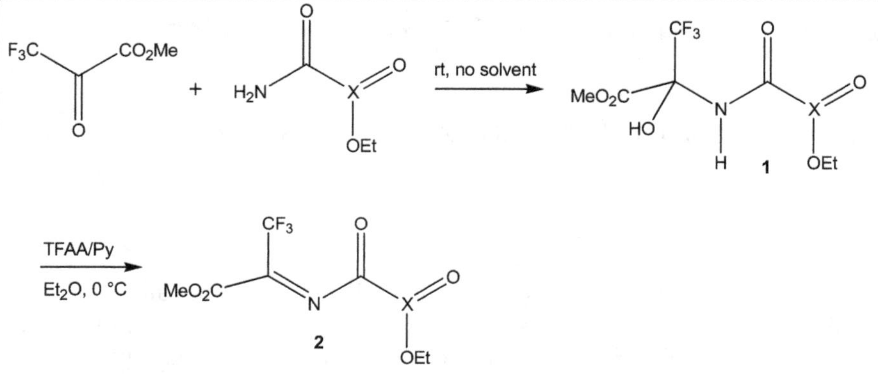

X = C (**1a** - 98%, **2a** - 82%)
X = P-OEt (**1b** - 97%, **2b** - 72%)

32.1 GENERAL PROCEDURE FOR THE SYNTHESIS OF HEMIAMIDALS (1) [12]

Apparatus A 100-mL flask equipped with a magnetic stirrer, nitrogen inlet, safety glasses, laboratory coat, protective gloves.

Chemicals MTFP, ethyl oxamate (for **1a**) or diethyl carbamoylphosphonate (for **1b**), petroleum ether.

Attention! Safety glasses and protective gloves must be used at all times.

Caution! MTFP is an irritant.

Experimental Procedure All solvents used in reactions were freshly distilled from appropriate drying agents before use. Reactions were performed under an atmosphere of dry nitrogen.

A mixture of MTFP (10 mmol) and ethyl oxamate (or diethyl carbamoylphosphonate; 10 mmol) was kept at room temperature for 16 h. The crude solid product was washed with petroleum ether to give analytically pure hemiamidals **1**.

Characterization Data For 2-(ethoxyoxalyl-amino)-3,3,3-trifluoro-2-hydroxy-propionic acid methyl ester (**1a**). Yield: 98% (white solid), mp 58–61 °C. ^{1}H NMR ($CDCl_3$): δ 8.11 (s, 1H), 5.42 (s, 1H), 4.41 (q, 2H, $^3J_{HH}$ = 7.1 Hz), 3.90 (s, 3H), 1.42 (t, 3H, $^3J_{HH}$ = 7.1 Hz) ppm. ^{19}F NMR ($CDCl_3$): δ −81.7 (s, CF_3) ppm. ^{13}C NMR ($CDCl_3$): δ 16.6, 55.2, 65.5, 80.5 (q, $^2J_{CF}$ = 29.5 Hz), 121.6 (q, CF_3, $^1J_{CF}$ = 285.0 Hz), 156.1, 158.3, 159.1 ppm. HRMS: [M^+] calcd for $C_8H_{10}F_3NO_6$, 273.0460; found, 273.0462.

For 2-(diethylphosphonoformamido)-3,3,3-trifluoro-2-propionic acid methyl ester (**1b**). Yield: 97% (white solid), mp 69–72 °C. ^{1}H NMR ($CDCl_3$): δ 8.51 (s, 1H), 5.32 (s, 1H), 4.20 (m, 4H), 3.93 (s, 3H), 1.4 (t, 6H, $^3J_{HH}$ = 7.0 Hz) ppm. ^{19}F NMR ($CDCl_3$): δ −75.5 (s, CF_3) ppm. ^{31}P NMR ($CDCl_3$): δ −1.9 (m) ppm. ^{13}C NMR ($CDCl_3$): δ 16.6, 16.7, 55.3, 65.6, 65.7, 80.5, 80.6 (both q, $^2J_{CF}$ = 29.0 Hz), 122.5 and 122.6 (both q, CF_3, $^1J_{CF}$ = 286.0 Hz), 165.5 (d, $^1J_{CP}$ = 123.0 Hz), 170.2 ppm. HRMS: [M^+] calcd for $C_9H_{15}F_3NO_7P$, 337.0538; found, 337.0539.

32.2 GENERAL PROCEDURE FOR THE SYNTHESIS OF IMINES (2) [12]

Apparatus Two-necked, 100-mL flask equipped with an internal thermometer, a dropping funnel, and a magnetic stirrer, N_2 inlet, Dewar flask (500 mL), filter funnel, safety glasses, laboratory coat, protective gloves.

Chemicals Trifluoroacetic anhydride, dry Et_2O, pyridine, petroleum ether, liquid N_2, ethanol for cooling bath.

Attention! Safety glasses and protective gloves must be used at all times.

Caution! Trifluoroacetic anhydride reacts violently with water. Pyridine is highly flammable.

Experimental Procedure All solvents used in the reactions were freshly distilled from appropriate drying agents before use. Reactions were performed under an atmosphere of dry nitrogen.

Trifluoroacetic anhydride (4.5 mL, 31.9 mmol) was added at 0 °C to a vigorously stirred solution of a hemiamidal (29.0 mmol) in dry ether (100 mL) over a period of 0.5 h. After stirring for 0.5 h, pyridine (5.2 mL, 64.0 mmol) was added slowly. Stirring was continued for additional 2 h. The reaction mixture was cooled to −20 °C and the precipitated pyridinium trifluoroacetate was filtered off under an inert gas atmosphere. The filtrate was concentrated in vacuum and triturated with petroleum ether (3 × 100 mL) to dissolve the imine and separate it from residual pyridinium trifluoroacetate. The combined petroleum ether solutions were evaporated to give compound **2**, which was additionally purified by distillation in the case of **2a**. Imine **2b** proved to be unstable under distillation conditions; therefore, it was used further as a crude product (purity c. 90% according the NMR data).

Characterization Data For methyl 2-[*N*-(2-ethoxyoxalyl)imino]-3,3,3-trifluoropropanoate (**2a**). Yield: 82% (colorless liquid), bp 95–97 °C/0.5 Torr. IR (thin layer) ν (cm^{-1}): 1034 (C–O–C), 1638 (C=N), 1756, 1760, and 1765 (C=O). ^{1}H NMR ($CDCl_3$): δ 4.41 (q, 2H, $^{3}J_{HH}$ = 6.9Hz), 4.08 (s, 3H), 1.42 (t, 3H, $^{3}J_{HH}$ = 6.9 Hz) ppm. ^{19}F NMR ($CDCl_3$): δ −71.3 (s, CF_3) ppm. ^{13}C NMR ($CDCl_3$): δ 14.2, 55.6, 64.4, 122.6 (q, CF_3, $^{1}J_{CF}$ = 279.0 Hz), 154.7, 156.3 (q, *C*=N–CF_3, $^{2}J_{CF}$ = 35.1 Hz), 160.4, 167.3 ppm. HRMS: [M^+] for calcd for $C_8H_8F_3NO_5$, 255.0457; found, 255.0461.

For methyl 2-[(*N*-diethylphosphonoformyl)]imino]-3,3,3-trifluoropropan oate (**2b**). Yield: 75% (pale yellow oil). IR (thin layer) ν (cm^{-1}): 1022, 1046 (P–O–C, C–O–C), 1270 (P=O), 1651 (C=N), 1759, and 1769 (C=O). ^{1}H NMR ($CDCl_3$): δ 4.33 (m, 4H), 4.07 (s, 3H), 1.69 (t, 6H, $^{3}J_{HH}$ = 6.8 Hz) ppm. ^{19}F NMR ($CDCl_3$): δ −71.3 (s, CF_3) ppm. ^{31}P NMR ($CDCl_3$): δ −2.2 (m) ppm. ^{13}C NMR ($CDCl_3$): δ 16.0 and 16.3, 55.3, 63.2 and 63.5, 122.9 and 123.2 (both q, CF_3, $^{1}J_{CF}$ = 282.0 Hz), 151.2, 155.1 and 155.4 (both q, $^{2}J_{CF}$ = 33.0 Hz), 164.8 (d, $^{1}J_{CP}$ = 120.0 Hz), 168.2 ppm. An analytically pure sample was not obtained.

Application Novel *N*-oxalyl and *N*-phosphonoformyl derivatives of α-Tfm-α-amino acids are potential inhibitors of medically important enzymes. In addition, the novel amino acid derivatives herein reported could find further applications as building blocks for the modification of other biologically active peptides. The new compounds will be examined for MMP inhibitory activity [12].

REFERENCES

1. (a) Myllyharju J, Kivirikko KI. Collagens and collagen-related diseases. *Ann Med* 2001;33:7–21. (b) Elkins JM, Hewitson KS, McNeill LA, Seibel JF, Schlemminger I, Pugh CW, Ratcliffe PJ, Schofield CJ. Structure of Factor-inhibiting Hypoxia-inducible Factor (HIF) Reveals mechanism of oxidative modification of HIF-1α. *J Biol Chem* 2003;278:1802–1806.
2. (a) Cunliffe CJ, Franklin TJ, Hales NJ, Hill GB. Novel inhibitors of prolyl 4-hydroxylase. 3. Inhibition by the substrate analog N-oxaloglycine and its derivatives. *J Med Chem* 1992;35:2652–2658. (b) Mole DR, Schlemminger I, McNeill LA, Hewitson KS, Pugh CW, Ratcliffe PJ, Schofield CJ. 2-Oxoglutarate analogue inhibitors of hif prolyl hydroxylase. *Bioorg Med Chem Lett* 2003;13:2677–2680.
3. Oberg B. Antiviral effects of phosphonoformate (pfa, foscarnet sodium). *Pharmacol Ther* 1989;40:213–285.
4. (a) Rosowsky A, Fu H, Mellors J, Richman DD, Hostetler KY. Synthesis and in Vitro Activity of Long-Chain 5′-*O*-[(Alkoxycarbonyl)phosphinyl]-3′-azido-3′-deoxythymidines against Wild-Type and AZT- and Foscarnet-resistant strains of HIV-1. *J Med Chem* 1997;40:2482–2491. (b) Hostetler KY, Kini GD, Beadle JR, Aldern KA, Gardner MF, Border R, Kumar R, Barshak L, Sridhar CN, Wheeler CJ, Richman DD. Lipid prodrugs of phosphonoacids: greatly enhanced antiviral activity of 1-*O*-octadecyl-*sn*-glycero-3-phosphonoformate in HIV-1, HSV-1 and HCMV-infected cells, in vitro *Antiviral Res*

1996;31:59–67. (c) Ferguson CG, Thatcher GRJ. A simple synthesis of phosphonoformamides. *Synlett* 1998;12:1325–1326.

5. (a) Salomon CJ, Breuer E. Spontaneous lossen rearrangement of (Phosphonoformyl)hydroxamates. The migratory aptitude of the phosphonyl group. *J Org Chem* 1997;62:3858–3861. (b) Breuer E, Frant J, Reich R. Recent non-hydroxamate matrix metalloproteinase inhibitors. *Expert Opin Ther Pat* 2005;15:253–269.
6. (a) Sewald N, Burger K. Synthesis of β-Fluorine-Containing Amino Acids. In: Kukhar VP, Soloshonok VA, editors. *Synthesis of Fluoro-Containing Amino Acids*. New York/Brisbane/Toronto/Singapore: John Wiley and Sons Ltd; 1995. p 139 and references cited therein. (b) Burger K, Muetze K, Hollweck W, Koksch B. Incorporation of α-trifluoromethyl substituted α-amino acids into C-and N-terminal position of peptides and peptide mimetics using multicomponent reactions. *Tetrahedron* 1998;54:5915–5928. (c) Sutherland A, Willis CL. Synthesis of fluorinated amino acids. *Nat Prod Rep* 2000;17:621–631. (d) Koksch B, Sewald N, Hofmann H-J, Burger K, Jakubke H-D. Proteolytically stable peptides by incorporation of α-Tfm amino acids. *J Pept Sci* 1997;3:157–167. (e) Horng J-C, Raleigh DP. Φ-Values beyond the ribosomally encoded amino acids: kinetic and thermodynamic consequences of incorporating trifluoromethyl amino acids in a globular protein. *J Am Chem Soc* 2003;125:9286–9287.
7. Banks RE, Taltow JC, Smart BE. *Organofluorine Chemistry: Principles and Commercial Applications*. New York: Plenum Press; 1994.
8. (a) Bilgicer B, Kumar K. Tetrahedron Synthesis and thermodynamic characterization of self-sorting coiled coils. 2002;58:4105–4112. (b) Wang P, Tang Y, Tirrell DA. Incorporation of trifluoroisoleucine into proteins in vivo. *J Am Chem Soc* 2003;125:6900–6906.
9. Carosati E, Sciabola S, Cruciani G. Hydrogen bonding interactions of covalently bonded fluorine atoms: from crystallographic data to a new angular function in the GRID force field. *J Med Chem* 2004;47:5114–5125.
10. Knunyants IL, Shokina VV, Tyuleneva VV. *Dokl Akad Nauk SSSR* 1996;169:594 (*Chem. Abstr.* 1996;65:15218e).
11. (a) Fluorine-containing ketimines. Osipov SN, Kolomiets AF, Fokin AV. *Russ Chem Rev* 1992;61:798–815. (b)Osipov SN, Golubev AS, Sewald N, Michel T, Kolomiets F, Fokin AV, Burger K. A new strategy for the synthesis of α-difluoromethyl-substituted α-hydroxy and α-amino acids. *J Org Chem* 1996;61:7521–7528.
12. Skarpos H, Vorob'eva DV, Osipov SN, Odinets IL, Breuer E, Röschenthaler G-V. Methyl-trifluoropyruvate imines possessing *N*-oxalyl and *N*-phosphonoformyl groups—precursors to a variety of α-CF_3-α-amino acid derivatives. *Org Biomol Chem* 2006;4:3669–3674.

CHAPTER 33

Rhodium-Mediated Synthesis of (3,3,3-Trifluoropropyl)trimethoxysilane and (3,3,3-Trifluoropropyl)triphenylsilane

FALK WEHMEIER AND THOMAS BRAUN

An important pathway for a transition metal-mediated synthesis of fluorinated compounds imparts a selective hydrodefluorination of highly fluorinated molecules. As an alternative, fluorinated building blocks can be derivatized in the coordination sphere of a metal atom. Hexafluoropropene can be activated by the rhodium hydrido complex $[Rh(H)(PEt_3)_3]$ to give the rhodium derivative $[Rh\{(Z)\text{-}CF{=}CFCF_3\}(PEt_3)_3]$ after cleavage of a carbon–fluorine bond [1]. A subsequent reaction with dihydrogen leads to the formation of 1,1,1-trifluoropropane. When tertiary silanes are used instead of dihydrogen, (3,3,3-trifluoropropyl)silanes are generated. The same products can also be synthesized by a rhodium-mediated hydrosilylation of 3,3,3-trifluoropropene with tertiary silanes. The formation of $R_3SiCH_2CH_2CF_3$ ($R_3{=}Ph_3$, $(OMe)_3$) can be carried out catalytically, with turnover numbers of up to 90 for $R_3{=}Ph_3$.

Apparatus All preparations should be carried out using standard Schlenk techniques under an argon atmosphere; all chemicals should be purified and dried by conventional methods and distilled under argon before use.

Chemicals $[Rh(H)(PEt)_3]$, hexafluoropropene or 3,3,3-trifluoropropene, trimethoxysilane, triphenylsilane, toluene, *n*-hexane.

Attention! Hexafluoropropene and 3,3,3-trifluoropropene are gases and should be handled appropriately.

Caution! Safety glasses and protective gloves must be used at all times. Trimethoxysilane is highly toxic. Inhalation and contact with skin and eyes should be avoided by handling under a well-functioning hood.

Experimental Procedure [2] About 5 mg (11 μmol) $[Rh(H)(PEt_3)_3]$ [3] is dissolved in 5 mL dry toluene. A slow stream of hexafluoropropene is then passed

Efficient Preparations of Fluorine Compounds, First Edition. Edited by Herbert W. Roesky.

through the solution for 3 min. A color change from red to yellow indicates the formation of the C–F activation product [Rh{(Z)-CF=CFCF$_3$}(PEt$_3$)$_3$] [4]. Then 0.5 mL (3.9 mmol) trimethoxysilane is added and the reaction mixture is stirred at room temperature overnight. The solution is then filtered over silica, and the silica gel is washed with *n*-hexane. The hexane filtrates are combined, and the solvent is removed under vacuum. The residue is then distilled under atmospheric pressure to yield (3,3,3-trifluoropropyl)trimethoxysilane as colorless liquid (TON=15, bp. 144 °C).

The same procedure can be used to synthesize (3,3,3-trifluorpropyl)triphenylsilane from hexafluoropropene and triphenylsilane. After filtration over silica gel and washing the silica gel with *n*-hexane, the solvent is removed in vacuo. The remaining solid is washed with *n*-hexane to give pure (3,3,3-trifluoropropyl)triphenylsilane (TON = 90).

Alternatively, a slow stream of 3,3,3-trifluoropropene is passed through a solution of 40 mg (70 μmol) [Rh(H)(PEt$_3$)$_3$] and 2.7 g triphenylsilane in 6 mL toluene. The reaction workup is the same as mentioned above to give 967 mg (3,3,3-trifluoropropyl)triphenylsilane (30%, not optimized).

Characterization Data For (3,3,3-Trifluoropropyl)trimethoxysilane: ^{1}H NMR (C_6D_6, 500 MHz): δ 0.87 (m, 2H, SiCH$_2$), 2.14 (m, 2H, CH$_2$CF$_3$), 3.39 (s, 9H, OCH$_3$) ppm. ^{19}F NMR (C_6D_6, 470 MHz): δ −68.2 (t, ^3J(H,F) = 10.8 Hz, CF$_3$) ppm. MS (EI, 70 eV, *m/z*): 141 ((MeO)$_3$SiHF$^+$), 121 ((MeO)$_3$Si$^+$).

For (3,3,3-Trifluoropropyl)triphenylsilane: ^{1}H NMR (C_6D_6, 500 MHz): δ 1.57 (m, 2H, SiCH$_2$), 2.13 (m, 2H, CH$_2$CF$_3$), 7.11–7.69 (m, 15H, Ph) ppm. ^{19}F NMR (C_6D_6, 470 MHz): δ −67.3 (t, ^3J(H,F) = 11 Hz, CF$_3$) ppm. ^{29}Si NMR (C_6D_6, 99 MHz): δ −10.8 ppm. MS (EI, *m/z*): 356 (M$^+$), 279 (Ph$_3$SiHF$^+$), 259 (Ph$_3$Si$^+$). Elemental analysis calcd (%) for $C_{21}H_{19}F_3Si$: C 70.76, H 5.37; found: C 71.32, H 5.87.

Application (3,3,3-Trifluoropropyl)trimethoxysilane can be used to produce trifluoropropylsilyl copolymers, which serve as precursors for poly(methylsilsesquioxane) films with low dielectric constants and low density [5, 6]. (3,3,3-(Trifluoropropyl)trimethoxysilane can also be applied in condensation reactions with tetraalkoxysilanes. The resulting gels and aerogels show a remarkable hydrophobicity [7].

REFERENCES

1. Wehmeier F, Braun T. C–F bond activation of highly fluorinated molecules at rhodium: from model reactions to catalysis. *Eur J Inorg Chem* 2011;613–625.
2. Braun T, Wehmeier F, Altenhöner K. Catalytic C-F activation of hexafluoropropene by rhodium: formation of 3,3,3-trifluoropropylsilanes. *Angew Chem* 2007;119:5415–5418; *Angew Chem Int Ed* 2007;46:5321–5324.
3. Braun T, Noveski D, Ahijado M, Wehmeier F. Hydrodefluorination of pentafluoropyridine at rhodium using dihydrogen: detection of unusual rhodium hydrido complexes. *Dalton Trans* 2007;5118–5132.

4. Braun T, Noveski D, Neumann B, Stammler HG. Conversion of hexafluoropropene into 1,1,1-trifluoropropane by C-F activation at rhodium. *Angew Chem* 2002;114:2870–2873; *Angew Chem Int Ed* 2002;41:2745–2747.
5. Mikoshiba S, Hayase S. Preparation of low density poly(methylsilsesquioxane)s for LSI interlayer dielectrics with low dielectric constant. Fabrication of Angstrom size pores prepared by baking trifluoropropylsilyl copolymers. *J Mater Chem* 1999;9:591–598.
6. Lee JH, Kim WC, Min SK, Ree H-W, Yoon DY. Synthesis of poly(methyl-*co*-trifluoropropyl)silsesquioxanes and their thin films for low dielectric application. *Macromol Mater Eng* 2003;288:455–461.
7. Tillotson TM, Foster KG, Reynolds JG. Fluorine-induced hydrophobicity in silica aerogels. *J Non-Cryst Solids* 2004;350:202–208.

CHAPTER 34

Rhodium-Mediated Synthesis of a Tetrafluoropyridyl-2-boronate Ester

MICHAEL TELTEWSKOI AND THOMAS BRAUN

Examples for C–F bond functionalization, that is, reactions by which a fluorine group is replaced by a new entity to access higher-value fluorinated compounds, are very limited, and most examples involve hydrodefluorination reactions. The *ortho* C–F bond in pentafluoropyridine can be activated by the rhodium boryl complex $[Rh(Bpin)(PEt_3)_3]$ [1]. The conversion yields the rhodium derivative $[Rh(2\text{-}C_5N(Bpin)F_4)(PEt_3)_3]$. Treatment of the latter with additional B_2pin_2 and pentafluoropyridine at 40 °C leads to the formation of a tetrafluoropyridyl-2-boronate ester. The product can be obtained catalytically with a turnover number of 16.

Apparatus PFA tube.

Chemicals $[Rh(F)(PEt_3)_3]$ or $[Rh(OPh)(PEt_3)_3]$, B_2pin_2, dry hexamethyldisilane.

Attention! Safety glasses and protective gloves must be used at all times.

Experimental Procedures *Synthesis of* $[Rh(Bpin)(PEt_3)_3]$:

(a) B_2pin_2 (9 mg, 0.036 mmol) is added to a solution of $[Rh(OPh)(PEt_3)_3]$ [1] (20 mg, 0.036 mmol) in hexamethyldisilane (0.5 mL) in a PFA tube. After 16 h, the NMR spectroscopic data of the reaction solution at 203 K reveal the formation of $[Rh(Bpin)(PEt_3)_3]$ and PhOBpin. (b) B_2pin_2 (6 mg, 0.024 mmol) is added to a solution of $[Rh(F)(PEt_3)_3]$ [2] (10 mg, 0.021 mmol) in hexamethyldisilane (0.5 mL) in a PFA tube. After 16 h, the NMR spectroscopic data of the reaction solution at 203 K reveal the formation of $[Rh(Bpin)(PEt_3)_3]$ and FBpin.

Characterization Data ^{1}H NMR (300.1 MHz, C_6D_6): δ 1.72 (apparent q, br, 18H, CH_2), 1.19 (apparent quint, br, 27H; CH_3) ppm. ^{11}B NMR (128.4 MHz, [D_8]toluene): δ 46.5 (s, br, $\Delta\nu_{\frac{1}{2}}$ =338 Hz) ppm. $^{31}P\{^1H\}$ NMR (121.5 MHz,

Efficient Preparations of Fluorine Compounds, First Edition. Edited by Herbert W. Roesky.

[D_8]toluene, 203 K): δ 20.5 (dd, ^1J(Rh,P) = 166, ^2J(P,P) = 30 Hz), 9.1 (dm, ^1J(Rh,P) = 110 Hz) ppm.

Preparation of the boronate ester 2-$C_5N(Bpin)F_4$: B_2pin_2 (10 mg, 0.04 mmol) is added to a solution of [Rh(F)(PEt_3)$_3$] (5 mg, 0.01 mmol) in hexamethyldisilane (0.5 mL). After 16 h, B_2pin_2 (110 mg, 0.43 mmol) and pentafluoropyridine (44 μL, 0.4 mmol) are added. The NMR spectroscopic data of the reaction solution reveal the conversion of pentafluoropyridine into 2-$C_5N(Bpin)F_4$ with 45% yield after 36 h at 40 °C.

Characterization Data ^{11}B NMR (128.4 MHz, C_6D_6): δ 28.5 (s, br) ppm. ^{19}F NMR (282.4 MHz, C_6D_6): δ −82.7 (d of apparent t, J(F,F) = 26, J(F,F) ≈ J(F,F) ≈ 18 Hz, 1F), −137.9 (ddd, J(F,F) = 26, J(F,F) = 18, J(F,F) = 7 Hz, 1F), −143.6 (apparent q, J(F,F) ≈ 18 Hz, 1F), −156.2 (ddd, J(F,F) = 26, J(F,F) = 18, J(F,F) = 7 Hz, 1F) ppm. Accurate HR-EI-MS (*m*/*z*): 277.08972 ($C_{11}H_{12}BF_4N_1O_2{}^+$).

Application The boronate ester 2-$C_5N(Bpin)F_4$ is a tetrafluoropyridyl derivative, which is functionalized at the 2-position. Note that it is very difficult to access compounds with such a substitution pattern [3]. In addition, the fluorinated boronate ester itself can provide a new fluorinated building block [4].

REFERENCES

1. Teltewskoi M, Panetier JA, Macgregor SA, Braun T. A highly reactive rhodium(I)–boryl complex as a useful tool for C–H bond activation and catalytic C–F bond borylation. *Angew Chem* 2010;122:4039–4043; *Angew Chem, Int Ed* 2010;49:3947–3951.
2. Noveski D, Braun T, Krückemeier S. Synthesis and reactivity of rhodium fluoro complexes. *J Fluorine Chem* 2004;125:959–966.
3. (a) Wehmeier F, Braun T. C–F bond activation of highly fluorinated molecules at rhodium: from model reactions to catalysis. *Eur J Inorg Chem* 2011;613–625. (b) Braun T, Noveski D, Ahijado M, Wehmeier F. Hydrodefluorination of pentafluoropyridine at rhodium using dihydrogen: detection of unusual rhodium hydrido complexes. *Dalton Trans* 2007;3820–3825. (c) Breyer D, Braun T, Penner A. Isolation and reactivity of palladium hydrido complexes: intermediates in the hydrodefluorination of pentafluoropyridine. *Dalton Trans* 2010;39:7513–7520; (d) Braun T, Perutz RN, Sladek MI. Catalytic C–F activation of polyfluorinated pyridines by nickel-mediated cross-coupling reactions. *Chem Commun* 2001;2254–2255; (e) Braun T, Perutz RN. Routes to fluorinated organic derivatives by nickel mediated C–F activation of heteroaromatics. *Chem Commun* 2002;2749–2757.
4. Ameduri B, Bardin VV, Boutevin B, Chambers RD, Dolbier WR Jr, Petrov UA, Vaughan JFS. Organofluorine chemistry: fluorinated alkenes and reactive intermediates. Topics in Current Chemistry, Chambers RD, editor; Springer: Berlin, 1997; Volume 192. Hiyama T, *Organofluorine Compounds*. Springer: Berlin; 2000.

CHAPTER 35

Palladium-Mediated Synthesis of 4-Vinyltetrafluoropyridine and 2,3,5,6-Tetrafluoropyridine

DAVID BREYER AND THOMAS BRAUN

An organometallic strategy to access fluoroorganic compounds involves the selective activation of highly fluorinated precursors followed by derivatization reactions in the coordination sphere of a transition metal. Some of the obtained products are often not accessible by any other means. Pentafluoropyridine can be selectively activated at the 4-position by $[Pd(PiPr_3)_2]$ to give the complex *trans*-$[Pd(F)(4\text{-}C_5NF_4)(PiPr_3)_2]$. Treatment of the latter with $Bu_3SnCH{=}CH_2$ leads to the formation of 4-vinyltetrafluoropyridine. The hydrido complex *trans*-$[Pd(H)(4\text{-}C_5NF_4)(PiPr_3)_2]$ can be prepared by reaction of *trans*-$[Pd(F)(4\text{-}C_5NF_4)(PiPr_3)_2]$ with HBpin (HBpin, 4,4,5,5-tetramethyl-1,3,2-dioxaborolane). Heating to 60 °C gives 2,3,5,6-tetrafluoropyridine. Both tetrafluoropyridine derivatives can be synthesized catalytically.

35.1 CATALYTIC DERIVATIZATION OF PENTAFLUOROPYRIDINE

Apparatus All preparations should be carried out using standard Schlenk techniques under an argon atmosphere; all chemicals should be purified and dried by conventional methods and distilled under argon before use.

Attention! Safety glasses and protective gloves must be used at all times. All reactions should be carried out in a well-ventilated hood.

35.2 PREPARATION OF *trans*-$[Pd(F)(4\text{-}C_5NF_4)(PiPr_3)_2]$

Chemicals $[PdMe_2(tmeda)]$ (tmeda, tetramethylethylenediamine), $PiPr_3$, pentafluoropyridine, toluene, hexane.

Efficient Preparations of Fluorine Compounds, First Edition. Edited by Herbert W. Roesky.

Experimental Procedure [$PdMe_2$(tmeda)] [1] (1.00 g, 4.0 mmol) is dissolved in toluene (25 mL) and P*i*Pr_3 (1.90 mL, 10.0 mmol) is added. After stirring the reaction mixture for 30 min, pentafluoropyridine (0.83 mL, 8.1 mmol) is added. The mixture is then stirred for another 15 min, after which the solution is heated for 1 h to 60 °C, 30 min to 80 °C, 90 min to 90 °C, and 5 h to 105 °C, during which the solution turned to dark orange. The solution is then cooled and filtered from a gray residue through a cannula, and the volatiles are removed from the toluene filtrate in the vacuum. The resulting orange residue is washed with hexane at −90 °C (3 ×), −60 °C (2 ×), and −30 °C (2 ×). The residue is extracted with hexane at room temperature. Evaporation of the solvent under vacuum yields a yellow powder (yield 1.90 g, 80%).

Characterization Data ^{1}H NMR (300.1 MHz, C_6D_6): δ 1.99 (m, 6H), 1.09 (dd, *J*(H,H) = 7.0 Hz, *J*(P,H) = 14.1 Hz, 36H) ppm. ^{31}P{^{1}H} NMR (121.5 MHz, C_6D_6): δ 40.1 (d, *J*(P,F) = 16.0 Hz) ppm.^{19}F NMR (282.4 MHz, C_6D_6): δ −98.7 (m, 2F), −116.1 (m, 2F), −324.6 (m, 1F) ppm. Elemental analysis calcd (%) for $C_{23}H_{42}F_5NP_2Pd$: C, 46.36; H, 6.94; N, 2.24. Found: C, 46.31; H, 7.10; N, 2.24.

35.3 PREPARATION 4-VINYLTETRAFLUOROPYRIDINE

Chemicals *trans*-[Pd(F)(4-C_5NF_4)(P*i*Pr_3)$_2$] [2], pentafluoropyridine, [D_8]thf, $Bu_3SnCH{=}CH_2$, 4-fluorotoluene.

Experimental Procedure In an NMR tube, $Bu_3SnCH{=}CH_2$ (104 μL, 0.46 mmol) is added to a solution of *trans*-[Pd(F)(4-C_5NF_4)(P*i*Pr_3)$_2$] (29 mg, 0.049 mmol) and pentafluoropyridine (51 μL, 0.46 mmol) in THF-d_8 (0.6 mL). The reaction mixture is then heated for 48 h to 50 °C. The yield of 4-vinyltetrafluoropyridine [3] is determined using an external standard of 4-fluorotoluene and is 60% (TON = 6).

Characterization Data ^{1}H NMR (500 MHz, [D_8]thf): δ 6.01 (dm, *J*(H,H) = 17.6 Hz, 1H), 5.71 (dm, *J*(H,H) = 17.8 Hz, 1H), 5.19 (dm, *J*(H,H) = 12.1 Hz, 1H) ppm. ^{19}F NMR (470.4 MHz, [D_8]thf): δ −91.6 (m, 2F), −144.2 (m, 2F) ppm. MS (EI): *m*/*z* 177 (M^+, 100%).

35.4 PREPARATION OF *trans*-[Pd(H)(4-C_5NF_4)(P*i*Pr_3)$_2$]

Chemicals *trans*-[Pd(F)(4-C_5NF_4)(P*i*Pr_3)$_2$], HBpin, benzene.

Experimental Procedure A solution of *trans*-[Pd(F)(4-C_5NF_4)(P*i*Pr_3)$_2$] (43 mg, 0.07 mmol) in C_6H_6 (4 mL) is treated with HBpin (13 μL, 0.087 mmol). The reaction mixture is subsequently stirred for 2 h at room temperature. The volatiles are then removed in vacuo to give a white solid (yield 41 mg, 98%).

Characterization Data ^{1}H NMR (300.1 MHz, C_6D_6): δ 1.72 (m, br, 6H), 0.96 (dd, *J*(H,H) = 7.1 Hz, *J*(P,H) = 14.3 Hz, 36H), −9.06 (m, *J*(P,H) = 0.74 MHz,

1H) ppm. $^{31}P\{^1H\}$ NMR (121.5 MHz, C_6D_6): δ 57.4 (t, *J*(P,F) = 1.5 Hz) ppm. ^{19}F NMR (282.4 MHz, C_6D_6): δ −100.5 (m, 2F), −115.3 (m, 2F) ppm. Elemental analysis calcd (%) for $C_{23}H_{43}F_4NP_2Pd$: C, 47.82; H, 7.51; N, 2.43. Found: C, 47.86; H, 7.45; N, 1.94.

35.5 PREPARATION OF 2,3,5,6-TETRAFLUOROPYRIDINE

Chemicals *trans*-[Pd(H)(4-C_5NF_4)(P*i*Pr_3)$_2$] [4], pentafluoropyridine, [D_8]thf, HBpin, fluorobenzene.

Experimental Procedure An NMR tube containing a capillary with fluorobenzene was charged with a solution of *trans*-[Pd(H)(4-C_5NF_4)(P*i*Pr_3)$_2$] (37 mg, 0.064 mmol) in [D_8]thf (0.6 mL), pentafluoropyridine (70 μL, 0.64 mmol), and HBpin (97 μL, 0.46 mmol). The reaction mixture was heated for 3 days to 60 °C. The yield of 2,3,5,6-tetrafluoropyridine was determined on using an external standard of fluorobenzene and is 44% (TON = 4).

Characterization Data ^1H NMR (300.1 MHz, [D_8]thf): δ 6.45 (m, 1H) ppm. ^{19}F NMR (282.4 MHz, [D_8]thf): δ −93.3 (m, 2F), −141.5 (m, 2F) ppm.

Application 4-Vinyltetrafluoropyridine and 2,3,5,6-tetrafluoropyridine are fluorinated compounds of higher value, and may be used as fluorinated organic building blocks.

REFERENCES

1. De Graaf W, Boersma J, Smeets WJJ, Spek AL, Van Koten G. Dimethyl(N,N,N′,N′-tetramethylethanediamine)palladium(II) and dimethyl[1,2-bis(dimethylphosphino)ethane]palladium(II): Syntheses, x-ray crystal structures, and thermolysis, oxidative-addition and ligand-exchange reactions. *Organometallics* 1989;8:2907–2917.
2. Jasim NA, Perutz RN, Whitwood AC, Braun T, Izundu J, Neumann B, Rothfeld S, Stammler H-G. Contrasting reactivity of fluoropyridines at palladium and platinum: C–F oxidative addition at palladium, alkyl transfer from phosphorus to platinum. *Organometallics* 2004;23:6140–6149.
3. Braun T, Izundu J, Steffen A, Neumann B, Stammler H-G. Reactivity of a palladium fluoro complex towards silanes and $Bu_3SnCH{=}CH_2$: catalytic derivatisation of pentafluoropyridine based on carbon-fluorine bond activation reactions. *Dalton Trans* 2006;5118–5123.
4. Breyer D, Braun T, Penner A. Hydrodefluorination of pentafluoropyridine at rhodium using dihydrogen: detection of unusual rhodium hydrido complexes. *Dalton Trans* 2010;39:7513–7520.

CHAPTER 36

Preparation of Polyfluoroaryl-1,2-difluorovinylsilanes

DONALD J. BURTON AND BA V. NGUYEN

36.1 PREPARATION OF 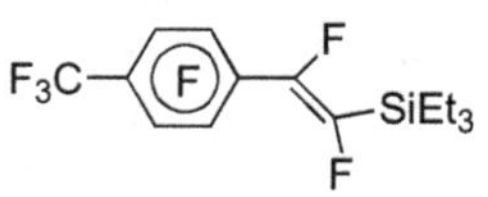

Apparatus A 150-mL, two-necked flask with Teflon-coated stir bar, septum, T-shaped N_2-inlet connected to a bubbler, syringe (10 mL), safety glasses, 50-mL flask, silica-gel chromatography column.

Chemicals Acid washed Cd powder, 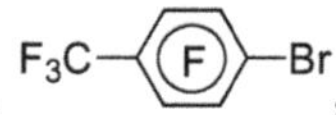, DMF (10 mL), Cu(I)Br medium-fritted Schlenk funnel, (*E*)-1-iodo-2-triethylsilyl-1,2-difluoroethene, pentane.

Attention! Safety glasses should be worn at all times.

Caution All reactions should be carried out in a well-ventilated hood.

Experimental Procedure A dry, two-necked, 150-mL, round-bottomed flask was equipped with a Teflon-coated magnetic stir bar, septum, and N_2 inlet connected to a bubbler. The flask was charged with acid-washed cadmium powder (3.3 g, 30.0 mmol), 1-bromo-2,3,5,6-tetrafluoro-4-(trifluoromethyl)benzene (5 g, 16.8 mmol), and 10 mL dry DMF. The mixture was stirred at room temperature for 20 min. ^{19}F NMR analysis of the reaction mixture indicated that all of the starting benzene had been consumed. The reaction mixture was then filtered through a medium-fritted Schlenk funnel under N_2 pressure into a 50-mL flask that contained Cu(I)Br (1 g, 7 mmol) and (*E*)-1-iodo-2-triethylsilyl-1,2-difluoroethene (5.1 g, 16.8 mmol) [1]. The mixture was stirred at room temperature for 2 h. The reaction mixture was introduced onto a silica-gel column and eluted with pentane. The eluate was collected and the

Efficient Preparations of Fluorine Compounds, First Edition. Edited by Herbert W. Roesky.

solvent removed via rotary evaporation. The yield of (Z)-1-(4-heptafluorotolyl)-2-triethylsilyl-1,2-difluoroethene was 6.4 g (96%).

Characterization Data ^{1}H NMR: δ 1.1 (t, $^3J_{HH}$ = 7.8 Hz, 9H), 0.8 (q, $^3J_{HH}$ = 7.9 Hz, 6H) ppm. ^{19}F NMR: δ −57.0 (t,....., $^4J_{FF}$ = 22.1 Hz, 3F), −135.0 (m, 2F), δ -140.4 (m, 2F), −147.5 (dm, $^3J_{FF}$ = 137.2 Hz, 1F), −153.1 (dt, $^3J_{FF}$ = 137.3 Hz, $^4J_{FF}$ = 13.2 Hz, 1F) ppm. ^{13}C NMR: δ 162.1 (dd, $^1J_{CF}$ = 279.2 Hz, $^2J_{CF}$ = 70.4 Hz), 148.9 (dd, $^1J_{CF}$ = 232.7 Hz, $^2J_{CF}$ = 43.6 Hz), 144.6 (dm, $^1J_{CF}$ = 261.9 Hz, δ 121.1 (q,....), 114.9 (dt, $^2J_{CF}$ = 28.3 Hz, $^2J_{CF}$ = 17.9 Hz), 111.7 (qt, $^2J_{CF}$ = 35.4 Hz, $^2J_{CF}$ = 12.6 Hz), 6.6 (s), 2.4 (s). FTIR (CCl_4, cm^{-1}): 1496.6 (s), 1344.7 (s), 1159.9 (s), 991.2 (m). MS (70 eV) *m/z* (relative intensity): 105 (100).

Waste Disposal The residues of the silica-gel column have to be collected in a labeled container for toxic metal waste that has to be properly deposited.

Application Fluorinated vinyl iodides are key intermediates in the preparation of polyfunctionalized derivatives. The vinyl iodides can be accessed via several different methods. Perfluoroolefins can be converted to perfluorovinyl phosphonium tetrafluoroborates via reaction with *n*-Bu_3P followed by reaction of the (Z)-perfluorovinyl phosphorane with $BF_3 \cdot Et_2O$. Subsequent cleavage of the (Z)-perfluorovinyl phosphonium salt with either KF or Na_2CO_3 gives stereospecifically the (Z)-perfluorovinyl iodide [2]. For aryl fluorine-containing vinyl iodides, the methodology involves Pd(0)-catalyzed stereospecific coupling of fluorine-containing vinylsilylzinc reagents with aryl iodides [3]. Polyfluorinated aryl halides do not readily undergo Pd(0)-catalyzed coupling with vinylsilylzinc reagents. However, polyfluorinated iodides or bromides readily react at room temperature with Cd powder to produce in situ the arylcadmium reagent [4–8]. Metathesis of the arylcadmium reagent with Cu(I)Br readily provides the polyfluoroarylcopper reagent, which can then be coupled with 1-iodo-1,2-difluorovinylsilanes or polyfluorinated vinyl iodides. Because all these reactions are carried out at room temperature, the methodology is easy to perform and avoids low-temperature preparation of unstable polyfluorovinyl lithium or magnesium reagents.

The corresponding vinyl iodide (from the vinylsilane) can be converted to a vinylzinc, cadmium, or copper reagent, can be coupled with vinylzinc reagents or vinyl copper reagents, or self-dimerized with Cu°/DMSO to provide symmetric dienes [9], or be converted to the vinylstannane [4], which can be coupled with Pd(0) catalysis or dimerized stereospecifically with $Cu(OAc)_2$ [10].

REFERENCES

1. Martinet P, Sauvetre R, Normant J-F. Preparation and reactivity of $Et_3SiCF{=}CFM$ (M = Li, ZnBr). *J Organomet Chem* 1989;367:1–10.
2. Heinz PL, Spawn TD, Burton DJ, Shin-Ya S. *J Fluorine Chem* 1988;38(1):131–134.

3. Davis CR, Burton DJ. Stereoselective preparation of Z-a,b-difluorostyrenes and stereospecific conversion to E-a,b-difluoro-b-iodostyrenes. *J Org Chem* 1997;26:9217–9222.
4. Xue L, Lu L, Pedersen SD, Liu Q, Narske RM, Burton DJ. A novel stereospecific route to E and Z-(2-substituted-1,2-difluoroethenyl)stannanes. *J Org Chem* 1997;62:1064–1071.
5. Heinze PL, Burton DJ. The preparation of F-cadmium reagents directly from F-alkyliodides and F-aryl bromide and cadmium metal. *J Fluorine Chem* 1985;29:359–361.
6. Nguyen BV, Burton DJ. Preparation and reactions of the 2,3,5,6-tetrafluoropyridylcopper reagent. *J Fluorine Chem* 1994;67:205–206.
7. MacNeil KJ, Burton DJ. The regiospecific syn addition of (polyfluoroaryl)copper reagents to fluorinated acetylenes: preparation and subsequent functionalization of internal vinylcopper reagents. *J Org Chem* 1993;58:4411–4417.
8. MacNeil KJ, Burton DJ. Preparation of pentafluorophenyl-substituted fluorinated dienes via internal vinyl- and dienyl copper reagents. *J Org Chem* 1995;60:4085–4089.
9. Burton DJ, Hansen SW, Morken PA, MacNeil KJ, Davis CR, Xue L. Coupling routes to hexafluoro-1,3-butadiene, substituted-1,3-fluorine-containing butadienes and fluorinated polyenes. *J Fluorine Chem* 2008;129:435–442.
10. Blumenthal EJ, Burton DJ. Copper (II) mediated homocoupling of 1,2-difluorovinylstannanes. *Israel J Chem* 1999;39:109–115.

CHAPTER 37

Preparation of (Z)-2-iodo-1,2-difluorostyrenes

DONALD J. BURTON AND LING XUE

37.1 PREPARATION OF 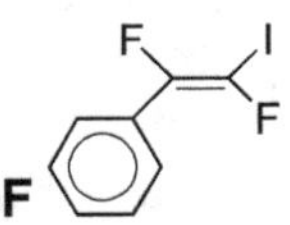

Apparatus A 250-mL, three-necked, round-bottom flask, Teflon-coated magnetic stir bar, N_2 inlet adapter, beaker, flash distillation apparatus.

Chemicals α,β,β-Trifluorostyrene, dry DMF, *n*-tributylphosphine, $BF_3 \cdot OEt_2$, I_2, anhydrous KF, aqueous sodium thiosulfate, 4 Å molecular sieves.

Attention! Safety glasses should be worn at all times.

Caution All reactions should be carried out in a well-ventilated hood.

Experimental Procedure A 250-mL, three-necked, round-bottom flask equipped with a Teflon-coated magnetic stir bar, two septum caps, and a nitrogen T-shaped adapter was charged with α,β,β-trifluorostyrene (19.1 g, 121 mmol) and 160 mL dry DMF. To the mixture, cooled with an ice–water bath, *n*-tributylphosphine (37 mL, 174 mmol) was slowly charged. The reaction mixture gradually turned from yellow to red. Then, $BF_3 \cdot OEt_2$ (18.2 mL, 148 mmol) was injected dropwise into the reaction mixture. The reaction mixture was allowed to warm to room temperature, and an orange solution was obtained. The flask was heated in an 80 °C oil bath for 2 h. To the reaction mixture, which was cooled again with an ice–water bath, I_2 (93.3 g, 367 mmol) and anhydrous KF (28.8 g, 497 mmol) were added. The reaction mixture was stirred at room temperature for 18 h. The ^{19}F NMR spectrum of the reaction mixture showed that the reaction was completed. Then, the contents of the flask were poured into a beaker containing 1500 mL of 5% aqueous sodium thiosulfate. The organic layer was extracted with ether, washed with H_2O, and dried

Efficient Preparations of Fluorine Compounds, First Edition. Edited by Herbert W. Roesky.

over 4 Å molecular sieves. After removal of the ether with a water aspirator, flash distillation gave (Z)-2-iodo-1,2-difluorostyrene (20.4 g, 63.4%), GLPC purity >98%.

Characterization Data ^{1}H NMR: δ 7.38 (m, 5H) ppm. ^{19}F NMR: δ −118.3 (d, $^3J_{FF}$ = 142 Hz, 1F), −133.2 (d, $^3J_{FF}$ = 142 Hz, 1F) ppm. ^{13}C NMR: δ 100.5 (dd, $^1J_{CF}$ = 321 Hz, $^2J_{CF}$ = 76 Hz), 122.8–130.0 (m), 153.1 (dd, $^1J_{CF}$ = 234 Hz, $^2J_{CF}$ = 34 Hz) ppm. MS (70 eV) *m/z* (relative intensity): 266 (M^+, 100), 139 (69), 119 (94), 99 (46). HRMS: calcd for $C_8H_5F_2I$, 265.9404; obsvd, 265.9392.

Waste Disposal The waste solvent and thiosulfate solution have to be collected in a labeled container for waste that has to be properly deposited.

Application Fluorinated vinyl iodides are key intermediates in the preparation of polyfunctionalized derivatives. The vinyl iodides, containing an aryl or substituted aryl group on the vinylic carbon, can be readily accessed via methodology initially developed for the conversion of perfluoroolefins to (Z)-perfluorovinyl iodides [1]. The tertiary phosphine, *n*-tributylphosphine, readily attacks a difluoromethylene olefin to initially produce (stereospecifically) the (Z)-phosphorane, which is easily converted (via $BF_3{\cdot}OEt_2$) to the (Z)-fluorinated vinyl phosphonium tetrafluoroborate. The tetrafluoroborate salt is readily cleaved in situ with I_2 and anhydrous KF or Na_2CO_3 to provide stereospecifically the (Z)-vinyl iodide [1]. The reaction described here is general for *o*-, *m*-, and *p*-substituted α,β,β-trifluorovinyl styrenes. The requisite styrenes are readily prepared via three methods: (a) Pd(0) cross-coupling of trifluorovinylzinc reagents with aryl iodides [2]. The trifluorovinylzinc reagent is easily prepared from bromo or iodotrifluoroethene and acid-washed zinc powder [2, 3]. (b) The trifluorovinylzinc reagent can also be efficiently prepared in situ via metallation of HFC-134A (CF_3CFH_2) in the presence of anhydrous zinc chloride [4,5]. Subsequent Pd(0)-catalyzed coupling of this in situ reagent with aryl iodides gives good to excellent yields of the respective styrenes. (c) A highly efficient non-organometallic route for the synthesis of α,β,β-trifluorostyrenes has also been reported [6, 7]. The (*E*)-aryl vinyl iodides can also be prepared from (Z)- α,β-difluorostyrenes [8]. The aryl vinyl iodide can be converted to a vinylzinc, cadmium, or copper reagent and functionalized further. Also, the aryl vinyl iodide can be coupled with other vinylzinc or vinylcopper reagents to provide polyfunctionalized compounds. Alternatively, the aryl vinyl iodide can be self-coupled with Cu°/DMSO to provide symmetrical dienes [9], or be converted to the vinylstannane [10], which can be coupled with Pd(0) catalysis or dimerized stereospecifically with $Cu(OAc)_2$ [11, 12].

REFERENCES

1. Heinz PL, Spawn TD, Burton DJ, Shin-Ya S. *J Fluorine Chem* 1988;38:131–134.
2. Heinz PL, Burton DJ. Palladium-catalyzed cross-coupling of perfluoroalkenylzinc reagents with aryl iodides. A new, simple synthesis of α,β,β-trifluorostyrenes and stereoselective preparation of 2-arylperfluoropropenes. *J Org Chem* 1988;53:2714–2720.

3. Hansen SW, Spawn TD, Burton DJ. The stereospecific preparation of fluorinated vinyl zinc reagents from polyfluorinated vinyl iodides or bromides and zinc metal. *J Fluorine Chem* 1987;26:415–420.
4. Anilkumar R, Burton DJ. A remarkable room temperature preparation of the trifluorovinylzinc reagent from HFC-134a. A cost-effective, high yield synthesis of α,β,β-trifluorostyrenes. *Tetrahedron Lett* 2002;43:2731–2733.
5. Raghavanpillai A, Burton DJ, Room temperature preparation of trifluoroethenylzinc reagent by metalation of the readily available halocarbon HFC-134a and an efficient, economically viable synthesis of 1,2,2-trifluorostyrenes. *J Org Chem* 2004;69:7083–7091.
6. Anilkumar R, Burton DJ. A highly efficient room temperature non-organometallic route for the synthesis of α,β,β-trifluorostyrenes by dehydrohalogenation. *Tetrahedron Lett* 2003;44:6661–6664.
7. Anilkumar R, Burton DJ. An efficient dehydrohalogenation method for the synthesis of 1,2,2-trifluorostyrenes, 1-chloro-2,2-difluorostyrenes and *E*-1-arylperfluoroalkenes. *J Fluorine Chem* 2005;126:1174–1184.
8. Davis CR, Burton DJ. Stereoselective preparation of *Z*-α,β-difluorostyrenes and stereospecific conversion to *E*-α,β-difluoro-β-iodostyrenes. *J Org Chem* 1997;62:9217–9222.
9. Burton DJ, Hansen SW, Morken PA, MacNeil KJ, Davis CR, Xue L. Coupling routes to hexafluoro-1,3-butadiene, substituted-1,3-fluorine-containing butadienes and fluorinated polyenes. *J Fluorine Chem* 2008;129:435–442.
10. Xue L, Lu L, Pedersen SD, Liu Q, Narske RM, Burton DJ. A novel stereospecific route to *E* and *Z*-(2-substituted-1,2-difluoroethenyl)stannanes. *J Org Chem* 1997;62:1064–1071.
11. Blumenthal EJ, Burton DJ. Copper (II) mediated homo-coupling of 1,2-difluorovinylstannanes. *Israel J Chem* 1999;39:109–115.
12. Burton DJ, Wesolowski CA, Liu Q, Davis CR. Approaches to the preparation of (Z)-1,2-difluorostilbenes. *J Fluorine Chem* 2010;131:989–995.

CHAPTER 38

Preparation of 1,4-Bis(*Z*-2-iodo-1,2-difluoroethenyl)benzene

DONALD J. BURTON AND CRAIG A. WESOLOWSKI

38.1 PREPARATION OF 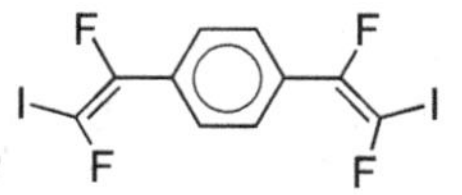

Apparatus A 50 mL, two-necked flask, Teflon-coated magnetic stir bar, rubber septum, N_2 inlet adapter, 250 mL separatory funnel, silica-gel chromatography column, safety glasses, I_2.

Chemicals Anhydrous KF, 1,4-bis(Z-1,2-difluoro-2-triethylsilylethenyl) benzene, DMSO, ethyl acetate, ethyl ether, H_2O, sodium bisulfite, brine, anhydrous $MgSO_4$, hexanes.

Attention! Safety glasses should be worn at all times.

Caution All reactions should be carried out in a well-ventilated hood.

Experimental Procedure A 50 mL, two-necked, round-bottom flask, equipped with a Teflon-coated magnetic stir bar, rubber septum, and nitrogen inlet adapter was charged with anhydrous KF (1.0 g, 17 mmol), I_2 (1.33 g, 5.25 mmol), 1,4-bis(Z-1,2-difluoro-2-triethylsilylethenyl)benzene (0.87 g, 2.02 mmol), and 10 mL DMSO. After stirring at 50 °C for 3 h, the reaction mixture was poured into a 250 mL separatory funnel and diluted with 30 mL ethyl acetate, 30 mL diethyl ether, and 30 mL water. The organic phase was washed sequentially with 30 mL aqueous 3% sodium bisulfite, 20 mL water, and 20 mL brine. The organic phase was then dried over anhydrous $MgSO_4$, filtered, and concentrated via rotary evaporation. The crude orange solid was chromatographed on a silica-gel column (hexane, R_f 0.56) to yield a light orange solid, which was further purified by recrystallization from hexanes to give 0.70 g (76%) of 1,4-bis(Z-2-iodo-1,2-difluoroethenyl)benzene as light orange flakes, mp 131–133 °C.

Efficient Preparations of Fluorine Compounds, First Edition. Edited by Herbert W. Roesky.

Characterization Data ^{1}H NMR: δ 7.63 (s, 4H) ppm. ^{19}F NMR: δ −115.4 (d, $^3J_{FF}$ = 142 Hz, 1F), −134.4 (d, $^3J_{FF}$ = 142.6 Hz, 1F) ppm. ^{13}C NMR: δ 152.3 (dd, $^1J_{CF}$ = 233.8 Hz, $^2J_{CF}$ = 34.8 Hz), 128.7 (ddm, $^2J_{CF}$ = 25.7 Hz, $^3J_{CF}$ = 6.6 Hz), 125.2 (ddm, $^3J_{CF}$ = 8.8 Hz, $^4J_{CF}$ = 6.9 Hz), 101.9 (dd, $^1J_{CF}$ = 323.4 Hz, $^2J_{CF}$ = 75.3 Hz) ppm. MS (70 eV) *m/z* (relative intensity): 453 (M^+ −1, 100). HRMS: calcd for $C_{10}H_4F_4I_2$, 453.8338; obsvd, 453.8364.

Waste Disposal The residues of the silica-gel column have to be collected in a labeled container for toxic waste that has to be properly deposited.

Application The requisite 1,4-bis(*Z*-1,2-difluoro-2-triethylsilylethenyl)benzene is readily prepared via Pd(0) cross-coupling of (*E*)-$Et_3SiCF{=}CFZnX$ [1], prepared from (*E*)-$ICF{=}CFSiEt_3$ [1], with 1,4-diiodobenzene. The 1,4-bis(Z-2-iodo-1,2-difluoroethenyl)benzene can be coupled with other zinc or copper reagents to afford polyfunctionalized derivatives. Alternatively, this bis-vinyliodide can be converted to a bis-vinylzinc reagent and be used in cross-coupling reactions. This bis-vinyliodide or bis-vinylzinc reagent is a key intermediate to the short path synthesis of multifunctionalized fluorine-containing compounds. For example, the bis-vinyliodide is readily converted to a (*Z*)-1,4-bis(2-carboalkoxy) benzene via the reported carboalkoxylation methodology [2].

REFERENCES

1. Martinet P, Sauvetre R, Normant J-F. Preparation and reactivity of $Et_3SiCF{=}CFM$ (M=Li, ZnBr). *J Organomet Chem* 1989;367:1–10.
2. Wesolowski CA, Burton DJ. Palladium-catalyzed stereospecific carboalkoxylation of 1,2-difluoro-1-iodoalkenes and α,β-difluoro-β-iodostyrenes. *Tetrahedron Lett* 1999;40:2243–2246.

CHAPTER 39

Stereospecific Carboamidation Route to (*Z*)-*N*-Phenyl-2,3-difluoro-3-(triethylsilyl) acrylamide

DONALD J. BURTON AND CRAIG A. WESOLOWSKI

39.1 PREPARATION OF

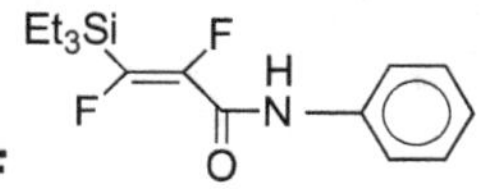

Apparatus Fischer–Porter glass pressure reactor, separatory funnel, silica gel chromatography column.

Chemicals (*E*)-1,2-difluoro-2-iodotriethylsilylethene, freshly distilled aniline, Et_3N, $Pd(PPh_3)_2Cl_2$, CO, hexanes, ethyl acetate.

Attention! Safety glasses should be worn at all times.

Caution CO is a toxic gas. All reactions should be carried out in a well-ventilated fume hood behind a safety shield.

Experimental Procedure A 100 mL Fischer–Porter glass pressure reactor was charged with (*E*)-1,2-difluoro-2-iodotriethylsilyl ethene (2.0 g, 6.6 mmol), freshly distilled aniline (0.92 g, 10 mmol), Et_3N (0.73 g, 7.3 mmol), and dichlorobis(triphenylphosphine) palladium(II). The Fischer–Porter glass pressure reactor was pressurized (behind a safety shield in a fume hood) to 80 psi with carbon monoxide, and the CO released in a fume hood. This pressurization was completed for four cycles to rid the system of air. Then, the reaction was heated to 80 °C for 10 h under CO pressure until the CO consumption ceases as noted by no further decrease in pressure. The reactor was then allowed to cool to room temperature, and the pressure (CO) was carefully released in a fume hood. The reaction mixture was transferred to a separatory funnel containing 40 mL ethyl acetate. The organic layer was washed sequentially with aqueous 10% HCl (2 × 30 mL), 20 mL saturated aqueous $NaHCO_3$, and 20 mL brine.

Efficient Preparations of Fluorine Compounds, First Edition. Edited by Herbert W. Roesky.

The organic layer was dried over anhydrous $MgSO_4$, filtered, and concentrated by rotary evaporation. The crude anilide was purified by silica gel chromatography, eluting first with 300 mL hexane, then with 10% ethyl acetate (10% ethyl acetate in hexanes, R_f 0.41, to yield (*Z*)-*N*-phenyl-2,3-difluoro-3-(triethylsilyl) acrylamide (1.30 g, 66%) as a viscous, yellow oil.

Characterization Data ^{19}F NMR: δ −149.7 (dd, $^3J_{FF}$ = 131.8 Hz, $^5J_{F\text{-}NH}$ = 12.4 Hz, 1F), −158.8 (d, $^3J_{FF}$ = 131.3 Hz, 1F) ppm. ^{1}H NMR: δ 8.05 (d, $^5J_{F\text{-}NH}$ = 11.9 Hz, 1H), 7.59 (dm, $^3J_{HH}$ = 8.4 Hz, 2H), 7.33 (tm, $^3J_{HH}$ = 7.4 Hz, 2H), 7.13 (tm, $^3J_{HH}$ = 7.4 Hz, 1H), 1.03 (t, $^3J_{HH}$ = 7.7 Hz, 9H), 0.82 (q = $^3J_{HH}$ = 7.7 Hz, 6H) ppm. ^{13}C NMR: δ 165.9 (dd, $^1J_{CF}$ = 287.0 Hz, $^2J_{CF}$ = 77.5 Hz), 156.1 (dd, $^2J_{CF}$ = 31.3 Hz, $^3J_{CF}$ = 2.3 Hz), 153.2 (dd, $^1J_{CF}$ = 232.7 Hz, $^2J_{CF}$ = 30.7 Hz), 136.9 (s), 129.1 (s), 125.0 (s), 120.3 (s), 7.0 (s), 2.1 (apparent t, *J* = 2.3 Hz). GC-MS (*m/z*) (relative intensity): 297 (M^+, 15), 268 (70), 120 (68), 105 (18), 77 (100). HRMS: calcd for $C_{15}H_{21}NOF_2Si$: 297.1360; obsvd: 297.1369.

Application This methodology illustrates a general stereospecific synthetic route to (*Z*)-1,2-difluoro-α,β-unsaturated amides and esters. The procedure is detailed here for an amide derivative. However, if an alcohol is used in place of the amine (aniline) in this procedure, a general palladium-catalyzed stereospecific carboalkoxylation of 1,2-difluoro-1-iodoalkenes and α,β-difluoro-β-iodostyrenes is easily achieved [1]. In the case of the described acrylamide derivative, the silyl group can easily be cleaved (iododesilylation) with KF/DMSO/I_2 to give stereospecifically the vinyl iodide product. The vinyl iodide is readily functionalized using Pd(0) catalysis and provides a stereospecific entry into polyfunctionalized compounds.

REFERENCE

1. Wesolowski CA, Burton DJ. Palladium-catalyzed stereospecific carboalkoxylation of 1,2-difluoro-1-iodoalkenes and α,β-difluoro-β-iodostyrenes. *Tetrahedron Lett* 1999;40:2243–2246.

CHAPTER 40

Preparation of 2-Trifluoromethyl-3,3,3-trifluoropropanal

DONALD J. BURTON AND DONALD A. WIEBE

40.1 PREPARATION OF

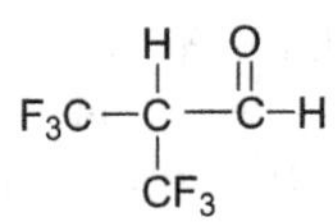

Apparatus A 50-mL flask, 6-in. Vigreux column, distillation head, 150 °C immersion thermometer, vacuum take-off, 20-mL receiver, Teflon-coated magnetic stir bar.

Chemicals Concentrated H_2SO_4 (25 mL), $(CF_3)_2C{=}CHOCH_3$.

Attention! Safety glasses should be worn at all times.

Caution All reactions and handling of polyfluorinated aldehydes should be carried out in a well-ventilated hood.

Experimental Procedure A 50-mL flask was equipped with a 6-in. Vigreux column topped with a distillation head, 150 °C immersion thermometer, vacuum take-off, 20-mL receiver, and a Teflon-coated magnetic stir bar. The flask was charged with 25 mL concentrated H_2SO_4, which was frozen with a liquid nitrogen bath. While the acid was frozen, 1-methoxy-2-trifluoromethyl-3,3,3-trifluoropropene (20.7 g, 107 mmol) was added. The pressure of the system was reduced to ~150 mm and the receiver was cooled in a liquid nitrogen bath. The reaction was warmed to 75 °C and the product (aldehyde) distilled as formed to give 2-trifluoromethyl-3,3,3-trifluoropropanal (16.7 g, 87%)

Characterization Data Bp 42–45 °C, IR 1765 cm^{-1} (s). ^1H NMR: δ 4.00 (heptet d, $^3J_{HF} = 9.0$ Hz, $^3J_{HH} = 1.5$ Hz, 1H), 9.65 (m, ^1H) ppm. ^{19}F NMR: δ −64.2 (dd, $^3J_{HF} = 9.0$ Hz, $^4J_{HF} = 2.0$ Hz) ppm. MS (70 eV) *m/z* (relative intensity): 180 (M^+, 27), 161 (91), 160 (89); 132 (52), 113 (100), 112 (83), 69 (87), 29 (77).

Efficient Preparations of Fluorine Compounds, First Edition. Edited by Herbert W. Roesky.

Waste Disposal The residues of the reaction mixture have to be collected in a labeled waste acid container that has to be properly deposited.

Application The starting vinyl ether $(CF_3)_2C{=}CHOCH_3$ is readily prepared by a Wittig reaction between Ph_3P^{+-}-$CHOCH_3$ and hexafluroacetone (polyfluorinated ketone). The Wittig reagent precursor, $[Ph_3P^+CH_2OCH_3]Cl^-$, is readily prepared by the procedure of Wittig and Schlosser [1] via the reaction of Ph_3P and $ClCH_2OCH_3$ [2]. The hydrolysis of the polyfluorinated vinyl ethers is a general reaction for a variety of polyfluorinated ketones, containing CF_3, CF_3CF_2, $CF_3CF_2CF_2$, CF_2Cl, C_6H_5, and n-C_4H_9 groups, and provides a useful chain extension reaction and conversion of the ketone to the chain-extended aldehyde. These 2-*H*-polyfluorinated aldehydes are easily converted to the α-halo (Cl, Br) aldehydes with Cl_2 or Br_2/acetic acid [3]. They are easily oxidized to the carboxylic acid [3]. Thus, they provide a useful addition to the organofluorine chemist's tool box for functional group transformation and an entry to aldehydes difficult to prepare via other methods.

REFERENCES

1. Wittig G, Schlosser M. Phosphinealkylenes as olefin-forming reagents. IV. Preparation of vinyl ethers, vinyl thioethers, and vinyl halides from ylids. *Chem Ber* 1961;94:1373–1383.
2. Marvel CS, Porter PK. Monochloromethyl ether. In: Gilman H, Blatt AH, editors. *Organic Synthesis, Collected Volume 1*. New York: Wiley; 1964. Vol. 94, p. 377–379.
3. Wiebe DA. Synthesis of polyfluorinated aldehydes, acyl halides, and carboxylic acids. PhD Thesis. 1973. Iowa City (IA): University of Iowa.

CHAPTER 41

Synthesis of Tetrafluorocatechol

SUDHARSANAM RAMANATHAN, DAYONG SANG, VIVEK KUMAR, AND DAVID M. LEMAL

Tetrafluorocatechol has excited considerable interest in connection with lithium ion batteries, which have an ever-expanding role in modern society. It has been found that borate [1,2], phosphate [3], and especially boronate complexes [4–6] of the catechol improve the performance of the electrolytes. As a result of the electron deficiency of the ligand, these electrochemically robust compounds are effective at complexing anions, thereby increasing the electrical conductivity of nonaqueous electrolytes. For the same reason, their anodic limits are high, which can be valuable for overcharge protection [4].

Oxidation of the catechol with nitric acid yields *o*-fluoranil (tetrafluoro-*o*-benzoquinone) [7]. This highly electron-deficient and reactive quinone is a versatile cycloaddend and potentially useful building block in the synthesis of organofluorine compounds [7–9].

Burdon et al. synthesized tetrafluorocatechol in three steps from hexafluorobenzene [10]. Reaction with ethylene glycol followed by cyclization gave tetrafluorobenzo-1,4-dioxene, which was cleaved with aluminum chloride in benzene to yield the catechol in an overall yield of 42%. Reproducing that yield has been difficult [11], however, and in a very recent paper that describes procedural modifications of the Burdon route, the authors claim that their overall yield of 5% represents an improvement [4]. Gores and coworkers developed a more practical variation of the Burdon route that proceeds in three steps from pentafluorophenol [2]. Conversion to the potassium salt, then high-temperature treatment with ethylene oxide gave the same dioxene, from which the catechol was obtained with aluminum chloride in toluene. The overall yield was 57%.

We have developed a variation on the Gores approach to the catechol, outlined next. The starting material is hexafluorobenzene instead of pentafluorophenol, and an easily handled liquid oxirane is substituted for the highly toxic gas ethylene oxide.

Efficient Preparations of Fluorine Compounds, First Edition. Edited by Herbert W. Roesky.

Aluminum iodide, prepared in situ, is used in hexanes for cleavage of the intermediate dioxene, a substitution that facilitates the workup of the catechol. Our overall yield is 64%.

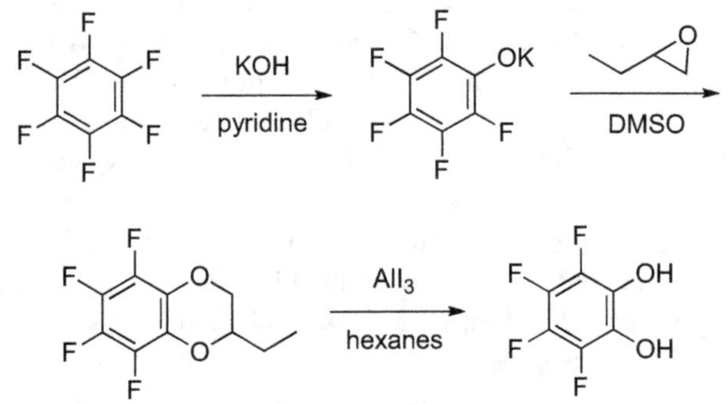

41.1 PREPARATION

Apparatus Three-necked flask (1 L), mechanical stirrer, reflux condenser with gas adapter, dropping funnels (60 and 125 mL), distillation heads (standard taper 24/40 and 14/20), separatory funnel (1 L), heat gun, filter funnel (6–8 cm diameter).

Chemicals Hexafluorobenzene, potassium hydroxide, pyridine, dimethyl sulfoxide, 1,2-epoxybutane, methylene chloride, sodium sulfate, aluminum powder, iodine, hexanes, hydrochloric acid, diethyl ether, sodium bisulfite.

Attention! Safety glasses and protective gloves should be worn at all times, and all operations should be conducted in an efficient fume hood.

Experimental Procedure *2-Ethyl-5,6,7,8-tetrafluorobenzo-1,4-dioxene.* A 1-L, three-necked flask fitted with a mechanical stirrer, reflux condenser connected to a nitrogen source, and stopper was charged with pyridine (100 mL) and freshly crushed KOH pellets (85%, 41 g, 620 mmol). When the stirred mixture had been brought to reflux, hexafluorobenzene (58 g, 310 mmol) was added dropwise over 10 min [12]. Refluxing was continued for 7 h, and then the pyridine was removed by distillation at a bath temperature of 120 °C and pressure down to 20 Torr. The potassium pentafluorophenolate thus formed was used as such for the next step. The distillation head was removed and the flask was again equipped with a reflux condenser connected to the nitrogen source. Dimethyl sulfoxide (100 mL) was added to the solid reaction mass. When the mixture had been heated with stirring to 120 °C, 1,2-epoxybutane (36 g, 500 mmol) was added drop by drop, and stirring at that temperature was continued for 8 h. After cooling, the reaction mixture was poured into ice-cold water (500 mL) and extracted with methylene chloride (5 × 30 mL). The combined organic extract was washed with brine (100 mL) and dried over anhydrous sodium sulfate. Methylene chloride was evaporated under vacuum and the product was purified by distillation at 130–140 °C and 20 Torr. The dioxene was obtained as yellowish oil (59 g, 81% yield).

Characterization Data ^{1}H NMR ($CDCl_3$): δ 4.35 (dd, J = 11.4, 2.4 Hz, 1H), 4.11 (m, 1H), 3.92 (dd, J = 11.4, 7.8 Hz, 1H), 1.78 (m, 2H), 1.09 (t, J = 7.5 Hz, 3H). ^{19}F NMR ($CDCl_3$): δ −165.5 (m, 1F), −165.6 (m, 1F), 171.2 (m, 1F), 171.6 (m, 1F). ^{13}C NMR ($CDCl_3$): δ 137.4 (dm, $^1J_{CF}$ ~ 245 Hz), 135.7 (dm, $^1J_{CF}$ ~ 237 Hz), 135.4 (dm, $^1J_{CF}$ ~ 242 Hz), 130.2 (dq, J = 12.0, 3.3 Hz), 129.9 (dq, J = 12.0, 3.3 Hz), 75.0, 67.8, 23.7, 9.0. Anal. calcd for $C_{10}H_8O_2F_4$: C, 50.85; H, 3.41; F, 32.17. Found: C, 50.83; H, 3.45; F, 32.16.

Tetrafluorocatechol Into a 1-L, three-necked flask fitted with a mechanical stirrer and reflux condenser connected to the nitrogen source, hexanes (280 mL), aluminum powder (200 mesh, 4.5 g, 170 mmol), and iodine (38.6 g, 152 mmol) were placed. The reaction mass was stirred for 30 min at room temperature and then refluxed for 6 h to complete formation of aluminum iodide. To the violet reaction mixture, 2-ethyl-5,6,7,8-tetrafluorobenzo-1,4-dioxene (20.0 g, 84.7 mmol) was added drop by drop, and refluxing was continued for an additional 9 h. The reaction mixture was allowed to cool, and then chilled in an ice bath. Hydrochloric acid (6 M, 100 mL) was added drop by drop with stirring while the temperature was maintained at 10–20 °C. After an additional 30 min of stirring, layers were separated and the organic layer was washed with water (2 × 50 mL). The combined aqueous layers were extracted with diethyl ether (3 × 100 mL). The violet ether layers were combined and washed with saturated sodium bisulfite solution (2 × 20 mL). Now orange, the ether solution was dried over anhydrous sodium sulfate. When the ether was evaporated under vacuum, the color again became an intense violet. Product was purified by vacuum distillation at 70–80 °C and 20 Torr. Because the distillate was a solid, a heat gun was occasionally needed for it to reach the receiver, which was cooled in ice. The catechol (12.8 g) was crushed and stirred with hexanes (50 mL) for 30 min. White solid was collected on a Buchner or sintered glass funnel and washed with hexanes (25 mL). The catechol was dried under vacuum (12.2 g, 79% yield; mp 68–69 °C, lit. [2] 68 °C).

Characterization Data ^{1}H NMR ($CDCl_3$): δ 5.68 (s, br, 2H). ^{19}F NMR ($CDCl_3$): δ −165.3 (m, 2F), −170.0 (m, 2F). ^{13}C NMR ($CDCl_3$): δ 137.8 (ddd, $^1J_{CF}$ = 241, $^2J_{CF}$ = 15.0, $^3J_{CF}$ = 6.9 Hz), 135.7 (dtd, $^1J_{CF}$ = 237, $^2J_{CF}$ = 14.3, $^3J_{CF}$ = 4.6 Hz), 129.9 (d, $^2J_{CF}$ = 12.9 Hz) [13].

ACKNOWLEDGMENT

This material is based on work supported by the National Science Foundation.

REFERENCES

1. Barthel J, Buestrich R. Procedure for production of new lithium-borate complex. German patent DE 19633027 A1; Feb 19, 1998.

2. Barthel J, Buestrich R, Carl E, Gores HJ. A new class of electrochemically and thermally stable lithium salts for lithium battery electrolytes. III. Synthesis and properties of some lithium organoborates. *J Electrochem Soc* 1996;143:3572–3575.
3. Eberwein M, Schmid A, Schmidt M, Zabel M, Burgemeister T, Barthel J, Kunz W, Gores HJ. Synthesis and electrochemical properties of some lithium chelatophosphates. *J Electrochem Soc* 2003;150:A994–A999.
4. Weng W, Zhang Z, Schlueter JA, Redfern PC, Curtiss LA, Amine K. Improved synthesis of a highly fluorinated boronic ester as dual functional electrolyte additive for lithium-ion batteries. *Power Sources* 2011;196:2171–2178.
5. Lee HS, Ma ZF, Yang XQ, Sun X, McBreen J. Synthesis of a series of fluorinated boronate compounds and their use as additives in lithium battery electrolytes. *J Electrochem Soc* 2004;151:A1429–A1435.
6. Lee HS, Sun X, Yang XQ, McBreen J. Synthesis and study of new cyclic boronate additives for lithium battery electrolytes. *J Electrochem Soc* 2002;149:A1460–A1465.
7. Shteingarts VD, Budnik AG. Aromatic fluoro derivatives. XXX. Tetrafluorobenzo-1,2-quinone. *Izv Sib Otd Akad Nauk SSSR, Ser Khim Nauk* 1967;124–127.
8. Lemal DM, Ramanathan S, Shellito J. *o*-Fluoranil chemistry: Diels-Alder versus Hetero Diels-Alder cycloaddition. *J Org Chem* 2008;73:3392–3396; Lemal DM, Sang D, Ramanathan S. *J Org Chem* 2009;74:7804–7811.
9. Kumar V, Ramanathan S, Sang D, Chen X, Lemal DM. The multifaceted reactivity of *o*-fluoranil. *J. Org. Chem* 2012;77:966–970; Kumar V, Lemal DM. Photochemistry of *o*-fluoranil. *J. Org. Chem* 2012;77:5017–5021
10. Burdon J, Damodaran VA, Tatlow JC. Polyfluoro-aromatic compounds. Part XV. The reaction of hexafluorobenzene with bifunctional nucleophiles. *J Chem Soc* 1964;763–765.
11. Macdonald C, Tomlinson AJ, Willis CJ. Fully fluorinated alkoxides. Part V. Derivatives of tetrafluorocatechol. *Can J Chem* 1971;49:2578–2580.
12. Pummer WJ, Wall LA. Reactions of hexafluorobenzene. *Science* 1958;127:643–644.
13. For the X-ray crystal structure of tetrafluorocatechol, see Ref. [4]. For its electrochemistry, see Stallings MD, Morrison MM, Sawyer DT. Redox chemistry of metal-catechol complexes in aprotic media. 1. Electrochemistry of substituted catechols and their oxidation products. *Inorg Chem* 1981;20:2655–2660.

CHAPTER 42

Preparation of an Unsymmetrical Bis((perfluoroalkyl)sulfonyl)imide

DARRYL D. DESMARTEAU AND THOMAS D. HICKMAN

Bis((perfluoroalkyl)sulfonyl)imides are very strong Brönsted acids and they can be prepared in large variety. In the general formula

$$R_fSO_2\underset{\displaystyle H}{\underset{|}{N}}SO_2R_f'$$

R_f and R_f' can be simple perfluoroalkyl groups such as CF_3 or C_nF_{2n+1}, and these can be equal or different [1–3]. R_f and R_f' can be connected to form a cyclic sulfonimide $-(CF_2)_n-$ [4]. R_f and R_f' may also contain functional groups such as $CF_2{=}CF{-}CF_2-$, and $CF_2{=}CFO-$ making possible perfluorinated polymeric materials containing the sulfonimide function as a pendant group on a polymeric chain [5]. Equally interesting are polyfunctional and polymeric materials where the sulfonimide function is in the polymer main chain [6,7].

$$R_fSO_2\left\{-\overset{\displaystyle H}{\overset{|}{N}}SO_2(CF_2)_nSO_2-\right\}_x\overset{\displaystyle H}{\overset{|}{N}}SO_2R_f'$$

Considering the many other structural variations possible, one can appreciate the large number of potential compounds and materials that can be prepared.

These fluorinated sulfonimides have many applications such as acid catalysis, selective fluorinations, fuel cell membranes, lithium ion batteries, and ionic liquids (ILs) [8–13]. Here we will describe the synthesis of the sulfonimide

Efficient Preparations of Fluorine Compounds, First Edition. Edited by Herbert W. Roesky.

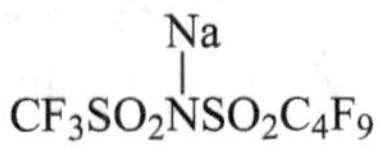

(**4**) and its use in preparing a low-melting, modest-viscosity IL.

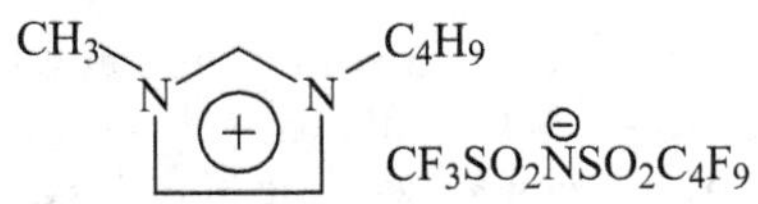

(**6**)

The synthesis route is outlined in Scheme 42.1.

42.1 SYNTHESIS OF $CF_3SO_2N(Na)SO_2C_4F_9$ (4)

General

All the syntheses in these procedures involve the use of corrosive and/or toxic chemicals. Adequate protective apparel (safety glasses, gloves, and laboratory coat) must be worn. All chemicals should be handled in a good fume hood or contained in a closed system. Many of these procedures are carried out using a simple vacuum system capable of producing a vacuum of 1.0×10^{-3} Torr or better. A typical two-stage mechanical pump protected by a liquid nitrogen trap and with appropriate connections is adequate. Teflon glass vacuum valves and ground glass connections lubricated by fluorinated grease are preferred.

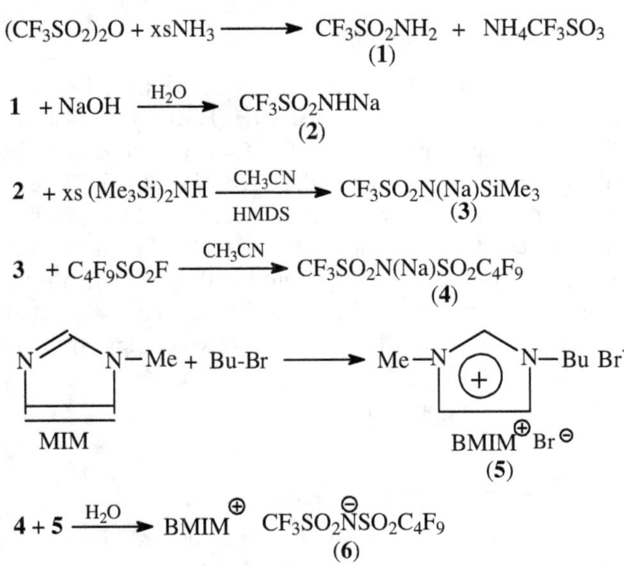

SCHEME 42.1 Synthesis route to **4** and **6**.

Chemicals The chemicals needed for these syntheses are all available from commercial sources. Except where noted, the reagent-grade chemicals can be used as supplied.

Preparation of Sulfonamide (1)

A dry, 250 mL, three-necked flask, fitted with a dropping funnel and containing a magnetic stir bar, is cooled to −78 °C while passing anhydrous NH_3 into the flask. After c. 100 mL liquid NH_3 is condensed into the flask, the NH_3 connection is replaced by a N_2 gas inlet. The triflic anhydride (50 g) is then added drop by drop over 30 min with vigorous stirring. (Caution! The reaction is very exothermic.) After this addition, a flow of dry N_2 (c. 100 sccm) is started and the flask with stirring is allowed to warm in the air to 22 °C. The N_2 flow is continued overnight. The flask is then connected to the vacuum line and evacuated for 20 min to remove any remaining NH_3. The white solid is then transferred to a sublimation apparatus and the amide **1** is vacuum sublimed at 80 °C onto a cold collector cooled to 0 °C (24.1 g, 91%).

NMR (CD_3CN): δ_F −79.4 ppm; δ_H 6.60 ppm.

Preparation of CF_3SO_2NHNa (2)

Compound **1** (16.8 g, 0.11 mmol) is placed in a 250 mL flask and neutralized with 1.0 M NaOH to a pH 8.4. The solution is stirred for 3 h at 22 °C and then the flask is connected to the vacuum line. Heating at 100 °C under dynamic vacuum for 12 h gives pure **2** (19.2 g, 95%).

NMR (CD_3CN): δ_F −79.3 ppm; δ_H 6.60 ppm.

Preparation of $CF_3SO_2N(Na)SiMe_3$ (3)

Into a 250 mL, round-bottom flask with a 24/40 ₮ connection, 19.2 g of compound **2** and 50 mL each of dry CH_3CN and $(Me_3Si)_2NH$ (HMDS) are placed. The flask is fitted with a reflux condenser connected to a $CaSO_4$ drying tube. The flask is heated for 24 h in an oil bath to maintain reflux of the CH_3CN at 80 °C. The excess CH_3CN and HMDS are then removed under dynamic vacuum and the remaining white to tan solid is vacuum dried at 100 °C for 12 h. The yield of compound **3** is typically about 95%, but it is best to use it directly from the flask in which it was prepared. Compound **3** is very moisture sensitive and must be handled in a good dry box. Very dry CD_3CN must be used for the NMR, or one will observe a spectrum for compound **2** and HMDS.

NMR (CD_3CN): δ_F −78.6 ppm; δ_H 0.02 ppm.

Preparation of $CF_3SO_2N(Na)SO_2C_4F_9$ (4)

In a dry box, compound **3** (3.17 g, 13 mmol) is added to a 250 mL, round-bottom flask fitted with an Ace-Thred® connector, a Teflon glass valve, and a Teflon-coated magnetic stir bar. Dried (P_4O_{10}) CH_3CN (50 mL) is then added followed by $C_4F_9SO_2F$

(5.90 g, 19.5 mmol) addition. The sealed flask is then heated in an oil bath at 110 °C for 48 h. The depth of the oil bath is such that one-half of the round-bottom part of the flask is immersed in the oil. (Caution! The flask has 2–3 atm pressure in the 110 °C bath.) After heating, the flask is attached to the vacuum line and the product $(CH_3)_3SiF$ and the CH_3CN and excess $C_4F_9SO_2F$ are removed. The remaining solid is then dried under vacuum at 80 °C for 12 h. The white solid **4** (5.17 g, 87%) is isolated from the reactor.

^{19}F NMR (CD_3CN): $CF_3{}^{A}SO_2N(Na)SO_2CF_2{}^{B}CF_2{}^{C}CF_2{}^{D}CF_3{}^{E}$: δ_A −78.9 (s, 3F), δ_E −80.3 (m, 3F), δ_B −112.7 (m, 2F), δ_C −120.4 (m, 2F), δ_D −125.2 (m, 2F) ppm.

42.2 SYNTHESIS OF IL (6)

Preparation of BMIM$^+$Br$^-$ (5)

The starting materials for this synthesis require purification. 1-Methylimidizole (1-MIM) (∼5 g) was added to a 25 mL flask containing 10 g KOH pellets. The flask was stoppered and allowed to stand at 22 °C for 12 h. The imidizole is then vacuum (10^{-2} Torr) distilled at 70 °C using a small distillation apparatus.

n-Butyl bromide (∼5 g) was added to a 25 mL flask and 98% H_2SO_4 (5 mL) is added. The mixture is shaken and the *n*-butyl bromide is decanted off into a 25 mL flask containing a magnetic stirring bar. A small portion (∼5 g) of P_4O_{10} is then added and the flask is closed with Teflon glass valve attached to a *** joint. The mixture was stirred at 22 °C overnight. The flask is connected to the vacuum line and the *n*-BuBr is vacuum transferred at 22 °C into a 25 mL flask fitted with a Teflon glass valve and cooled to −196 °C. After the transfer, the flask is closed, warmed to 22 °C, and dry N_2 is added to atmospheric pressure.

Into a 25 mL flask, 1-MIM (1.74 g, 21.2 mmol) is added and the flask is capped with a rubber septum. Then, *n*-BuBr (3.48 g, 25.4 mmol) is added by syringe. The flask is heated to 50 °C in a small ultrasound cleaning bath and sonicated for 12 h. The flask is then connected to vacuum line and heated to 70 °C overnight under dynamic vacuum to dry the BMIM ⊕ Br ⊖.

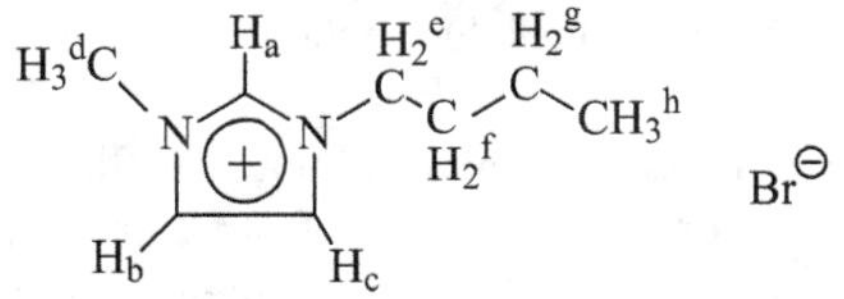

T_m 83 °C. ^{1}H NMR (CD_3CN): δ_a 9.28 (1H, s); δ_b 7.44 (1H, t); δ_c 7.48 (1H, t); δ_d 3.86 (3H, s); δ_e 4.18 (2H, m); δ_f 1.79 (2H, m); δ_g 1.30 (2H, m); δ_h 0.93 (3H, t) ppm.

Preparation of BMIM⊕ $CF_3SO_2N^{\ominus}SO_2C_4F_9$ (6) IL **5** (3.6 g, 16 mmol) is dissolved in 20 mL H_2O in a 100 mL flask. Compound **4** (6.3 g, 13 mmol) is dissolved in another 20 mL H_2O. This latter is added to IL **5** and the mixture is stirred

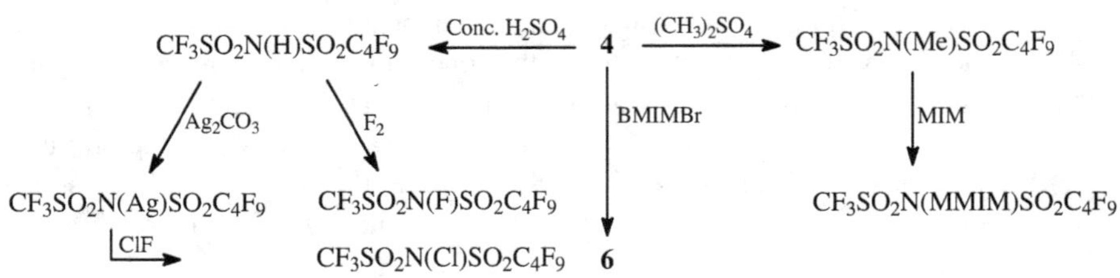

SCHEME 42.2 Illustrative reactions of compound **4**.

at 22 °C for 12 h. IL **6** is insoluble in H_2O and separates as a clear dense liquid. IL **6** is washed four times with 15 mL portions of distilled water and then dried under dynamic vacuum for 36 h at 70 °C.

$CF_3{}^ASO_2N(BMIM)SO_2CF_2{}^BCF_2{}^CCF_2{}^DCF_3{}^E$; ^{19}F NMR (CD_3CN, ppm): δ_A −78.9 (s, 3F), δ_E −80.3 (t, 3F), δ_B −112.7 (m, 2F), δ_C −120.4 (m, 2F), δ_D −125.2 (m, 2F). ^1H NMR (CD_3CN, ppm): δ 8.35 (s, 1H), δ 7.33 (m, 1H), δ 7.30 (m, 1H), δ 4.09 (5, 2H), δ 3.78 (2, 3H), δ 1.77 (m, 2H), δ 1.30 (m, 2H), δ 0.91 (t, 3H). DSC: T_g −86 °C. TGA (N_2) 10 °C/min: −1%, 300 °C; −20%, 400 °C.

42.3 DISCUSSION

Sulfonamide **4** is useful for preparing a variety of sulfonamide derivatives including **6**. Some of these are illustrated in Scheme 42.2 [2, 9, 12].

The *N*–Cl compound is a potent electrophilic chlorine source and the *N*–F derivative is a useful electrophilic-like, selective fluorination reagent. The *N*–Me compound is a potent alkylating agent and the halide-free synthesis of an MMIM derivative is shown.

ILs are a very large area of research and development. These unusual materials range from liquids at 22 °C and below to many with melting points considerably higher. Room temperature ionic liquids (RTILs) are usually any salt, liquid below 100 °C. ILs have very low or no vapor pressure over a wide range of temperatures and they often have very high thermal stability depending on the nature of the cation and the anion. ILs have many potential applications such as useful “green” solvents for chemical reactions, battery electrolytes due to their high conductivity, and lubricants [14–17].

There are a very large number of known and potential ILs because both the cation and anion can be varied over a wide range of structures that, in turn, influence their properties. One limitation to many ILs is their high viscosity. Viscosity is strongly influenced by both the structure of the cation and the anion. For example, a widely used anion for ILs is the bis((trifluoromethyl)sulfonyl)imide $(CF_3SO_2)_2N^{\ominus}$ or $TFSI^{\ominus}$. Viscosity values for $MMIM^{\oplus}TSFI^{\ominus}$ are reported from 44 to 100 cP at 20–25 °C, whereas $BMIM^{\oplus}TFSI^{\ominus}$ values range from 34 to 69 cP. It was hoped that the unsymmetrical anion in compound **4** would lower the viscosity but the opposite occurred. $MMIM^{\oplus}$ $NTFS^{\ominus}$ and $BMIM^{\oplus}NTFSI^{\ominus}$ have values at 25 °C of 92 and 125 cP, respectively. Clearly the issue of viscosity versus structure in ILs is complex, as are variations in other properties.

REFERENCES

1. Foropoulos J, Jr, DesMarteau DD. Synthesis, properties, and reactions of bis((trifluoromethyl)sulfonyl) imide, $(CF_3SO_2)2NH$. *Inorg Chem* 1984;23:3720–3723.
2. Singh S, DesMarteau DD. Chemistry of perfluoromethylsulfonyl perfluorobutylsulfonyl imide. *Inorg Chem* 1990;29:2982–2985.
3. Hu LQ, DesMarteau DD. Synthesis of perhaloalkanesulfonyl halides and their sulfonimide derivatives. *Inorg Chem* 1993;32:5079–5083.
4. DesMarteau DD, Zuberi SS, Pennington WT, Randolph BB. Structural studies of silver(I) salts of the cyclic sulfonimides cyclic $(CF_2)_n$ SO_2 $NHSO_2$ (n = 3,4). *Eur J Solid State and Inorg Chem* 1992;29:777–789.
5. Thomas BH, Shafer G, Ma JJ, Tu MH, DesMarteau DD. Synthesis of 3,6-dioxa-Δ7-4-trifluoromethyl perfluorooctyl trifluoromethyl sulfonimide: bis[(perfluoroalkyl)sulfonyl] superacid monomer and polymer. *J Fluorine Chem* 2004;125:1231–1240.
6. Appleby AJ, Velev OA, Parthasarthy JG, Srinivasan S, DesMarteau DD, Gillette MS, Ghosh JK. Polymeric perfluoro bis-sulfonimides as possible fuel cell electrolytes. *J Electrochem Soc* 1993;140:109–111.
7. Geiculescu OG, Yang J, Zhou S, Shafer G, Xie Y, Albright J, Creager SC, Pennington WT, DesMarteau DD. Solid polymer electrolytes from polyanionic lithium salts based on the LiTFSI anion structure. *J Electrochem Soc* 2004;151:1363–1368.
8. Singh S, DesMarteau DD, Zuberi SS, Witz M, Huang HN. N-Fluoroperfluoroalkyl sulfonimides. Remarkable new fluorination reagents. *J Am Chem Soc* 1987;109:7194–7196.
9. Zhang J, DesMarteau DD. Synthesis and comparison of N-fluorobis[(perfluoroalkyl)-sulfonyl]imides with different perfluoroalkyl groups. *J Fluorine Chem* 2001;111:253–257.
10. Savett SC, Atkins JR, Sides CR, Harris JL, Thomas BH, Creager SE, Pennington WT, DesMarteau DD. A comparison of Bis[(perfluoroalkyl)sulfonyl]imide ionomers and perfluorosulfonic acid ionomers for applications in PEM fuel-cell technology. *J Electrochem Soc* 2002;149:1527–1532.
11. Geiculescu OE, Creager SE, DesMarteau DD. Fluorinated electrolytes based on lithium salts of strong Bronsted acids. *Fluorinated Mater Energy Convers* 2005;223–251.
12. Zhang J, Marten GR, DesMarteau DD. Direct methylation and trifluoroethylation of imidazole and pyridine derivatives. *Chem Commun* 2003;18:2334–2335.
13. Xue H, Verma R, Shreeve JM. Review of ionic liquids with fluorine-containing anions. *J Fluorine Chem* 2006;127:159–176.
14. J. LeBideau, Viau L, Vioux A. Ionogels, ionic liquid based hybrid materials. *Chem Soc Rev* 2011;40:907–925.
15. Hallett JP, Welton T. Room-temperature ionic liquids: solvents for synthesis and catalysis. 2. *Chem Rev* 2011;111:3508–3576.
16. Hapiot P, Lagrost C. Electrochemical reactivity in room-temperature ionic liquids. *Chem Rev* 2008;108:2238–2264.
17. Greaves TL, Drummond CJ. Protic ionic liquids: properties and applications. *Chem Rev* 2008;108:206–237.

CHAPTER 43

Preparation of Perfluoroalkyl Sulfilimines and Sulfoximines

EMMANUEL MAGNIER

The incorporation of a fluoroalkylated group via an electrophilic pathway has recently grown in importance as a result of the increasing number of reagents proposed so far [1]. Recent innovative research in this area has highlighted the importance of the perfluoroalkyl sulfoximine structures as new electrophilic perfluoroalkylating agents as recently introduced by Shibata and coworkers for trifluoro- and monofluoromethylation [2] and Hu and coworkers as well as Prakash and coworkers for difluoromethylation [3].

We disclosed a Ritter-like reaction between nitriles and sulfoxides activated by trifluoromethanesulfonic anhydride. This scalable, flexible, and free of solvent procedure gave rise to a wide range of new acyl sulfilimines. We also described the preparation of sulfoximines, precursors of electrophilic fluoroalkylating reagents, by smooth oxidation of the sulfur(IV) precursors [4].

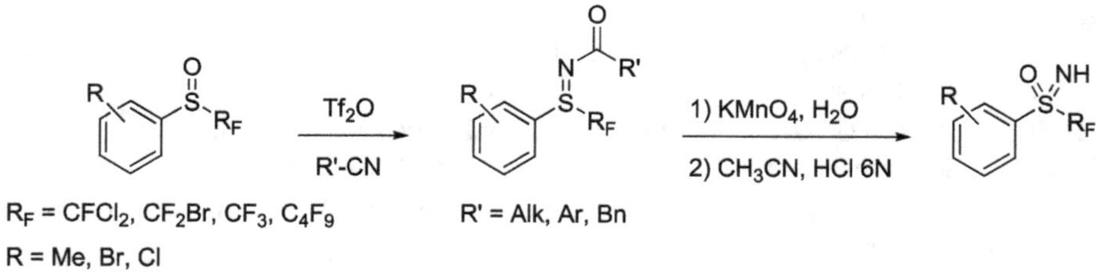

Efficient Preparations of Fluorine Compounds, First Edition. Edited by Herbert W. Roesky.

43.1 PREPARATION OF PERFLUOROALKYL SULFILIMINES AS EXEMPLIFIED BY THE PREPARATION OF *N*-ACETYL-TRIFLUOROMETHYLPHENYLSULFILIMINE

Apparatus A 100-mL Schlenk flask equipped with septum and magnetic stirrer, two disposable plastic syringes (5 and 2 mL), safety glasses, laboratory coat, protective gloves.

Chemicals $PhSOCF_3$, $(CF_3SO_2)_2O$, dry acetonitrile.

Attention! Safety glasses and protective gloves must be used at all times.

Caution! The reaction should be carried out in a well-ventilated hood.

Experimental Procedure A 100-mL Schlenk flask is charged with phenyl trifluoromethyl sulfoxide (3 g, 15 mmol) and cooled at −15 °C. Then, under argon, acetonitrile (1.2 mL, 22.5 mmol, 1.5 equiv) and trifluoromethanesulfonic anhydride (3.9 mL, 22.5 mmol, 1.5 equiv) are added through the septum with the help of syringes. The reaction mixture is stirred for 16 h at −15 °C, and then diluted with dichloromethane (20 mL). The solution is cooled at −40 °C, carefully hydrolyzed with water (15 mL), allowed to reach room temperature, and extracted with CH_2Cl_2 (3 × 45 mL). Combined organic phases are dried over $MgSO_4$ and concentrated under reduced pressure. The residue is purified by flash chromatography (ether/pentane, 7:3) to give 2.7 g (75%) of yellow oil.

Characterization Data ^{19}F NMR ($CDCl_3$, 188 MHz): δ −64.7 (s, 3F) ppm. ^{1}H NMR ($CDCl_3$, 300 MHz): δ 7.87 (d, J = 8.3 Hz, 2H), 7.72–7.57 (m, 3H), 2.22 (s, 3H) ppm. ^{13}C NMR ($CDCl_3$, 75 MHz): δ 183.7, 134.3, 130.2, 128.7, 127.3, 124.2 (q, J = 324 Hz, CF_3), 24.0 ppm. Pos. ESI MS (*m*/*z*): 189 ([MNa-CF_3]$^+$), 258 (MNa$^+$), 493 (2MNa$^+$). Elemental anal. (%): calcd for $C_9H_8F_3NOS$: C, 45.95; H, 3.43; N, 5.95. Found: C, 45.54; H, 3.53; N, 5.89.

Scope of the Reaction This methodology allows the variation of the functionalization of the aromatic ring (at *ortho*, *meta*, or *para* position), the nature of the nitrile (with a decrease of the yield with electron-withdrawing group close to the nitrogen atom), and the nature of the fluorinated chain ($CFCl_2$, CF_2Br, CF_3, C_4F_9).

43.2 PREPARATION OF PERFLUOROALKYL SULFOXIMINES AS EXEMPLIFIED BY THE PREPARATION OF *S*-(PHENYL)-*S*-(TRIFLUOROMETHYL)-SULFOXIMINE

Apparatus Round-bottom, 250- and 100-mL, flasks, magnetic stirrer, safety glasses, laboratory coat, protective gloves.

Chemicals Potassium permanganate, water, HCl 6 M, acetonitrile.

Attention Safety glasses and protective gloves must be used at all times.

Experimental Procedure A 250-mL, round-bottom flask is charged with *N*-acetyl-trifluoromethylphenylsulfilimine (6.5 g, 27 mmol). Then, water

(100 mL) and solid potassium permanganate (4.4 g, 27 mmol, 1 equiv) are added. The reaction is stirred at room temperature overnight. Solid $Na_2S_2O_4$ is added until disappearance of the color, then the reaction is diluted with water (50 mL) and extracted with CH_2Cl_2 (3 × 50 mL). The organic layers are dried over $MgSO_4$ and concentrated under reduced pressure to give the corresponding *N*-acyl sulfoximine (6.6 g). This compound is diluted in CH_3CN (26 mL) in a 100-mL flask and a solution of HCl 6 N (9 mL) is added. The reaction is stirred at room temperature for 18 h, and then water (50 mL) is added. The mixture is extracted with CH_2Cl_2 (3 × 50 mL). The combined organic layers are washed with a solution of $NaHCO_3$ 10%, dried over $MgSO_4$, and concentrated under reduced pressure. The residue is purified by column chromatography on silica gel (pentane/Et_2O, 9:1) to give 5.4 g (93%) of a white powder.

Characterization Data Mp: 89.9 ± 0.2 °C. ^{19}F NMR ($CDCl_3$, 188 MHz): δ −79.3 (3F, s) ppm. ^{1}H NMR ($CDCl_3$, 300 MHz): δ 8.16 (2H, d, $J = 7.3$ Hz), 7.83–7.75 (1H, m), 7.69–7.61 (2H, m), 3.62 (1H, br s for NH) ppm. ^{13}C NMR ($CDCl_3$, 75 MHz): δ 35.5, 131.6, 130.6, 129.5, 120.9 (q, $J = 332$ Hz, CF_3) ppm. Pos. ESI (*m*/*z*): 210 (MH^+), 232 (MNa^+), 441 ($2MNa^+$). Anal. calcd for $C_8H_8F_3NOS$: C, 43.05; H, 3.61; N, 6.27. Found: C, 43.03; H, 3.56; N, 6.31.

Scope of the Reaction This methodology can be nicely applied to a wide range of previously prepared sulfilimines (vide supra).

Application The easy *N*-functionalization of the sulfoximines allowed the introduction of a wide range of functional groups. The most promising derivatives were obtained with an electron-withdrawing function attached to the nitrogen for application as electron-withdrawing group [5], liquid crystals properties [6], and above all as electrophilic perfluoroalkylating reagent. The formation of an ammonium function by double methylation of the *S*-(phenyl)-*S*-(trifluoromethyl)-sulfoximine generates an activated species able to realize the trifluoromethylation of β-ketoesters as well as dicyanoalkylidene skeletons [2a]. We have recently showed that the trifluoromethanesulfonyl group was able to sufficiently activate dichlorofluoro- and bromodifluoromethyl sulfoximines, allowing then the transfer of these halogenated groups on an sp carbon [7].

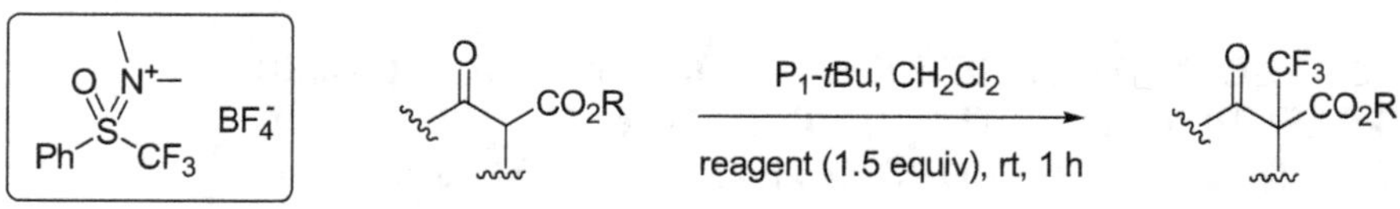

P$_1$-*t*Bu: *tert*butylimino-tris(dimethylamino)phosphorane

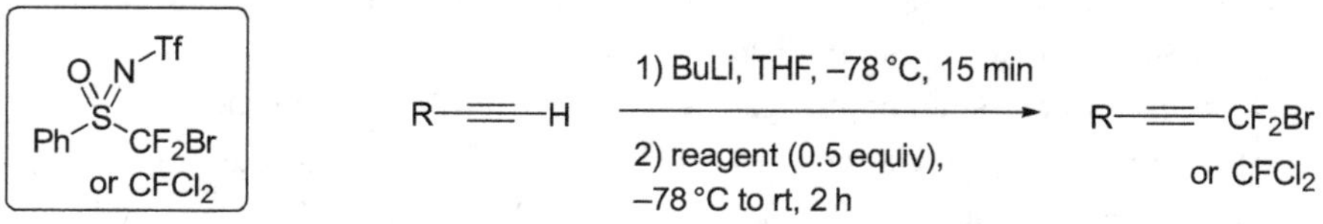

REFERENCES

1. (a) Hu J, Zhang W, Wang F. Selective difluoromethylation and monofluoromethylation reactions. *Chem Commun* 2009:7465–7478. (b) Shibata N, Matsnev A, Cahard D. Beilstein. Shelf-stable electrophilic trifluoromethylating reagents: a brief historical perspective. *J Org Chem* 2010;6. doi:10.3762/bjoc.6.65. (c) Macé Y, Magnier E. The new age of electrophilic perfluoroalkylation reactions. *Eur J Org Chem* 2012;2479–2494.
2. (a) Noritake S, Shibata N, Nakamura S, Toru T, Shiro M. Fluorinated Johnson reagent for transfer-trifluoromethylation to carbon nucleophiles. *Eur J Org Chem* 2008;3465–3468. (b) Nomura Y, Tokunaga E, Shibata N. Angew. Inherent oxygen preference in enolate monofluoromethylation and a synthetic entry to monofluoromethyl ethers. *Chem, Int Ed* 2011;50:1885–1889.
3. (a) Zhang W, Wang F, Hu J. N-Tosyl-S-difluoromethyl-S-phenylsulfoximine: a new difluoromethylation reagent for S-, N-, and C-nucleophiles. *Org Lett* 2009;11:2109–2112. (b) Zhang W, Huang W, Hu J. Angew. Highly stereoselective synthesis of monofluoroalkenes from α-fluorosulfoximines and nitrones. *Chem, Int Ed* 2009;48:9858–9861. (c) Prakash G. K. S, Zhang Z, Wang F, Ni C, Olah G. A. N,N-Dimethyl-S-difluoromethyl-S-phenylsulfoximinium tetrafluoroborate: a versatile electrophilic difluoromethylating reagent. *J Fluorine Chem* 2011;132:792–798.
4. (a) Macé Y, Urban C, Pradet C, Marrot J, Blazejewski J-C, Magnier E. Sulfilimines and sulfoximines by reaction of nitriles with perfluoroalkyl sulfoxides. *Eur J Org Chem* 2009;3150–3153. (b) Urban C, Macé Y, Cadoret F, Blazejewski J-C, Magnier E. Divergent preparation of fluoroalkylated sulfilimine and sulfilimino iminium salts. *Adv Synth Catal* 2010;352:2805–2814.
5. (a) Yagupolskii LM. Aromatic compounds with new fluorine-containing substituents. *J Fluorine Chem* 1987;36:1–28. (b) Terrier F, Magnier E, Kizilian E, Wakselman C, Buncel E. Superacidifiers: assessing the activation and the mode of charge transmission of the extraordinary electron-withdrawing SO_2CF_3 and $S(O)(=NSO_2CF_3)CF_3$ substituents in carbanion stabilization. *J Am Chem Soc* 2005;127:5563–5571.
6. Kirsch P, Lenges M, Kühne D, Wanczek K-P. Synthesis and structural characterization of highly fluorinated sulfimides and sulfoximides as functional building blocks for materials science. *Eur J Org Chem* 2005;797–802.
7. Urban C, Cadoret F, Blazejewski J-C, Magnier E. Sulfoximines as a versatile scaffold for electrophilic fluoroalkylating reagents. *Eur J Org Chem* 2011;4862–4867.

CHAPTER 44

Preparation of Trifluoromethylsulfonium Salts

EMMANUEL MAGNIER

Many recent progresses have been recently realized in the domain of electrophilic introduction of fluoroalkylated groups due to the availability of various and complementary reagents [1]. One family of molecule, inspired by the pioneering work of Yagupolskii et al. [2], is based on sulfonium skeletons. Many teams have contributed to the growth of this family, which now covers not only electrophilic trifluoromethylation reaction (Yang et al. [3], Umemoto and Ishihara [4], and Matsnev et al. [5]) but also mono- and difluoroalkylation (Prakash et al. [6]) as well as pentafluoroethylation [7].

Our group has recently proposed a one-pot and benchmark procedure that allowed the easy homemade preparation of biaryl- and dibenzothiophenium sulfonium salts [8]. We have shown that the simple treatment of an aromatic derivative (mono- or biphenylic) by the couple sodium trifluoromethanesulfinate/trifluoromethanesulfonic anhydride produced the corresponding sulfonium salts. The flexibility of our methodology gives the possibility to adapt the reactivity of the reagent to the nucleophile, by the variation of its structure and the modification of the substituents attached to the aromatic rings.

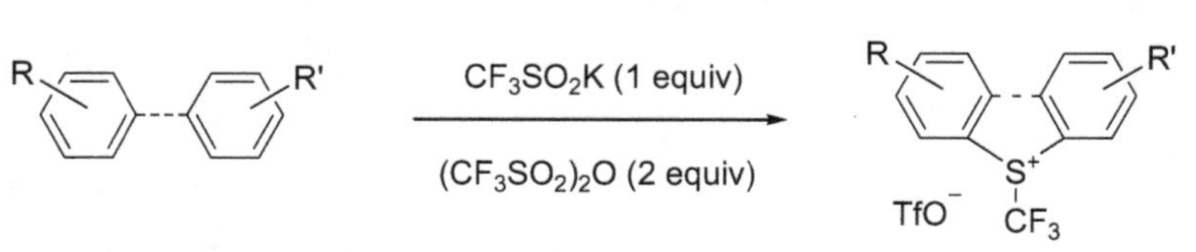

Efficient Preparations of Fluorine Compounds, First Edition. Edited by Herbert W. Roesky.

44.1 PREPARATION OF TRIFLUOROMETHYLSULFONIUM SALTS AS EXEMPLIFIED BY THE PREPARATION OF *S*-(TRIFLUOROMETHYL)DIPHENYLSULFONIUM TRIFLATE

Apparatus A 100-mL Schlenk flask equipped with septum and magnetic stirrer, two disposable plastic syringes (5 and 2 mL), safety glasses, laboratory coat, protective gloves.

Chemicals CF_3SO_2Na, $(CF_3SO_2)_2O$, benzene.

Attention! Safety glasses and protective gloves must be used at all times.

Caution! Benzene is toxic and volatile. Contact with skin and inhalation must be avoided. All reactions should be carried out in a well-ventilated hood.

Experimental Procedure A 100-mL Schlenk flask is charged with well-dried sodium trifluoromethanesulfinate (2.2 g, 12.8 mmol). Under argon, benzene (1.14 mL, 12.8 mmol, 1 equiv) and then trifluoromethanesulfonic anhydride (4.28 mL, 24.4 mmol, 2 equiv) are added through the septum with the help of syringes. After 16 h, the reaction mixture is filtered, diluted with CH_2Cl_2 (30 mL), and washed with water (3 × 10 mL). The organic phase is dried over $MgSO_4$ and concentrated under reduced pressure. The residue is purified by column chromatography on silica gel using dichloromethane/methanol (90:10) as eluent to give 1.76 g (70%) of a slightly colored powder. Recrystallization from pentane/ethyl acetate (2:8) affords 1.5 g (60%) of a white solid.

For large-scale reaction (4 g of sodium trifluoromethanesulfinate and more), the reaction mixture should be cooled to 0 °C before the introduction of trifluoromethanesulfonic anhydride and then slowly bring back to room temperature.

The purification by column chromatography on silica gel can be avoided by successive washing of the pasty crude mixture by diethyl ether until a solid is obtained.

Characterization Data Mp: 82 ± 0.2 °C. IR (KBr): 3098, 1270, 1250, 1152, 1090, 630 cm^{-1};.^{1}H NMR ($CDCl_3$, 200 MHz): δ 8.20 (d, $J = 8$ Hz, 4H), 7.95–7.77 (m, 6H) ppm. ^{19}F NMR ($CDCl_3$, 188 MHz): δ −50.6 (m, 3F, SCF_3), −79.0 (m, 3F, SO_2CF_3) ppm. ^{13}C NMR ($CDCl_3$, 50 MHz): δ 137.1, 132.9, 132.3, 123.2 (q, $J = 325.0$ Hz, CF_3), 120.7 (q, $J = 318.7$ Hz, CF_3), 117.6 ppm. Pos. ESI (*m/z*): 255 (M^+). Anal. calcd for $C_{14}H_{10}F_6O_3S_2$: C, 41.59; H, 2.49. Found: C, 41.69; H, 2.42.

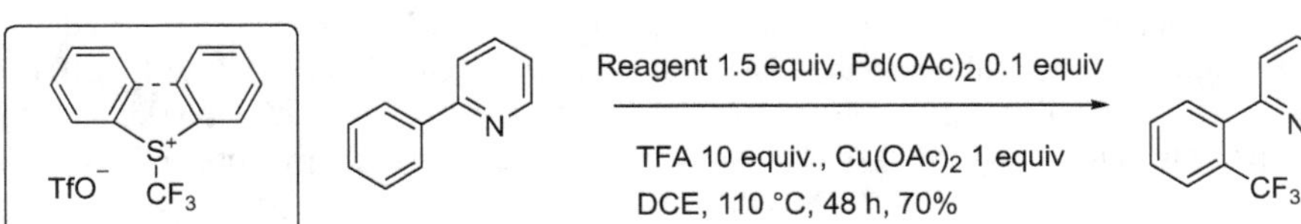

SCHEME 44.1 Metal catalysed trifluoromethylation through CH-activation.

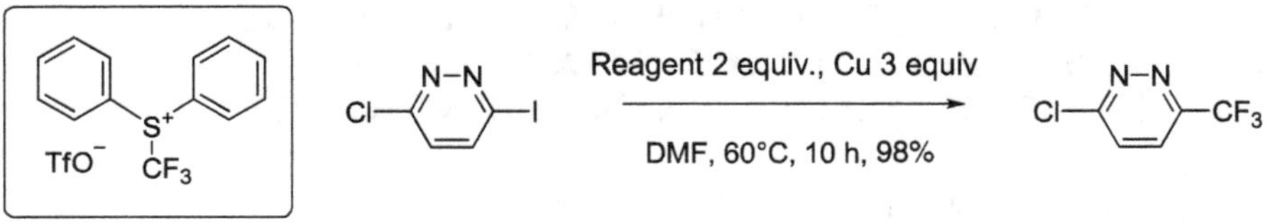

SCHEME 44.2 Aromatic trifluoromethylation.

Scope of the Reaction This methodology can be nicely applied to a set of aromatic and biphenylic compounds.

Application Trifluoromethylsulfonium salts have been widely used for the electrophilic trifluoromethylation of numerous soft nucleophiles: thiolates, enamines, enoxysilanes, aromatics, heteroaromatics, β-ketoesters, phosphines, etc. [1b, 4, 9].

Very recently, interesting work has described new reactivity of sulfonium salts due to the presence of metals. The Pd(II)-catalyzed *ortho*-trifluoromethylation of 2-phenylpyridine by Umemoto reagent was realized in the presence of copper(II) acetate and trifluoroacetic acid as promoters (Scheme 44.1) [10]. The scope of the reaction has been evaluated through the variation of the heterocyclic groups and the functionalization of the aromatic part.

Xiao and coworkers have shown the importance of "open" sulfonium salts as *S*-(trifluoromethyl)diphenylsulfonium triflate by its use for the trifluoromethylation of iodo heterocyclic compounds in the presence of copper [11]. The efficiency of this transformation has been proved by the description of impressive numbers of examples, including important variation of heteroaromatic rings (Scheme 44.2).

REFERENCES

1. (a) Hu J, Zhang W, Wang F. Selective difluoromethylation and monofluoromethylation reactions. *Chem Commun* 2009;7465–7478. (b) Shibata N, Matsnev A, Cahard D. Shelf-stable electrophilic trifluoromethylating reagents: a brief historical perspective. *Beilstein J Org Chem* 2010;6. doi:10.3762/bjoc.6.65. (c) Macé Y, Magnier E. The new age of electrophilic perfluoroalkylation reactions. *Eur J Org Chem* 2012; 2479–2494
2. (a) Yagupolskii LM, Kondratenko NV, Timofeeva GN. Fluoro(trifluoromethyl) arylsulfonium and (trifluoromethyl)diarylsulfonium salts. *J Org Chem USSR* 1984;20:103–105. (b) Yagupol'skii LM, Matsnev VA, Orlova RK, Deryabkin BG, Yagupolskii YL. A new method for the synthesis of trifluoromethylating agents - Diaryltrifluoromethylsulfonium salts. *J Fluorine Chem* 2008;129:131–136.
3. Yang JJ, Kirchmeier RL, Shreeve JM. New electrophilic trifluoromethylating agents. *J Org Chem* 1998;63:2656–2660.

4. Umemoto T, Ishihara S. Power-variable electrophilic trifluoromethylating agents. S-, Se-, and Te-(Trifluoromethy1)dibenzothio-, -seleno-, and -tellurophenium salt system. *J Am Chem Soc* 1993;115:2156–2164.
5. Matsnev A, Noritake S, Nomura Y, Tokunaga E, Nakamura S, Shibata N. Efficient access to extended Yagupolskii-Umemoto-type reagents: triflic acid catalyzed intramolecular cyclization of ortho-ethynylaryltrifluoromethylsulfanes. *Angew Chem, Int Ed* 2010;49:572–576.
6. (a) Prakash GKS, Weber C, Chacko S, Olah GA. New electrophilic difluoromethylating reagent. *Org Lett* 2007;9:1863–1866. (b) Prakash GKS, Ledneczki I, Chacko S, Olah GA. Direct electrophilic monofluoromethylation. *Org Lett* 2008;10:557–560.
7. Zhang C-P, Cao H-P, Wang Z-L, Zhang C-T, Chen Q-Y, Xiao J-C. New electrophilic bromodifluoromethylation and pentafluoroethylation reagents. *Synlett* 2010;7:1089–1092.
8. (a) Magnier E, Blazejewski J-C, Tordeux M, Wakselman C. Straightforward one-pot synthesis of trifluoromethyl sulfonium salts. *Angew Chem, Int Ed* 2006;45:1279–1282. (b) Macé Y, Raymondeau B, Pradet C, Blazejewski J-C, Magnier E. Benchmark and solvent-free preparation of sulfonium salt based electrophilic trifluoromethylating reagents. *Eur J Org Chem* 2009;1390–1397.
9. (a) Umemoto T. Electrophilic perfluoroalkylating agents. *Chem Rev* 1996;96:1757–1778. (b) Ma J-A, Cahard D. Mild electrophilic trifluoromethylation of β-ketoesters and silyl enol ethers with 5-trifluoro methyldibenzothiophenium tetrafluoroborate. *J Org Chem* 2003;68:8726–8729. (c) Macé Y, Pradet C, Popkin M, Blazejewski J-C, Magnier E. Mechanistical insight into 'electrophilic' trifluoromethylation with *S*-(trifluoromethyl)dibenzothiophenium salts. *Tetrahedron Lett* 2010;51:5388.
10. Wang X, Truesdale L, Yu J-Q. Pd(II)-catalyzed ortho-trifluoromethylation of arenes using tfa as a promoter. *J Am Chem Soc* 2010;132:3648–3649.
11. Zhang C-P, Wang Z-L, Chen Q-Y, Zhang C-T, Gu Y-C, Xiao J-C. Copper-mediated trifluoromethylation of heteroaromatic compounds by trifluoromethyl sulfonium salts. *Angew Chem, Int Ed* 2011;50:1896–1900.

CHAPTER 45

Preparation of Organometallic Fluorides of Main Group and Transition Elements

HERBERT W. ROESKY

Over the past 40 years, organometallic fluoride chemistry has developed in a fascinating way. An important factor was the application of new fluorinating reagents, which are efficient in metathesis reactions without cleavage of all bonds around the metal center. Otherwise, the thermodynamic favorable inorganic metal fluorides are formed. We have discovered that Me_3SnF and Ph_3SnF are reacting in many cases quantitatively with organometallic precursor under fluoride transformation.

Me_3SnF is quantitatively obtained from Me_3SnCl and NaF in water. It is separated by filtration and carefully dried in vacuum. In most of the fluorine transfer reactions, Me_3SnF can be used in stoichiometric quantities. Me_3SnF can be applied to metal complexes containing M–C, M–O, and M–N bonds (M = metal). In metathesis reactions with metal chlorides Me_3SnCl is formed and can be easily removed in vacuum from the reaction product and reconverted with aqueous NaF to Me_3SnF. Therefore, Me_3SnF is preferentially used when the fluorinated products are not volatile. The progress of the fluorination reaction can be easily observed when all of the Me_3SnF has been dissolved in the organic solvent. Me_3SnF itself is only slightly soluble in solvents such as benzene, toluene, tetrahydrofuran (THF), diethyl ether, or dichloromethane [1–11].

Ph_3SnF is mainly used for systems that produce volatile products such as SiF_4 and PF_3 [11].

Efficient Preparations of Fluorine Compounds, First Edition. Edited by Herbert W. Roesky.

45.1 PREPARATIONS OF TRANSITION AND MAIN GROUP METAL COMPOUNDS USING Me_3SnF AS A FLUORINATING AGENT

Preparation of (η^5-C_5Me_5)TiF_3 [1]

Apparatus A 100-mL Schlenk flask with septum, magnetic stirrer, T-shaped N_2 inlet, syringe (50 mL), safety glasses, laboratory coat, protective gloves.

Chemicals Me_3SnF [9], (η^5-C_5Me_5)$TiCl_3$, dry toluene.

Attention! Safety glasses and protective gloves must be used at all times.

Caution! Because of its toxicity and volatility, care should be taken to avoid inhalation of toluene, trimethyltin chloride, and trimethyltin fluoride or contact of their solution with the skin. All reactions should be carried out in a well-ventilated hood.

Experimental Procedure An oven-dried, two-necked, 100-mL Schlenk flask equipped with a magnetic stirrer, a septum, and a T-shaped N_2 inlet is charged in a dry box with freshly sublimed Me_3SnF (2.75 g, 15 mmol) and (η^5-C_5Me_5)$TiCl_3$ (1.45 g, 5 mmol). Toluene (50 mL) is added and the resulting mixture is stirred at room temperature for 4 h. During this time, the suspension disappeared and a clear solution has been formed. The disappearance of the Me_3SnF suspension is an indicator for the progress of the reaction. The solvent and the Me_3SnCl are removed in vacuum. The orange residue is sublimed at 110 °C/10^{-2} mbar to yield 1.1 g (92 %) of orange (η^5-C_5Me_5)TiF_3.

Characterization Data Mp 180 °C. ^{1}H NMR: δ 1.93 (s, C_5Me_5) ppm. ^{19}F NMR: δ 124.0 (s) ppm. IR (CsI, cm^{-1}): $\bar{\upsilon}$ 1072, 1023, 704, 648, 595, 590, 486, 341. MS (*m*/*z*): 240 (M^+, 38%), 135 (C_5Me_5, 100%).

Application Figure 45.1 compares the polymerization activities for styrene with $CpTiF_3$ and $CpTiCl_3$, and dependence on the temperature shows that the fluorine compound is much more active than the chlorine counterpart [12]. Table 45.1 summarizes the results of styrene polymerization. A comparison of the chlorine and fluorine derivatives indicates that the fluorine is also much more stable than the chlorine analogs. It may be noted that the higher activities for the fluorinated catalyst are reached at low Al to Ti ratio of 300, while that for the chlorinated system is achieved at a higher Al to Ti ratio of 900. This result clearly shows the enhanced Lewis acidity of the fluorine-containing metal centers, which makes them more electrophilic and hence more catalytically active.

The highest activity in styrene polymerization is reached with (η^5-C_5H_4Me)TiF_3 in combination with MAO. Activities of up to 140,000 kg of syndiotactic polystyrene in 1 h can be achieved by using a low Al to Ti ratio of 300. Obviously, a high activity is reached by subtle balance between the electronic and steric factors at the metal center. The low Al to Ti ratio reduces the amount of cocatalyst, which contributes to lower costs and sustainable chemistry [12].

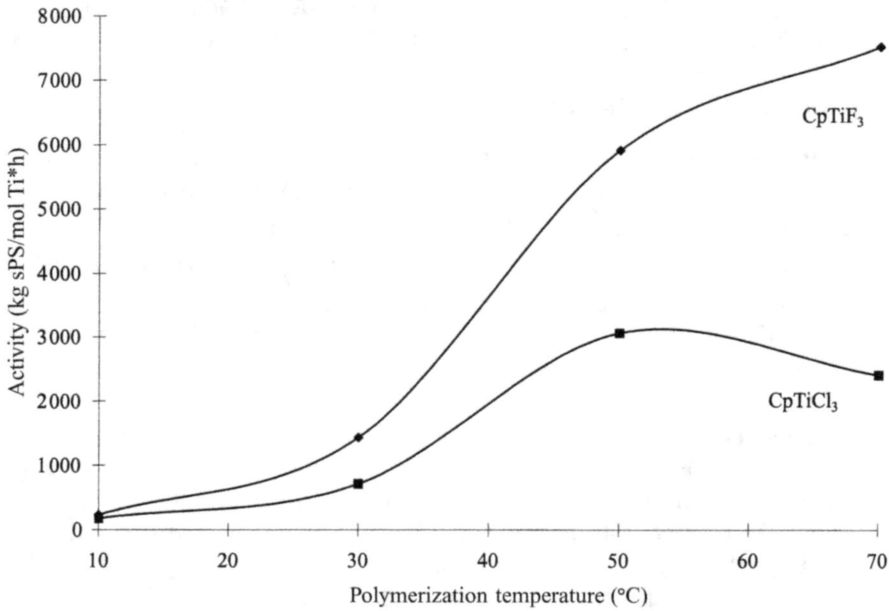

FIGURE 45.1 Activity of the polymerization of styrene with $CpTiX_3$/MAO (X = Cl and F) at different polymerization temperatures. [Ti] = 6.25×10^{-5}, Al/Ti = 900.

TABLE 45.1 Comparative Styrene Polymerization Data[a]: Effect of Fluorine

Catalyst	A[b] ($\times 10^6$)	M_w	M_w/M_n	mp (°C)
$CpTiCl_3$	1.1	140,000	1.9	258
$CpTiF_3$	3.0	100,000	2.0	265
Cp^*TiCl_3	0.015	169,000	3.6	275
Cp^*TiF_3	0.69	660,000	2.0	275

[a] Polymerization condition: toluene with 20-mL styrene at 50 °C, MAO:catalyst = 300.
[b] Activity (A) = gPS/mol cat·h.

Preparation of $[LCaF(thf)]_2$ L = CH[CMe(2,6-*i*$Pr_2C_6H_3N$)]$_2$ [13]

Apparatus A 100-mL Schlenk flask with septum, magnetic stirrer, syringe (50 mL), T-shaped N_2 inlet, safety glasses, laboratory coat, protective gloves.

Chemicals $[LCaN(SiMe_3)_2(thf)]$, Me_3SnF, dry THF, dry hexane.

Attention! Safety glasses and protective gloves must be used at all times.

Experimental Procedure The Schlenk flask is charged in a dry box with $[LCaN(SiMe_3)_2(thf)]$ (1.38 g, 2.0 mmol) and Me_3SnF (0.368 g, 2.0 mmol). After adding THF (40 mL), the mixture is stirred for 16 h. All the volatiles are removed in vacuum and the residue is extracted with hexane (60 mL). The solution on concentration in vacuum afforded $[LCaF(thf)]_2$ as a white solid. The solid can be recrystallized from a mixture of hot toluene and THF. Yield: 0.65 g (60%).

Characterization Data Mp 277–280 °C. ^{1}H NMR (500 MHz, C_6H_6): δ 7.09 (s, 12 H, *m*-, *p*-Ar-*H*), 4.72 (s, 2H, γ-*CH*), 3.54 (m, 8H, O-CH_2-CH_2) 3.14 (sept, 8H, *CH*$(Me)_2$), 1.64 (s, 12H, CH_3), 1.42 (m, 8H, O-CH_2-CH_2), 1.24–1.22 (d, 24H, $CH(CH_3)_2$), 1.02–1.01 (d, 24H, $CH(CH_3)_2$) ppm. ^{19}F NMR (188.77 MHz, C_6D_6): δ −78 ppm. MS (70 eV) (*m*/*z*): 202 (100%, 2,6-*i*$Pr_2C_6H_3$NCMe).

Waste Disposal The residues of Me_3SnCl, Me_3SnF, and the organometallic compounds have to be collected in a labeled container for toxic metal waste that have to be properly deposited [11].

Application The monoanionic β-diketiminate ligand is a versatile tool for addressing synthetic challenges, as amply shown with soluble $[LCaF(thf)]_2$ and $[LCaOH(thf)]_2$. The ability of CaF_2 in optics encouraged us to use the β-diketiminatocalcium compounds with HF·Py (Py = pyridine) for the preparation of CaF_2 dip coatings in common organic solvents. The CaF_2 coatings were obtained by dipping about 12 times unpolished pure silicon surface successively into toluene solutions of the calcium compound and HF·Py followed by toluene washing and drying at room temperature. The coating of CaF_2 on the silicon surface was investigated by scanning electron microscopy (SEM) and energy dispersive X-ray (EDX) spectroscopic analyses [13].

REFERENCES

1. Herzog A, Liu F-Q, Roesky HW, Demsar A, Keller K, Noltemeyer M, Pauer F. Trimethyltin fluoride: a new fluorinating reagent for the preparation of organometallic fluorides. *Organometallics* 1994;13:1251–1256.
2. Roesky HW, Herzog A, Liu F-Q. Organometallic fluorides. *J Fluorine Chem* 1995;71:161.
3. Roesky HW, Herzog A, Liu F-Q. Organometallic fluorides. *J Fluorine Chem* 1995;72:183–185.
4. Liu F-Q, Herzog A, Roesky HW, Usón I. Syntheses and properties of cyclopentadienyl-substituted imidotitanium fluorides. *Inorg Chem* 1996;35:741–744.
5. Murphy EF, Lübben T, Noltemeyer M, Schmidt H-G. First mixed fluoro-chloro Group 4 organometallics: synthesis and spectroscopic and structural characterization of $[\{(C_5Me_5)ZrF_2Cl\}_4]$, $[\{(C_5Me_5)HfF_2Cl\}_4]$, $[(C_5Me_5)_4Zr_4(\mu\text{-}F_2)_2(\mu\text{-}Cl)_2Cl_4]$, $[C_5Me_5)_4Hf_4(\mu\text{-}F)_2(\mu\text{-}F_2)_2(\mu\text{-}Cl)_2Cl_4]$, $[(C_5Me_4Et)_2ZrClF]$, and $[(C_5Me_5)_2HfClF]$. *Inorg Chem* 1996;35:23–29.
6. Murphy EF, Yu P, Dietrich S, Roesky HW, Parisini E, Noltemeyer M. Synthesis and spectroscopic characterization of a series of substituted cyclopentadienyl Group 4 fluorides; crystal structure of the acetylacetonato complex $[(acac)_2(\eta\text{-}C_5Me_5)Zr(\mu\text{-}F)SnMe_3Cl]$. *J Chem Soc Dalton Trans* 1996;1983–1987.
7. Shah SAA, Dorn H, Voigt A, Roesky HW, Parisini E, Schmidt H-G, Noltemeyer M. Group 4 metal amido fluorides and chlorides: molecular structures and the first comparison in ethylene polymerization catalysis. *Organometallics* 1996;15:3176–3181.
8. Walawalkar MG, Murugavel R, Roesky HW. Organometallic fluorides. *Eur J Solid State Inorg Chem* 1996;33:943–955.

9. (a) Krause E. Fluoride metallorganischer Verbindungen. I.: Trialkylzinnfluoride und Dialkylzinndifluoride. *Ber Dtsch Chem Ges* 1918;51:1447–1456. (b) Johnson WK. The alkylation of Tin(IV) chlorides with alkylaluminums. *J Org Chem* 1960;25:2253–2254; Levchuk LE, Sams JR, Aubke F. Synthesis and Moessbauer and vibrational spectra of some new tin(IV)-fluorine compounds. *Inorg Chem* 1972;11:43–50.
10. Cardoso AM, Clark RJH, Moorhouse S. Reactions of trimethylsilylcyclopentadiene derivatives with titanium, niobium, and tantalum halides. *J Chem Soc Dalton Trans* 1980;1156–1160.
11. Roesky HW, Herzog A, Keller K. Zinnorganische Fluoride als Fluorierungsreagenzien für Chloride von Hauptgruppenelementen - Quantitatives Recycling des Fluorierungsreagenzes. *Z Naturforsch* 1994;49b:981–982.
12. Kaminsky W, Lenk S, Scholz V, Roesky HW, Herzog A. Fluorinated half-sandwich complexes as catalysts in syndiospecific styrene polymerization. *Macromolecules* 1997;30:7647–7650.
13. Nembenna S, Roesky HW, Nagendran S, Hofmeister A, Magull J, Hildebrandt P-J, Hahn M. A well defined hydrocarbon-soluble calcium monofluoride, $[\{LCaF(thf)\}_2]$: the application of soluble calcium derivatives for surface coating. *Angew Chem* 2007;119:2564–2566; *Angew Chem, Int Ed* 2007;46:2512–2514.

CHAPTER 46

Preparation of Pentafluorosulfanyl Carbonyl Compounds

JOHN T. WELCH

The pentafluorosulfanyl (SF_5) group affords the agrichemical and medicinal chemist the unique opportunity to introduce a totally novel substituent, a substituent that is remarkably robust toward the majority of synthetic transformations.

The organic chemistry of the SF_5 group, previously reviewed [1] and extensively developed by Gard, has only recently come under more widespread investigation with the ready availability of previously difficultly accessible building blocks or reagents [2]. SF_5 groups are relatively hydrolytically and chemically stable [3]; for example, the hydrolytic stability of aromatic SF_5 group equals or exceeds that of the trifluoromethyl (CF_3) group.

The relative steric demand and symmetry of the SF_5 group can be compared and contrasted with both the *tert*-butyl and CF_3 groups. The unique octahedral geometry around sulfur enables ligands to interact with host molecules without many classic conformational constraints. The volume of the SF_5 group is slightly less than that of a *tert*-butyl group [1a] and therefore considerably greater than that of a CF_3 group. However, the electrostatic surface presented by SF_5 is comparable to that of CF_3; a pyramid of electron density as opposed to the inverted cone of density of the CF_3 group. The electron-withdrawing effects of the two groups are similar in magnitude as assessed by the carbon 1s photoelectron spectra [4]. The electronegativity of the SF_5 group has been proposed to be as high as 3.65 in comparison to the 3.36 value of the CF_3 group [5].

The unique properties of the SF_5 group have shown diverse utility in biological applications. With the unique octahedral geometry around sulfur and a square pyramidal array of fluorines, the SF_5 group has a reduced barrier to rotation and as such can optimize receptor interactions efficiently. The potential dehydrofluorination of a CF_3 group to form a Michael-type acceptor and hence a potential mechanism-based inhibitor is not possible with a SF_5 group. In environmental degradation studies of

Efficient Preparations of Fluorine Compounds, First Edition. Edited by Herbert W. Roesky.

SF_5-substituted molecules, degradation was shown to lead to environmentally benign products [6]. Finally, the remarkable hydrophobicity and steric demand of the SF_5 group profoundly influence molecular conformation in aqueous solutions as shown in heptapeptide studies.

46.1 PREPARATIONS OF PENTAFLUOROSULFANYL CARBONYL COMPOUNDS USING SF_5Br

Preparation of 2-Pentafluorosulfanyl Heptanal

Apparatus A 100-mL round flask with septum, magnetic stirrer, T-shaped N_2 inlet, syringe (50 mL), safety glasses, laboratory coat, protective gloves.

Chemicals SF_5Br, hept-1-enyl acetate, dry hexane, triethylboron.

Attention! Safety glasses and protective gloves must be used at all times.

Caution! Because the toxicity of pentafluorosulfanyl bromide is unknown, care should be taken to avoid inhalation of pentafluorosulfanyl bromide or contact of the solution with the skin. All reactions should be carried out in a well-ventilated hood.

Experimental Procedure *1-Bromo-2-pentafluorosulfanylalkyl acetates*

Into a round-bottom flask cooled to 0 °C, a SF_5Br stock solution in hexane (6.0 mL, 1.59 M, 9.5 mmol, 1.2 equiv) and triethylborane (0.80 mL, 1 M, 0.80 mmol, 0.1 equiv) were added. The enol acetate (7.7 mmol, 1.0 equiv) was then added drop by drop over approximately 5–10 min. The mixture was allowed to stir at 0 °C for 30 min and then quenched with saturated sodium bicarbonate solution. The mixture was extracted with diethyl ether and the organic fractions washed with brine, dried over magnesium sulfate, filtered, and concentrated. Purification by flash column chromatography using hexane/dichloromethane afforded the pure product.

2-Pentafluorosulfanyl aldehydes. A mixture of a 1-bromo-2-pentafluorosulfanylalkyl acetate (2.3 mmol, 1.0 equiv), aqueous hydrochloric acid (3.5 M, 4 mL, 14 mmol, 5.9 equiv), and acetic acid (17 M, 3 mL, 51 mmol, 22 equiv) was refluxed for 3 h. After cooling to room temperature, 10 mL diethyl ether was added and the mixture poured into a beaker containing 40 mL of 20 % sodium bicarbonate. The mixture was stirred until there was no visible bubbling observed then the layers were separated. The aqueous layer was extracted with diethyl ether and the combined organic fractions dried over sodium sulfate, filtered, and concentrated. Purification by flash column chromatography eluting with hexane/dichloromethane gives the pure product.

Characterization Data *1-Bromo-2-pentafluorosulfanylheptyl acetate*. Yield: 91%, colorless oil, two isomers (3.3:1). ^1H NMR (400 MHz, $CDCl_3$): δ 7.27 (bs, 1H), 7.11 (d, 3.4 Hz, 1H), 4.16 (m, 1H), 4.02 (m, 1H), 2.27 (m, 4H), 2.10 (s, 6H), 1.73 (m, 1H), 1.56 (m, 3H), 1.36 (m, 8H), 0.91 (t, 7.0 Hz, 6H) ppm. ^{13}C NMR (100 MHz, $CDCl_3$): δ 167.1 (C=O), 167.0 (C=O), 90.7 (C_{ipso}, qn,

8.2 Hz), 88.1 (C_{ipso}, qn, 8.2 Hz), 73.3 (C–Br, qn, 5.5 Hz), 71.8 (C–Br, qn, 5.5 Hz), 31.4, 31.3, 27.7 (bs), 27.5 (bs), 26.4, 22.2, 20.7, 20.5, 13.8 ppm. ^{19}F NMR (376 MHz, $CDCl_3$): δ 83.4 (9 peaks, 145 Hz, 1F), 83.3 (9 peaks, 142 Hz, 1F), 59.7 (dd, 144 Hz, 5.4 Hz, 4F), 58.8 (d, 143 Hz, 4F) ppm.

2-Pentafluorosulfanylheptanal. Yield: 77%, colorless oil. ^{1}H NMR (400 MHz, $CDCl_3$): δ 9.59 (m, 1H), 4.24 (m, 1H), 2.13 (m, 2H), 1.29 (m, 6H), 0.86 (t, 6.9 Hz, 3H) ppm. ^{13}C NMR (100 MHz, $CDCl_3$): δ 190.6 (C=O, qn, 5.1 Hz), 90.4 (C_O, qn, 7.3 Hz), 31.2, 27.2 (C_O, qn, 3.6 Hz), 26.1, 22.2, 13.7 ppm. ^{19}F NMR (376 MHz, $CDCl_3$): δ 82.5 (9 peaks, 144 Hz, 1F), 64.1 (dd, 145 Hz, 6.1 Hz, 4F) ppm. Anal. calcd for $C_7H_{13}F_5OS$: C, 35.00; H, 5.45. Found: C, 34.93; H, 5.51.

Application Little has been reported about the reactions of SF_5-containing aldehydes; however, these substances undergo many of the common reactions of aldehydes easily. Reduction with sodium borohydride gave the corresponding primary alcohols. The generally high yields of this process are indicative of the stability of the SF_5 group to the reaction conditions.

Nucleophilic addition to the carbonyl group using organolithium or Grignard reagents as anticipated gave the corresponding alcohols with high diastereoselectivity (>99%). In contrast to decomposition that occurred on the exposure of aromatic SF_5 compounds to *n*-butyllithium [7], the aliphatic SF_5 group was observed to be quite robust (Scheme 46.1).

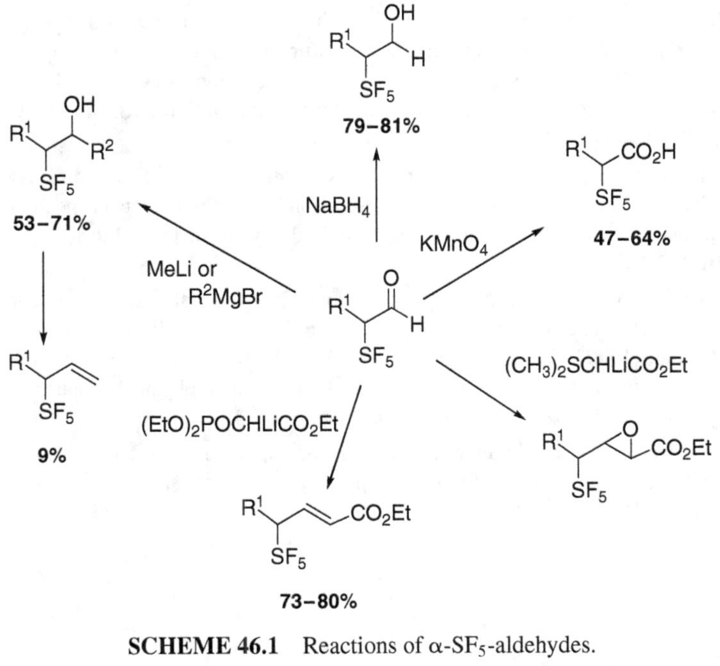

SCHEME 46.1 Reactions of α-SF_5-aldehydes.

REFERENCES

1. (a) Lentz D, Seppelt K, Akiba KY. The SF_5, SeF_5, and TeF_5 groups in organic chemistry. In: Akiba K-Y, editor. *Chemistry of Hypervalent Compounds*. New York, NY: Wiley-VCH; 1999. pp. 295–323. (b) Winter RW, Dodean RA, Gard GL, Soloshonok VA. SF_5 synthons: pathways to organic derivatives of SF_6. In: Soloshonok VA, editor. *Fluorine-Containing Synthons*. Washington, DC: American Chemical Society; 2005. pp. 87–118. (c) Gard GL. Recent milestones in SF5-chemistry Chem Today 2009;27:10–13. (d) Lal GS, Syvret RG. Enabling Chemistry Innovation with New SF5 Synthons, in Focus on Fluorine Chemistry a Special Supplement to Chem Today 2008;26. (e) Welch JT. Applications of pentafluorosulfanyl substitution in life sciences research. In Governeur V , Müller K, editors. Fluorine in Pharmaceutical and Medicinal Chemistry: From *Biophysical Aspects to Clinical Applications*. Imperial College Press, 2011. pp. 171–203

2. Kirsch P. *Modern Fluoroorganic Chemistry. Synthesis, Reactivity and Applications*. Weinheim: Wiley-VCH; 2004.

3. (a) Kirsch P, Bremer M. Nematic liquid crystals for active matrix displays: molecular design and synthesis. *Angew Chem, Int Ed* 2000;39:4216–4235. (b) Kirsch P, Bremer M, Heckmeier M, Tarumi K. Liquid crystals with pentafluorosulfuranyl as a polar terminal group. *Mol Cryst Liq Cryst* 2000;346:29–33. (c) Kirsch P, Bremer M, Taugerbeck A, Wallmichrath T. Difluorooxymethylene-bridged liquid crystals. A novel synthesis based on the oxidative alkoxydifluorodesulfuration of dithianylium salts. *Angew Chem, Int Ed* 2001;40:1480–1484. (d) Bowden RD, Comina PJ, Greenhall MP, Kariuki BM, Loveday A, Philp D. A new method for the synthesis of aromatic sulfur pentafluorides and studies of the stability of the sulfur pentafluoride group in common synthetic transformations. *Tetrahedron* 2000;56:3399–3408.

4. (a) Brant P, Berry AD, DeMarco RA, Carter FL, Fox WB, Hashmall JA. Gas phase and solid state X-ray photoelectron spectra of the substituted acetylenes $(SF_5)_nC_2(CF_3)_2$-n(n = 0, 1, 2): a comparison of the pentafluorosulfur and trifluoromethyl groups. *J Electron Spectrosc Relat Phenom* 1981;22:119–129. (b) True JE, Thomas D, Winter RW, Gard GL. Electronegativities from core-ionization energies: electronegativities of SF_5 and CF_3. *Inorg Chem* 2003;42:4437–4441.

5. Saethre LJ, Berrah N, Bozek JD, Boerve KJ, Carroll TX, Kukk E, Gard GL, Winter R, Thomas TD. Chemical insights from high-resolution X-ray photoelectron spectroscopy and *ab initio* theory: propyne, trifluoropropyne, and ethynylsulfur pentafluoride. *J Am Chem Soc* 2001;123:10729–10737.

6. (a) Jackson DA, Mabury SA. Environmental properties of pentafluorosulfanyl compounds: physical properties and photodegradation. Abstracts of Papers. 236th ACS National Meeting; 2008 Aug 17–21; Philadelphia, PA. pp. AGRO-216. (b) Jackson DA, Mabury SA. Environmental properties of pentafluorosulfanyl compounds: physical properties and photodegradation. *Environ Toxicol Chem* 2009;28:1866–1873.

7. Kirsch P, Bremer M, Heckmeier M, Tarumi K. Liquid crystals based on hypervalent sulfur fluorides: pentafluorosulfuranyl as polar terminal group. *Angew Chem, Int Ed* 1999;38:1989–1992.

CHAPTER 47

Preparation of Power-Variable Electrophilic Trifluoromethylating Agents, *S*-(Trifluoromethyl) dibenzothiophenium Salts Series

TERUO UMEMOTO

Electrophilic trifluoromethylating agents, useful for the preparation of trifluoromethyl compounds, *S*-, *Se*-, and *Te*-(trifluoromethyl)dibenzo-, -seleno-, and -tellurophenium salts, have been developed by the author and coworkers as power-variable electrophilic trifluoromethylating agents [1–5].

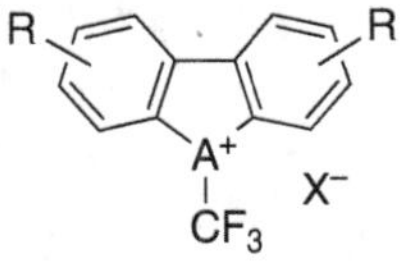

Power order
A: Te < Se < S
R: Alkyl < H < NO_2

The trifluoromethylating power of the reagents increases in the order of Te $<$ Se $<$ S, and furthermore increases by electron-withdrawing substituents, while it decreases by electron-donating ones. Powerful reagents trifluoromethylate less reactive nucleophiles well, while less powerful reagents do more reactive nucleophiles well. Intermediately powerful reagents trifluromethylate intermediately reactive nucleophiles well. Because of this variation, trifluoromethylations of a wide range of nucleophilic substrates are possible.

Efficient Preparations of Fluorine Compounds, First Edition. Edited by Herbert W. Roesky.

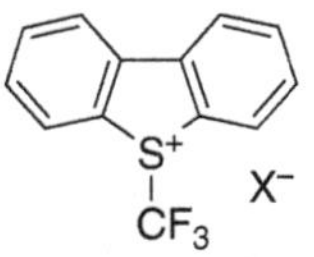

X=BF_4 **1**
OTf **2**

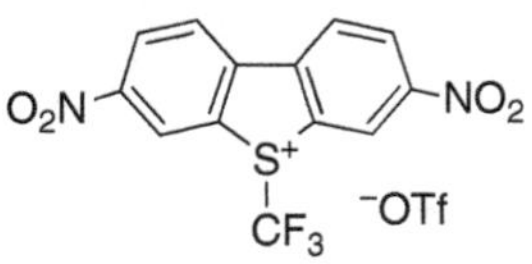

3

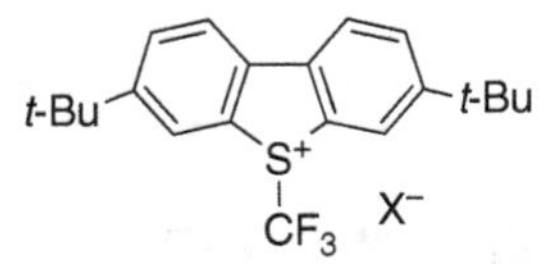

X=BF_4 **4**
OTf **5**

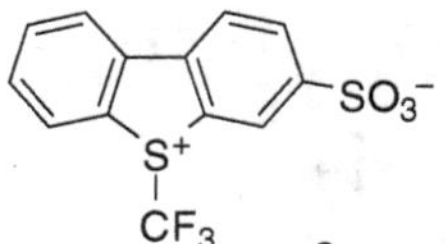

6

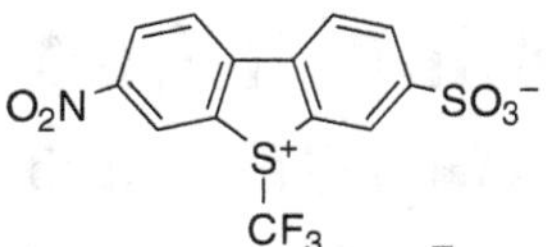

7

Among the heteroatoms, the *S*-series reagents are the most useful because of broad applications. *S*-(Trifluoromethyl)dibenzothiophenium tetrafluoroborate (**1**) and triflate (**2**) are particularly useful because of satisfactory reactivity, stable crystalline solid, and relatively easy preparation. 3,7-Dinitro-*S*-(trifluoromethyl)dibenzothiophenium triflate (**3**), *S*-(trifluoromethyl)dibenzothiophenium-3-sulfonate (**6**), and 7-nitro-*S*-(trifluoromethyl)dibenzothiophenium-3-sulfonate (**7**) were derived from the same intermediate [2, 3]. 3,7-Bis(*tert*-butyl)dibenzothiophenium tetrafluoroborate (**4**) and triflate (**5**) were also prepared [1].

Salts **1** and **2** have essentially the same trifluoromethylating power, but salt **1** can be prepared at less cost. Dinitro **3** is more powerful and bis(*tert*-butyl) **4** is less powerful. Thus, the power order is $\mathbf{4} \approx \mathbf{5} < \mathbf{1} \approx \mathbf{2} < \mathbf{3}$. A zwitterionic type of **6** and **7** has an advantage of easy posttreatment following trifluoromethylation. The power order is $\mathbf{6} < \mathbf{7}$.

47.1 PREPARATION OF *S*-(TRIFLUOROMETHYL) DIBENZOTHIOPHENIUM TETRAFLUOROBORATE (1)

Apparatus 1-L and 500-mL glass flasks, mechanical and magnetic stirrers, T-shaped N_2 inlet, 50-mL dropping funnel.

Chemicals 2-(Trifluoromethanesulfinyl)biphenyl (**8**) [1,4], 60% fuming sulfuric acid, methanol, diisopropyl ether, dry dichloromethane, sodium tetrafluoroborate, $NaHCO_3$.

Attention! Safety glasses, protective gloves, and laboratory coat must be used at all times.

Caution! Because of its toxicity, care should be taken to handle 60% fuming sulfuric acid. All reactions should be carried out in a well-ventilated hood.

Experimental Procedure

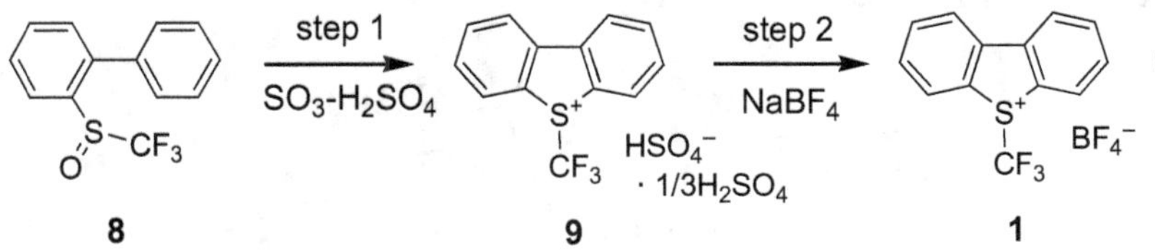

(47.1)

Step 1: An oven-dried, three-necked, 1-L glass flask equipped with a 50-mL dropping funnel, a mechanical stirrer, and a T-shaped N_2 inlet is charged with 2-(trifluoromethanesulfinyl)biphenyl (**8**) (64 g, 237 mmol) and dry dichloromethane (160 mL). The flask is cooled on an ice bath and then 60% fuming sulfuric acid (29 mL; SO_3, 426 mmol) is slowly dropped through a dropping funnel over 30 min to the stirred mixture cooled on an ice bath. After addition, the ice bath is removed and the mixture is stirred for 1.5 h at room temperature. The mixture is then cooled on an ice bath. Methanol (17 mL, 421 mmol) is carefully dropped through a dropping funnel to the cooled and stirred reaction mixture (*caution*: a vigorous exothermic reaction occurs on mixing). After that, diisopropyl ether (~500 mL) is carefully added to the cooled and stirred reaction mixture (*caution*: a vigorous exothermic reaction may still occur). The reaction mixture is filled with precipitates, which are collected by filtration and recrystallized from methanol–ethyl acetate and dried. The product (~72 g, ~79% yield) is assigned as *S*-(trifluoromethyl)dibenzothiophenium hydrogen sulfate $1/3(H_2SO_4)$ adduct **9**: mp 122–127 °C (with dec.). ^{19}F NMR: δ −52.8 (s) ppm. MS (*m*/*z*): 253 (M^+-$HSO_4 \cdot 1/3(H_2SO_4)$).

Step 2: An oven-dried, three-necked, 500-mL glass flask equipped with a magnetic stirrer and a T-shaped N_2 inlet is charged with sodium tetrafluoroborate (14.1 g, 126 mmol) and acetonitrile (200 mL). The mixture is heated on an oil bath of 60 °C to give a clear solution. Salt **9** (48 g, 126 mmol) obtained in step 1 is added in portions to the stirred solution over 5 min. The mixture is stirred at 60 °C for 1.5 h and cooled to room temperature. Powdered $NaHCO_3$ (~48 g, ~572 mmol) is added in portions to the mixture (*caution*: carbon dioxide generates) and the mixture is stirred at room temperature for 3 h. Insoluble materials are removed by filtration. The solvent is evaporated from the filtrate and the residue is recrystallized from acetonitrile–diethyl ether to give product **1** (~41 g, ~95% yield).

Characterization Data of Product 1 Mp 171–172 °C. ^1H NMR: δ 7.79–8.22 (m, 2, 3, 7, and 8H), 8.36–8.52 (m, 1, 4, 6, and 9H) ppm. ^{19}F NMR: δ −52.5 (s, CF_3), −150.1 (s, BF_4) ppm. MS (SIMS): *m*/*z* 253 (M^+-BF_4).

47.2 PREPARATION OF *S*-(TRIFLUOROMETHYL) DIBENZOTHIOPHENIUM TRIFLATE (2)

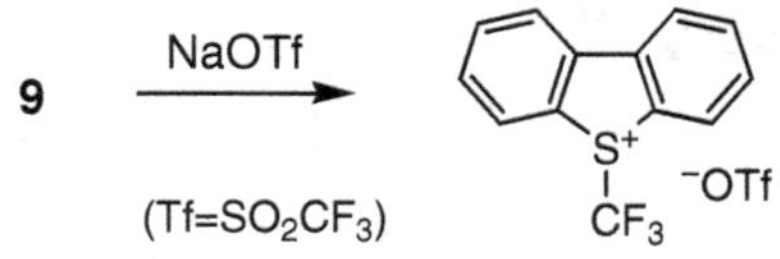

2

(47.2)

Triflate **2** is prepared in ~92% yield with sodium triflate instead of sodium tetrafluoroborate in the procedure of step 2 for tetrafluoroborate **1**. Triflate **2** is recrystallized from acetonitrile–diethyl ether.

Characterization Data of Triflate 2 Mp 155 °C. ^{1}H NMR: δ 7.78–8.18 (m, 2, 3, 7, and 8H), 8.38–8.55 (m, 1, 4, 6, and 9H) ppm. ^{19}F NMR: δ −52.6 (s, SCF_3), −78.1 (s, SO_2CF_3) ppm. MS (*m/z*): 184 (M^+-OSO_2CF_3-CF_3).

47.3 PREPARATION OF 3,7-DINITRO-*S*-(TRIFLUOROMETHYL) DIBENZOTHIOPHENIUM TRIFLATE (3)

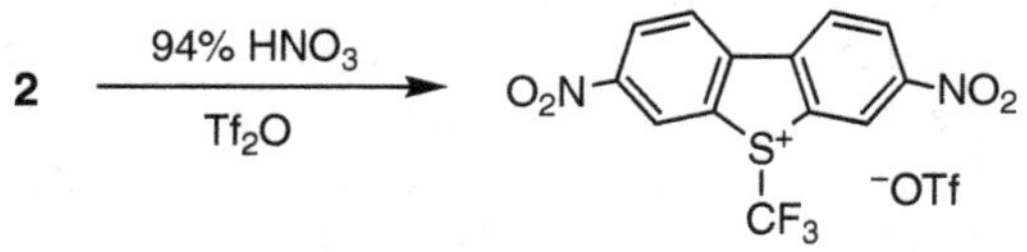

3

(47.3)

Apparatus A 50-mL glass flask, magnetic stirrer, T-shaped N_2 inlet, syringe (10 mL).

Chemicals *S*-(Trifluoromethanesulfinyl)dibenzothiophenium triflate (**2**), 94% concentrated nitric acid, trifluoromethanesulfonic anhydride (triflic anhydride), diethyl ether.

Attention! Safety glasses, protective gloves, and laboratory coat must be used at all times.

Caution! Because of its toxicity, care should be taken to handle 94% concentrated nitric acid and triflic anhydride. All reactions should be carried out in a well-ventilated hood.

Experimental Procedure An oven-dried, two-necked, 50-mL glass flask equipped with a magnetic stirrer, a septum, and a T-shaped N_2 inlet is charged with 94% concentrated nitric acid (1.41 mL, 33.6 mmol). Triflic anhydride (6.86 mL, 40.8 mmol) is slowly added to the nitric acid at room temperature with a syringe while stirring. After an additional stirring for 1 h, powdered **2** (4.5 g, 11.2 mmol) is added and the reaction mixture is stirred for 2 days at room temperature. The liquid layer is removed from the resulting solid as much as possible by decantation. Diethyl ether is slowly and carefully added to the solid layer while stirring under cooling with an ice bath (*caution*: a vigorous exothermic reaction occurs). After an adequate amount of diethyl ether is added, the resulting pale yellow crystals are collected by filtration and dried at room temperature. Yield: ~3.9 g (~70% yield).

Characterization Data of Salt 3 dec 130–135 °C. ^{1}H NMR: δ 8.69 (d, $J = 9$ Hz, 1 and 9H), 8.91 (dd, $J = 9.0$ Hz, 1.8 Hz, 2 and 8H), 9.34 (d, $J = 1.8$ Hz, 4 and 6H) ppm. ^{19}F NMR: δ −48.4 (s, CF_3), −78.1 (s, SO_2CF_3) ppm. MS (*m*/*z*): 343 (M^+-OTf).

47.4 PREPARATION OF S-(TRIFLUOROMETHYL) DIBENZOTHIOPHENIUM-3-SULFONATE (6)

8

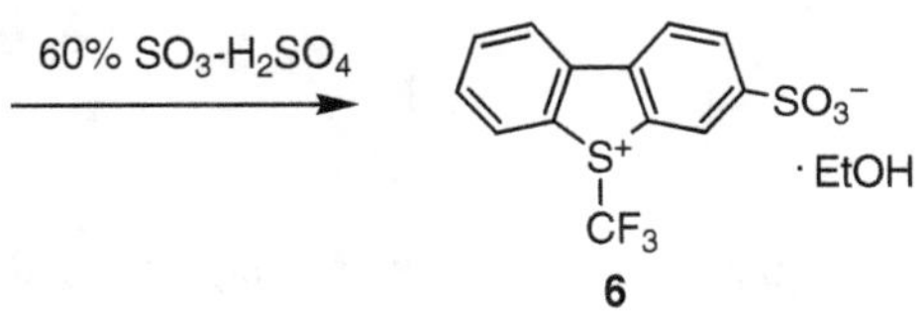

(47.4)

Apparatus A 1-L glass flask, mechanical stirrer, T-shaped N_2 inlet, 100-mL dropping funnel.

Chemicals 2-(Trifluoromethanesulfinyl)biphenyl (**8**) [1,4], 60% fuming sulfuric acid, dry dichloromethane, diisopropyl ether, ethanol.

Attention! Safety glasses, protective gloves, and laboratory coat must be used at all times.

Caution! Because of its toxicity, care should be taken to handle 60% fuming sulfuric acid. All reactions should be carried out in a well-ventilated hood.

Experimental Procedure An oven-dried, three-necked, 1-L glass flask equipped with a mechanical stirrer, a dropping funnel, a condenser, and a T-shaped N_2 inlet is charged with **8** (52.5 g, 195 mmol) and dry dichloromethane (150 mL). The T-shaped N_2 inlet is set at the top of the condenser. The solution is cooled on an ice bath, and 60% fuming sulfuric acid (65.1 mL; SO_3, 960 mmol) is added dropwise over 30 min through the dropping funnel into the solution while stirring. After the addition, the reaction mixture is stirred at 42 °C for 20 h. The reaction mixture is then cooled on an ice bath and then diisopropyl ether (~500 mL) is carefully added to the reaction mixture while stirring (*caution*: a vigorous exothermic reaction occurs on mixing with the ether). The resulting gummy oil is separated from the solution by decantation and then washed a few times with diisopropyl ether. Ethanol (~165 mL) is added to the gummy oil and the mixture is stirred until all the oily material becomes white powdered precipitate. It may take close to 2 h. The precipitate is collected by filtration, washed with ethanol (2 times), and dried under vacuum for about 6 h at 60 °C. Product **6** (~62 g, ~84% yield) is identified as an adduct with one molecule of ethanol.

Characterization Data of Product 6 *EtOH adduct*: Mp 142–147 °C (with dec.). ^{1}H NMR: δ 1.17 (t, $J = 7$ Hz, CH_3CH_2OH), 3.60 (q, $J = 7$ Hz, CH_3CH_2OH), 7.92 (td, $J = 8$ Hz, 1 Hz, 7H), 8.10 (td, $J = 8$ Hz, 1 Hz, 8H), 8.42 (dd, $J = 8$ Hz, 1 Hz, 2H), 8.46–8.57 (m, 1, 6, and 9H), 8.90 (d, $J = 1$ Hz, 4H) ppm. ^{19}F NMR: δ −53.9 (s, CF_3) ppm. MS (*m/z*): 333 ($M^+ + 1$).

47.5 PREPARATION OF 7-NITRO-*S*-(TRIFLUOROMETHYL) DIBENZOTHIOPHENIUM-3-SULFONATE (7)

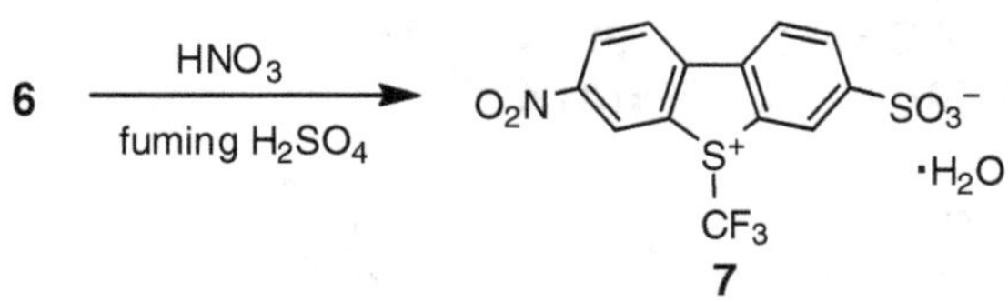

(47.5)

Apparatus A 50-mL glass flask, magnetic stirrer, T-shaped N_2 inlet.

Chemicals *S*-(Trifluoromethyl)dibenzothiophenium-3-sulfonate (**6**), concentrated nitric acid, 20% fuming sulfuric acid, diethyl ether, methanol, dichloromethane.

Attention! Safety glasses, protective gloves, and laboratory coat must be used at all times.

Caution! Because of toxicity, care should be taken to handle fuming sulfuric acid and concentrated nitric acid. All reactions should be carried out in a well-ventilated hood.

Experimental Procedure An oven-dried, 50-mL glass flask equipped with a magnetic stirrer and a T-shaped N_2 inlet is charged with 2.0 mL of 20% fuming sulfuric acid and 0.6 mL of concentrated nitric acid. Into the stirred mixture, salt **6** (2.0 g, 6.0 mmol) is added in several portions. The reaction mixture is stirred at room temperature for 16 h. Then, the mixture is cooled on an ice bath and diethyl ether is added slowly and carefully to the reaction mixture while stirring (*caution*: a vigorous exothermic reaction occurs). The ether is removed from the resulting precipitate by decantation. Again, some diethyl ether is added to the precipitate and the ether is removed by decantation. While stirring, a small amount of methanol and then a large amount of diethyl ether are added, and the precipitate is collected by filtration. The precipitate is washed with a 10:1 mixture of dichloromethane and methanol and dried to give product **7** (~1.7 g, ~75% yield).

Characterization Data of Product 7 dec 196–210 °C as a monohydrate. ^{1}H NMR: δ 8.27 (dd, $J = 1.2$ Hz, 8.1 Hz, 2H), 8.65 (d, $J = 8.1$ Hz, 1H), 8.75 (d, $J = 8.6$ Hz, 9H), 8.86 (dd, $J = 2.1$ Hz, 8.6 Hz, 8H), 9.05 (d, $J = 1.2$ Hz, 4H), 9.64 (d, $J = 2.1$ Hz, 6H) ppm. ^{19}F NMR: δ −50.9 (s, CF_3) ppm. MS (*m/z*): 378 ($M^+ + 1$).

47.6 TRIFLUOROMETHYLATION OF SODIUM SALT OF 2-METHYL-1,3-CYCLOPENTANEDIONE WITH *S*-(TRIFLUOROMETHYL)DIBENZOTHIOPHENIUM TRIFLATE (2)

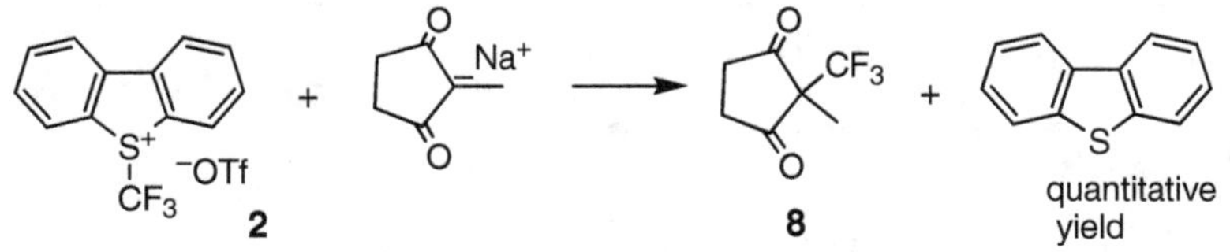

(47.6)

Apparatus A 50-mL glass flask, magnetic stirrer, T-shaped N_2 inlet, septum, syringe (500 μL).

Chemicals *S*-(Trifluoromethyl)dibenzothiophenium triflate (**2**), 2-methyl-1,3-cyclopentanedione, 60% sodium hydride in oil, dry dimethylformamide (DMF), diethyl ether, dry hexane.

Attention! Because of strong base, care should be taken to handle sodium hydride. Safety glasses, protective gloves, and laboratory coat must be used at all times.

Caution! All reactions should be carried out in a well-ventilated hood.

Experimental Procedure An oven-dried, 50-mL flask equipped with a magnetic stirrer, a septum, and a T-shaped N_2 inlet is charged with sodium hydride (60% in oil) (104 mg, 2.6 mmol), which is then washed with dry hexane to remove oil (*the detailed procedure for removing the oil*: a few milliliters of dry hexane is added to the flask, the mixture is stirred with a magnetic stirrer and left, and the upper clear hexane solution is removed by a syringe. This procedure is repeated three times and finally the mixture is dried by means of a vacuum pump). Dry DMF (5 mL) is added to the flask and the mixture is cooled at 0 °C. 2-Methyl-1,3-cyclopentanedione (291 mg, 2.6 mmol) is slowly added to the mixture while stirring (*note*: hydrogen gas evolves). After stirring for an additional 0.5 h, the mixture is cooled to −45 °C. Then, salt **2** (804 mg, 2 mmol) is added to the mixture while stirring. The reaction mixture is warmed to room temperature over a period of 1 h and maintained for 1 h at room temperature. ^{19}F NMR of the reaction mixture shows a singlet at −69.8 ppm (CF_3) that corresponds to product **8**. The yield of the product is calculated to be around 84% by using a CF_3 signal (a singlet at −78 ppm) of a triflate anion part of salt **2** in this reaction as an internal standard. To isolate the product, diethyl ether is added to the reaction mixture, which is then washed with water and then aqueous NaCl solution, dried with anhydrous magnesium sulfate, and filtered. Evaporation of solvent from the filtrate gives a residue that includes the product, dibenzothiophene, and some others, which are separated by column chromatography on silica gel.

TABLE 47.1 Trifluoromethylation of Nucleophiles with *S*-(Trifluoromethyl)dibenzothiophenium Salts

Entry	Nucleophile	Salt	Solvent	Reaction conditions	Product	Yield
1	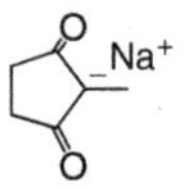	**1**	DMF	$-45\,^\circ$C -> rt, 1 h	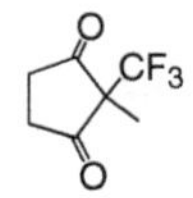	70%
2		**2**	DMF	$-60\,^\circ$C -> rt, 1 h		84%
3		**6**	DMF	$-45\,^\circ$C -> rt, 1 h		86%
4	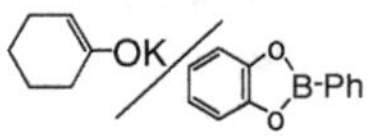	**1**	THF	$-78\,^\circ$C -> 0, 2.5 h	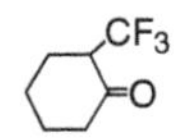	65%
5		**2**	THF	$-78\,^\circ$C -> 0, 2.5 h		86%
6	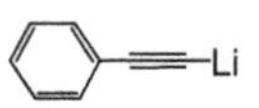	**2**	THF	$-78\,^\circ$C -> rt, 1.3 h	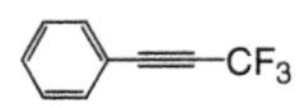	58%
7		**5**	THF	$-78\,^\circ$C -> rt, 3 h		67%
8	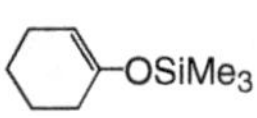	**2**	DMF	Pyridine (1 equiv), 80 °C, overnight	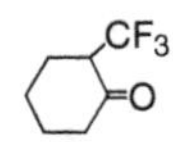	65%
9	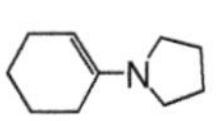	**2**	DMF	DMPy (1 equiv), 0 °C, 2 h	2-CF_3-cyclhexanone 2,6-Bis(CF_3) cyclohexanone	57% 26%
10	Aniline	**3**	DMF	rt, 0.5 h	2-CF_3-aniline 4-CF_3-aniline	54% 20%
11	Aniline	**7**	DMF	80 °C, 0.5 h	2-CF_3-aniline 4-CF_3-aniline	37% 18%
12	2-Naphthol	**3**	DMF	DMPy (1 equiv) $-20\,^\circ$C -> rt, 1.25 h	1-CF_3-2-naphthol 8-CF_3-2-naphthol	52% 6%
13	R-SNa	**2**	THF	rt, 0.5 h	R-SCF_3	47%
14	R-SNa	**5**	THF	0 °C -> rt, 0.5 h	R-SCF_3	76%
15	Ar-SNa	**2**	DMF	$-30\,^\circ$C -> rt, 1 h	Ar-SCF_3	78%
16	Ph_3P	**3**	CH_3CN	rt, 5 h	$Ph_3P^+CF_3\ ^-OTf$	78%

rt, room temperature; DMPy, 4-(*N*,*N*-dimethylamino)pyridine; R, n-$C_{12}H_{25}$; Ar, 2-phenylphenyl.

Application

According to Equation 47.7, an *S*-(trifluoromethyl)dibenzothiophenium salt reacts with a nucleophile (Nu^-) to give a trifluoromethyl compound along with the quantitative formation of dibenozothiophene [1–3]. Many electrophilic trifluoromethylations are exemplified in Table 47.1.

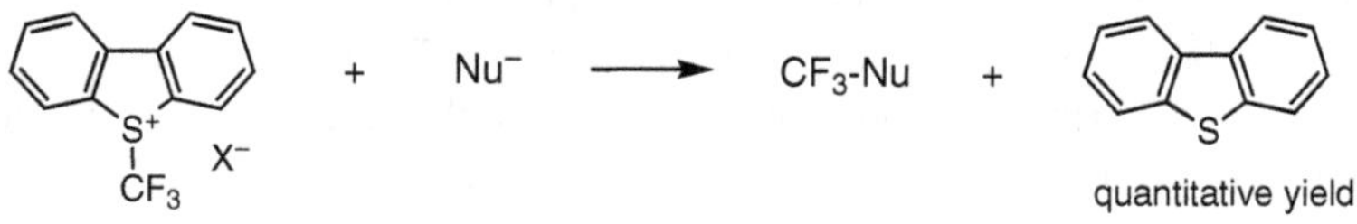

(47.7)

Salts **1** and **2** have basically the same trifluoromethylating power. As salt **2** is more soluble in organic solvent than **1**, it often provides better results. Moderately powerful salts **1** and **2** satisfactorily react with nucleophiles such as carbanions of active methylene compounds (entries 1 and 2, Table 47.1), lithium acetylides (entries

4 and 5), silyl enol ethers (entry 8), enamines (entry 9), and thiolates (entries 13 and 15), while powerful salt **3** does so with less reactive nucleophiles such as aromatics (entries 10 and 12) and triphenylphosphine (entry 16). As metal enolates of ketones are too reactive, salts **1** and **2** do not fit with them. However, salts **1** and **2** can trifluoromethylate them well when their reactivity is moderated by combining with a suitable Lewis acid (entries 4 and 5) [5]. Less powerful salt **5** provides better yields in trifluoromethylating the acetylide (67%, entry 7) and the alkanethiolate (76%, entry 14) than salt **2**. With these acetylide and alkanethiolate, less powerful *Se*-(trifluoromethyl)dibenzothiophenium triflate gives furthermore improved yields (89% and 87%) [1].

When zwitterionic **6** and **7** are used for trifluoromethylation, trifluoromethylated compounds are produced together with dibenzothiophene-3-sulfonic acids or their salts, which are soluble in water or insoluble in organic solvents. Therefore, the by-products are easily removed from products by washing with water or an organic solvent. Their reactions are exemplified in entries 3 and 11 [Table 47.1].

These trifluoromethylating agents have been applied to the first enantioselective electrophilic trifluoromethylation [5] and many other reactions [6–8]. In industrial applications, Amgen reported the synthesis of a potent drug candidate for treatment of metabolic syndrome as well as type 2 diabetes using **6** (Eq. 47.8) [9]. Merck [10], Eli Lilly [11], Sumitomo Pharma [12], Takeda Pharma [13], and Novartis [14] have also used the trifluoromethylating agents in their drug researches.

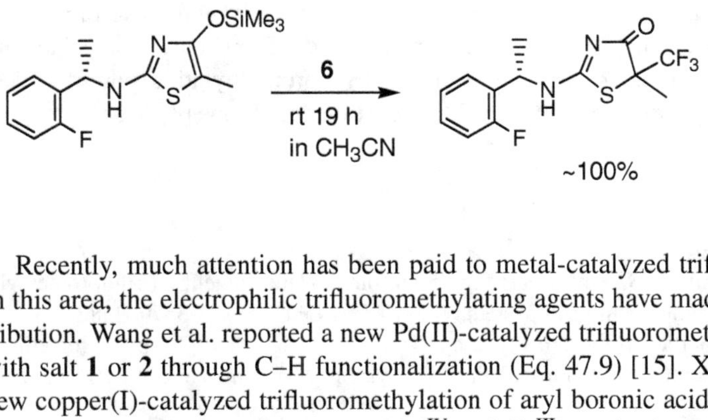

(47.8)

Recently, much attention has been paid to metal-catalyzed trifluoromethylation. In this area, the electrophilic trifluoromethylating agents have made significant contribution. Wang et al. reported a new Pd(II)-catalyzed trifluoromethylation of arenes with salt **1** or **2** through C–H functionalization (Eq. 47.9) [15]. Xu et al. reported a new copper(I)-catalyzed trifluoromethylation of aryl boronic acids using salt **1** or **2** (Eq. 47.10) [16]. High valence metal Pd^{IV} and Cu^{III} species have been suggested as a possible reaction mechanism.

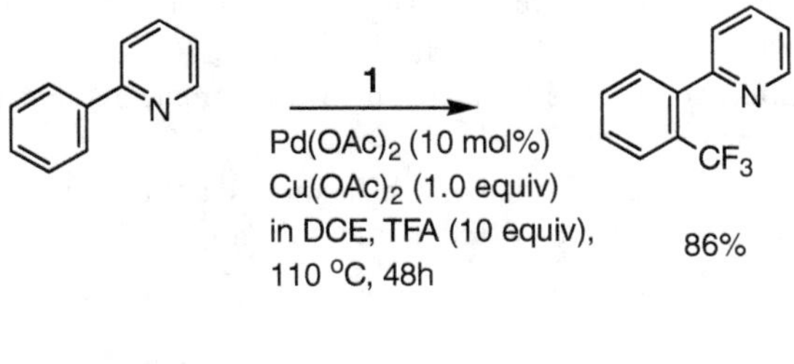

(47.9)

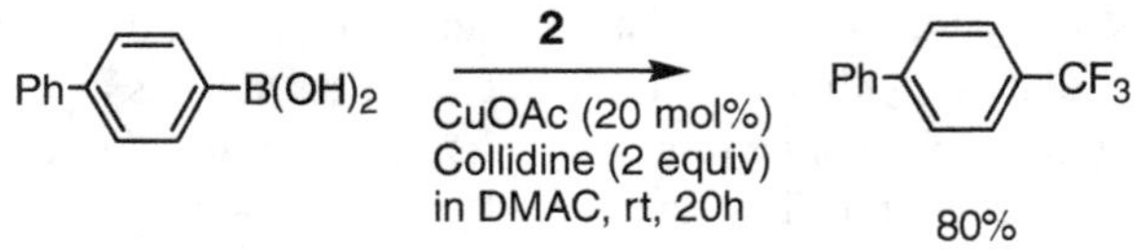

(47.10)

It is worthwhile to compare the *S*-CF_3 salts with the corresponding *O*-CF_3 salts, *O*-(trifluoromethyl)dibenzofuranium salts [17]. There is a complete difference in the reaction manner between them. As seen in Equation 47.11, the *O*-CF_3 salts exclusively give *N*- and *O*-CF_3 products with aniline and phenol.

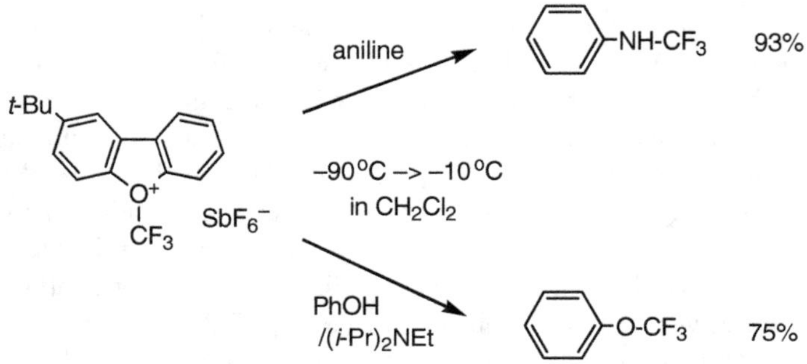

(47.11)

The *S*-salts form *C*-products with aniline and phenol. However, when the powerful dinitrated **3** is heated in phenol, the product is an *O*-CF_3 product, α,α,α-trifluoroanisole [17]. Thus, while the *O*-CF_3 salts exclusively undertake the CF_3^+ formation mechanism, the *S*-CF_3 salts undertake the different reaction mechanism varying from $CF_3\cdot$ to CF_3^+ depending on the trifluoromethylating power of the *S*-salts, the reactivity of nucleophiles, and the reaction conditions [3,17,18].

REFERENCES

1. Umemoto T, Ishihara S. Power-variable trifluoromethylating agents, (trifluoromethyl) dibenzo- and selenophenium salt system. *Tetrahedron Lett* 1990;31:3579–3582; Power-variable electrophilic trifluoromethylating agents, *S*-, *Se*-, and *Te*-(trifluoromethyl) dibenzothio-, -seleno- and –tellurophenium salt system. *J Am Chem Soc* 1993;115:2156–2164.
2. Umemoto T, Ishihara S, Adachi K. Useful electrophilic trifluoromethylating agents; *S*-, Se-, and *Te*-(trifluoromethyl)dibenzo-thio-, -seleno-, and -telluro-phenium-3-sulfonates. *J Fluorine Chem* 1995;74:77–82.
3. Ono T, Umemoto T. Kinetic study of trifluoromethylation with *S*-(trifluoromethyl)dibenzothiophenium salts. *J Fluorine Chem* 1996;80:163–166.
4. Umemoto T, Ishihara S. Effective methods for preparing *S*-(trifluoromethyl)dibenzothiophenium salts. *J Fluorine Chem* 1999;98:75–81.
5. Umemoto T, Adachi K. New method for trifluoromethylation of enolate anions and applications to regio-, diastereo-, and enantio-selective trifluoromethylation. *J Org Chem* 1994;59:5692–5699.

6. Ma J-A, Cahard D. Strategies of nucleophilic, electrophilic, and radical trifluoromethylations. *J Fluorine Chem* 2007;128:975–996.
7. Noritake S, Shibata N, Nomura Y, Huang Y, Matsnev A, Nakamura S, Toru T, Cahard D. Enantioselective electrophilic trifluoromethylation of β-keto esters with Umemoto reagents induced by chiral nonracemic guanidines. *Org Biomol Chem* 2009;7:3599–3604.
8. Pégot B, Macé Y, Urban C, Diter P, Blazejewski J-C, Magnier E. Ionic liquids as new media for electrophilic trifluoromethylation reactions. *J Fluorine Chem* 2012;134; 156–159.
9. St. Jean DJ, Jr, Yuan C, Bercot EA, Cupples R, Chen M, Fretland J, Hale C, Hungate RW, Komorowski R, Veniant M, Wang M, Zhang X, Fotsch C. 2-(*S*)-Phenethylaminothiazolones as potent, orally efficacious inhibitors of 11β-hydroxysteroid dehydrogenase type 1. *J Med Chem* 2007;50:429–432.
10. Kothandaraman S, Donnely KL, Butora G, Jiao R, Pasternak A, Morriello GJ, Goble SD, Zhou C, Mills SG, Maccoss M, Vicario PP, Ayala JM, Demartino JA, Struthers M, Cascieri MA, Yang L. Design, synthesis, and structure-activity relationship of novel CCR2 antagonists. *Bioorg Med Chem Lett* 2009;19:1830–1834; Jiao R, Morriello G, Yang L, Moyes C, inventors; Merck, assignee. Tetrahydropyranyl cyclopentyl tetrahydropyridopyridine modulators of chemokine receptor activity. WO/2003/093266. 2003 Nov 13.
11. Aicher TD, Chen Z, Le Hueron Y, Martin FM, Pineiro-Nunez MM, Rocco VP, Ruley KM, Schaus JM, Spinazze PG, Tupper DE, inventors; Eli Lilly, assignee. Piperazine substituted aryl benzodiazepines. WO/2004/014895. 2004 Feb 19.
12. Imazaki N, Kitano M, Fujibayashi T, Asano S, inventors; Sumitomo Pharmaceuticals, assignee. Benzamide derivatives. WO/2003/082808. 2003 Oct 9.
13. Ishikawa T, Miwa K, Seto M, Banno H, Kawakita Y, inventors; Takeda Pharmaceutical, assignee. Fused heterocyclic compound. WO/2007/064045. 2007 June 7.
14. Adams C, Hu Q-Y, Mcquire LW, Papillon J, inventors; Novartis, assignee. Organic compounds. WO/2009/156462. 2009 Dec 30.
15. Wang X, Truesdale L, Yu J-Q. Pd(II)-Catalyzed *ortho*-trifluoromethylation of arenes using TFA as a promoter. *J Am Chem Soc* 2010;132:3648–3649.
16. Xu J, Luo D-F, Xiao B, Liu Z-J, Gong T-J, Fu Y, Liu L. Copper-catalyzed trifluoromethylation of aryl boronic acids using a CF_3^+ reagent. *Chem Commun* 2011;47:4300–4302.
17. Umemoto T, Adachi K, Ishihara S. CF_3 oxonium salts, *O*-(trifluoromethyl)dibenzofuranium salts: in situ synthesis, properties, and application as a real CF_3^+ species reagent. *J Org Chem* 2007;72:6905–6917.
18. Mace Y, Pradet C, Popkin M, Blazejewski J-C, Magnier E. Mechanistical insight into 'electrophilic' trifluoromethylation with *S*-(trifluoromethyl)dibenzothiophenium salts. *Tetrahedron Lett* 2010;51:5388–5391.

CHAPTER 48

Synthesis of Fluorine-Containing Heterocycles from α,α-Dihydropolyfluoroalkylsulfides and Fluorinated Thiocarboxylic Acids Derivatives

VADIM M. TIMOSHEMKO AND YURIY G. SHERMOLOVICH

Heterocyclic compounds constitute the essential part of all pesticides and medical products. The replacement of H atoms in heterocyclic systems with fluorine atoms or perfluoroalkyl groups exerts great influence on the physical, chemical, and biological properties of these molecules. The main methods for preparing these compounds can be divided into two groups.

The first one consists of procedures where the fluorine atom or perfluoroalkyl groups are introduced into the existent heterocyclic molecule.

The second method which was more intensively studied during the last years consists of the synthesis of heterocyclic compounds from fluorine-containing building blocks. The most known and used precursors are fluorine-containing olefines, vinyl ketones, acids, and aromatic compounds.

In the present chapter we will demonstrate the use of α,α-dihydropolyfluoroalkyl sulfides, esters, and amides of polyfluorinated aliphatic thiocarboxylic acids for the synthesis of fluorine-containing heterocycles.

Efficient Preparations of Fluorine Compounds, First Edition. Edited by Herbert W. Roesky.

48.1 PREPARATION OF 1,1-DIHYDROPOLYFLUOROALKYLSULFIDES

1 **2**

$R_F = CF_3$, R = CH_2Ph (**a**);
$R_F = CF_3$, R = *p*-Tol (**b**);
$R_F = CF_3CF_2$, R = *p*-Tol (**c**);
$R_F = CF_3(CF_2)_2$, R = CH_2Ph (**d**);
$R_F = H(CF_2)_4$, R = *p*-Tol (**e**)

Apparatus Three-necked flask, mechanical stirrer, N_2-inlet, dropping funnel, safety glasses, laboratory coat, protective gloves.

Chemicals Benzylmercaptane, *p*-toluenemercaptane, dry dimethyl sulfoxide, KOH, $(C_2H_5)_2O$.

Attention! Safety glasses and protective gloves must be used at all times.

Caution! Mercaptanes are harmful.

Common Procedure for Benzyl Sulfides 2a, d [1, 2]

Experimental Procedure 2a To a suspension of fine-powdered KOH (22.4 g, 0.4 mol) in dry DMSO (160 mL), benzylmercaptane (49.7 g, 0.4 mol) in DMSO (30 mL) was added under N_2 atmosphere keeping the temperature below 30 °C. The mixture was stirred for 3–4 h until it became clear and a solution of 2,2,2-trifluoroethyl tosylate (50.8 g, 0.2 mol) in DMSO (50 mL) was added keeping the temperature below 30 °C by regulation of the rate of addition. The mixture was stirred for 5 h at 30 °C and 8 h at room temperature. Then 200 mL of H_2O was added and the product was extracted with ether (4 × 100 mL). The etheral extracts were combined and washed with H_2O (300 mL), and then with a NaOH solution (50 mL, 10% in water), and finally with water to neutralize the reaction. Lacmus indicator was added to show the neutral state. After drying over Na_2SO_4 ether was distilled off, and the residue was fractionalized in vacuum. Yield: 37.10 g (90%).

Characterization Data bp 41–42 °C (0.05 mm Hg). ^{1}H NMR, δ: 2.89 (2H, q, $J = 9.9$ Hz, $CF_3\underline{CH_2}$), 3.83 (2H, s, CH_2), 7.25–7.37 (5H, m, H_{Ar}) ppm. ^{19}F NMR, δ: –67.0 (s, CF_3) ppm.

Common Procedure for the Preparation of Tolyl Sulfides 2b,c,e

Experimental Procedure 2c To a suspension of fine-powdered KOH (3.37 g, 0.06 mol) in dry DMSO (50 mL) a solution of *p*-tolylmercaptane (7.45 g, 0.06 mol) in DMSO (30 mL) was added under N_2 atmosphere keeping the temperature below 30 °C. The mixture was stirred for 1.5–2 h until it became clear, and a solution of pentafluoropropyl tosylate (15.2 g, 0.05 mol) in DMSO (20 mL) was added drop by drop. The reaction mixture was heated to 65 °C to start the exothermic reaction. The reaction temperature increased to 80 °C, and the mixture was stirred at 75–80 °C for additional 2–3 h, monitoring the full conversion of tosylate by ^{19}F NMR. The mixture was diluted with water (100 mL) and stirred for 0.5 h. Then the solution was transferred to a separating funnel, an additional amount of water (100 mL) was added, and the mixture was extracted with diethyl ether (3 × 100 mL). The combined etheral solution was washed with water (75 mL), dried over $MgSO_4$, and finally the ether was distilled off. The residue was separated by fractional distillation in vacuum. Yield: 10.2 g (80%).

Characterization Data bp 50–54 °C (0.09 mm Hg). ^1H NMR, δ: 2.35 (3H, s, CH_3), 3.40 (2H, t, $J = 16.5$ Hz, $CF_2\underline{CH_2}$), 7.15 (2H, d, $J = 8.0$ Hz, ArH-*para*), 7.41 (2H, d, $J = 8.0$ Hz, ArH-*para*) ppm. ^{19}F NMR, δ: –85.61 (3F, s, CF_3), –118.02 (2F, $J = 8.0$ Hz, CF_2) ppm.

Waste Disposal All liquid and solid residues of the reactions have to be collected in a proper labeled container for toxic waste for utilization by incineration.

48.2 PREPARATION OF SULFONES [1]

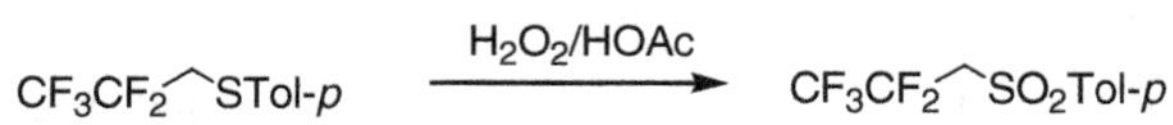

3

Experimental Procedure To a solution of $CF_3CF_2CH_2STol$-*p* (10.2 g, 0.04 mol) in HOAc (50 mL) was added H_2O_2 (10 g of 50% water solution), and the mixture was carefully warmed to 60 °C and then stirred for 4 h at this temperature. The reaction mixture was poured into water (400 mL) and left for 1–1.5 h for precipitating the sulfone. Crystals were filtered off, washed with water (100 mL), and dried. Yield: 11.1 g (96%).

Characterization Data mp 61–63 °C. ^1H NMR, δ 2.47 (3H, s, CH_3), 3.80 (2H, t, $J = 15.5$ Hz, $CF_2\underline{CH_2}$), 7.39 (2H, d, $J = 8.0$ Hz, ArH-*para*), 7.85 (2H, d, $J = 8.0$ Hz, ArH-*para*) ppm. ^{19}F NMR, δ –86.02 (3F, s, CF_3), –117.43 (2F, $J = 8.0$ Hz, CF_2) ppm.

48.3 PREPARATION OF 1,1-DICHLOROPOLYFLUOROALKYLSULFIDES [3,4]

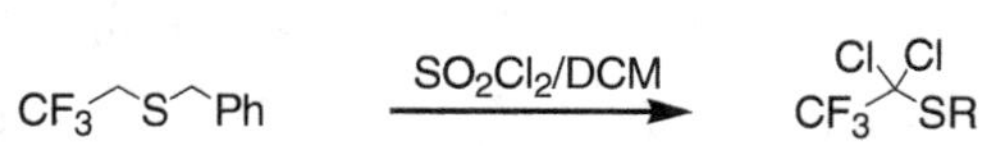

4

Apparatus Two-hundred-and-fifty-milliliter flask, with septum, magnetic stirrer, T-shaped N_2-inlet, syringe (50 mL), safety glasses, laboratory coat, protective gloves.

Chemicals Benzyl-2,2,2-trifluoroethylsulfide, SO_2Cl_2, DCM.

Caution! Hydrogen chloride evolved has to be trapped with NaOH water solution.

Attention! Safety glasses and protective gloves must be used at all times.

Experimental Procedure To a solution of benzyl-2,2,2-trifluoroethylsulfide (15.09 g, 73.2 mmol) in dry DCM (100 mL) was added a solution of SO_2Cl_2 (20.25 g, 150 mmol) in DCM (30 mL) under stirring for 1 h. The mixture was refluxed for 3 h and DCM was removed in vacuum to give as a residue 1,1-dichloro-2,2,2-trifluoroethylbenzyl sulfide, which can be used without additional purification. Yield: 19.5 g (97%).

Characterization Data ^{1}H NMR ($CDCl_3$), δ 4.25 (s, 2H, CH_2), 7.32 (m, 5H, H_{Ar}) ppm. ^{19}F NMR ($CDCl_3$), δ –77.80 (s, CF_3) ppm.

48.4 PREPARATION OF DITHIOESTERS

Preparation of Trifluorothioacetyl Chloride [4]

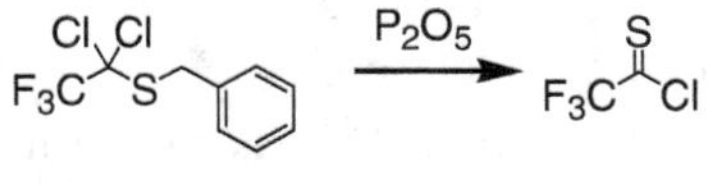

Apparatus One-hundred-milliliter flask with septum, magnetic stirrer, T-shaped N_2-inlet, syringe (50 mL), safety glasses, laboratory coat, protective gloves.

Chemicals Benzyl dichlorosulfide, P_2O_5.

Attention! Safety glasses and protective gloves must be used at all times.

Experimental Procedure A mixture of 1,1-dichloro-2,2,2-trifluoroethyl benzyl sulfide (0.1 mol) and P_2O_5 (0.2–0.3 mol) was prepared by shaking the mixture in a airtight flask. Then the flask was connected with a cooled receiver, and slowly heated on an oil bath to 150–180 °C. The volatile product was collected in the cooled receiver. Yield: 81%.

Characterization Data bp 28–29 °C. ^{19}F NMR ($CDCl_3$), δ –77.7 ppm.

Waste Disposal To the residues in the flask the solution of NaOH (20% in water) is carefully added and the mixture is kept for 24 h at room temperature. Finally, the mixture is collected in a labeled container for waste.

48.5 PREPARATION OF *P*-TOLYLTRIFLUORODITHIOACETATE [3]

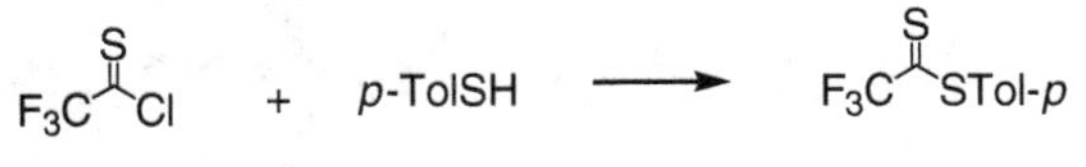

Experimental Procedure *p*-Toluenethiol (2.48 g, 20 mmol) was added to a solution of trifluorothioacetyl chloride (10.39 g, 70 mmol) in 20 mL of diethyl ether at room temperature. After stirring for 1 h the next portion of *p*-toluenethiol (3.73 g, 30 mmol) was added, and the mixture was stirred for 8 h. The mixture was distilled in vacuum at 12 mm Hg. Yield: 8 g (58%).

Characterization Data bp 110–112 °C (12 mm Hg). ^{19}F NMR ($CDCl_3$, 188 MHz) δ –65.98 (s, CF_3) ppm. ^{1}H NMR ($CDCl_3$, 300 MHz) δ 2.44 (3H, s, CH_3), 7.29 (2H, d, $J = 8.1$ Hz, H_{Ar}), 7.35 (2H, d, $J = 8.4$ Hz, H_{Ar}) ppm.

48.6 PREPARATION OF *N*-BUTYL TRIFLUORODITHIOACETATE [5]

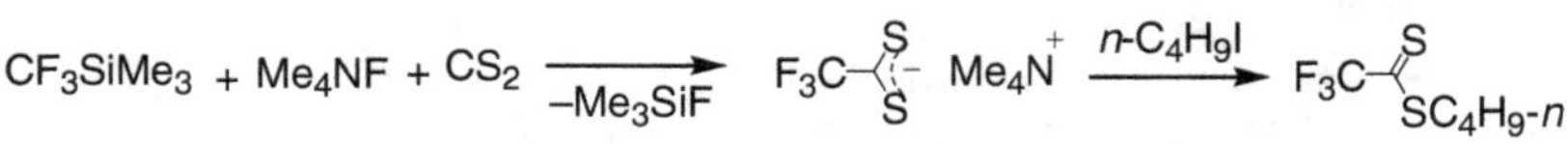

Experimental Procedure (1) To a solution of 0.152 g (2 mmol) of CS_2 in monoglyme (4 mL) at –60 °C, $[Me_4N]F$ (0.186 g, 2 mmol) and Me_3SiCF_3 (0.3 g, 2.1 mmol) were added. The reaction mixture was stirred for 1 h at –40 °C, and then for 1 h at room temperature. The solvent and the volatile compounds were evaporated; the residue was washed with Et_2O and dried at 0.02 Torr. The resulting salt of composition $Me_4NCF_3CS_2$ (0.43 g, 98% yield) was obtained as hygroscopic pink solid.

Characterization Data mp 225–228 °C (decomposition). ^{1}H NMR (CD_3CN), δ d 3.11 (12H, s, Me_4N) ppm; ^{19}F NMR (CD_3CN): δ –63.2 (3F, s) ppm; ^{13}C NMR (Me_2SO-d_6): δ 118.2 (q, $^1J_{CF} = 279$ Hz, CF_3), 229 (q, $^2J_{CF} = 40$ Hz, CS), 54.5 (q, $^1J_{CH} = 144$ Hz, CH_3) ppm. Anal. calcd. for $C_6H_{12}F_3NS_2$: C 32.8; H 5.5; S 29.2. Found: C 32.5; H 5.75; S 29.0.

Experimental Procedure (2) To a solution of $Me_4NCF_3CS_2$ 0.55 g (2.5 mmol) in CH_3CN (6 mL) was added 0.55 g (3 mmol) of *n*-butyliodide. The reaction mixture was stirred for 1 h at room temperature and the precipitate of Me_4NI was filtered off. The filtrate was concentrated and the residue was purified by distillation. *n*-Butyl trifluorodithioacetate $CF_3C(S)Sn$-C_4H_9 was obtained as red liquid, bp 73 °C at 12 Torr. Yield 95% (0.5 g).

Characterization Data ^{1}H NMR, δ 3.25 (2H, t, $^3J_{HH} = 7.3$ Hz, CH_2), 1.71 (2H, quintet, $^3J_{HH} = 7.6$ Hz, CH_2), 1.44 (2H, sept, $^3J_{HH} = 7.3$ Hz CH_2), 0.95 (3H, t, $^1J_{CF} = 7.3$ Hz, CH_3) ppm; ^{19}F NMR ($CDCl_3$): δ –65.6 (3F, s). Anal. calcd. for $C_6H_9F_3S_2$: C 35.6; H 4.5; S 31.7. Found: C 35.8; H 4.4; S 31.0.

48.7 PREPARATION OF BENZYL PERFLUOROBUTANEDITHIOATE [2]

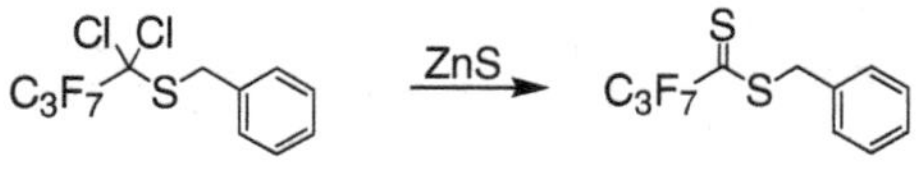

Experimental Procedure A mixture of dichlorosulfide (0.01 mol) and zinc sulfide (0.2 mol) in acetonitrile (20 mL) was refluxed for 8 h. The mixture was cooled to room temperature and filtered, acetonitrile distilled off at atmospheric pressure, and the residue separated by fractional distillation in vacuum. Yield: 76%.

Characterization Data bp 90 °C (0.1 mbar). ^{19}F NMR ($CDCl_3$): δ –80.6 (t, J = 11 Hz, CF_3), –103.9 (tm, J = 11 Hz, CF_2), –124.5 (m, CF_2) ppm. ^{1}H NMR, δ 4.45 (s, CH_2), 7.30 (m, C_6H_5) ppm. ^{13}C NMR ($CDCl_3$): δ 41.8 (CH_2), 104–121 (m, $CF_2CF_2CF_3$), 128.1, 129.0, 130.1, 132.7 (C_6H_5), 211 (t, J = 25.5 Hz, C=S).

48.8 PREPARATION OF KETOSULFONES [6, 7]

A. [6]

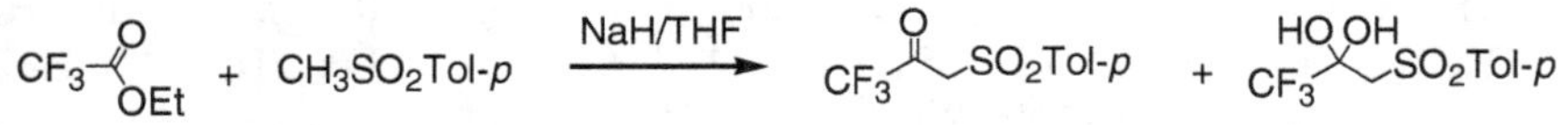

Apparatus A three-necked flask (250 mL), mechanical stirrer, N_2-inlet, dropping funnel (50 mL), safety glasses, laboratory coat, protective gloves.

Chemicals NaH (60% oil dispersion) methyl *p*-tolyl sulfone, ethyl trifluoroacetate.

Attention! Safety glasses and protective gloves must be used at all times.

Experimental Procedure To a stirred solution of methyl *p*-tolyl sulfone (1.7 g, 10 mmol) in THF (20 mL) under N_2 at 0 °C was added NaH (60% oil dispersion; 0.6 g, 15 mmol) in portions after which the resulting suspension was stirred at 0 °C for 10 min, and then treated drop by drop with ethyl trifluoroacetate (3.6 mL, 30 mmol) at 0 °C. After 2 h under reflux, the resulting solution was poured into a saturated aqueous NaCl solution (250 mL). Then the mixture was acidified by addition of diluted HCl and extracted with Et_2O (4 × 100 mL). The combined extracts were dried ($MgSO_4$) and evaporated under reduced pressure to give crude ketosulfone, which was recrystallized from hexane to give a mixture of ketone : hydrate $CF_3C(O)CH_2SO_2$*p*-Tol and $CF_3C(OH)_2CH_2SO_2$*p*-Tol in the ratio of 1 : 2. Yield: 78%.

Characterization Data mp 74–76 °C. *Hydrate*: ^{1}H NMR ($CDCl_3$), δ 7.39 and 7.85 (dd, 4H, $^{3}J_{HH}$ = 8.4 Hz, C_6H_4), 4.79 (br, 2OH), 3.58 (s, 2H, CH_2), 2.47 (s, 3H, CH_3) ppm; ^{19}F NMR ($CDCl_3$), δ –88.23 (s, CF_3) ppm; ^{13}C NMR ($CDCl_3$), δ 145.93, 136.67, 130.01, 128.43 ($4C_{Ar}$), 121.60 (q, CF_3, J_{CF} = 287.6 Hz), 92.55 (q, $CF_3\underline{C}$, $^{2}J_{CF}$ = 33.5 Hz), 56.59 (CH_2), 21.73 (CH_3) ppm. *Ketone*: ^{1}H NMR ($CDCl_3$), δ 7.41 and 7.83 (dd, 4H, $^{3}J_{HH}$ = 8.4 Hz, C_6H_4), 4.46 (s, 2H, CH_2), 2.48 (s, 3H, CH_3) ppm; ^{19}F NMR ($CDCl_3$), δ –79.81 (s, CF_3) ppm; ^{13}C NMR ($CDCl_3$), δ 180.23 (q, C=O, $^{2}J_{CF}$ = 39.0 Hz), 146.42, 135.41, 130.29, 128.68 ($4C_{Ar}$), 114.68 (q, CF_3, J_{CF} = 290.0 Hz), 60.99 (CH_2), 21.78 (CH_3) ppm.

B. [7]

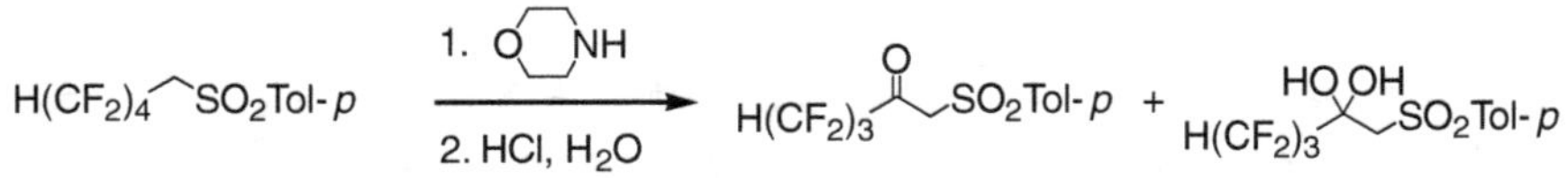

Experimental Procedure A mixture of sulfone (3.7 g, 10 mmol) and morpholine (2.6 g, 30 mmol) in benzene (30 mL) was refluxed for 1 h, then the solvent was removed in vacuum, and the residue was heated with conc. HCl (10 mL) at 90 °C for 10 min. After cooling the mixture to room temperature, the product was isolated by extraction with $CHCl_3$. Evaporation of $CHCl_3$ and recrystallization from benzene gave a mixture of ketone to hydrate (1:3) in 75% yield.

Characterization Data mp 48–50 °C. *Hydrate*: ^{1}H NMR ($CDCl_3$), δ 7.86 (d, 2H, ArH, $^2J_{HH}$ = 8.0 Hz), 7.39 (d, 2H, ArH, $^2J_{HH}$ = 8.0 Hz), 6.26 (tt, 1H, HCF_2, $^2J_{HF}$ = 52.2, $^3J_{HF}$ = 5.8 Hz), 5.23 (br, 2OH), 3.63 (s, 2H, CH_2), 2.47 (s, 3H, CH_3) ppm. ^{13}C NMR ($CDCl_3$), δ 146.06 (s, $\underline{C}_{Ar}$-CH_3), 135.51 (s, $\underline{C}_{Ar}$-SO_2), 130.12 (M-Ar), 128.37 (s, *o*-Ar), 110.03 (tt, $\underline{C}F_2$C-O, J_{CF} = 262.9 Hz), 109.37 (tm, $\underline{C}F_2$C=O, J_{CF} = 263.0 Hz), 107.78 (tt, HCF_2, J_{CF} = 254.2 Hz), 94.56 (t, C-OH, $^2J_{CF}$ = 27.5 Hz), 57.18 (s, CH_2), 21.76 (s, CH_3) ppm. ^{19}F NMR ($CDCl_3$), δ –126.80 M (2F, CF_2), –130.63 (m, 2F, CF_2), –138.14 (dm, 2F, HCF_2, $^2J_{FH}$ = 52.2 Hz) ppm. *Ketone*: ^{1}H NMR ($CDCl_3$), δ 7.83 (d, 2H, ArH, $^2J_{HH}$ = 8.0 Hz), 7.41 (d, 2H, ArH, $^2J_{HH}$ = 8.0 Hz), 6.05 (tt, 1H, HCF_2, $^2J_{HF}$ = 52.0, $^3J_{HF}$ = 5.0 Hz), 4.51 (s, 2H, CH_2), 2.48 (s, 3H, CH_3) ppm. ^{13}C NMR ($CDCl_3$), δ 184.30 (t, C=O, $^2J_{CF}$ = 29.5 Hz), 146.43 (s, $\underline{C}_{Ar}$-CH_3), 136.88 (s, $\underline{C}_{Ar}$-SO_2), 130.29 (M-Ar), 128.74 (s, *o*-Ar), 112.84 (tt, $\underline{C}F_2$C=O, J_{CF} = 264.3 Hz), 111.10 (tm, $\underline{C}F_2$C=O, J_{CF} = 267.1 Hz), 108.22 (tt, HCF_2, J_{CF} = 252.8 Hz), 61.65 (s, CH_2), 21.80 (s, CH_3) ppm. ^{19}F NMR ($CDCl_3$), δ –122.07 (m, 2F, CF_2), –131.52 (m, 2F, CF_2), –137.83 (dm, 2F, HCF_2, $^2J_{FH}$ = 52.0 Hz).

48.9 PREPARATION OF HETEROCYCLES

Synthesis of Fluorinated Thiopyran Derivatives [2, 3, 8]

Preparation of 2-chloro-2-(trifluoromethyl)-3,6-dihydro-2H-thiopyran (2) [8]

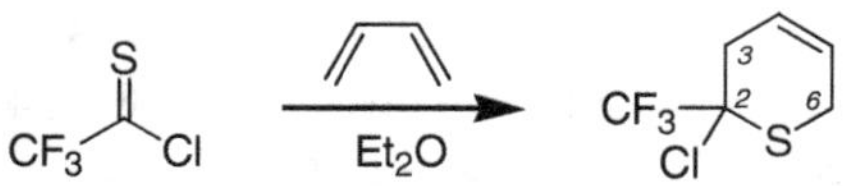

1 **2**

Apparatus Twenty-five-milliliter flask, magnetic stirrer, safety glasses, laboratory coat, protective gloves.

Chemicals Trifluorothioacetyl chloride, 1,3-butadiene, diethyl ether.

Attention! Safety glasses and protective gloves must be used at all times.

Caution! Trifluorothioacetyl chloride is corrosive lachrymator.

Experimental Procedure 1,3-Butadiene gas was bubbled into a solution of trifluorothioacetyl chloride (**1**) (1.48 g, 10 mmol) in 5-mL diethyl ether under stirring and cooling in an ice-water bath until decoloration of the solution. Then the solvent was distilled off at atmospheric pressure to give a pure residue of composition $C_5H_4(CF_3)_2S$ (**2**) as a pale-yellow liquid. Yield: 1.86 g (92%).

Characterization Data ^{1}H NMR, δ 2.74 (1H, dm, AB, J_{AB} = 17.7 Hz, H-3*eq*), 2.98 (1H, dm, AB, J_{AB} = 17.7 Hz, H-3*ax*), 3.31 (1H, dm, AB, J_{AB} = 17.7 Hz, H-6*eq*), 3.67 (1H, dm, AB, J_{AB} = 17.7 Hz, H-6*ax*), 5.82 (1H, dm, J = 10.8 Hz), 6.00 (1H, dm, J = 10.8 Hz) ppm. ^{19}F NMR, δ –78.46 (s, CF_3) ppm. ^{13}C NMR, δ 25.64 (s, C-6), 35.75 (q, J = 1.5 Hz, C-3), 72.51 (q, J = 31.5 Hz, C-2), 122.52 (s), 123.41 (s), 124.31 (q, J = 282.0 Hz, CF_3) ppm.

48.10 PREPARATION OF 6-(TRIFLUOROMETHYL)-2*H*-THIOPYRAN (3) [8]

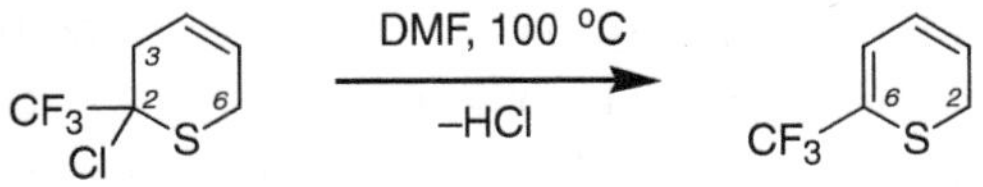

2 **3**

Apparatus Twenty-five-milliliter flask, magnetic stirrer, outlet tube, safety glasses, laboratory coat, protective gloves.

Chemicals Adduct **2**, dry DMF, hexane.

Attention! Safety glasses and protective gloves must be used at all times.

Caution! Eliminated hydrogen chloride must be trapped with a NaOH water solution.

Experimental Procedure A solution of adduct **2** (1.97 g, 10 mmol) in 10 mL of anhydrous DMF was heated at 100 °C by means of a boiling water bath for 2 h. The resulting dark-brown solution was diluted with 50-mL cold water and extracted with hexane (5 × 10 mL). The hexane extracts were combined, washed with water (2 × 20 mL), dried over Na_2SO_4, evaporated, and the residue was distilled in vacuum. Yield: 1.32 g (81%), brownish liquid.

Characterization Data bp 80–82 °C (130 mm Hg) ^{1}H NMR, δ 3.34 (2H, dq, J = 5.4, 0.7 Hz, CH_2), 5.78 (1H, dt, J = 9.8, 5.4 Hz, H-3), 6.13 (1H, ddq, J = 9.8, 5.9, $^5J_{HF}$ = 1.2 Hz, H-4), 6.66 (1H, dq, J = 5.9, $^4J_{HF}$ = 1.5 Hz, H-5) ppm. ^{19}F NMR, δ –66.16 (m, CF_3) ppm. ^{13}C NMR, δ 24.52 (s, C-2), 119.73 (q, J = 1.5 Hz, C-4), 122.39 (q, J = 271.5 Hz, CF_3), 123.84 (q, J = 6.0 Hz, C-5), 125.41 (s, C-3), 125.61 (q, J = 34.0 Hz, C-6) ppm.

GC/MS (*m*/*z*): 166 (M^+, 56%), 165 (M^+–H, 63%), 97 (100%), 69 (CF_3^+, 33%).

48.11 PREPARATION OF 2-TRIFLUOROMETHYLTHIOPYRYLIUM TETRAFLUOROBORATE (4) [8]

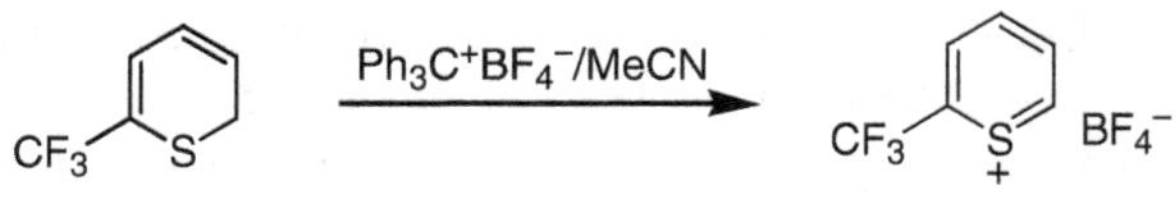

3 **4**

Apparatus Twenty-five-milliliter flask, magnetic stirrer, safety glasses, laboratory coat, protective gloves.

Chemicals 6-Trifluoromethyl-2*H*-thiopyran (**3**), trityl tetrafluoroborate, dry CH_3CN, diethyl ether.

Attention! Safety glasses and protective gloves must be used at all times.

Experimental Procedure 6-Trifluoromethyl-2*H*-thiopyran (**3**) (1.67 g, 10 mmol) was added to a suspension of trityl tetrafluoroborate (3.4 g, 10.3 mmol) in dry CH_3CN (10 mL) at 0 °C under stirring. The solution was stirred at room temperature for further 30 min, and then diethyl ether (30 mL) was added. A precipitate of thiopyrylium salt was formed, filtered off, and finally washed with ether and dried. Yield: 2.05 g (81%).

Characterization Data mp 150–160 °C. ^1H NMR, δ 9.09 (1H, t, $J = 8.3$ Hz, H-5), 9.21 (1H, d, $J = 8.3$ Hz, H-3), 9.30 (1H, t, $J = 8.3$ Hz, H-4), 10.40 (1H, d, $J = 8.3$ Hz, H-2) ppm. ^{19}F NMR, δ –59.84 (3F, s, CF_3), –149.26 (4F, s, BF_4) ppm. ^{13}C NMR (CD_3CN), δ 121.35 (q, $J = 277$ Hz, CF_3), 137.93 (q, $J = 3$ Hz, C-3), 140.58 (s, C-5), 153.58 (s, C-4), 156.41 (q, $J = 38$ Hz, C-2), 160.82 (s, C-6) ppm.

48.12 PREPARATION OF 2-METHOXY-6-TRIFLUOROMETHYL-2*H*-THIOPYRAN (5) [8]

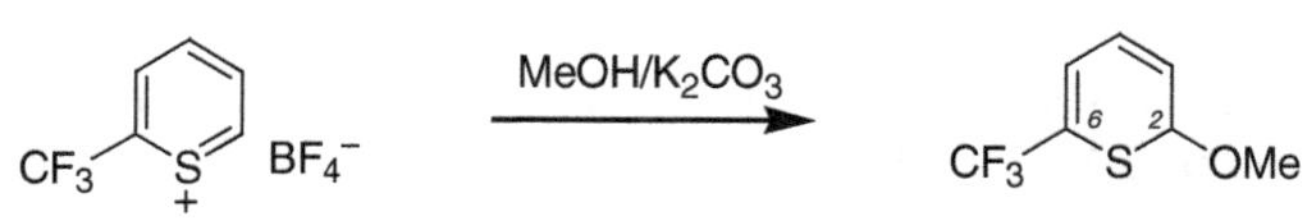

4 **5**

Apparatus Twenty-five-milliliter flask, magnetic stirrer, safety glasses, laboratory coat, protective gloves.

Chemicals 2-Trifluoromethylthiopyryliumtetrafluoroborate (**4**), dry CH_3OH, K_2CO_3.

Attention! Safety glasses and protective gloves must be used at all times.

Experimental Procedure The suspension of fine-powdered K_2CO_3 (1.1 g, 8 mmol) in 5 mL of methanol was stirred, and then 2-trifluoromethylthiopyrylium tetrafluoroborate (**4**) (1.0 g, 4 mmol) was added. The mixture was vigorously stirred at room temperature for 1 h and excess of methanol was removed in vacuum. The residue was treated with diethyl ether (5 mL), insoluble materials were filtered off, and the ether was evaporated to yield 0.48 g (61%) of **5** as yellow liquid.

Characterization Data ^{1}H NMR, δ 3.27 (3H, s, CH_3O), 5.21 (1H, dq, J = 6.4 Hz, 0.8 Hz, H-2), 6.11 (1H, ddq, J = 10.0 Hz, 6.4 Hz, 0.2 Hz, H-3), 6.53 (ddq, J = 10.0 Hz, 6.6 Hz, $^5J_{HF}$ = 1.0 Hz, H-4), 6.91 (dq, J = 6.6 Hz, $^4J_{HF}$ = 1.4 Hz, H-5) ppm. ^{19}F NMR, δ –66.12 (m, CF_3) ppm. ^{13}C NMR (CD_3CN), δ 53.09 (s, CH_3O), 75.62 (s, C-2), 119.22 (q, J = 1.5 Hz, C-4), 121.20 (q, J = 5.5 Hz, C-5), 122.52 (q, J = 272 Hz, CF_3), 123.86 (q, J = 34 Hz, C-6) ppm.

GC/MS (*m*/*z*): 196 (M^+, 20%), 165 (M^+–CH_3O, 100%), 97 (M^+–CF_3, 18%), 69 ($CF_3{}^+$, 12%).

48.13 PREPARATION OF 2-METHOXY-6-(TRIFLUOROMETHYL)-3,4-DIHYDRO-2*H*-THIOPYRAN-3,4-DIOL (6) [8]

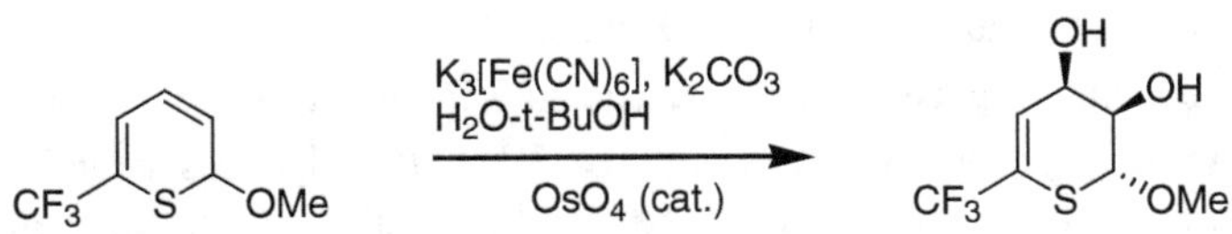

6

5

Apparatus Twenty-five-milliliter flask, magnetic stirrer, safety glasses, laboratory coat, protective gloves.

Chemicals Thiopyran **5**, OsO_4 (Merck, 10 mM solution in 0.05 M sulfuric acid), *t*-BuOH, $K_3[Fe(CN)_6]$, Na_2SO_3, ethyl acetate.

Attention! Safety glasses and protective gloves must be used at all times.

Caution! OsO_4 solution is a highly toxic oxidizer. The reaction has to be done in a well-ventilated hood because of the volatility of OsO_4.

Experimental Procedure The mixture of $K_3[Fe(CN)_6]$ (1.48 g, 4.5 mmol) and K_2CO_3 (0.63 g, 4.5 mmol), 1.5 mL (15 μmol) of osmium(VIII)-oxide solution (Merck, 10 mM solution in 0.05 M sulfuric acid) in H_2O (2 mL) and *t*-BuOH (2 mL) was vigorously stirred for 30 min at room temperature, then compound **5** (0.3 g, 1.5 mmol) was added and the reaction mixture was stirred for 3 days at room temperature. Consumption of **5** was monitored by ^{19}F NMR spectra of the nonaqueous phase. After the reaction was complete, Na_2SO_3 (0.19 g, 1.5 mmol) was added, stirring was continued for additional 10 min, and insoluble material was filtered off and washed with CH_3OH. The filtrate was extracted with ethyl acetate (5 × 10 mL), evaporated, and the residue was exposed to column chromatography on SiO_2 (CH_2Cl_2–MeOH, 10:1). Fraction with R_f = 0.50 was collected to give 35 mg (10%) of compound **6** as one diastereomer.

Characterization Data Viscous oil. ^{1}H NMR, δ 3.52 (3H, s, CH_3), 4.10 (1H, dd, $J = 4.8$ Hz, 4.1 Hz, H-3), 4.41 (1H, dm, $J = 4.1$ Hz, H-4), 4.94 (1H, d, $J = 4.8$ Hz, H-2), 5.17 (2H, OH), 6.36 (1H, dq, $J = 3.6$ Hz, $^4J_{HF} = 1.2$ Hz, H-5) ppm. ^{19}F NMR, δ –67.21 (m, CF_3) ppm. ^{13}C NMR (CD_3CN), δ 57.55 (s, CH_3), 64.82 (s, CH-OH), 66.16 (s, CH-OH), 87.14 (s, C-2), 121.61 (q, $J = 274$ Hz, CF_3), 123.59 (q, $J = 33.3$ Hz, C-6), 127.48 (q, $J = 5.0$ Hz, C-5) ppm.

Waste Disposal All liquid residues of reactions have to be collected in a proper labeled container for toxic waste disposal.

48.14 PREPARATION OF ENAMINONE [9]

$R_F = H(CF_2)_3$ (**a**), CF_3 (**b**)　　**2a,b**

Apparatus One-hundred-milliliter flask, magnetic stirrer, condenser.

Chemicals Keto sulfon hydrate, $POCl_3$, dry DMF.

Attention! Safety glasses and protective gloves must be used at all times.

Experimental Procedure A sample of DMF (6 mL) was added to $POCl_3$ (2.8 mL, 30 mmol) under stirring for 1 h. Then, a solution of keto sulfone hydrate **1a** or **b** (5 mmol) in DMF (6 mL) was added, stirred for 2 h at 20 °C, and then poured onto wet ice. The crystalline precipitate was filtered off, washed with water, and dried. Yield: 88% for **2a** and 82% for **2b**. The recovered enaminones were used in the subsequent syntheses without further purification. Analytical pure samples were obtained by recrystallization.

Characterization Data **2a**: bp 28–29 °C. IR spectrum (neat), ν, cm^{-1}: 1610 (C=C), 1660 (C=O). ^{1}H NMR, δ 8.26 (1H, s, CH=N), 7.28 and 7.75 (4H, dd, $^3J_{HH} = 8.0$ Hz, C_6H_4), 6.16 (1H, tt, $^2J_{HF} = 53.0$ Hz, $^3J_{HF} = 6.0$ Hz, HCF_2), 3.44 (3H, s, NCH_3), 3.04 (3H, s, NCH_3), 2.42 (3H, s, CH_3) ppm. ^{19}F NMR, δ −118.14 (2F, m, CF_2), −133.56 (2F, m, CF_2), −133.42 (2F, dm, $J_{FH} = 53.0$, HCF_2) ppm. Mass spectrum, *m/z*: 404 $[M]^+$. Found, %: C, 44.53; H, 3.74; N, 3.77; S, 8.07. $C_{15}H_{15}F_6NO_3S$. Calcd. %: C, 44.67; H, 3.75; N, 3.47; S, 7.95.

2b: mp 155–157 °C (methanol). ^{1}H NMR, δ 8.16 (1H, s, CH=N), 7.25 and 7.73 (4H, dd, $3J_{HH} = 8.0$ Hz, C_6H_4), 3.42 (3H, s, NCH_3), 2.92 (3H, s, NCH_3), 2.40 (3H, s, CH_3) ppm. ^{19}F NMR, δ −73.04 (s, CF_3) ppm.

48.15 PREPARATION OF ALKANOYL VINYLIC ETHER [10]

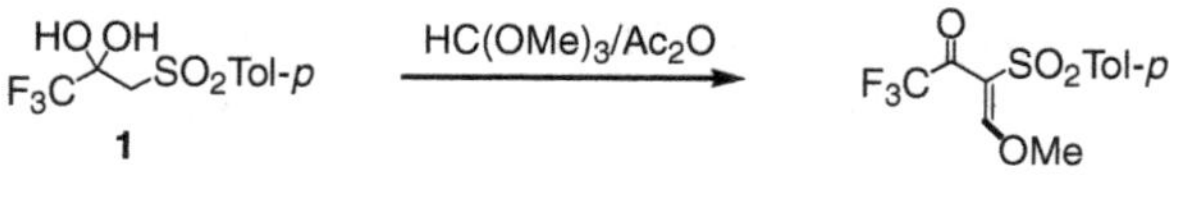

2

Apparatus One-hundred-milliliter flask, magnetic stirrer, condenser.

Chemicals Keto sulfon hydrate, trimethylformate, acetic anhydride.

Attention! Safety glasses and protective gloves must be used at all times.

Experimental Procedure A mixture of ketosulfone **1** (10 mmol), and trimethylformate (20 mmol) in acetic anhydride (30 mmol) was refluxed for 4 h. After cooling, the reaction mixture compound **2** was filtered off, washed with ether, and can be used without additional purification. Yield: 72%.

Characterization Data mp 153–155 °C. ^{1}H NMR ($CDCl_3$), δ 8.37 (s, 1H, =CH), 7.32 and 7.86 (dd, 4H, C_6H_4, $^3J_{HH}$ = 8.4 Hz), 4.24 (s, 3H, OCH_3), 2.43 (s, 3H, CH_3) ppm; ^{13}C NMR ($CDCl_3$), δ 174.48 (q, C=O, $^2J_{CF}$ = 40.0 Hz), 171.21 (CH=), 144.81 (C_{Ar}), 137.55 (C_{Ar}), 129.67 (CH_{Ar}), 128.52 (CH_{Ar}), 117.24 (<u>C</u>=CH), 114.95 (q, CF_3, J_{CF} = 290.0 Hz), 66.03 (OCH_3), 21.72 (CH_3) ppm, ^{19}F NMR ($CDCl_3$), δ –78.21 (s, CF_3) ppm.

48.16 PREPARATION OF PYRAZOLES [9, 10]

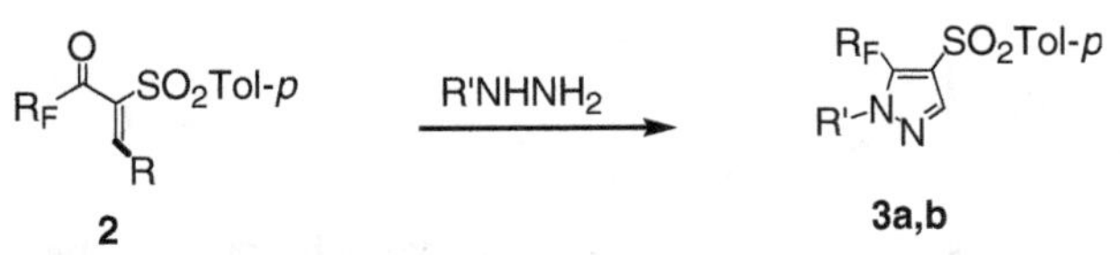

$R_F = H(CF_2)_3$, R' = Ph (**a**), $R_F = CF_3$, R' = H (**b**)
R = NMe_2, OMe

Apparatus One-hundred-milliliter flask, magnetic stirrer, condenser.

Chemicals Phenyl hydrazine, hydrazine sulfate, acetonitrile.

Attention! Safety glasses and protective gloves must be used at all times.

Experimental Procedure 3a Phenyl hydrazine (1 mmol) was added to a solution of **2a** (1 mmol) in acetonitrile (7 mL) and stirred for 2 h at 20 °C. The reaction mixture was filtered and the filtrate was evaporated at 50 °C (10–15 mm Hg). The residue was recrystallized from ethanol. Yield: 88%.

Characterization Data mp 144–146 °C (ethanol). ^{1}H NMR, δ 8.18 (1H, s, HC=N), 7.33 and 7.86 (4H, dd, $^3J_{HH}$ = 8.0 Hz, C_6H_4), 7.36–7.56 (5H, m, C_6H_5), 6.24 (1H, tt, $^2J_{HF}$ = 52.0 Hz, $^3J_{HF}$ = 6.0 Hz, HCF_2), 2.46 (3H, s, CH_3) ppm. ^{19}F NMR, δ −103.71 (2F, m, CF_2), −129.61 (2F, m, CF_2), −138.74 (2F, dm, J_{FH} = 52.0 Hz, HCF_2) ppm. ^{13}C NMR, δ 144.89 [s, C(Py)SO_2], 142.15 (s, CH=N), 138.70 [s, C(Ar)CH_3], 138.47 [C(Ar)SO_2], 130.59 [t, J_{CF} = 32.5 Hz, C(Py)CF_2], 130.48 [s, C(Ph)H], 129.85 [s, C(Ar)H], 128.67 [s, C(Ar)H], 127.91 [s, C(Ph)H], 127.23 [s, C(Ph)H], 112.52 (tm, J_{CF} = 253.0 Hz, $HCF_2CF_2CF_2$), 109.60 (tt, J_{CF} = 258.5 Hz, $^2J_{CF}$ = 32.0 Hz, $HCF_2CF_2CF_2$), 107.67 (tt, J_{CF} = 253.0 Hz, $^2J_{CF}$ = 29.5 Hz, $HCF_2CF_2CF_2$), 21.60 [s, CH_3(Ar)] ppm.

Mass spectrum, *m/z*: 449.2 $[M]^+$. Found, %: C, 50.88; H, 3.11; N, 6.32; S, 7.22. $C_{19}H_{14}F_6N_2O_2S$. Calcd. %: C, 50.89; H, 3.15; N, 6.25; S, 7.15.

Experimental Procedure 3b Hydrazine sulfate (1 mmol) and K_2CO_3 (2 mmol) were added to a solution of enaminone **2b** (1 mmol) in acetonitrile (7 mL). The mixture was heated under reflux with stirring for 8 h, followed by filtration and evaporation to dryness. The residue was purified by chromatography using ethyl acetate as the eluent to give pyrazole **3b** in 55% yield as yellow oil, R_f 0.9 (Silufol UV-254 plate, ethyl acetate as eluent, developed with iodine vapor).

Characterization Data ^{1}H NMR, δ 11.53 (1H, br. s, NH), 8.30 (1H, s, CH=N), 7.31 and 7.80 (4H, dd, $^3J_{HH} = 8.0$ Hz, C_6H_4), 6.29 (1H, tt, $^2J_{HF} = 52.5$ Hz, $^3J_{HF} = 6.0$ Hz, HCF_2); 2.46 (3H, s, CH_3) ppm. ^{19}F NMR, δ −107.32 (2F, m, CF_2), −131.94 (2F, m, CF_2), −138.14 (2F, dm, $^2J_{FH} = 52.5$ Hz, HCF_2) ppm.

48.17 PREPARATION OF PYRIMIDINES [9]

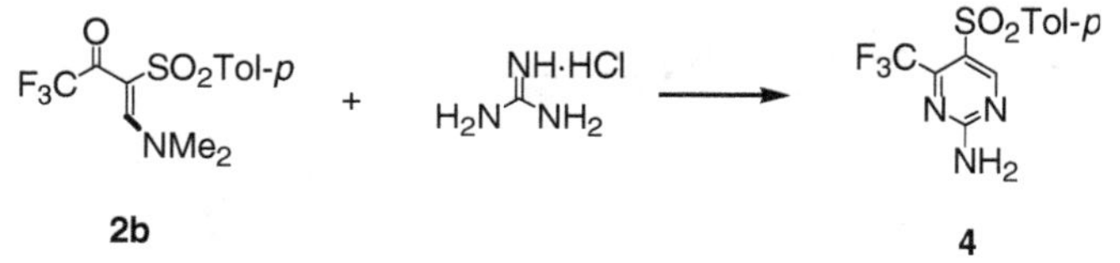

Experimental Procedure for 4 Guanidine hydrochloride (1 mmol) and K_2CO_3 (2 mmol) were added to a solution of enaminone **2b** in acetonitrile (7 mL). The reaction mixture was heated at reflux with stirring for 2 h, cooled, and the precipitate of **4** formed was filtered off, washed on the filter plate with water, and dried. Yield: 84%.

Characterization Data mp 260–262°C (acetonitrile). ^{1}H NMR, δ 9.09 (1H, s, CH=N), 8.43 (2H, br. s, NH_2), 7.41 and 7.78 (4H, dd, $^3J_{HH} = 8.0$ Hz, C_6H_4), 2.38 (3H, s, CH_3) ppm. ^{19}F NMR spectrum, δ −64.36 (s, CF_3) ppm.

48.18 PREPARATION OF PYRIMIDINE-2,4-ONES [11]

Apparatus Fifty-milliliter flask, magnetic stirrer, condenser.

Chemicals Sodium cyanate, triethylamine, acetonitrile.

Attention! Safety glasses and protective gloves must be used at all times.

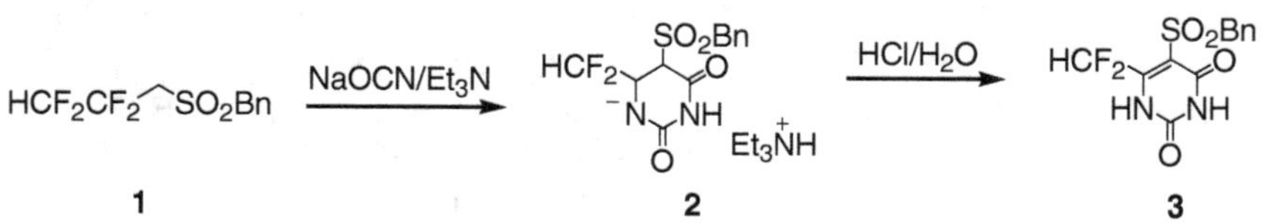

Experimental Procedure Et_3N (1.1 mL, 7.8 mmol) was added to a solution of sulfone **1** (7.4 mmol) in 15 mL of acetonitrile, the mixture was stirred for 5 min, and powdered NaOCN (1.0 g, 15.4 mmol) was added. The suspension was stirred at 70 °C for 4–6 h (monitoring by ^{19}F NMR), and the hot mixture was filtered through a layer of Celite®. After cooling, a quantity of triethylammonium salt **2** precipitated, and was taken for analysis. The warm solution of salt **2** in acetonitrile was acidified with conc. HCl (5 mL) and, after this, stored for 12 h. During this time crystals of **3** precipitated which were filtered off, washed with water, dried, and recrystallized from ethanol. Yield: 81%.

Characterization Data for 2 mp 230–232 °C (decomp.). ^1H NMR ([D$_6$]DMSO): δ 10.43 (broad s, 1 H, NH), 7.30 (m, 3 H, ArH), 7.20 (m, 2 H, ArH), 6.70 (t, $^2J_{H,F}$ = 55.2 Hz, 1 H, CHF_2), 4.56 (s, 2 H, CH_2SO_2), 3.39 (broad, NH–H_2O), 3.07 (q, 6 H, NCH_2), 1.16 (t, 9 H, CH_3) ppm. ^{19}F NMR (CH_3CN): δ −120.10 (d, $^2J_{H,F}$ = 55.2 Hz, 2 F, CF_2H). ^{13}C NMR ([D$_6$]DMSO): δ 163.19 (s, C4), 161.29 (t, $^2J_{C,F}$ = 21.1 Hz, C6), 158.77 (s, C2), 130.50 (s, ArC-*ortho*), 129.97 (s, ArC*para*), 128.07 (s, ArC-*meta*), 127.97 (s, ArC-*ipso*), 108.31 (t, $J_{C,F}$ = 240.0 Hz, CF_2H), 103.07 (s, C5), 59.55 (s, CH_2SO_2), 45.75 (s, CH_2N), 8.55 (s, CH_3) ppm. $C_{18}H_{25}F_2N_3O_4S$ (417.464): Calcd.: C, 51.79; H, 6.04; N, 10.07; S, 7.68; found: C, 51.62; H, 6.00; N, 10.14; S, 7.34.

Characterization Data for 3 mp 257–260 °C. ^1H NMR ([D$_6$]DMSO): δ 12.15 (s, 1 H, NH), 12.4 (s, 1 H, NH), 7.37 (m, 3 H, ArH), 7.30 (m, 2 H, ArH), 7.17 (t, $^2J_{H,F}$ = 52.5 Hz, 1 H, CHF_2), 4.76 (s, 2 H, CH_2), 3.35 (broad, NH−H_2O) ppm. ^{19}F NMR (CH_3CN): δ −120.09 (d, $^2J_{H,F}$ = 52.5 Hz, 2 F, CF_2H) ppm. ^{13}C NMR ([D$_6$]DMSO): δ 160.15 (s, C4), 150.62 (t, $^2J_{C,F}$ = 24.5 Hz, C6), 149.52 (s, C2), 131.04 (s, ArC-*ortho*), 128.83 (s, ArC-*para*), 128.65 (s, ArC-*meta*), 127.92 (s, ArC-*ipso*), 110.76 (t, $^3J_{C,F}$ = 4.1 Hz, C5), 106.65 (t, $J_{C,F}$ = 245.3 Hz, CF_2), 60.20 (s, CH_2). $C_{12}H_{10}F_2N_2O_4S$ (316.284): calcd.: C, 45.57; H, 3.19; N, 8.86; S, 10.14; found: C, 45.23; H, 3.30; N, 9.06; S, 10.13.

48.19 PREPARATION OF TRIAZOLES [12]

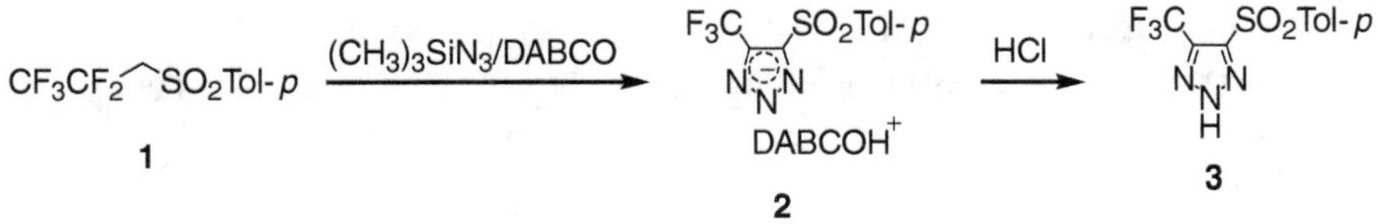

Apparatus Fifty-milliliter flask, magnetic stirrer, condenser.

Chemicals Trimethylsilyl azide, DABCO, benzene.

Attention! Safety glasses and protective gloves must be used at all times.

Experimental Procedure Solution of trimethylsilyl azide (0.28 mL, 2.1 mmol) was added to solution of sulfone **1** (0.58 g, 2 mmol) and 1,4-diazabicyclo-[2.2.2]octane (0.45 g, 4 mmol) in benzene (10 mL) which was heated to 80 °C. After cooling to 20 °C the precipitate of salt **2** was filtered off.

Yield: 97%. The salt **2** was dissolved in 5 mL of water and the solution was acidified with conc. HCl. After solidification of the separated oil the triazole **3** was filtered off and crystallized from water. Yield: 82%.

Characterization Data for 3 mp 150–151 °C. ^1H NMR ($CDCl_3$): δ 7.37, 7.93 (4H, dd, C_6H_4), 2.45 (3H, s, CH_3) ppm. ^{19}F NMR ($CDCl_3$), δ –60.87 (3F, c, CF_3) ppm.

48.20 PREPARATION OF TRIAZOLES

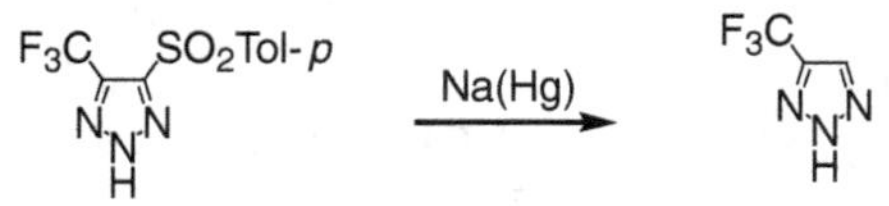

3 4

Apparatus Five-hundred-milliliter flask, magnetic stirrer, argon, condenser.

Chemicals Sodium amalgam (3%), methanol.

Attention! Safety glasses and protective gloves must be used at all times.

Caution! Because of its toxicity and volatility, care should be taken to avoid inhalation of Hg vapor. All reactions should be carried out in a well-ventilated hood.

Experimental Procedure To a mixture of triazole (0.3 mol) and Na_2HPO_4 (17 g, 0.12 mol) in absolute CH_3OH (150 mL) under argon and stirring, sodium amalgam (3%, 125 g) was added and the mixture was stirred at 30 °C for 72 h. Methanol solution was decanted and Hg was washed with CH_3OH (2 × 70 mL). Combined methanolic solutions were evaporated in vacuum (20 mm Hg) and the residue was dissolved in 200 mL of H_2O. Water solution was acidified with conc. HCl to pH 2–2.5, extracted with ether (4 × 100 mL), dried over $MgSO_4$, evaporated, and the residue was crystallized from $CHCl_3$. Yield: 60%.

Characterization Data mp 78–79 °C. ^1H NMR ($CDCl_3$), δ 14.80 (br, 1H, NH), 8.45 (s, 1H, CH) ppm. ^{19}F NMR (MeOH), δ –57.88 (s, CF_3) ppm.

Waste Disposal The residues of Hg have to be collected in a labeled container for toxic metal waste which has to be properly deposited.

48.21 PREPARATION OF TRIFLUOROTHIOACETAMIDE [13]

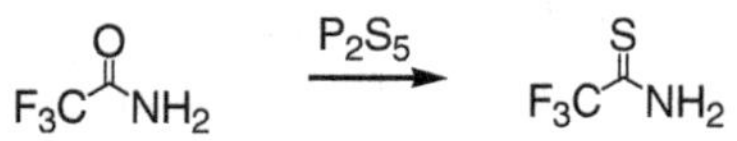

Apparatus Five-hundred-milliliter flask, magnetic stirrer, argon, condenser.

Chemicals Trifluoroacetamide, P_2S_5, toluene, hexamethyldisiloxane.

Attention! Safety glasses and protective gloves must be used at all times.

Caution! Because of its toxicity and smelling, reactions with P_2S_5 should be carried out in a well-ventilated hood.

Experimental Procedure Phosphorus pentasulfide (110.7 g, 0.249 mol) was added to a suspension of trifluoroacetamide (47 g, 0.416 mol) in a mixture of hexamethyldisiloxane (134.8 g, 0.830 mol) and 200 mL of toluene. The reaction mixture was stirred at 45 °C for 15 h. Solvents were removed in vacuum (20 mm Hg) up to one-half of its volume and the residue was diluted with 50 mL of diethyl ether. Organic solution was consecutively washed with $NaHCO_3$ solution (4 × 50 mL), saturated aqueous NaCl solution (2 × 100 mL), and water (2 × 100 mL). The water phase was additionally washed with diethyl ether (2 × 100 mL). The combined ethereal solutions were dried over Na_2SO_4 and the solvents were removed in vacuum. The residue was crystallized from hexane. Yield: 23.64 g (44%).

Characterization Data mp 42–44 °C. ^{1}H NMR ($CDCl_3$): δ 7.40 (1H, broad, NH), 7.63 (1H, broad, NH) ppm. ^{19}F NMR ($CDCl_3$): δ −71.18 (s, CF_3) ppm.

48.22 PREPARATION OF THIAZOLIDINONE [13]

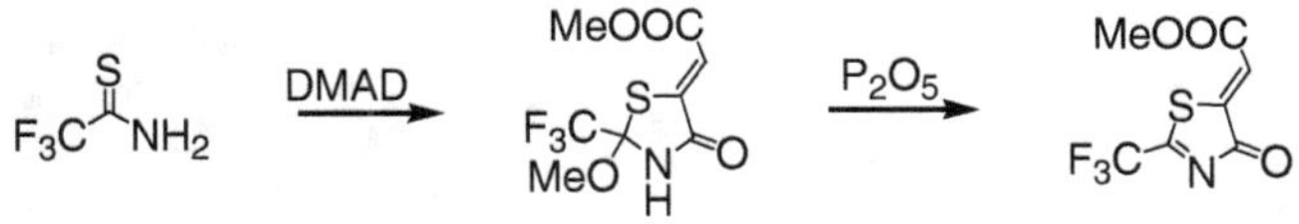

1 2 3

Apparatus Flask, magnetic stirrer.

Chemicals Trifluorothioacetamide, dimethyl acetylenedicarboxylate.

Attention! Safety glasses and protective gloves must be used at all times.

Experimental Procedure The equimolar mixture of trifluorothioacetamide and DMAD was kept at room temperature until reaction was complete (72 h). The solidified mass was dissolved in a mixture of petroleum ether–diethyl ether (1 : 1). The solution was stirred at −20 °C for 40 h to obtain the methanol adduct **2**. Yield: 95%.

Characterization Data mp 115–117 °C. ^{1}H NMR ($CDCl_3$): δ 3.47 (3H, s, CH_3), 3.85 (3H, s, CH_3), 6.92 (1H, s, CH=), 7.72 (1H, broad, NH) ppm. ^{19}F NMR ($CDCl_3$): δ −82.00 (s, CF_3) ppm. ^{13}C NMR with C–H coupling ($CDCl_3$): δ 51.3 (q, CH_3O, J_{CH} = 146.3 Hz), 52.7 (q, CH_3O, J_{CH} = 148.9 Hz), 96.85 (qm, C-2, $^2J_{CF}$ = 35.0 Hz), 115.6 (d, C-6, J_{CH} = 173.0 Hz), 121.7 (q, CF_3, J_{CF} = 283.5 Hz), 143.7 (s, C-5), 166.0 (d, C-4, $^3J_{C^4H^6}$ = 5.4 Hz), 166.8 (q, C-7, $^2J_{C^7H^6}$ = 5.0 Hz) ppm. Anal. Calcd. for $C_8H_8F_3NO_4S$: S, 11.82; N, 5.16. Found: S, 12.25; N, 4.68.

48.23 PREPARATION OF 2-TRIFLUOROMETHYLIMIDAZOLINE [14]

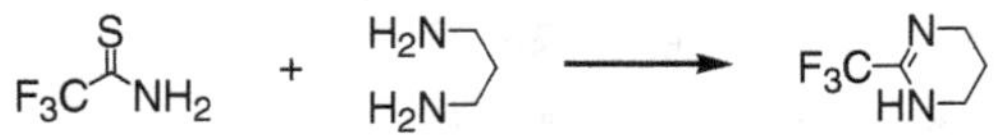

Apparatus Flask, magnetic stirrer.

Chemicals Trifluorothioacetamide, propane 1,3-diamine, diethyl ether.

Attention! Safety glasses and protective gloves must be used at all times.

Experimental Procedure To a stirred solution of thioamide (5 mmol, 1 equiv) in Et_2O (20 mL), propane 1,3-diamine (10 mmol, 2 equiv) was added and the mixture was stirred at 15–20 °C for 22 h. The reaction mixture was diluted with 20 mL of diethyl ether and washed with water (2 × 20 mL). The water layer was extracted with ether (2 × 20 mL). Combined ethereal solution was dried over Na_2SO_4 overnight and ether was removed in vacuum. The remaining crude product was purified by crystallization from hexane–$CHCl_3$ (4 : 1). Yield: 78%.

Characterization Data mp 110–111 °C. ^{1}H NMR ($CDCl_3$), δ: 1.82 (m, $^2J_{HH}$ = 11.4 Hz, $^3J_{HH}$ = 5.4 Hz, 2 H, C^5H_2), 3.44 (t, $^3J_{HH}$ = 5.4 Hz, 4 H, C^4H_2, and C^6H_2), 4.95 (s, 1 H, NH) ppm. ^{19}F NMR ($CDCl_3$), δ –74.76 (s, 3 F, CF_3) ppm. MS (*m/z*): 153 $[M + H]^+$ ppm.

REFERENCES

1. Timoshenko VM, Listvan VV, Rusanov EB, Shermolovich YuG, Markovskii LN. 2,2,3,3,4,4,5,5-Octafluoropentyl benzyl sulfide, sulfoxide, and sulfone and their dehydrofluorination under the action of bases. *Russian J Org Chem* 1997;33:60–65; *Zh Organ Khim* 1997;33:70–76.
2. Portella C, Shermolovich YuG., Tschenn O. Synthesis of alkyl perfluoroalkanedithiocarboxylates and some aspects of their reactivity in cycloaddition reactions. *Bull Soc Chim Fr* 1997;134:697–702.
3. Timoshenko VM, Siry SA, Rozhenko AB, Shermolovich YuG. Asymmetric induction in thia-Diels-Alder reactions of chiral polyfluoroalkylthionocarboxylates. *J Fluorine Chem* 2010;131:172–183.
4. Sizov AY, Kovregin AN, Serdyuk RN, Vorob'ev MV, Porosyatnikov VA, Tsvetkov AA, Korneev DO, Ermolov AF. Synthesis and reactivity of fluorine-containing thiols and thioacyl halides. *Russ Chem Bull* 2006;55:1200–1208.
5. Badadzhanova LA, Kirij NV, Yagupolskii YL. The synthesis of salts and esters of perfluorinated dithio- and thiocarboxylic acids from carbon disulfide and carbonyl sulfide reacting with R_FSiMe_3/F^-. *J Fluorine Chem* 2004;125:1095–1098.
6. Takahashi M, Kotajima H. 1,1,1-Trifluoro-3-phenylsulfonyl-2,2-propanediol: a building block for trifluoromethylated and sulfonylated heterocycles. Synthesis of 1-alkyl-3-trifluoromethyl-4-phenylsulfonyl-pyrazoles. *Synlett* 1990;353–354.

7. Shermolovich YuG, Timoshenko VM, Listvan VV, Markovskii LN. Synthesis of fluorocontaining ketones from 1,1-dihydropolyfluoroalkyl sulfides. *Russian J Org Chem* 1998;34:1112–1116; *Zh Organ Khim* 1998;34:1167–1171.
8. Siry SA, Timoshenko VM. 2-Polyfluoroalkyl thiopyrylium salts: synthesis and reactions with nucleophiles. *Tetrahedron Lett* 2010;51:6406–6409.
9. Kanischev OS, Bandera YuP, Timoshenko VM, Rusanov EB, But SA, Shermolovich YuG. Synthesis of 5-polyfluoroalkyl-4-(p-tolyl-sulfonyl)pyrazoles and 4-polyfluoroalkyl-5-(p-tolylsulfonyl)pyrimidines from 1-dimethylamino-2-(p-tolylsulfonyl)-polyfluoro-1-alken-3-ones. *Chem Heterocycl Compd* 2007;43:887–892; *Khim Geterotsikl Soedin (Russ)* 2007;7:1052–1058.
10. Markitanov YuM, Timoshenko VM, Shermolovich YuG. β-Polyfluoroalkyl-β-oxosulfones and –sulfamides as methylene components in reactions with orthoesters. *Zh Org Pharm Khim (Ukr)* 2010;8:68–78.
11. Timoshenko VM, Nikolin YV, Chernega AN, Shermolovich YuG. Treatment of polyfluoro-1,1-dihydroalkyl sulfones with sodium cyanate and triethylamine: a new method for the synthesis of 6-(polyfluoroalkyl)uracils. *Eur J Org Chem* 2002;10:1619–1627.
12. Timoshenko VM, Nikolin YV, Rusanov EB, Chernega AN, Shermolovich YuG. Synthesis of 4-benzylsulfonyl-5-polyfluoroalkyl-ν-triazoles from 1,1-dihydropolyfluoroalkyl benzyl sulfones. *Chem Heterocycl Compd* 2001;37: 470–476; *Khim Geterotsikl Soedin (Russ)* 2001;4:518–524.
13. Rudnichenko AV, Timoshenko VM, Shermolovich YuG. Cycloaddition reactions of polyfluoroalkylthioncarboxylic acid derivatives with dimethyl acetylenedicarboxylate (DMAD). *J Fluorine Chem* 2004;125:439–444.
14. Rudnichenko AV, Shermolovich YuG. New synthesis of 2-polyfluoroalkyl substituted 1,4,5,6-tetrahydro-pyrimidines and 4,5,6,7-tetrahydro-1H-1,3-diazepines. *Syn Commun* 2007;37:459–465.

CHAPTER 49

Synthesis of Octafluorocyclooctatetraene

JOHN M. BUZBY, ALDOS C. BAREFOOT, III, MICHAEL W. GRAYSTON, EVAN D. LAGANIS, ROY F. WALDRON, AND DAVID M. LEMAL

Octafluorocyclooctatetraene (**1**) was first prepared in 1975 by vapor-phase photolysis of tetrafluorocyclobutenedicarboxylic anhydride (**2**) [1]. Its formation apparently proceeded via the cycloadduct (**3**) of two highly reactive intermediates, tetrafluorocyclobutadiene and tetrafluorocyclopentadienone [2].

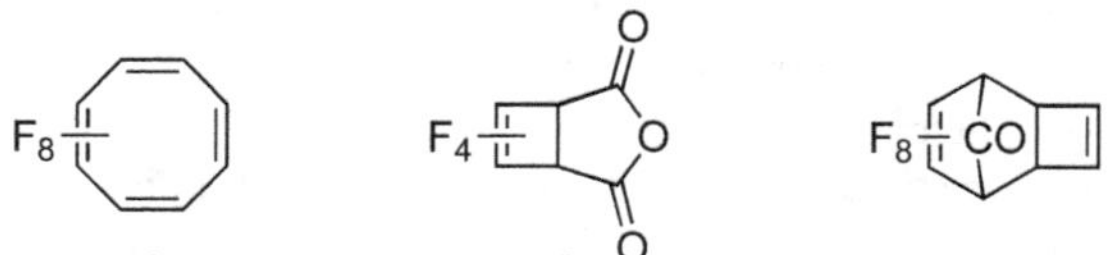

1 **2** **3**

Later a more practical synthesis was developed, starting with photocycloaddition of hexafluorobenzene to 1,2-dichlorodifluoroethylene [3]. The initially formed bicyclic adducts (**4**) photocyclized to tricyclooctenes (**5**) under the reaction conditions, but heating returned the bicyclooctadienes (**4**). Reductive dechlorination with zinc in acetic acid gave cyclooctatetraene (**1**) via its labile bicyclooctatriene valence isomer (**6**) [4]. It was subsequently found that zinc in dimethyl sulfoxide assisted with ultrasound smoothly dechlorinated the mixture of tricyclic adducts to yield *anti*-tricyclooctadiene (**7**) which ring opened quantitatively to tetraene **1** at 150 °C. This variation on the synthesis, gives cleaner tetraene **1** in higher yield and is the method of choice. Tetraene **1** has also been prepared by three other routes: photolysis of tetrafluorocyclopentadienone [5], photorearrangement of perfluorobarrelene [6], and pyrolysis of 3,4-diiodocyclobutene [7].

Efficient Preparations of Fluorine Compounds, First Edition. Edited by Herbert W. Roesky.

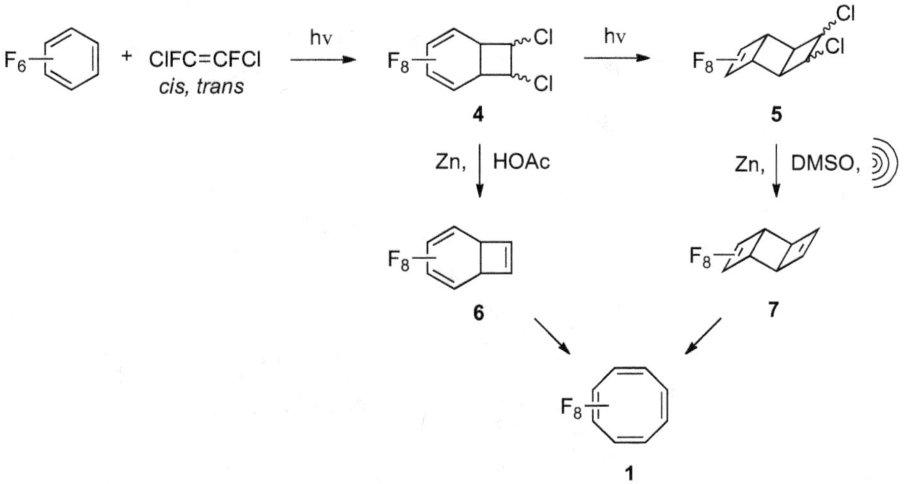

The tetraene exists in equilibrium with a tiny amount of its bicyclic isomer **6** (K_{eq} = 0.003 at 20 °C) [8], and like the parent cyclooctatetraene, it undergoes Diels–Alder reactions via the bicyclic form. Treatment of the compound with boron trifluoride at 250 °C results in rearrangement to octafluorobenzocyclobutene, again via isomer **6**. Photolysis of tetraene **1** results in cyclization to *anti*-tricyclooctadiene (**7**) accompanied by a small amount (20:1) of its *syn* isomer [9]. Multiphoton ionization spectroscopy [10] and multiphoton photochemistry [11] of tetraene **1** have been analyzed. Explored extensively by the Hughes group, the rich organometallic chemistry of octafluorocycloooctatetraene features novel structures and transformations [12–21].

49.1 PREPARATION

Apparatus Photolysis apparatus (see description later), three-necked flasks (100 and 500 mL) with gas adapters, dry ice condenser, mechanical stirrer, small pressure-equalizing dropping funnel, vacuum distillation assembly with short Vigreux column, ultrasonic bath (e.g., 40 KHz), U-trap with Dewar flask, manometer, drying tube, small steel or Monel bomb, bomb heater.

Chemicals Hexafluorobenzene, 1,2-dichlorodifluoroethylene, zinc dust, dimethyl sulfoxide, isopentane, isopropanol, dry ice.

Attention! Safety glasses and protective gloves should been worn at all times, and all operations should be conducted in an efficient fume hood.

Experimental Procedure *7,8-Dichlorooctafluoro-*anti*-tricyclo[4.2.0.0.^{2,5}]oct-3-enes* [3]. Photolysis apparatus was assembled as depicted in Figure 49.1. A 450-W medium pressure mercury lamp surrounded by a Vycor filter sleeve was contained in a water-cooled quartz immersion well with a 60/40 male joint. The immersion well was mounted in a borosilicate vessel that has two

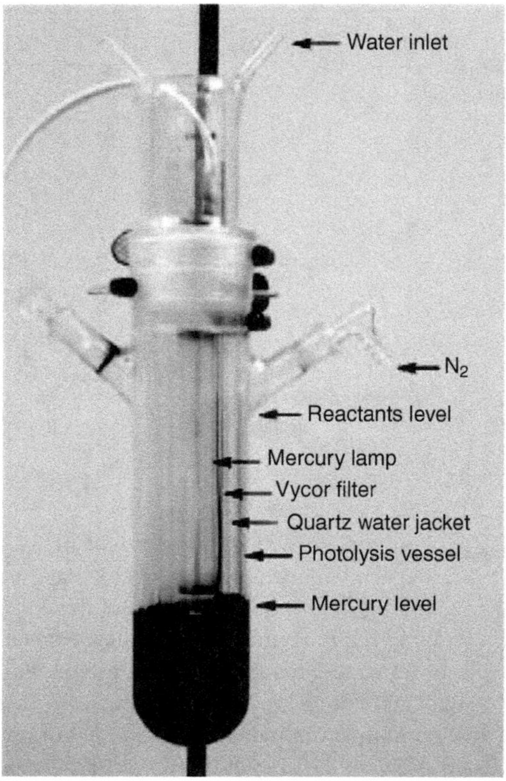

FIGURE 49.1 Apparatus for the photocycloaddition. (For a color version of the figure, please see color plates.)

side arms with 14/20 male joints and an I.D. only ~2 mm larger than the O.D. of the well. Enough clean mercury was placed in the vessel to reach almost to the bottom of the arc when the well was in place. This way all of the mixture to be photolyzed, contained in the narrow annulus between well and vessel, was efficiently irradiated. To prevent freezing of the large immersion well joint by photolysis of its grease, the joint was wrapped with a layer of Teflon tape before silicone grease was applied. One of the side arms was connected to a nitrogen source and the other was sealed after the reactants were introduced.

A 100-mL, three-necked flask fitted with a dry ice condenser and stopper was flushed with nitrogen through the condenser, and hexafluorobenzene (25 mL, 40 g, 220 mmol) was poured in. The open neck was fitted with a gas inlet connected by flexible tubing to a preweighed canister containing (*E,Z*)-1,2-dichlorodifluoroethylene (bp 21–22 °C) and mounted on a balance. With dry ice/isopropanol in the condenser and the flask cooled in an ice bath, the alkene (25.55 g, 192 mmol) was introduced, most by pouring. After the flask was swirled to complete mixing, the transfer tubing end was clamped shut and the condenser was replaced with a stopper. When the photolysis apparatus had been flushed with nitrogen, the cooling water started, and the whole assembly placed in a bath of ice–water, the mixture of reactants was poured in through the open side arm via the transfer tubing. The side arm was stoppered and the apparatus was maintained under nitrogen. Reactants filled the annular space up to the bottom of the side arms. Irradiation was begun, and external cooling by a bath of flowing water was continued throughout the photolysis. After 70 h, ^{19}F NMR showed that no (*Z*)-alkene remained, but there was still some (*E*)-alkene; so irradiation was continued for another ~30 h. The golden brown product was distilled through a short Vigreux column under reduced pressure to preclude ring opening of the tricyclic adducts. Unreacted hexafluorobenzene (14.73 g) was collected up to 44 °C at 115 Torr, then the pressure was lowered and tricyclooctenes distilled up to 50 °C at 2 Torr (26.32 g, 82.5 mmol, 43% yield) [22].

Characterization Data ^{19}F NMR ($CDCl_3$, isomers in decreasing order of abundance): *trans*, δ −113.5, −113.6 (F_7 *endo*, F_8 *exo*); −118−119 (F_3, F_4); −166.1 (F_6); −178.3 (F_1), −185.7 (F_5); −188.7 (F_2); *endo, cis*, δ −118−119 (F_3, F_4); −127.0 (F_7, F_8); −176.2 (F_1, F_6); −188.5 (F_2, F_5); *exo, cis*, δ~ −119.4 (F_3, F_4, F_7, F_8); −167.2 (F_1, F_6); −185.7 (F_2, F_5) ppm.

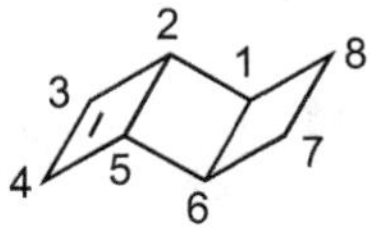

Octafluoro-anti-*tricyclo[4.2.0.0.^{2,5}]octa-3,7-diene* [23]. A 500-mL, three-necked flask was fitted with a mechanical stirrer, small pressure-equalizing dropping funnel, and a gas takeoff adapter. A length of polyethylene tubing connected to the tip of the dropping funnel extended close to the bottom of the flask. The assembly was mounted in an ultrasonic bath that was maintained at ~22 °C. The adapter was connected to a U-trap contained in a Dewar flask, and the trap was connected in turn to an aspirator via a drying tube, with a manometer in the line. Tricyclooctenes from the above procedure (10.08 g, 31.6 mmol) were introduced into the dropping funnel, and dimethyl sulfoxide (150 mL) and activated zinc dust (7.90 g, 120 mmol) were placed in the flask. The system was evacuated to 17 Torr, the trap was cooled with dry ice/isopropanol, and then the tricyclooctenes were added during 1 h with stirring and sonication. The mixture darkened in 15–20 min and a white solid accumulated in the trap. Reaction was allowed to proceed for 45 min after addition was complete. Recrystallization of the wet crystals from isopentane at −28 °C gave 4.58 g (58% yield) of dry, colorless crystals of the diene, mp 41–42 °C. On a half-gram scale, diene with mp 40–42 °C was obtained in 64% yield without recrystallization.

Characterization Data ^{19}F NMR ($CDCl_3$): δ −119.7 (s, 4 vinyl F), −183.8 (s, 4 bridgehead F) ppm. ^{13}C NMR ($CDCl_3$): δ 131.9 (d of m, $^1J_{CF}$ ~343 Hz), 99.0 (d of m, $^1J_{CF}$ ~283 Hz) ppm. IR (vapor): 1744, 1369, 1214, 1101, 891, 814 cm^{-1}. MS (*m/z*): 248 (M^+), 179 (base, M^+ −CF_3) [9].

Octafluorocyclooctatetraene. A small sample of the diene was sealed in a heavy-walled glass ampoule and heated in an oil bath at 150 °C. The half-life for quantitative ring opening to the tetraene was $\sim$ 0.5 h [9]. On a large scale, the use of a steel or Monel bomb for the isomerization is advisable.

Characterization Data Mp 40–41.5 °C. ^{19}F NMR ($CDCl_3$): δ −123.1 (narrow s, 8F) ppm. ^{13}C NMR ($CDCl_3$): δ 138.2 (d of m, $^1J_{CF} \sim 272$ Hz) ppm. IR (CH_2Cl_2): 1724, 1692, 1333 cm^{-1}. MS (*m/z*): 248 (M^+) [1]. X-ray crystal structure [24].

ACKNOWLEDGMENT

This material is based on work supported by the National Science Foundation.

REFERENCES

1. Gerace MJ, Lemal DM, Ertl H. Tetrafluorocyclobutadiene. *J Am Chem Soc* 1975;97:5584–5586.
2. The initial surmise that **1** was formed via dimerization of tetrafluorocyclobutadiene was proven incorrect (Barefoot AC III. On the synthesis and chemistry of octafluorocyclooctatetraene. Masters thesis. Dartmouth College; Hanover, NH 03755, USA, 1978). Decarbonylation of **3**, then ring opening yields **1**.
3. Lemal DM, Buzby JM, Barefoot AC III, Grayston MW, Laganis ED. A new synthesis of perfluorocyclooctatetraene. *J Org Chem* 1980;45:3118–3120.
4. Waldron RF, Barefoot AC III, Lemal DM. Perfluorobicyclo[4.2.0]octa-2,4,7-triene, a Labile (CF)8 valence isomer. *J Am Chem Soc* 1984;106:8301–8302.
5. Grayston MW, Saunders WD, Lemal DM. Perfluorocyclopentadienone. *J Am Chem Soc* 1980;102:413–414.
6. Ralli P, Zhang Y, Lemal DM. Perfluorobarrelene. *Tetrahedron Lett* 2008;49:7349–7351.
7. Barlow MG, Crawley MW, Haszeldine RN. Fluoro olefin chemistry. Part 12. Some reactions of 3,4-dihalotetrafluorocyclobutenes: a synthesis of perfluorocyclooctatetraene. *J Chem Soc, Perkin Trans I* 1980;1972–1999.
8. Rahman MM, Secor BA, Morgan KM, Shafer PR, Lemal DM. Valence isomerization equilibria of highly fluorinated cycloocta-1,3,5-trienes. *J Am Chem Soc* 1990;112:5986–5990.
9. Barefoot AC III, Saunders WD, Buzby JM, Grayston MW, Lemal DM. Configurations and chemistry of the perfluorotricyclo[4.2.0.0.^{2,5}]octa-3,7-dienes. *J Org Chem* 1980;45:4292–4295.
10. BelBruno JJ, Raiche GA. Laser ionization spectroscopy of octafluorocyclooctatetraene. *Spectrosc Lett* 1989;22:747–761.
11. BelBruno JJ, Greenfield SR, Carl RT, Hughes RP. Competition among collisional deactivation, ionization, and dissociation in the multiphoton excitation of octafluorocyclooctatetraene. *J Phys Chem* 1988;92:2480–2484.

12. Hughes RP, Carl RT, Doig SJ, Hemond RC, Samkoff DE, Smith WL, Stewart LC, Davis RE, Holland KD. Effect of polyfluorination on ring inversion barriers for cyclooctatetraenes. Transition-metal compounds of unsaturated, polyfluorinated cycloaliphatics. Crystal and molecular structures of [Fe(η-C_5R_5)(η^1-heptafluorocycloocta-1,3,5,7-tetraenyl)$(CO)_2$] (R = H, Me), {[Fe(η-C_5H_5)$(CO)_2$]$_2$(μ^2-1η,5η)-hexafluorocycloocta-1,3,5,7-tetraenediyl)}, and [Mn ((3η)-heptafluorotricyclo[4.2.0.0^{2,5}]octa-3,7-dienyl)$(CO)_5$]. *Organometallics* 1990;9:2732–2745.

13. Hughes RP, Doig SJ, Hemond RC, Smith WL, Davis RE, Gadol SM, Holland KD. Synthesis, structures, and conformational dynamics of dicobalt complexes containing the hexafluorodidehydrocyclooctatetraene (hexafluorocycloocta-3,5,7-trien-1-yne) ligand. Crystal and molecular structures of [(Co(L)$(CO)_2$)(μ^2-(1η,2η)-C_8F_6] (L = CO, PPh_3, $PPhMe_2$, PMe_3). *Organometallics* 1990;9:2745–2753.

14. Carl RT, Corcoran EW Jr, Hughes RP, Samkoff DE. Nickel, palladium, and platinum complexes derived from octafluorocyclooctatetraene. Synthesis of 1-2:5-6-η-octafluorocyclooctatetraene complexes of nickel (0) and η^2-octafluorobicyclo[3.3.0]octa-2,7-diene-4,6-diyl complexes of nickel (II), palladium (II), and platinum (II). *Organometallics* 1990;9:838–844.

15. Hemond RC, Hughes RP, Rheingold AL. Synthesis of η^6-octafluorocyclooctatetraene and η^6-cyclooctatetraene complexes of manganese (I). Molecular structures of [Mn(η^5-C_5R_5)(η^6-C_8X_8)] (R = H, Me, X = F; R = Me, X = H). *Organometallics* 1989;8:1261–1269.

16. Carl RT, Hughes RP, Rheingold AL, Marder TB, Taylor NJ. Synthesis, structures, and solution dynamics of mononuclear and dinuclear (η^5-indenyl)rhodium complexes of octafluorocyclooctatetraene. Crystal and molecular structures of [Rh(η^5-C_9H_7)(1,2,5,6-η-C_8F_8)], {[Rh(η^5-C_9H_7)]$_2$[μ-(1,5,6-η:2-4-η-C_8F_8)](*Rh-Rh*)}, {[Rh(η^5-C_9H_7)]$_2$[μ-(1,5,6-η:2-4-η-C_8F_7H)](*Rh-Rh*)}, and {[Rh(η^5-C_9H_7)(CO)][Rh(η^5-(C_9H_7)][μ-(1,2,5,6-η:7,8-η-C_8F_8)]}. *Organometallics* 1988;7:1613–1624.

17. Carl RT, Doig SJ, Geiger WE, Hemond RC, Hughes RP, Kelly RS, Samkoff DE. Pentamethylcyclopentadienyl cobalt and rhodium complexes of octafluorocyclooctatetraene. Photochemical and thermal interconversion of 1,2,5,6-η- and 1,2,3,6-η-C_8F_8 isomers. Electrochemical and ESR characterization of the 19-electron radical anion [Co(η-C_5Me_5)(1,2,5,6-η-C_8F_8)]$^-$. *Organometallics* 1987;6:611–616.

18. Doherty NM, Ewels BE, Hughes RP, Samkoff DE, Saunders WD, Davis RE, Laird BB. Cyclopentadienyl cobalt and rhodium complexes containing monocyclic and bicyclic valence isomers of octafluorocyclooctatetraene: Crystal and molecular structures of the exo and endo isomers of (η-cyclopentadienyl)(2-5-η-octafluorobicyclo[4.2.0]octa-2,4,7-triene)rhodium (I). *Organometallics* 1985;4:1606–1611.

19. Doig SJ, Hughes RP, Patt SL, Samkoff DE, Smith WL. Octafluorocyclooctatetraene transition metal compounds: displacement of fluoride by metal carbonyl anions. *J Organomet Chem* 1983;250:C1–C4.

20. Hughes RP, Samkoff DE, Davis RE, Laird BB. Octafluorocyclooctatetraene transition metal compounds. Novel transannular ring closures and a formal intramolecular redox equilibrium between 1,2,5,6-η and 1,2,3,6-η ligands. *Organometallics* 1983;2:195–197.

21. Barefoot AC III, Corcoran EW Jr., Hughes RP, Lemal DM, Saunders WD, Laird BB, Davis RE. Octafluorocyclooctatetraene transition metal chemistry: 1,2-η and 1,2,3,6-η complexes of iron and platinum. *J Am Chem Soc* 1981;103:970–972.

22. Yield here is based on the alkene as the limiting reagent; it is currently much more expensive than hexafluorobenzene. The 60% yield reported in reference 3 was based on hexafluorobenzene as the limiting reagent, but that based on unrecovered alkene was only ∼34%.

23. Waldron RF. The valence isomerism of octafluorocyclooctatetraene. Senior Thesis. Dartmouth College; 1984.
24. Laird BB, Davis RE. Octafluorocyclooctatetraene at 238 K. *Acta Crystallogr B* 1982;B38:678–680.

CHAPTER 50

Preparation of Fluoroolefins

MORITZ F. KUEHNEL AND DIETER LENTZ

Fluoroolefins are important starting materials for the synthesis of a variety of polymers such as PTFE, FEP, PFA, and Nafion®, which have numerous applications ranging from nonadhesive coatings and chemically resistant lubricants to fuel-cell membranes. While many simple olefins such as tetrafluoroethene and trifluoropropene are commercially available, the access to more sophisticated fluoroalkenes is very limited. Apart from a few examples, olefin metathesis is not applicable to fluoroolefins.

On a lab scale, fluoroolefins are often synthesized by base-induced elimination reactions. Typically, alkyllithiums (e.g., *n*-butyllithium) are necessary for dehydrofluorination reactions, while alkali hydroxides (e.g., KOH) and nitrogen bases (e.g., DBU) are sufficient for dehydrobrominations. Precursors are conveniently accessible by palladium-catalyzed cross-coupling of vinyl synthons. The choice of a suitable precursor, solvent, and base is crucial to a successful synthesis, especially differences in volatility are of particular concern to allow for an efficient separation of the product from the reaction mixture. Fractional condensation and low-temperature distillation are the most commonly used purification techniques.

Fluoroolefins are volatile and subject to polymerization, nucleophilic addition, and substitution reactions. They should be stored under inert conditions at low temperatures.

50.1 PREPARATION OF CYCLOHEXYLLITHIUM [1–3]

Apparatus A 1000-mL, two-necked Schlenk flask, magnetic stirrer, reflux condenser, 100-mL dropping funnel, glass frit, PTFE cannula, argon-filled glove box, heat gun, safety glasses, laboratory coat, protective gloves.

Chemicals Cyclohexyl chloride, powdered lithium (0.6% Na, washed with dry pentane and dried in vacuo before use), dry toluene.

Efficient Preparations of Fluorine Compounds, First Edition. Edited by Herbert W. Roesky.

Attention! Safety glasses and protective gloves must be used at all times.

Caution! All manipulations have to be performed under an atmosphere of dry argon using standard Schlenk techniques. Powdered lithium reacts with water to release highly flammable hydrogen gas. Cyclohexyllithium is pyrophoric and reacts violently with air and water. All reactions should be carried out in a well-ventilated hood to avoid inhalation of toluene.

Experimental Procedure In a glove box, a 1000-mL, two-necked Schlenk flask equipped with a magnetic stirrer, a dropping funnel, and a reflux condenser is charged with powdered lithium (3.50 g, 504 mmol, 2.2 equiv) and removed from the glove box. After addition of toluene (500 mL), cyclohexyl chloride (27.21 g, 229 mmol) is added drop by drop over a period of 1 h while the reaction is initiated by heating with a heat gun until a temperature slightly above room temperature is maintained by itself. After 3 h, the purple reaction mixture is filtered over a glass frit and the yellow filtrate is cooled to −80 °C overnight to precipitate cyclohexyllithium as colorless crystals. The supernatant is removed with a cannula; the crystals are washed with cold toluene and dried in vacuo overnight. Yield: 12.47 g (60%).

Characterization Data IR (NaCl, cm^{-1}): $\bar{\upsilon}$ 2800, 2720, 1450, 1340, 1260, 1160, 1080, 1030, 1015, 980, 900, 850, 830, 790, 770, 675, 600, 570.

Waste Disposal Residues of cyclohexyllithium in the supernatant are most conveniently deactivated by drop-by-drop addition of 2-propanol under argon.

Application Cyclohexyllithium is a strong secondary alkyllithium base similar to *sec*-butyllithium. Its conjugate acid cyclohexane has a high boiling point of 81 °C; it is therefore highly suitable for the synthesis of low-boiling fluoroalkenes by dehydrohalogenation.

50.2 PREPARATION OF TETRAFLUOROALLENE [3,4]

Apparatus A 1000-mL, two-necked Schlenk flask, magnetic stirrer, two stoppers, septum, low-temperature thermometer, fractionation column (2 cm × 50 cm, packed with Raschig rings) equipped with a column head cooled to −80 °C using a suitable cryostat, collection flask cooled to −80 °C, two cold traps, Dewar flasks, vacuum line, argon-filled glove box, safety glasses, laboratory coat, protective gloves.

Chemicals 1,1,3,3,3-Pentafluoropropene, cyclohexyllithium, dry diethyl ether, dry ice, ethanol, liquid N_2.

Attention! Safety glasses and protective gloves must be used at all times.

Caution! All manipulations have to be performed under an atmosphere of dry argon using standard Schlenk techniques. Due to its high reactivity toward nucleophiles, tetrafluoroallene is potentially toxic. Care should be taken to avoid inhalation of tetrafluoroallene, pentafluoropropene, and diethyl ether; all reactions should be carried out in a well-ventilated hood. Cyclohexyllithium is pyrophoric and reacts violently with air and water. Liquefied gases may expand explosively when allowed to warm to room temperature without proper pressure release. Tetrafluoroallene slowly dimerizes at room temperature and should be stored at −80 °C or −196 °C. Liquid nitrogen and dry ice can cause skin burns.

Experimental Procedure In a glove box, a 1000-mL, two-necked Schlenk flask equipped with a magnetic stirrer and two stoppers is charged with cyclohexyllithium (4.52 g, 50.2 mmol, 1.1 equiv). After removal from the glove box, cold diethyl ether (500 mL) is added at −80 °C, the resulting suspension is cooled to −196 °C and degassed. 1,1,3,3,3-Pentafluoropropene (6.06 g, 45.9 mmol) is condensed into the flask. After allowing the reaction mixture to warm to approximately −100 °C, the flask is vented with argon, a stopper is replaced by a septum, and a low-temperature thermometer is inserted into the suspension. The mixture is stirred for 3 h at −75 °C and subsequently allowed to warm to −20 °C, resulting in a cloudy red solution. A stopper is replaced by a fractionation column with its head cooled to −80 °C and the reaction mixture is heated to reflux. Distillation (−39 °C to + 15 °C) gives a mixture of tetrafluoroallene and diethyl ether. Fractional condensation of the distillate via two subsequent traps kept at −125 °C and −196 °C, respectively, yields 3.49 g (69 %) tetrafluoroallene in the second trap.

On a smaller scale or if a cryostat is not available, the fractionation column can be replaced by a dry ice condenser during the reaction. Isolation of tetrafluoroallene is achieved by repeated fractional condensation of the reaction mixture via two subsequent traps kept at −125 °C and −196 °C.

Characterization Data Mp −165 °C, bp −39 °C. ^{19}F NMR (376 MHz, $CDCl_3$): δ −63.9 (s) ppm. ^{13}C NMR (100 MHz, $CDCl_3$): δ 117.9 (qi, J = 39 Hz), 140.1 (tdd, J = 276, 14, 5 Hz) ppm. IR (gas, cm^{-1}): $\bar{\upsilon}$ 2071 (s), 1250 (s), 1040 (s).

Application Tetrafluoroallene is highly reactive toward various nucleophilic and electrophilic reagents (Figure 50.1). It can be used as a versatile building block in [2 + 2] cycloadditions and 1,3-dipolar cycloadditions to yield fluorinated heterocycles. Nucleophilic addition of alcohols and amines leads to ethers and amides, respectively. Tetrafluoroallene forms η^2 complexes with electron-rich transition metals and can dimerize in the presence of metal carbonyls to give μ-η^3:η^3 complexes.

50.3 PREPARATION OF OCTAFLUORO-1,2-DIMETHYLENECYCLOBUTANE (TETRAFLUOROALLENE DIMER) [17–19]

Apparatus Glass ampoule 8 mm OD × 200 mm, three cold traps, oven (used for drying glassware), vacuum line.

Chemicals Tetrafluoroallene, α-terpinene, liquid N_2, acetone.

Attention! Safety glasses and protective gloves must be used at all times.

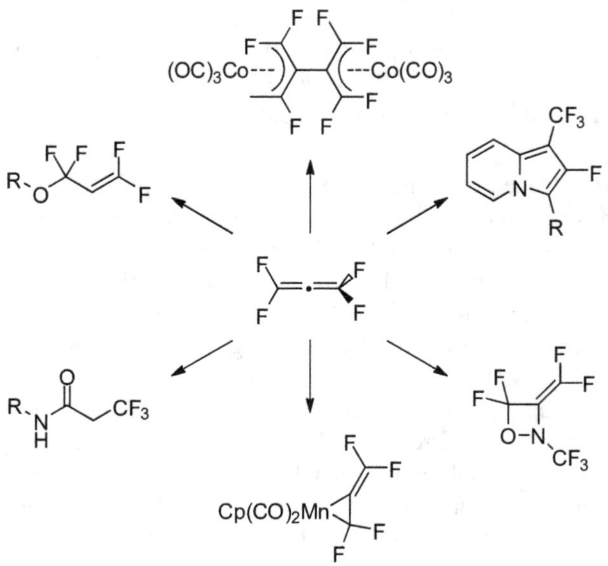

FIGURE 50.1 Synthetic potential of tetrafluoroallene [5–16].

Caution! All manipulations have to be performed under an atmosphere of dry argon using standard Schlenk techniques. Tetrafluoroallene and octafluoro-1,2-dimethylenecyclobutane are potentially toxic. Care should be taken to avoid inhalation; all reactions should be carried out in a well-ventilated hood. Pressurized glass ampoules can explode violently, causing serious injuries. Adequate precautions such as protective shields may be necessary.

Experimental Procedure A few drops of α-terpinene are added to a glass ampoule and tetrafluoroallene (2.48 g, 20.6 mmol) is added via vacuum transfer. The ampoule is flame-sealed under vacuum, placed in an oven and heated to 55 °C for 65 h. Subsequently, the ampoule is cooled to −196 °C, broken open inside a PVC hose connected to a cold trap, and all volatiles are removed in vacuo. Fractional condensation of the volatiles via two traps kept at −100 °C (acetone/liquid N_2) and −196 °C, respectively, gives 2.0 g (81%) octafluoro-1,2-dimethylenecyclobutane in the first trap.

Characterization Data Mp −35 °C, bp 64 °C. ^{19}F NMR (376 MHz, $CDCl_3$): δ −72.0 (m, 2F), −73.7 (m, 2F), −113.4 (br s, 4F) ppm. ^{13}C NMR (100 MHz, $CDCl_3$): δ 82.4, 112.2 (t, $^1J_{CF} = 289$ Hz), 150.2 (t, $^1J_{CF} = 306$ Hz) ppm. IR (gas, cm^{-1}): $\tilde{\upsilon}$ 1788 (s), 1734 (vs), 1394 (m), 1359 (vs), 1335 (s), 1282 (vw), 1249 (w), 1195 (w), 1179 (s), 1172 (s), 1131 (m), 896 (vs), 785 (vw), 668 (vw), 621 (w), 567 (m).

Application The chemistry of octafluoro-1,2-dimethylenecyclobutane is very limited. Examples include isomerization, cycloaddition, and oxidation reactions; an η^4 manganese complex has been reported (Figure 50.2) [17, 19, 20].

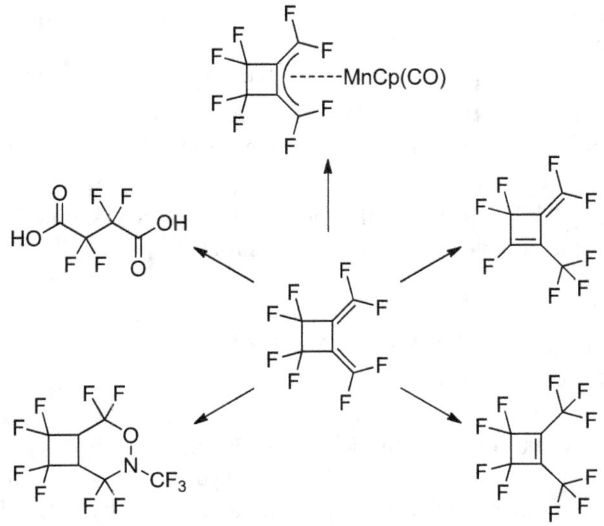

FIGURE 50.2 Synthetic potential of octafluoro-1,2-dimethylenecyclobutane.

50.4 PREPARATION OF 2,3-DIBROMO-1,1,1-TRIFLUOROPROPANE [21,22]

Apparatus A 250-mL, three-necked flask with magnetic stirrer, reflux condenser, thermometer and gas dispersion tube, cryostat, UV lamp, ice–water bath, two cold traps, vacuum line, safety glasses, laboratory coat, protective gloves.

Chemicals 3,3,3-Trifluoropropene, bromine, dry ice, ethanol, liquid N_2.

Attention! Safety glasses and protective gloves must be used at all times.

Caution! All manipulations have to be performed under an atmosphere of dry argon using standard Schlenk line techniques. Bromine is toxic and corrosive. Care should be taken to avoid inhalation or contact with skin, all reactions should be carried out in a well-ventilated hood. Cold liquefied gases may expand explosively when allowed to warm to room temperature without proper pressure release. Dry ice can cause skin burns.

Experimental Procedure The three-necked flask equipped with stirrer, reflux condenser, thermometer, and gas dispersion tube is charged with bromine (310 g, 1.94 mol), and placed inside an ice–water bath. The condenser is cooled to −45 °C with a suitable cryostat and 3,3,3-trifluoropropene (200 g, 2.08 mol, 1.1 equiv) is introduced through the dispersion tube at a slow rate so that the temperature of the reaction mixture remains below 25 °C, while being irradiated with a UV lamp. After 4.5 h, the obtained yellow solution is purified by fractional condensation via two traps kept at −78 °C and −196 °C, respectively, to give 2,3-dibromo-1,1,1-trifluoropropane (485 g, 98%) in the first trap.

If a cryostat is not available, the reflux condenser can be replaced by a dry ice condenser.

Characterization Data bp 115–120 °C. ^{19}F NMR (376 MHz, $CDCl_3$): δ −86.3 (d, 3F, CF_3) ppm. ^{1}H NMR (400 MHz, $CDCl_3$): δ 3.5–4.1 (m, CH_2Br, 2H), 4.2–4.6 (m, CHBr, 1H) ppm. IR (neat, cm^{-1}): $\bar{\upsilon}$ 2990 (w), 1340, 1275, 1250, 1180, 1110, 910, 895.

Waste Disposal The residues of bromine are toxic and should be deactivated using an aqueous solution of sodium thiosulfate.

50.5 PREPARATION OF 2-BROMO-3,3,3-TRIFLUOROPROPENE [22, 23]

Apparatus A 500-mL, three-necked flask, mechanical stirrer, dropping funnel, fractionation column (2 cm × 50 cm, packed with Raschig rings) equipped with a column head cooled to 0 °C using a suitable cryostat, collection flask cooled to 0 °C, drying tube, safety glasses, laboratory coat, protective gloves.

Chemicals 2,3-Dibromo-1,1,1-trifluoropropane, potassium hydroxide pellets, ice–water bath, dry ice–ethanol bath.

Attention! Safety glasses and protective gloves must be used at all times.

Caution! Hot potassium hydroxide is highly corrosive! Care should be taken to avoid inhalation of 2,3-dibromo-1,1,1-trifluoropropane and 2-bromo-3,3,3-trifluoropropene; all reactions should be carried out in a well-ventilated hood.

Experimental Procedure The three-necked flask equipped with stirrer, dropping funnel, and fractionation column is charged with potassium hydroxide pellets (161 g, 2.87 mol, 3.9 equiv); the column head is cooled to 0 °C, the collection flask is equipped with a drying tube and placed in an ice–water bath. The reaction mixture is heated to 100 °C, and 2,3-dibromo-1,1,1-trifluoropropane (150 g, 0.59 mol) is added drop by drop during a period of 2.5 h. 2-Bromo-3,3,3-trifluoropropene is collected by fractional distillation. Yield: 84.3 g (482 mmol, 82%).

On a smaller scale or if a cryostat is not available, the fractionation column can be replaced by a standard distillation apparatus. The yield decreases to approximately 60%.

Characterization Data Bp 33–35 °C. ^{19}F NMR (376 MHz, $CDCl_3$): δ −69.3 (dd, $^4J_{FH}$ = 1.35 Hz, $^4J_{FH}$ = 0.90 Hz, 3F, CF_3) ppm. ^{1}H NMR (400 MHz, $CDCl_3$): δ 6.0 (m, 1H), 6.4 (m, 1H) ppm. IR (gas, cm^{-1}): $\bar{\upsilon}$ 1640 (w), 1285, 1185, 1090.

Application 2-Bromo-3,3,3-trifluoropropene is a valuable fluorinated C_3 building block for the synthesis of styrenes, arylimines, and indoles by palladium-catalyzed cross-coupling reactions (Figure 50.3) [24–27]. In addition, it can be used as a useful precursor to the in situ generation of trifluoropropynyllithium and trifluoropropenyllithium to access fluorinated propargylic and allylic amines and alcohols [21, 28–32].

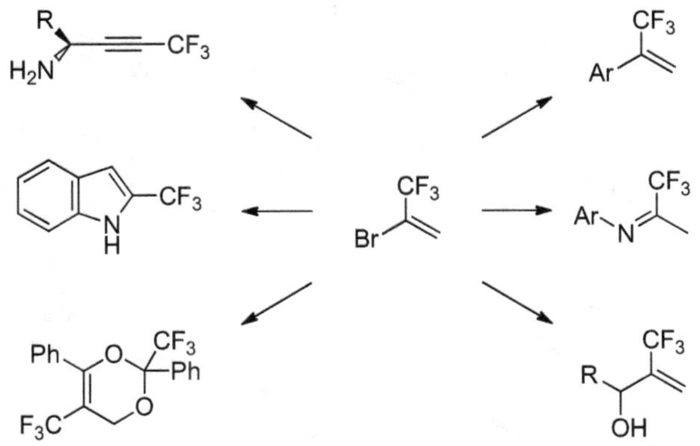

FIGURE 50.3 Synthetic potential of 2-bromo-3,3,3-trifluoropropene.

50.6 PREPARATION OF 1,1-DIFLUOROALLENE [22]

Apparatus A 1000-mL, three-necked flask with magnetic stirrer, dropping funnel, septum, low-temperature thermometer, fractionation column (2 cm × 50 cm, packed with Raschig rings) equipped with a column head cooled to −80 °C using a suitable cryostat, collection flask cooled to −80 °C, Dewar flasks, safety glasses, laboratory coat, protective gloves.

Chemicals 2-Bromo-3,3,3-trifluoropropene, *n*-butyllithium 2.5-M solution in hexanes, dry hexane, hexane, dry ice, ethanol.

Attention! Safety glasses and protective gloves must be used at all times.

Caution! All manipulations have to be performed under an atmosphere of dry argon using standard Schlenk line techniques. Care should be taken to avoid inhalation of 1,1-difluoroallene and hexane; all reactions should be carried out in a well-ventilated hood. *n*-Butyllithium is pyrophoric and reacts violently with air and water. Liquefied gases may expand explosively when allowed to warm to room temperature without proper pressure release. 1,1-Difluoroallene should be stored at −80 °C or −196 °C. Liquid nitrogen and dry ice can cause skin burns.

Experimental Procedure The three-necked flask equipped with magnetic stirrer, low-temperature thermometer, and dropping funnel is charged with dry hexane (250 mL) and 2-bromo-3,3,3-trifluoropropene (20.0 g, 114 mmol). The flask is cooled with a hexane slush bath by addition of liquid nitrogen until the solution reaches −85 °C. A solution of *n*-butyllithium (50 mL, 125 mmol, 1.1 equiv) is added drop by drop over a period of 30 min to the reaction mixture at such a rate that the temperature remains below −80 °C. After additional 15 min, the dropping funnel is replaced by the fractionation column with its head cooled to −80 °C, and the reaction mixture is allowed to warm slowly to room temperature and subsequently heated to reflux. Distillation gives pure 1,1-difluoroallene in the collection flask. Yield: 72%.

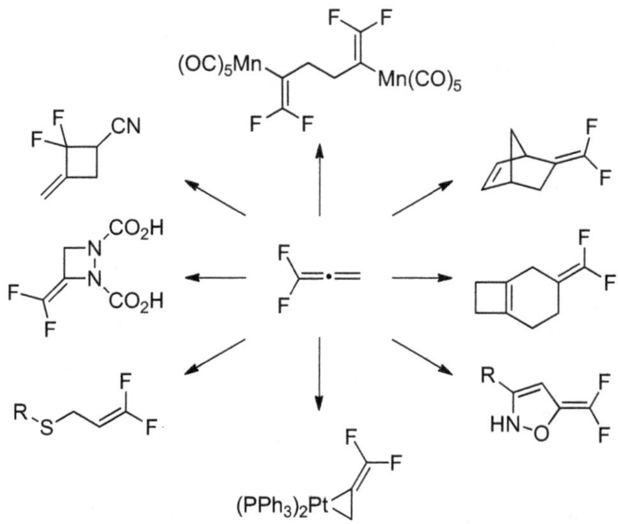

FIGURE 50.4 Synthetic potential of 1,1-difluoroallene.

On a smaller scale or if a cryostat is not available, the fractionation column can be replaced by a dry ice condenser cooled to 0 °C and connected to a trap kept at −78 °C. Fractional condensation of its content via two subsequent traps kept at −100 °C and −196 °C yields 1,1-difluoroallene in up to 97% yield.

Characterization Data Mp −137 °C, bp −23 °C. ^{19}F NMR (376 MHz, $CDCl_3$): δ −106.9 (t, $^4J_{FH} = 3.5$ Hz, 2F, CF_2) ppm. ^{1}H NMR (400 MHz, $CDCl_3$): δ 6.04 (t, $^4J_{HF} = 3.5$ Hz, 2H, CH_2) ppm. IR (gas, cm^{-1}): $\bar{\upsilon}$ 3289 (w), 3049 (m), 2976 (m), 2907 (m), 2793 (w), 2532 (w), 2439 (m), 2315 (w), 2268 (w), 2119 (m), 2020 (vs), 1835 (w), 1754 (w), 1618 (w), 1515 (m), 1488 (vs), 1479 (vs), 1471 (vs), 1395 (m), 1339 (vs), 1330 (vs), 1321 (vs), 1247 (vs), 1232 (vs), 1010 (m), 992 (m), 935 (s), 920 (s), 876 (w), 870 (w), 862 (w), 854 (w), 846 (w), 810 (s), 802 (s), 668 (w).

Application 1,1-Difluoroallene is an interesting fluorinated substrate for [2 + 2], [4 + 2], and 1,3-dipolar cycloaddition reactions (Figure 50.4) [22, 33–37]. Nucleophilic addition reactions have also been reported [38]. In a number of transition-metal complexes, difluoroallene was shown to coordinate preferentially via the non-fluorinated side [3, 11–16, 39].

50.7 PREPARATION OF 1,1-DIFLUORO-2-IODOETHENE [40]

Apparatus A 2000-mL, single-necked flask with a PFTE needle valve, magnetic stirrer, vacuum line, cold trap, distillation apparatus.

Chemicals 1,1-Difluoroethene, iodine monochloride, 2,6-lutidine, 1,8-diazabicyclo[5.4.0]undec-7-ene (DBU), ice–water bath, liquid N_2.

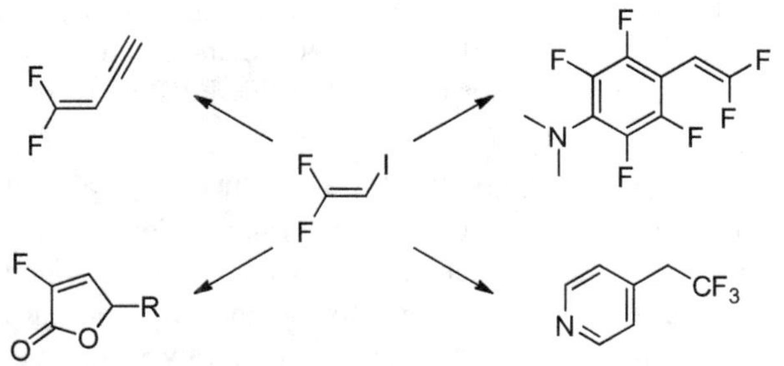

FIGURE 50.5 Synthetic applications of 1,1-difluoro-2-iodoethene.

Attention! Safety glasses and protective gloves must be used at all times.

Caution! Iodine monochloride is corrosive. Care should be taken to avoid inhalation of iodine monochloride, 1,1-difluoroethene, and 1,1-difluoro-2-iodoethene and contact with the skin. All manipulations should be performed in a well-ventilated hood.

Experimental Procedure The flask equipped with a stirrer is charged with iodine monochloride (100 g, 600 mmol), cooled to −196 °C and evacuated. After placing it in an ice–water bath, difluoroethene (40 g, 620 mmol) is introduced in several portions into the flask via the vacuum line, the consumption is being monitored by the decrease in pressure. When no more difluoroethene is consumed, 2,6-lutidine (300 mL) and DBU (110 g, 720 mmol) are added. After stirring overnight, all volatiles are removed from the reaction mixture into a trap cooled to −196 °C. Water (3 × 50 mL) is added to the residue, stirred for 5 min and again, all volatiles are removed under vacuum. Distillation of the combined volatiles yields 1,1-difluoro-2-iodoethene (97 g, 511 mmol, 85%).

Characterization Data Bp 35 °C. ^{1}H NMR (400 MHz, $CDCl_3$): δ 4.82 (dd, $^3J_{HF}$ = 22.8 Hz, $^3J_{HF}$ = 2.1 Hz, 1H, CHI) ppm. ^{19}F NMR (376 MHz, $CDCl_3$): δ −71.4 (dd, $^2J_{FF}$ = 26.2 Hz, $^3J_{FH}$ = 2.1 Hz, 1F, CF_2), −75.8 (dd, $^2J_{FF}$ = 26.2 Hz, $^3J_{FH}$ = 23.6 Hz, 1F, CF_2) ppm. ^{13}C NMR (100 MHz, $CDCl_3$): δ 156.5 (dd, $^1J_{CF}$ = 296 Hz, $^1J_{CF}$ = 286 Hz, CF_2), 24.2 (dd, $^2J_{CF}$ = 37 Hz, $^2J_{CF}$ = 29 Hz, CHI) ppm. IR (gas, cm^{-1}): $\bar{\nu}$ 3126, 3117, 2662, 2270, 1726, 1717, 1700, 1469, 1317, 1308, 1144, 1136, 953, 946, 747, 738, 731.

Application 1,1-Difluoro-2-iodoethene is a useful difluorovinyl building block for palladium-catalyzed cross-coupling reactions to give fluorinated styrenes and enynes (Figure 50.5) [40–42].

50.8 PREPARATION OF 1,1,4,4-TETRAFLUOROBUTA-1,3- DIENE [40, 43]

Apparatus Two 100-mL Schlenk flasks with magnetic stirrer, two cold traps, PTFE cannula (ID = 1 mm), rubber septum.

Chemicals 1,1-Difluoro-2-iodoethene, activated zinc powder (treated with dilute hydrochloric acid, washed with water, ethanol, and acetone, and dried in vacuo), dry *N*,*N*-dimethylformamide (DMF), $Pd(PPh_3)_4$, dry ice, ethanol, liquid N_2.

Attention! Safety glasses and protective gloves must be used at all times.

Caution! All manipulations have to be performed under an atmosphere of dry argon using standard Schlenk techniques. DMF is toxic. Care should be taken to avoid contact with skin or inhalation. Organozinc reagents can react violently with water. Liquefied gases may expand explosively when allowed to warm to room temperature without proper pressure release. Liquid nitrogen and dry ice can cause skin burns. 1,1,4,4-Tetrafluorobuta-1,3-diene should be stored at −80 °C.

Experimental Procedure In the Schlenk flask, 1,1-difluoroiodoethene (13.0 g, 68.4 mmol) is added drop by drop to a suspension of activated zinc powder (8.0 g, 122 mmol, 1.8 equiv) in DMF (50 mL) at room temperature. An exothermic reaction occurs while the mixture is stirred for 1 h. Subsequently, the excess zinc is allowed to settle at the bottom of the flask and the supernatant is transferred via cannula to another Schlenk flask containing 1,1-difluoroiodoethene (13.0 g, 68.5 mmol, 1.0 equiv) and $Pd(PPh_3)_4$ (2.0 g, 1.7 mmol, 2.5 mol%). The flask is connected to a trap kept at −78 °C and heated to 75 °C for 4 h, during which a colorless liquid condenses in the trap. Fractional condensation via two traps kept at −120 °C (ethanol/N_2 slush) and −196 °C (liquid N_2) gives 1,1,4,4-tetrafluorobuta-1,3-diene (5.2 g, 41.3 mmol, 60%) in the first trap.

Characterization Data Bp 3–4 °C. ^{19}F NMR (376 MHz, $CDCl_3$): δ −86.8 (m, 2F, CF_2), −88.1 (m, 2F, CF_2) ppm. ^1H NMR (400 MHz, $CDCl_3$): δ 4.52 (m, 2H, CH) ppm. ^{13}C NMR (100 MHz, $CDCl_3$): δ 72.3 (m, CH), 155.6 (m, CF_2) ppm. IR (gas, cm^{-1}): $\bar{\upsilon}$ 3122 (vw), 1740 (sh), 1715 (vs), 1321 (s), 1172 (m), 1140 (w), 949 (m), 922 (s), 826 (w), 739 (vw), 548 (vw).

50.9 PREPARATION OF 1,4-DIBROMO-1,1,2,2-TETRAFLUOROBUT-2-ENE [40, 43]

Apparatus A 100-mL Schlenk flask with magnetic stirrer, vacuum line, separation funnel, two traps.

Chemicals 1,1,4,4-Tetrafluorobuta-1,3-diene, bromine, dichloromethane, ethanol, liquid N_2.

Attention! Safety glasses and protective gloves must be used at all times.

Caution! Bromine is toxic and corrosive. Care should be taken to avoid contact with skin and inhalation of bromine, 1,1,4,4-tetrafluorobuta-1,3-diene, 1,4-dibromo-1,1,4,4-tetrafluorobut-2-ene, and dichloromethane; all reactions should be carried out in a well-ventilated hood.

Experimental Procedure In the Schlenk flask, bromine (6.7 g, 42 mmol) is dissolved in dichloromethane (10 mL) and 1,1,4,4-tetrafluorobuta-1,3-diene (5.2 g, 42 mmol) is added by vacuum transfer. After stirring for 5 h at room temperature, sodium thiosulfate (5 mL saturated solution in water) is added and the reaction mixture is stirred for an additional 5 min. Subsequently, it is transferred to the separation funnel, the phases are separated, and the aqueous layer is extracted with dichloromethane (3 × 2 mL). Fractional condensation of the combined organic layers via two traps kept at −60 °C (ethanol/N_2 slush) and −196 °C (liquid N_2) gives 1,4-dibromo-1,1,4,4-tetrafluorobut-2-ene (9.87 g, 34.5 mmol, 82 %) in the first trap.

Characterization Data Bp 105–107 °C. ^{19}F NMR (376 MHz, $CDCl_3$): δ −50.3 (m, 4F, CF_2Br) ppm. ^{1}H NMR (400 MHz, $CDCl_3$): δ 6.33 (m, CH, 2H) ppm. ^{13}C NMR (100 MHz, $CDCl_3$): δ 114.2 (t, $^1J_{CF}$ = 301 Hz, CF_2Br), 129.3 (tt, $^2J_{CF}$ = 27 Hz, $^3J_{CF}$ = 7 Hz, CH) ppm.

50.10 PREPARATION OF 1,1,4,4-TETRAFLUOROBUTA-1,2,3-TRIENE

Apparatus Three traps, U-shaped tube (ID = 1 cm), oil bath, vacuum line, Dewar flasks.

Chemicals 1,4-Dibromo-1,1,4,4-tetrafluorobut-2-ene, potassium hydroxide (technical grade flakes), dry ice, ethanol, liquid N_2.

Attention! Safety glasses and protective gloves must be used at all times.

Caution! All manipulations must be performed under strict exclusion of air using standard vacuum-line techniques. 1,1,4,4-Tetrafluorobuta-1,2,3-triene has been reported to explode violently above −5 °C or on contact with air; it polymerizes even at −80 °C and must be stored at −196 °C; it can be safely transferred under reduced pressure using a vacuum line. Hot potassium hydroxide is highly corrosive.

Experimental Procedure A reservoir trap containing 1,4-dibromo-1,1,4,4-tetrafluorobut-2-ene (1.0 g, 3.5 mmol) is cooled to −196 °C and connected to the vacuum line via a U-shaped tube charged with potassium hydroxide and two traps kept at −78 °C and −196 °C, respectively. The U-shaped tube is placed in an oil bath and maintained at 88 °C. Under dynamic vacuum, the reservoir is allowed to warm very slowly to room temperature inside an empty Dewar flask, passing a slow stream of starting material over the hot KOH. It is important to keep the pressure below 10^{-2} mbar. 1,1,4,4-Tetrafluorobuta-1,2,3-triene (40–99%) collects in the last trap, unreacted starting material can be recovered from the second trap.

Characterization Data mp −130 °C, bp −5 °C. ^{19}F NMR (376 MHz, CD_2Cl_2, −80 °C): δ −96.1 (s, 4F, CF_2) ppm. ^{13}C NMR (100 MHz, CD_2Cl_2, −80 °C): δ 147.8, 158.9 ppm. IR (gas, cm^{-1}): $\bar{\upsilon}$ 1732, 1281, 973, 939, 536.

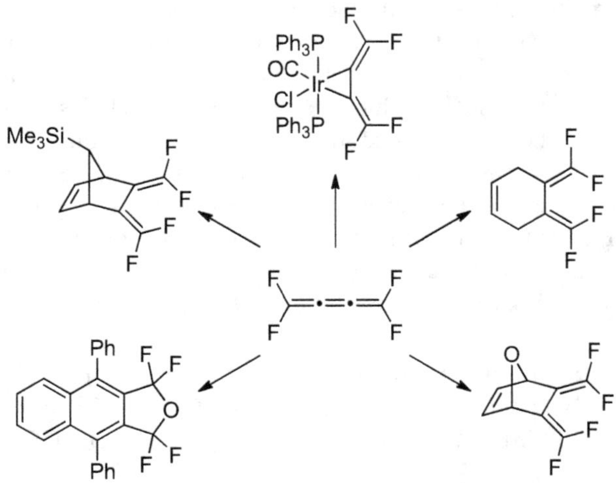

FIGURE 50.6 Synthetic potential of 1,1,4,4-tetrafluorobuta-1,2,3-triene.

Application Tetrafluorobutatriene is a powerful dienophile for [4 + 2] cycloadditions with a number of 1,3-dienes. It forms air-stable transition-metal complexes with rhodium and iridium (Figure 50.6) [44–46].

50.11 PREPARATION OF HEXAFLUOROBUTA-1,3-DIENE [47]

Apparatus A 50-mL, single-necked flask with a PTFE needle valve, magnetic stirrer, three traps, Dewar flasks, vacuum line.

Chemicals Tri(*n*-butyl)(trifluorovinyl)tin, trifluorobromoethene, palladium acetate, Me-Phos (dicyclohexyl-2-(2′-methylbiphenyl)-phosphine), dry DMF, ethanol, liquid N_2.

Attention! Safety glasses and protective gloves must be used at all times.

Caution! All manipulations must be performed under exclusion of air using standard Schlenk and vacuum line techniques. Tributyltin compounds and DMF are toxic; care should be taken to avoid contact with skin and inhalation of tributyl(trifluorovinyl)tin, trifluorobromoethene, hexafluorobutadiene, and DMF. Cold liquefied gases may expand explosively when allowed to warm to room temperature without proper pressure release. Liquid nitrogen can cause skin burns. Hexafluorobuta-1,3-diene should be stored at −80 °C.

Experimental Procedure The single-necked flask is charged with DMF (12 mL), tributyl(trifluorovinyl)tin (11.13 g, 30 mmol), palladium acetate (14 mg, 0.063 mmol, 0.25 mol %), and Me-Phos (114 mg, 0.31 mmol, 1.25 mol%). After degassing at −196 °C, trifluorobromoethene (4.05 g, 25 mmol) is added by vacuum transfer and the mixture is stirred at 60 °C

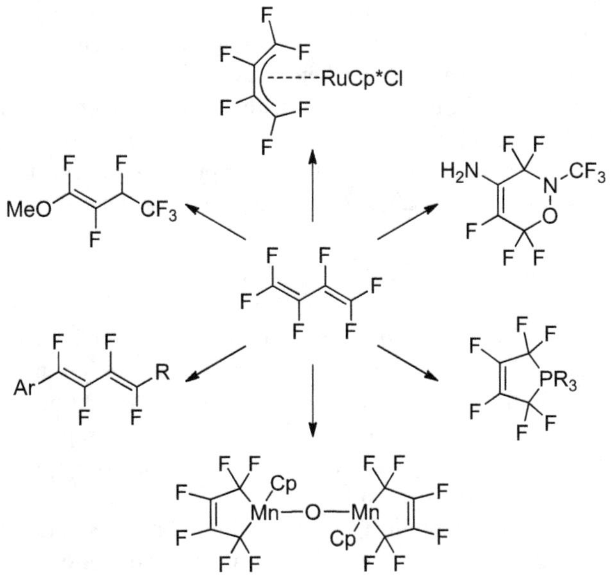

FIGURE 50.7 Synthetic applications of hexafluorobuta-1,3-diene.

for 40 h. Fractional condensation of the reaction mixture via three subsequent traps kept at −50 °C, −130 °C, and −196 °C yields hexafluorobuta-1,3-diene (3.24 g, 20 mmol, 80 %) in the second trap.

Characterization Data Mp −132 °C, bp 5–10 °C. ^{19}F NMR (376 MHz, $CDCl_3$): δ −93.3 (m, 2F, CF_2), −107.8 (m, 2F, CF_2), −180.2 (m, 2F, CF) ppm. ^{13}C NMR (100 MHz, $CDCl_3$): δ 116.0 (CF), 152.3 (CF_2) ppm. IR (gas, cm^{-1}): $\tilde{\upsilon}$ 1795 (m), 1765 (vs), 1330 (vs), 1191 (s), 1141 (vs), 1136 (vs), 972 (vs).

Application Hexafluorobutadiene is used for the synthesis of fluoropolymers and as an etching agent. It can also be used to synthesize fluorinated heterocycles by cycloaddition reactions and undergoes substitution and addition reactions. Transition-metal complexes with η^1-η^1, η^4, and η^2 coordination modes have been synthesized (Figure 50.7) [16, 48–55].

50.12 PREPARATION OF 1,1,2,4,4-PENTAFLUOROBUTA-1,3-DIENE [47]

Apparatus A 50-mL, single-necked flask with a PTFE needle valve, magnetic stirrer, three traps, Dewar flasks, vacuum line.

Chemicals Tri(*n*-butyl)(2,2-difluorovinyl)tin, trifluorobromoethene, palladium acetate, X-Phos (dicyclohexyl-2-(2′,4′,6′-triisopropylbiphenyl)-phosphine), dry DMF, ethanol, liquid N_2.

Attention! Safety glasses and protective gloves must be used at all times.

Caution! All manipulations must be performed under exclusion of air using standard Schlenk and vacuum line techniques. Tributyltin compounds and DMF are toxic; care should be taken to avoid contact with skin and inhalation of tri-butyl(1,1-difluorovinyl)tin, trifluorobromoethene, pentafluorobutadiene, and DMF. Cold liquefied gases may expand explosively when allowed to warm to room temperature without proper pressure release. Liquid nitrogen can cause skin burns. 1,1,2,4,4-Pentafluorobuta-1,3-diene should be stored at −80 °C.

Experimental Procedure The single-necked flask is charged with DMF (3 mL), tributyl(1,1-difluorovinyl)tin (3.20 g, 9.06 mmol), palladium acetate (5 mg, 0.022 mmol, 0.25 mol%), and X-Phos (54 mg, 0.11 mmol, 1.26 mol%). After degassing at −196 °C, trifluorobromoethene (1.29 g, 8.0 mmol) is added by vacuum transfer and the mixture is stirred at 60 °C for 39 h. Fractional condensation of the reaction mixture via three subsequent traps kept at −50 °C, −130 °C, and −196 °C yields 1,1,2,4,4-pentafluorobuta-1,3-diene (1.12 g, 7.78 mmol, 97 %) in the second trap.

The product may contain up to 5% trifluorobromoethene, which can be removed by reaction with activated zinc in dry DMF followed by fractional condensation (yield after purification approximately 30%).

Characterization Data Mp −110 °C, bp 19 °C. ^{1}H NMR (400 MHz, $CDCl_3$): δ 4.9 (m, 1H, CH) ppm. ^{19}F NMR (376 MHz, $CDCl_3$): δ −78.4 (m, 1F, CF_2), −79.5 (m, 1F, CF_2), −101.9 (m, 1F, CF_2), −118.8 (m, 1F, CF_2), −179.3 (m, 1F, CF) ppm. ^{13}C NMR (100 MHz, $CDCl_3$): δ 70.2 (m, CH), 121.8 (m, *C*F=CF_2), 152.5 (m, CF_2), 155.6 (m, CF_2) ppm. IR (gas, cm^{-1}): $\bar{\upsilon}$ 1790 (w), 1728 (vs), 1389 (m), 1323 (s), 1300 (m), 1246 (m), 1184 (m), 1146 (m), 1084 (m), 935 (s), 845 (w), 789 (w).

50.13 PREPARATION OF 1,1-DIFLUOROBUTA-1,3-DIENE [47]

Apparatus A 250-mL, single-necked flask with a PTFE needle valve, magnetic stirrer, three traps, Dewar flasks, vacuum line.

Chemicals Tri(*n*-butyl)(vinyl)tin, 1,1-difluoro-2-iodoethene, $Pd(PPh_3)_4$, dry DMF, ethanol, liquid N_2.

Attention! Safety glasses and protective gloves must be used at all times.

Caution! All manipulations must be performed under exclusion of air using standard Schlenk and vacuum line techniques. Tributyltin compounds and DMF are toxic; care should be taken to avoid contact with skin and inhalation of tributyl(vinyl)tin, difluoroiodoethene, difluorobutadiene, and DMF. Cold liquefied gases may expand explosively when allowed to warm to room temperature without proper pressure release. Liquid nitrogen can cause skin burns. 1,1-Difluorobuta-1,3-diene should be stored at −80 °C.

Experimental Procedure The single-necked flask is charged with DMF (30 mL), tributyl(vinyl)tin (10.10 g, 53.4 mmol), 1,1-difluoro-2-iodoethene

(17.46 g, 55.1 mmol), and $Pd(PPh_3)_4$ (400 mg, 0.34 mmol, 0.65 mol%). After degassing at −196 °C the mixture is stirred at 60 °C for 23 h. Fractional condensation of the reaction mixture via three subsequent traps kept at −50 °C, −130 °C, and −196 °C yields 1,1-difluorobuta-1,3-diene (4.23 g, 47.0 mmol, 88%) in the second trap.

Characterization Data Bp 3–5 °C. ^{1}H NMR (400 MHz, $CDCl_3$): δ 6.24 (ddddd, $^3J_{HH}$ = 17.2 Hz, $^3J_{HH}$ = 11 Hz, $^3J_{HH}$ = 11 Hz, $^4J_{HF}$ = 1 Hz, $^4J_{HF}$ = 1 Hz, 1H, CH=CH_2), 5.15 (m, $^3J_{HH}$ = 17.2 Hz, 1 H, –*CH*=CH_2), 5.02 (m, $^3J_{HH}$ = 11 Hz, 1H, CH_2), 4.96 (m, $^3J_{HH}$ = 11 Hz, $^3J_{HF}$ = 24 Hz, 1H, –CH=CF_2) ppm. ^{19}F NMR (376 MHz, $CDCl_3$): δ −86.09 (dd, $^2J_{FF}$ = 26 Hz, $^3J_{FH}$ = 24 Hz, 1F, =CF_2), −88.61 (d, $^2J_{FF}$ = 26 Hz, 1F, =CF_2) ppm. ^{13}C NMR (100 MHz, $CDCl_3$): δ 156.7 (dd, $^1J_{CF}$ = 296.5 Hz, $^1J_{CF}$ = 290.7 Hz, =CF_2), 126.1 (dd, $^4J_{CF}$ = 5.0 Hz, $^4J_{CF}$ = 1.3 Hz, CH_2), 116.4 (dd, $^3J_{CF}$ = 11.0 Hz, $^3J_{CF}$ = 2.7 Hz, *C*H=CH_2), 82.8 (dd, $^2J_{CF}$ = 26.7 Hz, $^2J_{CF}$ = 16.8 Hz, *C*H=CF_2) ppm. IR (gas, cm^{-1}): $\tilde{\upsilon}$ 3110 (w), 3097 (w), 1734 (vs), 1731 (vs), 1725 (vs), 1437 (m), 1425 (m), 1339 (s), 1327 (s), 1215 (s), 1210 (s), 1205 (s), 990 (m), 904 (s), 895 (m), 658 (w).

50.14 PREPARATION OF 1,1,2-TRIFLUOROPENTA-1,4-DIENE [47]

Apparatus A 100-mL Schlenk flask, 100-mL, single-necked flask with PTFE needle valve, magnetic stirrer, three traps, PTFE cannula (ID = 1 mm), rubber septum, Dewar flasks, vacuum line.

Chemicals Trifluorobromoethene, activated zinc powder (treated with dilute hydrochloric acid, washed with water, ethanol, and acetone, and dried in vacuo), allyl bromide, dry DMF, $Pd(PPh_3)_4$, ethanol, liquid N_2.

Attention! Safety glasses and protective gloves must be used at all times.

Caution! All manipulations have to be performed under exclusion of air and moisture using standard Schlenk and vacuum line techniques. DMF is toxic. Care should be taken to avoid contact with skin or inhalation; all reactions should be carried out in a well-ventilated hood. Organozinc reagents can react violently with water. Cold liquefied gases may expand explosively when allowed to warm to room temperature without proper pressure release. Liquid nitrogen can cause skin burns. 1,1,2-Trifluoropenta-1,4-diene should be stored at −80 °C.

Experimental Procedure In the Schlenk flask, trifluorobromoethene (2.41 g, 15 mmol) is added to a suspension of activated zinc powder (2.94 g, 45 mmol, 3.0 equiv) in DMF (10 mL) by vacuum transfer. An exothermic reaction occurs while the mixture is stirred at room temperature for 1 h and at 60 °C for another 2 h. Subsequently, the excess zinc is allowed to settle at the bottom of the flask and the supernatant is transferred via cannula to the PTFE valve flask containing allyl bromide (1.81 g, 15 mmol, 1.0 equiv) and $Pd(PPh_3)_4$ (173 mg, 0.15 mmol, 1.0 mol%). After degassing at −196 °C, the mixture is stirred at 80 °C for

6 h. Fractional condensation of the reaction mixture via three subsequent traps kept at −50 °C, −130 °C, and −196 °C yields 1,1,2-trifluoropenta-1,4-diene (1.37 g, 11.2 mmol, 86%) in the second trap.

Characterization Data bp 37–38 °C. ^{1}H NMR (400 MHz, $CDCl_3$): δ 5.77 (ddt, 1H, $^3J_{HH}$ = 16.9 Hz, $^3J_{HH}$ = 10.4 Hz, $^3J_{HH}$ = 6.2 Hz, 1H, CH), 5.12–5.22 (m, 2H, CH_2), 2.97–3.03 (m, 2H, CH_2) ppm. ^{19}F NMR (376 MHz, $CDCl_3$): δ −105.9 (ddt, $^2J_{FF}$ = 87.2 Hz, $^3J_{FF}$ = 32.7 Hz, $^4J_{FH}$ = 2.7 Hz, 1F, CF_2), −124.7 (ddt, $^3J_{FF}$ = 114.4 Hz, $^2J_{FF}$ = 87.2 Hz, $^4J_{FF}$ = 4.2 Hz, 1 F, CF_2), −174.1 (ddt, $^3J_{FF}$ = 114.4, $^3J_{FF}$ = 32.7 Hz, $^3J_{FH}$ = 22.2 Hz, 1F, CF) ppm. ^{13}C NMR (100 MHz, $CDCl_3$): δ 153.2 (ddd, $^1J_{CF}$ = 286.2 Hz, $^1J_{CF}$ = 270.4 Hz, $^2J_{CF}$ = 46.3 Hz, CF_2), 134.0 (s, CH), 127.2 (ddd, $^1J_{CF}$ = 235.3 Hz, $^2J_{CF}$ = 53.0 Hz, $^2J_{CF}$ = 15.8 Hz, CF), 118.4 (s, CH_2), 30.1 (dd, $^3J_{CF}$ = 22.3 Hz, $^4J_{CF}$ = 2.5 Hz, CH_2) ppm. IR (gas, cm^{-1}): $\bar{\nu}$ 3096, 3025, 2997, 2919, 1801, 1652, 1646, 1436, 1431, 1430, 1424, 1304, 1267, 1218, 1141, 1105, 1072, 993, 929, 923, 792.

REFERENCES

1. Bartlett PD, Tauber SJ, Weber WP. Ethylenation of secondary and tertiary alkyllithiums. I. Conditions and limits of the reaction. *J Am Chem Soc* 1969;91:6362–6366.
2. Zerger R, Rhine W, Stucky G. Stereochemistry of polynuclear compounds of the main group elements. Bonding and the effect of metal-hydrogen-carbon interactions in the molecular structure of cyclohexyllithium, a hexameric organolithium compound. *J Am Chem Soc* 1974;96:6048–6055.
3. Kühnel MF, Lentz D. Hydrometalation of fluoroallenes. *Dalton Trans* 2009;4747–4755.
4. Lu H, Friedrich HB, Burton DJ. A facile route to tetrafluoroallene. *J Fluorine Chem* 1995;75:83–86.
5. Banks RE, Haszeldine RN, Taylor DR. 1044. Polyhalogeno-allenes. Part II. Reaction of tetrafluoroallene with trifluoronitrosomethane. *J Chem Soc (Resumed)* 1965;5602–5612.
6. Banks RE, Deem WR, Haszeldine RN, Taylor DR. Polyhalogenoallenes. Part IV. Thermal co-dimerisation of tetrafluoroallene with hexafluorobut-2-yne. *J Chem Soc C* 1966;2051–2052.
7. Blackwell GB, Haszeldine RN, Taylor DR. Polyhalogeno-allenes and -acetylenes. Part 15. Dipolar cycloadditions of N-phenylsydnone and aryl azides to perfluoropropadiene and perfluoropropyne. *J Chem Soc, Perkin Trans 1* 1982;2207–2210.
8. Blackwell GB, Haszeldine RN, Taylor DR. Polyhalogeno-allenes and -acetylenes. Part 16. Further 1,3-dipolar cycloadditions to perfluoropropadiene. *J Chem Soc, Perkin Trans 1* 1983;1–5.
9. Xiao J-C, Chen Q-Y. The chemistry of tetrafluoroallene: One-pot synthesis of trifluoromethylindolizines from 1,3-diiodo-1,1,3,3-tetrafluoropropane by 1,3-dipolar cycloaddition. *Chin J Chem* 2003;21:898–903.
10. Xiao J-C, Chen Q-Y. The chemistry of tetrafluoroallene: nucleophilic addition reactions with phenols and amines. *J Fluorine Chem* 2003;123:189–195.

11. Lentz D, Willemsen S. Formation of novel unsaturated fluorinated ligands by metal induced C-C coupling reaction of difluoroallene with fluoromethylidyne ligands of Fe3(CO)9(μ3-CF)2. *J Organomet Chem* 2002;641:215–219.

12. Lentz D, Willemsen S. Fluorinated bis(enyl) ligands by metal-induced dimerization of fluorinated allenes. *Angew Chem, Int Ed* 2001;40:2087–2091; *Angew Chem* 2001;113:2142–2146.

13. Lentz D. Organometallic chemistry of fluorinated propadienes and butadienes. *J Fluorine Chem* 2004;125:853–861.

14. Lentz D, Willemsen S. Coordination chemistry of fluorinated allenes. *Organometallics* 1999;18:3962–3964.

15. Lentz D, Nickelt N, Willemsen S. Organometallic chemistry of fluorinated allenes. *Chem Eur J* 2002;8:1205–1217.

16. Kühnel MF, Lentz D. Fluorinated dienes in transition-metal chemistry—the rich chemistry of electron-poor ligands. *Dalton Trans* 2010;39:9745–9759.

17. Banks RE, Haszeldine RN, Taylor DR. 176. Polyhalogeno-allenes. Part I. Tetrafluoroallene, its dimer, and the reaction of the dimer with trifluoronitrosomethane. *J Chem Soc (Resumed)* 1965;978–990.

18. Jacobs TL, Bauer RS. The synthesis and polymerization of tetrafluoroallene. *J Am Chem Soc* 1959;81:606–610.

19. Lentz D, Patzschke M. Octafluor-1,2-dimethylencyclobutan, ein perfluorierter Dien-Komplexligand—Carbonyl(η5-cyclopentadienyl)(η4-octafluor-1,2-dimethylencyclobutan)mangan. *Z Anorg Allg Chem* 2004;630:973–976.

20. Banks RE, Barlow MG, Deem WR, Haszeldine RN, Taylor DR. Polyhalogenoallenes. Part III. Isomerisation of the dimer of tetrafluoroallene, perfluoro-(1,2-dimethylenecyclobutane), to perfluoro-(2-methyl-3-methylenecyclobutene). *J Chem Soc C* 1966;981–984.

21. Drakesmith FG, Stewart OJ, Tarrant P. Preparation and reactions of lithium derivatives of trifluoropropene and trifluoropropyne. *J Org Chem* 1968;33:280–285.

22. Dolbier WR, Burkholder CR, Piedrahita CA. Fluorinated allenes: the syntheses of 1-fluoropropadiene, 1,1-difluoropropadiene and 1,1-difluoro-3-methyl-1,2-butadiene. *J Fluorine Chem* 1982;20:637–647.

23. Henne AL, Nager M. Trifluoropropyne. *J Am Chem Soc* 1951;73:1042–1043.

24. Pan R-q, Liu X-x, Deng M-Z. A novel and convenient synthetic method for producing α-(trifluoromethyl)styrenes (3). *J Fluorine Chem* 1999;95:167–170.

25. Kobayashi O, Uraguchi D, Yamakawa T. Synthesis of α-trifluoromethylstyrene derivatives via Pd-catalyzed cross-coupling of 2-bromo-3,3,3-trifluoropropene and arylmagnesium bromides. *J Mol Catal A: Chem* 2009;302:7–10.

26. Kobayashi O, Uraguchi D, Yamakawa T. Synthesis of α-trifluoromethylstyrene derivatives via Ni-catalyzed cross-coupling of 2-bromo-3,3,3-trifluoropropene and aryl Grignard reagents. *J Fluorine Chem* 2009;130:591–594.

27. Kino T, Nagase Y, Horino Y, Yamakawa T. Pd-catalyzed coupling of arylamines and 2-bromo-3,3,3-trifluoropropene. *J Mol Catal A: Chem* 2008;282:34–51.

28. Xiong Z, Qiu X-L, Huang Y, Qing F-L. Regioselective synthesis of 5-trifluoromethyl-1,2,3-triazole nucleoside analogues via TBS-directed 1,3-dipolar cycloaddition reaction. *J Fluorine Chem* 2011;132:166–174.

29. Nadano R, Fuchibe K, Ikeda M, Takahashi H, Ichikawa J. Regioselective synthesis of 5-trifluoromethyl-1,2,3-triazole nucleoside analogues via TBS-directed 1,3-dipolar cycloaddition reaction. *Chem Asian J* 2010;5:1875–1883.

30. Jeon SL, Choi JH, Cho JA, Kim BT, Jeong IH. Regioselective synthesis of 5-trifluoromethyl-1,2,3-triazole nucleoside analogues via TBS-directed 1,3-dipolar cycloaddition reaction. *J Fluorine Chem* 2008;129:1018–1028.

31. Chen X-Y, Qiu X-L, Qing F-L. Asymmetric synthesis of trifluoromethylated propargylamines via 1,2-additions of trifluoromethylacetylide to N-tert-butanesulfinyl imines. *Tetrahedron* 2008;64:2301–2306.

32. Nadano R, Ichikawa J. A facile synthesis of N-[2-(trifluoromethyl)allyl]amides and their transformation into angularly trifluoromethylated bicyclic cyclopentenones. *Chem Lett* 2007;36:22–23.

33. Knoth WH, Coffman DD. Synthesis and chemistry of 1,1-difluoroallene. *J Am Chem Soc* 1960;82:3873–3875.

34. Dolbier WR, Piedrahita C, Houk KN, Strozier RW, Gandour RW. Cycloadditions of 1,1-difluoroallene. A novel probe of mechanism. *Tetrahedron Lett* 1978;19:2231–2234.

35. Dolbier WR, Burkholder CR. The [2 + 2] and [2 + 4] cycloadditions of fluorinated allenes. *J Org Chem* 1984;49:2381–2386.

36. Dolbier WR, Burkholder CR, Winchester WR. The [2 + 2] and [2 + 4] cycloadditions of fluorinated allenes. *J Org Chem* 1984;49:1518–1522.

37. Dolbier WR, Purvis GD, III, Seabury MJ, Wicks GE, Burkholder CR. The regiochemistry and stereochemistry of 1,3-dipolar cycloadditions of 1-fluoro- and 1,1-difluoroallene. *Tetrahedron* 1990;46:7991–8004.

38. Xu Y-Y, Jin F-Q, Huang W-Y. Facile nucleophilic addition of thiols to 1,1-difluoroallene: a convenient approach to γ,γ-difluoroallyl sulphides. *J Fluorine Chem* 1995;70:5–6.

39. Ristic-Petrovic D, Anderson DJ, Torkelson JR, Ferguson MJ, R. McDonald, M. Cowie. Binuclear activation of cumulenes: roles of the adjacent metals and the cumulene binding mode in the activation process. *Organometallics* 2005;24:3711–3724.

40. Ehm C, Akkerman FA, Lentz D. Fluorinated butatrienes. *J Fluorine Chem* 2010;131:1173–1181.

41. Tellier F, Sauvêtre R, Normant J-F. Synthèse et reactivité d'énynes fluorés. *J Organomet Chem* 1987;328:1–13.

42. Nguyen BV, Burton DJ. A new route for the preparation of substituted 2,2-difluorostyrenes and a convenient route to substituted (2,2,2-trifluoroethyl)benzenes. *J Org Chem* 1997;62:7758–7764.

43. Bach A, Lentz D, Luger P, Messerschmidt M, Olesch C, Patzschke M. Crystal structure analysis of 1,1,4,4-tetrafluorobutadiene and experimental determination of the charge density of 1,1,4,4-tetrafluorobutatriene. *Angew Chem, Int Ed* 2002;41: 296–299; *Angew Chem* 2002;114:311–314.

44. Servis KL, Roberts JD. Nuclear magnetic resonance spectroscopy. Studies of 1,1,4,4-tetrafluoro-1,3-alkadienes. *J Am Chem Soc* 1965;87:1339–1344.

45. Ehm C, Lentz D. Diels-Alder reactions of 1,1,4,4-tetrafluorobutatriene. *Chem Commun* 2010;46:2399–2401.

46. Akkerman FA, Lentz D. Stabilization of tetrafluorobutatriene by complex formation. *Angew Chem, Int Ed* 2007;46:4902–4904; *Angew Chem* 2007;119:4989–4991.

47. Akkerman FA, Kickbusch R, Lentz D. Synthesis of fluorinated dienes by palladium-catalyzed coupling reactions. *Chem Asian J* 2008;3:719–731.

48. Fokin AV, Uzun AT, Dement'eva OV. Reactivity of 1-nitro-2-nitrosotetrafluoroethane. IV. Reaction with hexafluorobutadiene. *Zh Org Khim* 1967;37:834–836.

49. Denney DB, Denney DZ, Hsu YF. Condensation of trivalent phosphorus compounds with hexafluorobutadiene. *Phosphorus* 1974;4:217–219.

50. Kaz'mina NB, Knunyants IL, Mysov EI, Kuz'yants GM, Lutsenko AI. New reaction of sulfur dioxide. *Izv Akad Nauk SSSR Ser Khim* 1975;2379.

51. Bludssus W, Mews R. 1λ4,2-Thiazacyclohexa-1,4-dienes. *J Chem Soc, Chem Commun* 1979;35–36.

52. Carson PA, Roberts DW. Nuclbophilic substitution reactions of the heterocyclic fluoro-olefin perfluoro-(3,6-dihydro-2-methyl-2-1,2-oxazine). *Tetrahedron* 1986;42:6495–6510.

53. Kazmina NB, Mysov EI, Aerov AF, Kagramanova EM. Thermal reactions of hexafluorobutadiene and hexafluorocyclobutene with elemental sulphur. *J Fluorine Chem* 1993;65:203–208.

54. Shtarev AB, Kremlev MM, Chvátal Z. 1-Aryl-F-1,3-butadienes and unsymmetrical α,ω-Diaryl-F-polyenes. *J Org Chem* 1997;62:3040–3045.

55. Suzuta T, Abe T, Sekiya A. Methyl hypofluorite (MeOF) reactions with various fluoro-olefins. *J Fluorine Chem* 2003;119:3–8.

CHAPTER 51

Preparation of Ionic Liquids of Fluorocomplex and Oxofluorocomplex Anions by Fluoroacid–Base Reactions

KAZUHIKO MATSUMOTO AND RIKA HAGIWARA

Ionic liquids are attractive materials for various applications such as electrolytes and reaction media because of their unique properties such as nonflammability, negligible vapor pressure, and wide temperature range of the liquid phase [1–4]. Many ionic liquids containing fluoroanions have been prepared because of their higher hydrolytic stability than that of chloro or bromo analogs. One of the problems in synthesis of ionic liquids is halide impurities [5]. Fluorohydrogenate ionic liquids, which often exhibit high conductivities, low viscosities, and low melting points, contain $(FH)_nF^-$ as anionic species and are usually prepared by reaction of chloride salts with excess HF [6, 7]. One application of fluorohydrogenate ionic liquids is their use as precursors of ionic liquids based on fluorocomplex anions [8]. In this reaction, Lewis acidic fluorides (MF_m) react with $(FH)_nF^-$ to give MF_{m+1}^- and the by-product HF. This method provides ionic liquids with a low content of halide impurity and water, because the halide impurities are removed in synthesis of the starting material, fluorohydrogenate ionic liquid, and water is not involved in this reaction. The following describes the details of this method, taking [EMIm][BF_4] and [EMIm][NbF_6] (EMIm, 1-ethyl-3-methylimidazolium) for example. The first procedure is applied to the cases of volatile Lewis acidic fluorides such as PF_5, AsF_5, and WF_6 [8,9], and the second procedure is applied to the cases of non- or low-volatile Lewis acidic fluorides or oxofluorides such as TaF_5, UF_5, VOF_3, $MoOF_4$, and WOF_4 [8–12]. The precursor ionic liquid, [EMIm][$(FH)_{2.3}F$], is prepared according to a literature method [6, 7]. Alkylpyridinium and alkylpyrrolidinium salts are also prepared in the same manner [9, 10,12].

Efficient Preparations of Fluorine Compounds, First Edition. Edited by Herbert W. Roesky.

51.1 PREPARATION OF 1-ETHYL-3-METHYLIMIDAZOLIUM TETRAFLUOROBORATE ([EMIm][BF_4])

Apparatus PFA (tetrafluoroethylene–perfluoroalkylvinylether copolymer) reactor (o.d. 1/2″, 20 cm in length) with a stainless steel valve, magnetic stirrer, fluoride-resistant vacuum line, safety glasses, laboratory coat, protective gloves.

Chemicals BF_3, [EMIm][$(FH)_{2.3}F$].

Attention! Safety glasses and protective gloves must be used at all times.

Caution! HF and BF_3 should be carefully treated because they are highly corrosive and toxic. Appropriate protective equipment should be used as well. The pressure of the reactor and vacuum line should be always monitored to avoid an unexpected burst by high pressure. The following Lewis acid–base reaction is highly exothermic and temperature controlling with a coolant may be necessary if the reaction is performed on a large scale.

Experimental Procedure A PFA reactor is charged in a dry box with [EMIm][$(FH)_{2.3}F$] (1.74 g, 9.88 mmol). The reactor is connected to the vacuum line and evacuated for 10 min. The valve of the reactor is closed and BF_3 gas is introduced to the reaction line (~2 atm). The reactor is cooled to −196 °C and BF_3 is condensed onto frozen [EMIm][$(FH)_{2.3}F$]. The reaction proceeds by warming up the mixture to room temperature. The valve of the reactor should be opened during this process to monitor the pressure inside. The liquid phase in the reactor is stirred for 30 min to promote the reaction. Introduction of BF_3 is repeated several times until the uptake of BF_3 gas ceases. All the volatiles are roughly pumped off at room temperature for 3 h through a soda lime chemical trap and a liquid nitrogen trap in series (10–100 Pa). The final pumping at 70 °C for 2 days through a liquid nitrogen trap (<1 Pa) resulted in the product [EMIm][BF_4] (1.95 g, 9.84 mmol).

Characterization Data Anal. calcd for $C_6H_{11}BF_4N_2$: C, 36.38; H, 5.56; N, 14.15. Found: C, 36.38; H, 5.42; N, 14.30. ^{1}H NMR (neat): δ 8.35 (s, 1H, H(2) ring), 7.30 (s, 1H, H(4) ring), 7.23 (s, 1H, H(5) ring), 3.97 (q, 2H, $J = 21.6$ Hz, H(6) methylene), 3.66 (s, 3H, H(8) methyl), 1.19 (t, 3H, $J = 14.7$ Hz, H(7) methyl). ^{19}F NMR (neat): δ 150.35 (s, BF_4^-). Raman (cm^{-1}): $\bar{v}$ 1078 (vw, BF_4^-), 767 (s, BF_4^-), 519 (w, BF_4^-), 384 (vw, BF_4^-). IR (AgCl, cm^{-1}): $\bar{v}$ 1070 (vs, br, BF_4^-), 522 (vs, BF_4^-).

51.2 PREPARATION OF 1-ETHYL-3-METHYLIMIDAZOLIUM HEXAFLUORONIOBATE ([EMIm)[NbF_6])

Apparatus T-shaped PFA (tetrafluoroethylene–perfluoroalkylvinylether copolymer)) reactor (o.d. 1/2″, 15 cm in length for each arm) with a stainless valve, magnetic stirrer, fluoride-resistant vacuum line, safety glasses, laboratory coat, protective gloves.

Chemicals NbF_5, [EMIm][$(FH)_{2.3}F$].

Attention! Safety glasses and protective gloves must be used at all times.

Caution! HF and NbF_5 should be carefully treated because they are highly corrosive and toxic. Appropriate protective equipment should be used as well. The following Lewis acid–base reaction is highly exothermic and temperature controlling with a coolant may be necessary if the reaction is performed on a large scale.

Experimental Procedure A T-shaped PFA reactor is charged in a dry box with [EMIm][$(FH)_{2.3}F$] (0.919 g, 5.22 mmol) at one end and NbF_5 (0.990 g, 5.27 mmol) at the other end. The reactor is connected to the vacuum line and evacuated for 10 min. Niobium pentafluoride at one end is slowly added to [EMIm][$(FH)_{2.3}F$] at the other end and is cooled to 0 °C. As the reaction proceeds, the pressure in the reactor increases because HF is produced as a by-product. The ionic liquid–HF mixture is stirred until the product becomes a homogeneous liquid (resulting in the disappearance of the NbF_5 powder). When the reaction ceases, the part that originally contained NbF_5 is carefully washed three times with the resulting liquid. All the volatiles are roughly pumped off at room temperature for 3 h through a soda lime chemical trap and a liquid nitrogen trap in series (10–100 Pa). The final pumping at 70 °C for 2 days through a liquid nitrogen trap (<1 Pa) results in the product [EMIm][NbF_6] (1.622 g, 5.10 mmol).

Characterization Data Anal. calcd for $C_6H_{11}F_6N_2Nb$: C, 22.65; H, 3.46; N, 8.81. Found: C, 22.95; H 3.27; N, 8.80%. ^1H NMR (neat): δ 8.99 (br s, 1H, H(2) ring), 7.70 (br s, 1H, H(4) ring), 7.64 (br s, 1H, H(5) ring), 4.40 (br s, 2H, H(6)methylene), 4.09 (br s, 3H, H(8) methyl), 1.64 (br s, 3H, H(7) methyl). ^{19}F NMR (neat): δ 127.25 (br s, NbF_6^-). Raman (cm^{-1}): $\bar{\upsilon}$ 681 (vs, NbF_6^-), 558 (w, NbF_6^-), 379 (vw, NbF_6^-). IR (AgCl, cm^{-1}): $\bar{\upsilon}$ 606 (vs, br).

TABLE 51.1 Selected Physical Properties of EMIm Salts of Fluoro- and Oxofluorocomplex Anions [8, 10–12]

Catalyst	$T_m(T_g)$ (°C)	ρ ($g\ cm^{-3}$)	η (cP)	σ ($mS\ cm^{-1}$)	Λ ($S\ cm^2\ mol^{-1}$)
[EMIm][BF_4]	15(−79)	1.28	32	13.6	2.1
[EMIm][PF_6]	60	1.56	–	–	–
[EMIm][AsF_6]	53	1.78	–	–	–
[EMIm][SbF_6]	10	1.85	67	6.2	1.16
[EMIm][NbF_6]	−1	1.67	49	8.5	1.62
[EMIm][TaF_6]	2	2.17	51	7.1	1.33
[EMIm][WF_7]	−15	2.27	171	3.2	0.60
[EMIm][UF_6]	11	2.38	59	7.9	1.54
[EMIm][VOF_4]	75(−30)	–	–	–	–
[EMIm][$MoOF_5$]	(−83)	1.76	86	5.1	0.92
[EMIm][WOF_5]	−20(−91)	2.25	105	3.0	0.54

T_m, melting point; ρ, density at 25 °C; η, viscosity at 25 °C; σ, ionic conductivity at 25 °C; Λ, molar conductivity at 25 °C.

Application Ionic liquids are widely studied as potential electrolytes [4]. Some important physical properties of ionic liquids composed of $EMIm^+$ and fluoro- or oxofluorocomplex anions are compared in Table 51.1. Melting point of $[EMIm][MF_6]$ decreases with increase in size of the anion except for $[EMIm][UF_6]$. The molar conductivity (Λ) and viscosity (η) of these ionic liquids at 298 K roughly fulfill the Walden's rule ($\Lambda \cdot \eta$ = constant) as in the case of many other ionic liquids [4].

REFERENCES

1. Welton T. Room-temperature ionic liquids. Solvents for synthesis and catalysis. *Chem Rev* 1999;99:2071–2083.
2. Wasserscheid P, Kein W. Ionic liquids—new "solutions" for transition metal catalysis. *Angew Chem, Int Ed* 2000;39:3772–3789.
3. Hagiwara R, Ito Y. Room temperature ionic liquids of alkylimidazolium cations and fluoroanions. *J Fluorine Chem* 2000;105:221–227.
4. Ohno H, editor. *Electrochemical Aspects of Ionic Liquids*. Hoboken, NJ: John Wiley & Sons; 2005.
5. Seddon KR, Stark A, Torres MJ. Influence of chloride, water, and organic solvents on the physical properties of ionic liquids. *Pure Appl Chem* 2000;72:2275–2287.
6. Hagiwara R, Matsumoto K, Nakamori Y, Tsuda T, Ito Y, Matsumoto H, Momota K. Physicochemical properties of 1,3-dialkylimidazolium fluorohydrogenate room-temperature molten salts. *J Electrochem Soc* 2003;150:D195–D199.
7. Hagiwara R, Nakamori Y, Matsumoto K, Ito Y. The effect of anion fraction on the physicochemical properties of $EMIm(HF)_nF$ ($n = 1.0{\sim}2.6$). *J Phys Chem B* 2005;109:5445–5449.
8. Matsumoto K, Hagiwara R, Yoshida R, Ito Y, Mazej Z, Benkič P, Žemva B, Tamada O, Yoshino H, Matsubara S. Syntheses, structures and properties of 1-ethyl-3-methylimidazolium salts of fluorocomplex anions. *Dalton Trans* 2004;144–149.
9. Kanatani T, Ueno R, Matsumoto K, Nohira T, Hagiwara R. Thermal properties of *N*-alkyl-*N*-methylpyrrolidinium and *N*-butylpyridinium fluorometallates and physicochemical properties of their melts. *J Fluorine Chem* 2009;130:979–984.
10. Kanatani T, Matsumoto K, Hagiwara R. Syntheses and physicochemical properties of new ionic liquids based on hexafluorouranate (V) anion. *Chem Lett* 2009;7:714–715.
11. Matsumoto K, Hagiwara R. A new room temperature ionic liquid of oxyfluorometallate anion:1-ethyl-3-methylimidazolium oxypentafluorotungstate ($EMImWOF_5$). *J Fluorine Chem* 2005;126:1095–1100.
12. Kanatani T, Matsumoto K, Hagiwara R. Syntheses and physicochemical properties of low melting salts based on VOF_4^- and $MoOF_5^-$, and the molecular geometries of the dimeric $(VOF_4^-)_2$ and $Mo_2O_4F_6^{2-}$ anions. *Eur J Inorg Chem* 2010;1049–1055.

CHAPTER 52

Synthesis of Difluorocyclopropyl Building Blocks: 2,2-Difluorocyclopropylmethanol and 2-(Bromomethyl)-1,1-difluorocyclopropane

WEI XU AND WILLIAM R. DOLBIER JR.

52.1 INTRODUCTION

It is now well recognized that the introduction of fluorine into potential pharmaceutical or agrochemical candidate molecules often imparts significant beneficial properties to these compounds. As reported recently by Hagmann and O'Hagan [1,2], over the past 60 years about 15–20% of new chemical entities that have been licensed for the clinical market contain fluorine, including about 30% of the top-selling pharmaceutical products.

In the past decade, there has been a noticeable interest in potential pharmaceuticals and agrochemicals containing the 2,2-difluorocyclopropyl group. No doubt this interest was generated as a result of the previous recognition that 2,2-dimethylcyclopropyl groups can impart useful activity (pyrethroid insecticides) [3,4]. This fact combined with recent advances in the synthesis of 2,2-difluorocyclopropyl building blocks [5–7] inevitably led to incorporation of 2,2-difluorocyclopropyl entities in a number of particularly agrochemical, but also pharmaceutical candidate compounds. A recent SciFinder search for the two title building block compounds (Figure 52.1) indicated 12 patents listed for use of the former compound **1** and 9 patents for use of the latter compound **2** since 2006.

Although a preparation of alcohol **1** has been reported [8], improvements have been made on its synthesis, and the preparation and characterization of bromide **2** has

Efficient Preparations of Fluorine Compounds, First Edition. Edited by Herbert W. Roesky.

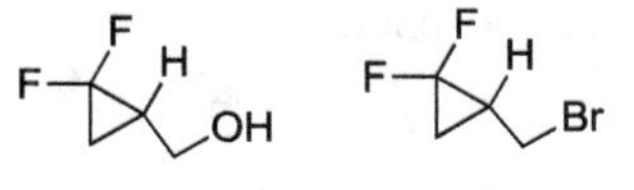

1 2

FIGURE 52.1 Title building block compounds: 2,2-difluorocyclopropylmethanol (**1**) and 2-(bromomethyl)-1,1-difluorocyclopropane (**2**).

not been previously reported. Therefore, these syntheses, as well as the currently preferred preparation of difluorocarbene precursor, TFDA (trimethylsilyl 2,2-difluoro-2-(fluorosulfonyl)acetate) [6], are presented here for the benefit of organic chemists in need of these compounds (Scheme 52.1).

52.2 EXPERIMENTAL PROCEDURES

Preparation of TFDA (Trimethylsilyl 2,2-difluoro-2-(fluorosulfonyl)acetate) [6, 7]

$$FSO_2CF_2COOH + Me_3SiCl \xrightarrow{\text{reflux}} FSO_2CF_2COOSiMe_3 + HCl \uparrow$$

A dry, 1-L, three-necked, round-bottomed flask is equipped with water condenser and dropping funnel. A two-hose adapter is fitted at the top of the condenser, with one hose being an inlet for nitrogen and the other an outlet for venting gaseous HCl that is formed during the reaction. The vented HCl is neutralized by passing through a trap containing aqueous sodium hydroxide solution. The flask is then charged with 2,2-difluoro-2-(fluorosulfonyl)acetic acid (178 g, 1 mol), and chlorotrimethylsilane (216 g, 3 mol) is added drop by drop at room temperature. After addition, the mixture is heated to reflux for 4 h. At this point, around 90% of acid has been consumed. Most of the excess of chlorotrimethylsilane is then distilled out at atmospheric pressure and retained for use in a future run. To the residue, fresh chlorotrimethylsilane (108 g, 1 mol) is added and the mixture is again heated to reflux. After 4 additional hours, the purity of TFDA product reaches approximately 98%. After removal of most

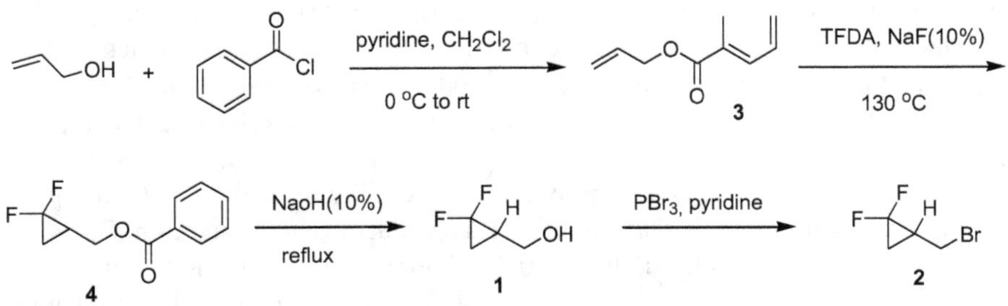

SCHEME 52.1 Reactions leading to preparation of 2,2-difluorocyclopropyl building blocks, **1** and **2**.

of the chlorotrimethylsilane at atmospheric pressure, distillation of the TFDA is performed under reduced pressure, and 200 g TFDA (80% yield) is collected, bp 62 °C/ 30 mm Hg.

Preparation of Allyl Benzoate (3)

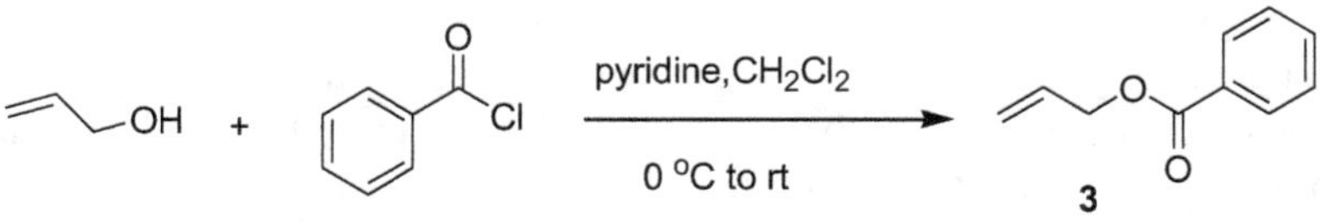

A 2-L, one-necked, round-bottomed flask is charged with allyl alcohol (116 g, 2 mol), pyridine (162 g, 2.1 mol), and methylene chloride (1 L), the mixture is cooled to 0 °C with an ice bath, and benzoyl chloride (280 g, 2 mol) is added drop by drop. After addition, the mixture is warmed to room temperature slowly and stirred overnight. The precipitated salt is then filtered off and washed with methylene chloride (200 mL). The combined methylene chloride extract is then washed with three 200 mL portions of 10% hydrochloric acid, followed by three 500 mL portions of water, and the solution then dried over sodium sulfate. The product is distilled under reduced pressure after removal of solvent to obtain 275 g (85%) of allyl benzoate: bp 118–119 °C/10 mmHg. ^{1}H NMR ($CDCl_3$): δ 4.82 (d, $J = 5.6$ Hz, 2H), 5.28 (dd, $J = 10.3$, 1.0 Hz, 1H), 5.41 (dd, $J = 17.2$, 1.3 Hz, 1H), 6.05 (ddt, $J = 17.2$, 10.5, 5.6 Hz, 1H), 7.40–7.46 (m, 2H), 7.51–7.58(m, 1H), 8.08 (d, $J = 7.6$ Hz, 2H) ppm [9].

Preparation of (2,2-difluorocyclopropyl)methyl benzoate (4) [7]

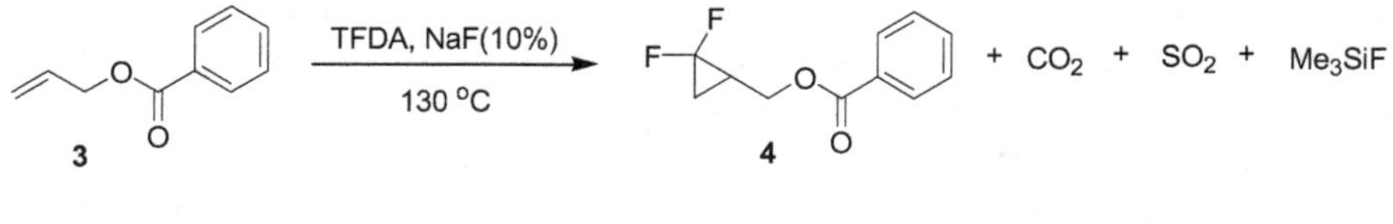

A dry, 500 mL, three-necked, round-bottomed flask is equipped with water condenser and dropping funnel. A two-hose adapter is fitted at the top of the condenser, with one hose being an inlet for nitrogen and the other an outlet for venting gases (CO_2, SO_2, and Me_3SiF) that are formed during the reaction. These vented gases are neutralized by passing through a trap containing aqueous sodium hydroxide solution. The flask is then charged with allyl benzoate (162 g, 1 mol) and sodium fluoride (4.2 g, 0.1 mol), and the mixture heated to 130 °C (bath temperature). No solvent is added. TFDA (400 g, 1.6 mol) is then added drop by drop to the mixture, with the addition proceeding over a period of 2 h. Gas evolution is apparent during the addition. Once the addition of TFDA is completed, the reaction mixture is stirred for additional 1 h. At this time, the conversion of allyl benzoate is confirmed by proton NMR and that of TFDA by fluorine NMR. The product is used in the next step without further purification being required. ^{1}H NMR ($CDCl_3$): δ 1.29 (m, 1H), 1.56 (m, 1H), 2.08 (m, 1H),

4.45 (m, 1H), 4.29 (m, 1H), 7.44 (m, 2H), 7.56 (m, 1H), 8.04 (d, J = 7.6 Hz, 2H) ppm. ^{19}F NMR ($CDCl_3$): δ −129.6 (dt, J = 160 Hz and 13 Hz, 1F), −143.8 (dd, J = 160 Hz and 13 Hz, 1F) ppm.

Preparation of 2,2-difluorocyclopropylmethanol (1) [8]

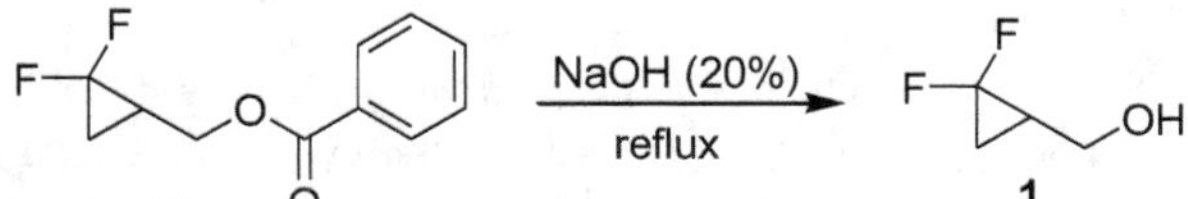

4 1

A 2-L, round-bottomed flask is charged with 20% aqueous sodium hydroxide solution (the amount of sodium hydroxide is 120 g, 3 mol), and ester **4** is added in one portion. The mixture is heated to reflux and stirred for 8 h to form a dark, clear solution, after which the solution is cooled to 0 °C and acidified to a pH value of 4 using concentrated hydrochloric acid. The precipitated benzoic acid is filtered off and washed with water. The combined aqueous phase is extracted three times with 500 mL portions of diethyl ether, and the ether phase dried over sodium sulfate. After removing the ether, distillation provided alcohol product **1** (70 g, 65% yield for two steps) as a colorless liquid: bp 131 °C. ^{1}H NMR ($CDCl_3$): δ 1.14 (m, 1H), 1.45 (m, 1H), 1.87 (m, 1H), 2.45 (bs, 1H), 3.62 (m, 1H), 3.75 (m, 1H) ppm. ^{19}F NMR ($CDCl_3$): δ −129.10 (dt, J = 160 Hz and 12.4 Hz, 1F), −144.76 (dd, J = 161 Hz and 12.4 Hz, 1F) ppm.

Preparation of 2-(bromomethyl)-1,1-difluorocyclopropane (2)

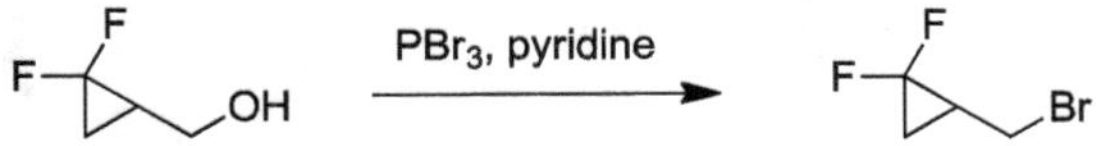

1 2

A 250 mL, round-bottomed flask is charged with phosphorus tribromide (73 g, 0.26 mol), and after cooling to −10 °C with a ice–acetone bath, a mixture of 2,2-difluorocyclopropylmethanol (**1**) (70 g, 0.65 mol) and dry pyridine (14.3 g, 0.182 mol) is added drop by drop with stirring. After addition, the mixture is warmed to room temperature slowly and stirred overnight. The reaction mixture is then distilled at atmospheric pressure, and the fraction above 100 °C collected. This crude product is washed with 2 N aqueous sodium hydroxide, and then with water. After drying the crude product over calcium chloride, a final distillation gives 110 g (80%) of 2-(bromomethyl)-1,1-difluorocyclopropane as a colorless liquid: bp 108–109 °C. ^{1}H NMR ($CDCl_3$): δ 1.17 (m, 1H), 1.62 (m, 1H), 2.07 (m, 1H), 3.36 (m, 1H), 3.48 (m, 1H) ppm. ^{19}F NMR ($CDCl_3$): δ −127.50 (dt, J = 160 Hz and 12.4 Hz, 1F), −144.74 (dm, J = 160 Hz, 1F) ppm.

REFERENCES

1. Hagmann WK. The many roles for fluorine in medicinal chemistry. *J Med Chem* 2008;51:4359–4369.
2. O'Hagan D. Fluorine in health care organofluorine containing blockbuster drugs. *J Fluorine Chem* 2010;131:1071–1081.
3. Katsuda Y. Development of and future prospects for pyrethroid chemistry. *Pestic Sci* 1999;55:775–782.
4. Khambay BPS. Pyrethroid insecticides. *Pestic Outlook* 2002;13:49–54.
5. Tian F, Kruger V, Bautista O, Duan J-X, Li A-R, Dolbier WR, Jr, Chen Q-Y. A novel and highly efficient synthesis of gem-difluorocyclopropanes. *Org Lett* 2000;2:563–564.
6. Dolbier WR, Jr, Duan J-X, Tian F, Chen Q-Y. *n*-Butyl 2,2-difluorocyclopropanecarboxylate. Preparation and use of a new difluorocarbene reagent. *Org Syn* 2003;80:172–176.
7. Dolbier WR, Jr., Tian F, Duan J-X, Li AR, Ait-Mohand S, Bautista O, Buathong S, Baker JM, Crawford J, Anselme P, Cai X-H, Modzelewska A, Koroniak H, Battiste MA, Chen Q-Y. Trimethylsilyl fluorosulfonyldifluoroacetate (TFDA). A new, highly efficient difluorocarbene reagent. *J Fluorine Chem* 2004;125:459–469.
8. Battiste MA, Tian F, Baker JM, Bautista O, Villalobos J, Dolbier WR, Jr. Reactivity and regiochemical behavior in the solvolysis reactions of 2,2-difluorocyclopropylcarbinyl tosylates. *J Fluorine Chem* 2003;119:39–51.
9. Guo Q-X, Miyaji T, Hara R, Shen B, Takahashi T. Group 5 and group 6 metal halides as very efficient catalysts for acylative cleavage of ethers. *Tetrahedron* 2002;58:7327–7334.

CHAPTER 53

Preparation of 1,1,2,2,9,9,10,10-Octafluoro[2.2]paracyclophane and Perfluoro[2.2]paracyclophane

WILLIAM R. DOLBIER JR., JIAN-XI DUAN, ALEX J. ROCHE, AND LIANHAO ZHANG

The unique structural features, chemical reactivities, and synthetic challenges associated with [2.2]paracyclophanes (PCPs) have fascinated organic chemists since Brown and Farthing first prepared the parent molecule in 1949 [1], and Cram and Steinberg embarked on his classic studies of this molecule [2, 3]. Current interest in this very interesting class of compounds derives largely from their unique three-dimensional structure, which is characterized by layered, face-to-face, bent benzene rings. In addition to the fundamental interest in these strained compounds, their spectroscopy, and their chemistry, there is also much interest in the many potential applications that derive from their unique structures, including their use as chemical vapor deposition (CVD) precursors of parylene polymers, and their unique chiral properties, which make them useful as templates in stereoselective synthesis. A recent book, edited by Gleiter and Hopf, provides a reasonably up-to-date summary of the structural aspects [4], chemical and physical properties, and current applications of PCPs, as does an earlier book by Vögtle [5].

The two fluorinated PCPs, 1,1,2,2,9,9,10,10-octafluoro[2.2]paracyclophane (the bridged fluorinated PCP commonly known as AF4) [6] and perfluoro[2.2]paracyclophane (commonly known as F8) [7], exemplify the synthetic challenges, the unique chemical behavior, and potential for future application as well as any other PCPs. Both are readily prepared, but their syntheses are unique [7, 8], different from one another in important ways, and not applicable to preparation of other, less fluorinated PCPs.

In terms of reactivity, AF4 and F8 are very different, each having distinct advantages over their hydrocarbon counterparts. As can be discerned from examination of

Efficient Preparations of Fluorine Compounds, First Edition. Edited by Herbert W. Roesky.

Cram's early papers on the chemistry of the parent hydrocarbon [3, 9], there are many problems associated with the use of PCP as a substrate for electrophilic aromatic substitution, and the yields of simple reactions such as bromination and nitration are relatively low. These low yields most likely derive from mechanistic aspects of the reaction that lead to destruction of the PCP.

In contrast, AF4, in spite of having highly electron-deficient benzene rings, and in spite of its requiring harsher conditions for its electrophilic aromatic substitution reactions, provides very good yields for monobromination, mononitration, and dinitration [10, 11]. Moreover, reduction of the nitro group and subsequent diazonium chemistry proceed quite efficiently, thus allowing the preparation of a diverse number of mono- and disubstituted AF4 molecules.

Perfluoro[2.2]paracyclophane (F8), bearing no aromatic hydrogen, of course cannot undergo electrophilic substitution, but it is highly susceptible to nucleophilic substitution by the S_NAr mechanism, allowing the preparation of a wide variety of mono- and multisubstituted F8 derivatives [12].

These two fluorinated PCPs might well become of interest as synthetic building blocks if they were available commercially. Because they are not available from catalogs of the usual specialty chemical companies, if they are to be used, they must be prepared in house. Although preparations of both of these molecules appear in the chemical journal literature [7, 8], we provide them here in their most up-to-date versions for the convenience and benefit of the organic/fluoroorganic synthetic community.

53.1 PREPARATION OF AF4

The precursor to AF4, 1,4-bis-(chlorodifluoromethyl)benzene (**1**), had first been prepared by Chow in 1961 by the procedure given in Scheme 53.1 [13, 14]. This method was not ideal for large-scale, inexpensive synthesis of precursor **1**. For that reason, a one-step synthesis from 1,4-bis-(trichloromethyl)benzene via reaction with anhydrous HF was devised (Scheme 53.2) [15].

AF4 is then prepared by heating precursor **1** with powdered Zn in *N,N*-dimethylacetamide (DMA) at 100 °C (Scheme 53.3) [8]. All previous methods of preparing AF4 had required high dilution methods, and none of them proved useful for large-scale preparations. However, the Zn method shown in Scheme 53.3 was highly scalable, and it can readily be used for preparing kilogram quantities of AF4.

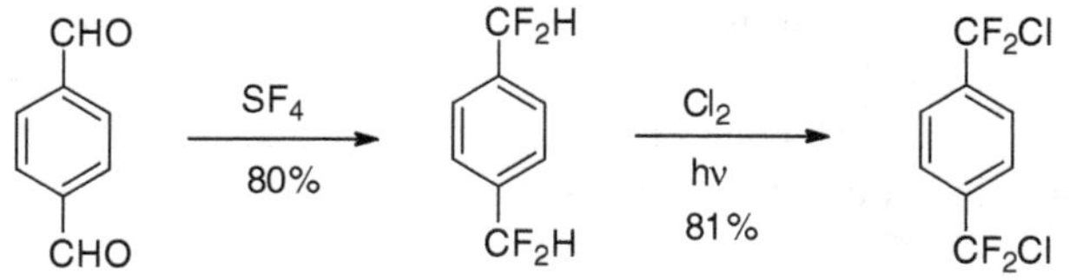

1

SCHEME 53.1 Lab-scale method for making AF4 precursor **1**.

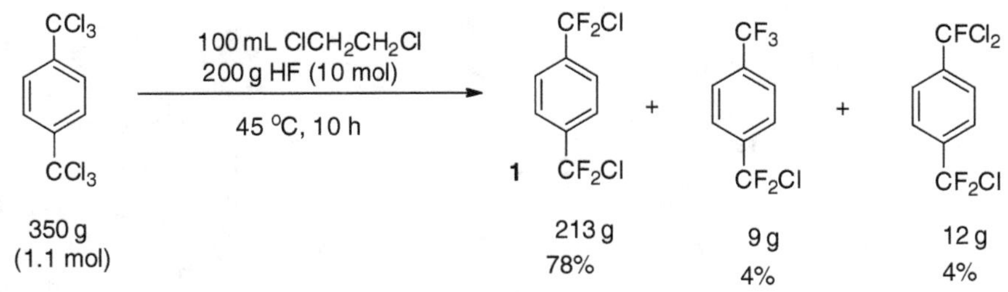

SCHEME 53.2 Large-scale, commercial method for making AF4 precursor **1**.

Experimental Procedure

Into a 1-L, three-necked, round-bottomed flask, 600 mL DMA, *p*-bis-(chlorodifluoromethyl)benzene (**1**) (60 g, 0.24 mol), and zinc dust (61 g, 0.96 mol) are added. The mixture is heated over a period of 40 min to 100 °C under vigorous stirring and an N_2 atmosphere, and it was maintained at that temperature for 4 h, after which no precursor **1** remained and the solution had become yellow. The reaction mixture was then cooled to room temperature and filtered, washing the resulting oligomeric solid residue with two 50 mL portions of DMA to extract small amounts of AF4. Combining these washings with the original DMA filtrate, impurities such as the bridge-unsaturated $C_8H_8F_6$ compound (~6% yield) and heptafluoro[2.2]paracyclophane (<1%) were removed by stirring the combined DMA solution at room temperature with 5 g $KMnO_4$ overnight, followed by again filtering the mixture to obtain a clear, colorless DMA solution. The crude AF4 product was then obtained by evaporating the DMA filtrate at 100 °C under full vacuum. After washing the resulting residue with water and cold methanol, the crude product was extracted with benzene; the benzene was evaporated, and the residue was recrystallized from hexane to give >99% pure AF4 (25.6 g, 60.5%).

53.2 PREPARATION OF PERFLUORO[2.2]PARACYCLOPHANE (F8)

Although a method of preparation of F8 precursor, 1,4-bis-(chlorodifluoromethyl)-2,3,5,6-tetrafluorobenzene (**2**), was given in the original F8 paper, an improved, more convenient procedure that is outlined in Scheme 53.4 has proved superior

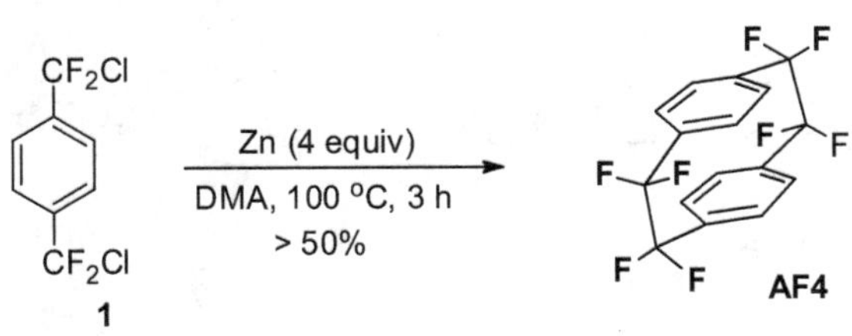

SCHEME 53.3 Method for preparation of AF4.

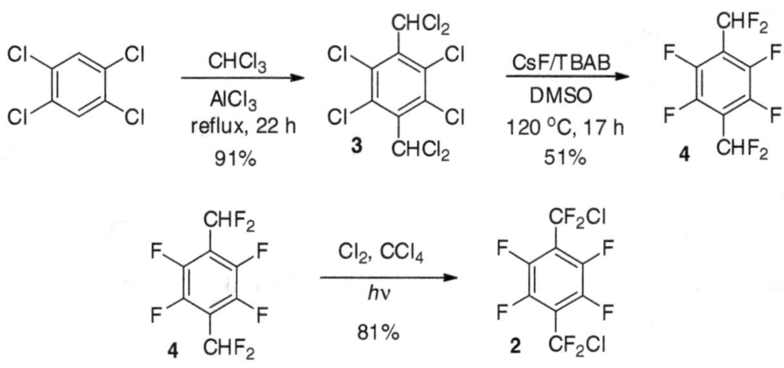

SCHEME 53.4 New preparation of F8 precursor **2**.

for making 100 g quantities of high-purity precursor **2**. This method was based on the reported preparation of 1,4-bis-(dichloromethyl)-2,3,5,6-tetrachlorobenzene (**3**) by Castaner and Riera by an $AlCl_3$-catalyzed condensation of chloroform with the inexpensive 1,2,4,5-tetrachlorobenzene [16]. This compound underwent chlorine–fluorine exchange on heating with CsF in DMSO to form 1,4-bis-(difluoromethyl)-2,3,5,6-tetrafluorobenzene, which could then be chlorinated to form precursor **2**.

When the Zn/DMA conditions that led to formation of AF4 were applied to precursor **2**, no desired F8 product was formed. Instead it appeared that the reduction conditions were too severe for this system, and over-reduction occurred to form a bis-Zn intermediate. However, going to a less polar solvent, acetonitrile, inhibited the over-reduction, and F8 could be formed and isolated. Optimization led to the given procedure that allowed the preparation of F8 in 39% yield (Scheme 53.5).

Experimental Procedure

1,4-Bis(dichloromethyl)-2,3,5,6-tetrachlorobenzene (3) A mixture of 2,3,5,6-tetrachlorobenzene (22.6 g, 0.1 mol) and aluminum chloride (30 g, 0.225 mol) in anhydrous chloroform (300 mL) was refluxed for 22 h in a 1-L, round-bottomed flask. The reaction mixture was cooled to room temperature, diluted with trichloromethane (200 mL), and poured into a beaker containing a mixture

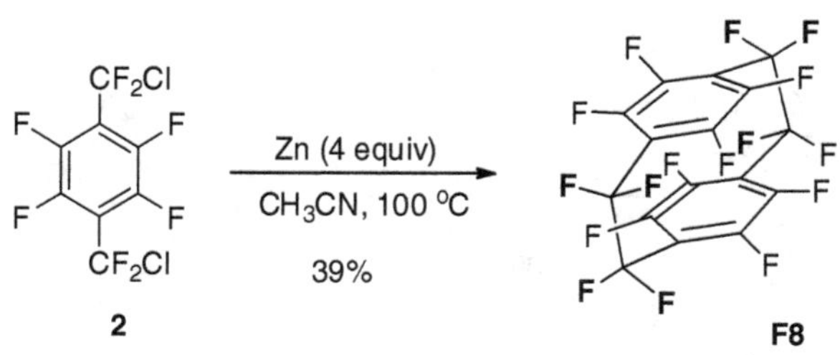

SCHEME 53.5 Method for preparation of perfluoro[2.2]paracyclophane (F8).

of hydrochloric acid (30 mL) and ice water (300 mL). The organic layer was separated, dried over magnesium sulfate, and concentrated to give crude product (45 g, wet), which was recrystallized from hexanes (225 mL) to obtain compound **3** (31.6 g) as a yellow solid. The mother liquor was concentrated to a volume of 45 mL, after which a second crop of product (4.1 g) was obtained. The total yield was 91.1%: mp 134–136 °C (lit. 127–129 °C) [16]. ^{1}H NMR ($CDCl_3$): δ 7.59 (br s), 7.63 (br s) ppm (equal intensity, due to atropisomers deriving from restricted rotation). ^{13}C NMR: δ 66.52, 137.29 ppm (C1/C4 carbons not observed).

1,4-Bis(difluoromethyl)-2,3,5,6-tetrafluorobenzene (4) A mixture of compound **3** (23 g, 60 mmol), cesium fluoride (91.2 g, 600 mmol) (powdered, high purity), and tetrabutylammonium bromide (1.2 g, 3.7 mmol) in anhydrous DMSO (110 mL) was heated to 120 °C with vigorous stirring in a 500 mL, round-bottomed flask for 17 h. The reaction mixture was then cooled down to room temperature and poured into ice water (330 mL), extracted with diethyl ether (2 × 250 mL). The combined organic layers were washed with water (500 mL), dried with magnesium sulfate, filtered, and concentrated to remove solvent. The residue was distilled to give crude product (8.4 g), which was recrystallized from hexanes (15 mL) to furnish (**4**) (7.6 g, yield 50.6%, mp 70–72 °C) as white crystals. ^{1}H NMR ($CDCl_3$): δ 6.97 (t, $J = 53$ Hz) ppm. ^{19}F NMR ($CDCl_3$): δ −115.2 (d, $J = 52$ Hz), −142.2 (s) ppm. ^{13}C NMR ($CDCl_3$): δ 108.22 (t, $J = 242$ Hz), 115.97 (br s), 144.79 (d, $J = 262$ Hz) ppm. Anal. calcd for $C_8H_2F_8$: C, 38.42; H, 0.81. Found: C, 38.07; H, 0.68.

1,4-Bis-(chlorodifluoromethyl)-2,3,5,6-tetrafluorobenzene (2) Molecular Cl_2 was introduced to a solution of 1,4-bis-(difluoromethyl)-2,3,4,5-tetrafluoroben zene (**4**) (18.7 g, 74.8 mmol) in CCl_4 (250 mL) while irradiating with a sunlamp for 20 h. The reaction mixture was slowly evaporated to remove CCl_4, and the residue distilled under reduced pressure (85–87 °C/20 mm Hg) to give 1,4-bis-(chlorodifluoromethyl)-2,3,4,5-tetrafluorobenzene (**3**) (19.4 g, yield 81.3%) as a colorless oil. ^{13}C NMR ($CDCl_3$): δ 121.10 (t, $^1J_{FC} = 295$ Hz), 143.61 (d, $^1J_{FC} = 268$ Hz) ppm (non-fluorine-substituted C-1/C-4 not observed). ^{19}F NMR ($CDCl_3$): δ −47.6 (m, 4F), −137.9 (m, 4F) ppm. HRMS: Calcd for $C_8F_8Cl_2$, 317.9249. Found, 317.9239.

Perfluoro[2.2]paracyclophane (F8) A mixture of 1,4-bis(chlorodifluoro methyl)-2,3,5,6-tetrafluorobenzene (**2**) (10 g, 31.3 mmol) and zinc (8.2 g, 125.2 mmol) (99.7%, activated by 2% HCl) in anhydrous acetonitrile (100 mL) was heated to 100 °C (oil bath temperature) under N_2 atmosphere. The reaction mixture was refluxed gently for 38 h and then cooled to room temperature, filtered, and washed with acetone (3 × 30 mL). The combined filtrates were concentrated to dryness. The residue was purified by column chromatography (silica gel, hexanes) to give crude product (3.4 g) as a white powder. This crude product was recrystallized from chloroform (40 mL) to furnish 2.7 g of pure product as white needles. The mother liquor was concentrated to dryness and recrystallized from chloroform (10 mL) to obtain a second crop of pure product (0.3 g) as white needles. The yield is 38.6% based on isolated pure F8: mp 195–196 °C. ^{19}F NMR ($CDCl_3$): δ −102.8 (s), −132.4 (s) equal intensities ppm. ^{13}C NMR ($CDCl_3$): δ 118.0 (tt, $J = 283.29$ Hz), 147.4 (dd, $J = 267.22$ Hz) ppm, bridgehead carbon not seen. HRMS: Calcd for $C_{16}F_{16}$, 495.9739; Found, 495.9719. Anal. calcd for $C_{16}F_{16}$: C, 38.73; H, 0.00; N, 0.00. Found: C, 39.07; H, 0.00; N, 0.04.

REFERENCES

1. Brown CJ, Farthing AC. Preparation and structure of di-p-xylylene. *Nature* 1949;164:915–915.
2. Cram DJ, Steinberg HJ. Macro rings .1. Preparation and spectra of the paracyclophanes. *J Am Chem Soc* 1951;73:5691–5740.
3. Cram DJ, Cram LM. Cyclophane chemistry - bent and battered benzene rings. *Acc Chem Res* 1971;4:204–213.
4. Gleiter R, Hopf H, editors. *Modern Cyclophane Chemistry*. Weinheim: Wiley-VCH; 2004.
5. Vögtle F. *Cyclophane Chemistry*. Chichester: John Wiley & Sons; 1993.
6. Dolbier WR, Jr, Beach WF. Parylene-AF4: A polymer with exceptional dielectric and thermal properties. *J Fluorine Chem* 2003;122:97–104.
7. Dolbier WR. Jr, Xie P, Zhang L, Xu W, Chang Y, Abboud KA. Synthesis of perfluoro[2.2]paracyclophane. *J Org Chem* 2008;73:2469–2472.
8. Dolbier WR, Jr, Duan J-X, Roche AJ. A novel, non-high dilution method for preparation of 1,1,2,2,9,9,10,10-Octafluoro[2.2]paracyclophane. *Org Lett* 2000;2:1867–1869.
9. Cram DJ. [2.2]Paracyclophanes. In: Keehn PM, Rosenfeld SM, editors. *Cyclophanes*. New York: Academic Press; 1983. Vol. 1, p 1.
10. Roche AJ, Dolbier WR, Jr. Electrophilic substitution of 1,1,2,2,9,9,10,10-octafluoro [2.2] paracyclophane. *J Org Chem* 1999;64:9137–9143.
11. Roche AJ, Dolbier WR, Jr. Multiple electrophilic substitution of 1,1,2,2,9,9,10,10-octafluoro[2.2]paracyclophane. *J Org Chem* 2000;65:5282–5290.
12. Zhang L, Ogawa K, Ghiviriga I, Dolbier WR, Jr. Reactions of nucleophiles with perfluoro[2.2]paracyclophane. *J Org Chem* 2009;74:6831–6836.
13. Fuqua SA, Parkhurst RM, Silverstein RM. Synthesis and chemistry of several fluorinated p-xylenes designed as precursors for a,a,a,a,-tetrafluoro-p-xylene. *Tetrahedron* 1964;20:1625–1632.
14. Chow SW, Pilato LA, Wheelwright WL. Synthesis of 1,1,2,2,9,9,10,10-octafluoro [2.2]paracyclophane. *J Org Chem* 1970;35:20–22.
15. Dolbier WR, Jr, Duan J-X, Rong XX. Efficient synthesis of p-bis-(chlorodifluoromethyl)benzene. *J Fluorine Chem* 2007;128:1091–1093.
16. Castaner J, Riera J. Highly crowded perchloropolyphenyl-para-xylylenes with exceptional thermal-stability. *J Org Chem* 1991;56:5445–5448.

CHAPTER 54

Synthesis and Application of gem-Difluoromethylenated and Trifluoromethylated Building Blocks

FENG-LING QING AND XIAO-LONG QIU

As the most electronegative element, fluorine is a small-sized atom, its size is between hydrogen and oxygen but closer to oxygen [1,2]. When bound to carbon, it forms the strongest bond in organic chemistry and the resultant C–F bond is polarized with significant dipolar character [3]. Although fluorine is extremely rare in natural biological compounds, organic chemists, medicinal chemists, and pharmaceutical chemists synthesized and widely used enormous numbers of synthetic fluorine-containing compounds because the introduction of fluorine atom(s) into many biologically active molecules can bring about remarkable and profound changes in their physical, chemical, and biological properties. In pharmaceutical research, fluorine has been considered as a suitable bioisostere for hydrogen on steric grounds; meanwhile, the electronegative properties of fluorine make it an "isoelectronic" replacement for the hydroxyl group. Quite often, fluorine is introduced to improve metabolic stability [4,5] and modulate physicochemical properties, such as lipophilicity or basicity [6,7]. Furthermore, the strong gauche and antiperiplanar effects of fluorine atom due to its high electronegativity have profound stereoelectronic effect on neighboring groups, thereby the fluorine substituent can lead to a change in the preferred molecular conformation [8–13], which could affect protein binding affinity and selectivity at the molecular level [14,15]. Therefore, fluorinated compounds are nowadays synthesized in pharmaceutical research on a routine basis [16]. As an effort to introduce the fluorine atom into medicinal molecules, an increasing number of fluorinated agents and methods have been developed because most terrestrial F is bound in insoluble form and no haloperoxidase exists to incorporate F owing to the high oxidation potential of F, hindering uptake by bioorganisms.

Efficient Preparations of Fluorine Compounds, First Edition. Edited by Herbert W. Roesky.

In an attempt to explore the fundamental aspects of the fluorine atom and C–F bond in chemical reactions and biological activities, our group has focused on the development of efficient methodologies for introduction of fluorine-containing groups into organic molecules and their applications in synthesis of fluorinated analogs of natural bioactive products. Herein reported is the synthesis and application of some trifluoromethylated and *gem*-difluoromethylenated building blocks conducted in our group.

54.1 PREPARATIONS OF *GEM*-DIFLUOROMETHYLENATED BUILDING BLOCKS

Preparation of (2*R*)-1,2-di-*O*-isopropylidene-4,4-difluoro-5-hexen-1,2,3-triol (Scheme 54.1) [17]

Apparatus A 2000-mL Schlenk flask with septum, magnetic stirrer, silica gel column, safety glasses, laboratory coat, protective gloves.

Chemicals 1-(*R*)-glyceraldehyde acetonide, 3-bromo-3,3-difluoropropene, powdered indium, DMF, 1 N HCl, ether, anhydrous Na_2SO_4.

Attention! Safety glasses and protective gloves must be used at all times.

Caution! Because of its toxicity and/or volatility, care should be taken to avoid inhalation of ether, 3-bromo-3,3-difluoropropene, and 1 N HCl and avoid contact of 1-(*R*)-glyceraldehyde acetonide and powdered indium. All reactions should be carried out in a well-ventilated hood.

Experimental Procedure To a 2-L Schlenk flask, freshly distilled 1-(*R*)-glyceraldehyde acetonide (15.0 g, 115.4 mmol), 3-bromo-3,3-difluoropropene (27.2 g, 173.1 mmol), powdered indium (19.9 g, 173.1 mmol), and DMF (830 mL) were added. The resulting suspension was stirred for 22 h at room temperature. After that, the reaction mixture was quenched with 1 N HCl and extracted with ether. The combined organic extracts were washed with brine, dried over anhydrous Na_2SO_4, filtered, and the solvent was removed in vacuo. The residue was purified by silica gel column chromatography (petroleum ether/EtOAc = 3:1) to give (2*R*)-1,2-di-*O*-isopropylidene-4,4-difluoro-5-hexen-1,2,3-triol as a clear oil (21.6 g, 90%). The ratio of *anti*/*syn* was 7.7:1 determined by ^{19}F NMR.

Characterization Data ^{1}H NMR (300 MHz, $CDCl_3$): δ 6.09–5.95 (m, 1H), 5.75 (dt, J = 19.2 Hz, 1H), 5.55 (d, J = 11.7 Hz, 1H), 4.33–4.27 (m, 1H), 4.09–3.90 (m, 3H), 2.56 (br, 1H), 1.45 (s, 3H), 1.40 (s, 3h) ppm. ^{19}F NMR (282 MHz, $CDCl_3$): δ −108.14 (dt, J = 252.2, 11.4 Hz, 0.12F), −108.5 (dt = 252.2, 11.4

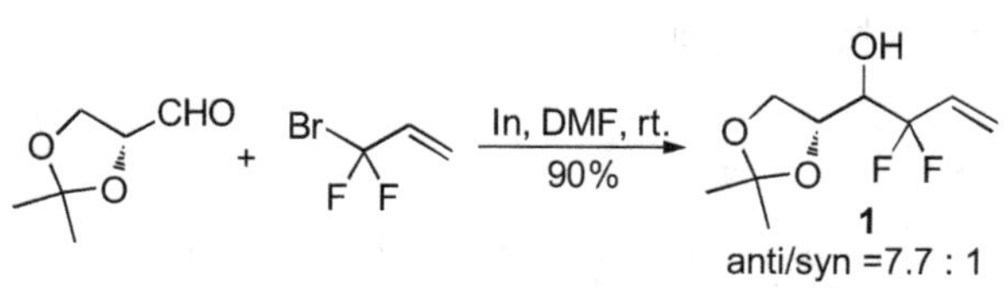

SCHEME 54.1 Preparations of *gem*-difluoromethylenated building block **1**.

TABLE 54.1 Kinetic Resolution of (2*R*)-1,2-di-*O*-isopropylidene-4,4-difluoro-5-hexen-1,2,3-triol

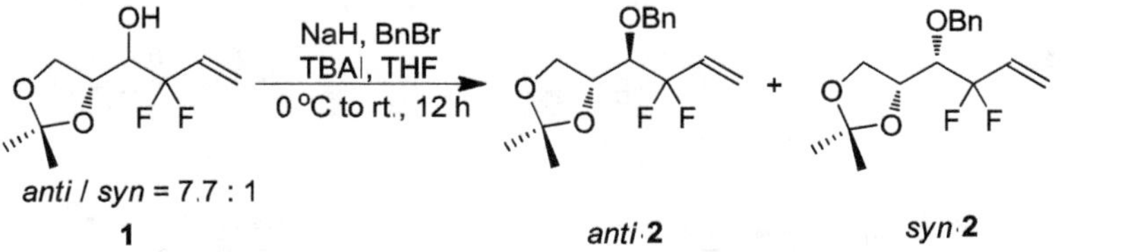

Entry	NaH (equiv)	Product **2** (*anti*/*syn*)^a	Recovered **1** (*anti*/*syn*)^a	Yield^b (%)
1	1.2	6.1:1		87.4^c
2	1.4	5.6:1		90.0^c
3	1.6	5.7:1		93.1^c
4	1.0	8.2:1		80.0^c
5	0.6	21.0:1	2.8:1	(35.6)^d; 75.5
6	0.8	21.8:1	1.8:1	(23.0)^d; 78.5

^{a} Determined by ^{19}F NMR before column chromatography purification.
^{b} Isolated yield based on the NaH.
^{c} Isolated yield based on the compound **1**.
^{d} Recovery of the substrate **1**.

Hz, 0.88F), −110.4 (dt, J = 252.2, 12.2 Hz, 0.12F), −111.0 (dt, J = 252.2, 0.88 Hz, 0.88F) ppm. IR (thin film, cm^{-1}): $\bar{\upsilon}$ 3446, 1652, 1219. MS (*m/z*): 193 (58%), 101 (35%), 43 (100%).

Application Table 54.1 shows the epimerization of (2*R*)-1,2-di-*O*-isopropylidene-4,4-difluoro-5-hexen-1,2,3-triol in benzylation [17]. Kinetic resolution takes place when it was treated with sodium hydride. When 1.2 equiv of sodium hydride was used, the epimerization started, and the ratio of *anti*/*syn* was 6.1:1 (entry 1). When 1.6 equiv of sodium hydride was used, further epimerization happened, but *syn*-product was isolated in the highest yield (entry 3). No doubt, the epimerization at the C3 position did not occur until benzylated compound was formed. The more excess base was used, the more epimeric *syn*-product was obtained (entries 1–3). More surprisingly, when the *gem*-difluorohomoallyl alcohol reacted with less than 1.0 equiv of sodium hydride, only one isomer *anti*-**2** was obtained (entries 5–6). Furthermore, the more sodium hydride was used, the less compound was recovered, and the more *syn*-**1** was enriched in the recovered product. Notably, based on the yield and the ratio of *anti*/*syn*-recovered **1** and the formed **2**, respectively (entries 5–6), no epimerization happened in the recovered **1** and the product **2**. These results suggested that the reactivity of *anti*-**1** seemed to be much more active than that of *syn*-**1**. Thus, the single *anti*-**2** could be obtained by controlling the amount of sodium hydride used in the benzylation of compound **1**.

So far, many potentially bioactive compounds were synthesized starting from (2*R*)-1,2-di-*O*-isopropylidene-4,4-difluoro-5-hexen-1,2,3-triol, such as L- and D-β-3′-deoxy-3′,3′-difluoronucleosides [18], N^1-(3-deoxy-3,3-difluoro-D-arabinofuranosyl)cytosine, [17], 2′,3′-dideoxy-6′,6′-difluoro-3′-thionucleoside

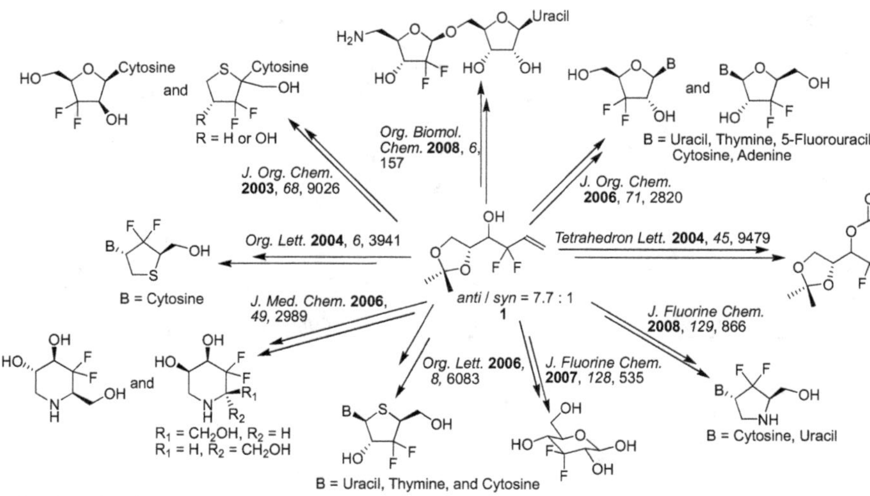

FIGURE 54.1 The potentially bioactive compounds prepared from *gem*-difluoromethylenated building block **1**.

[19], 2′,3′-dideoxy-6′,6′-difluoro-3′-azanucleosides [20], *gem*-difluoromethylenated nojirimycin analogs [21], *gem*-difluoromethylenated α,β-unsaturated-δ-lactone [22], L-β-3′-deoxy-3′,3′-difluoro-4′-thionucleosides [23], 3-deoxy-3,3-difluoro-D-ribohexose [24], *gem*-difluorinated nucleoside analog of the liposidomycins [25] (Figure 54.1), etc.

Preparation of [[difluoro(trimethylsilyl)methyl]seleno]benzene (Scheme 54.2) [26]

Apparatus A 25-mL Schlenk flask with septum, magnetic stirrer, syringe, flask, silica gel column, safety glasses, laboratory coat, protective gloves.

Chemicals Mg turnings, bromodifluoromethyl phenyl selenide, TMSCl, THF, silica gel, *n*-pentane.

Attention! Safety glasses and protective gloves must be used at all times.

Caution! Because of its toxicity and/or volatility, care should be taken to avoid inhalation of TMSCl, bromodifluoromethyl phenyl selenide, and avoid contact of powdered silica gel. All reactions should be carried out in a well-ventilated hood.

Experimental Procedure To a dry 25-mL Schlenk flask under nitrogen atmosphere, Mg turnings (30 mg, 1.25 mmol) and a solution of bromodifluoromethyl

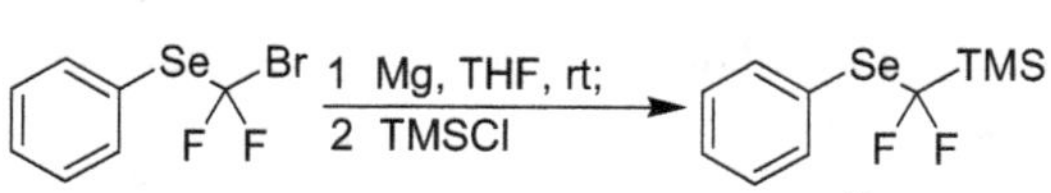

3

SCHEME 54.2 Preparations of *gem*-difluoromethylenated building block **3**.

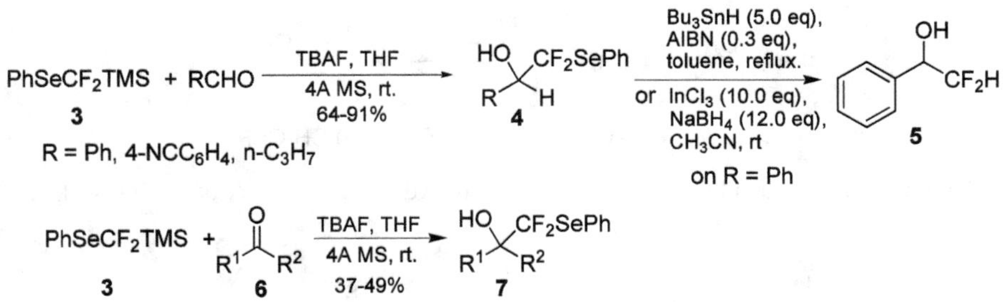

SCHEME 54.3 Nucleophilic addition of **3** with aldehydes and ketones in the presence of TBAF.

phenyl selenide (143 mg, 0.5 mmol) in THF (2.5 mL) at room temperature were added. Then, TMSCl (0.26 mL, 2.0 mmol) was added slowly via a syringe. The reaction mixture was stirred at room temperature for 30 min and the starting substrate was completely consumed by TLC. Then, the reaction mixture was transferred to a flask via a syringe and the excess of TMSCl was removed under reduced pressure. The residue was purified by silica gel chromatography (*n*-pentane) to give [[difluoro(trimethylsilyl)methyl]seleno]benzene as a pale yellow oil (113 mg, 81%).

Characterization Data ^{1}H NMR (300 MHz, acetone-D_6): δ 7.71–7.68 (m, 2H), 7.43–7.40 (m, 3H), 0.20 (s, 9H) ppm. ^{19}F NMR (282 MHz, acetone-D_6): δ −85.24 (s, 2F) ppm. IR (neat, cm^{-1}): $\bar{\upsilon}$ 3036, 2964, 1579, 1477, 1440, 1255, 1081, 1066. HRMS (*m/z*): 280.0022.

Application The *gem*-difluoromethylenation of aldehydes with [[difluoro(trimethylsilyl)methyl]seleno]benzene (**3**) proceeded smoothly in the presence of TBAF, and the corresponding α,α-difluoro-β-hydroxy selenides were obtained in moderate to high yields (Scheme 54.3) [26]. It was reported that the existence of a small amount of water in TBAF did not create any serious problems in the trifluoromethylation of aldehydes and ketones with CF_3TMS. The use of anhydrous TBAF did not offer any advantages [27,28]. However, the existence of a small amount of water in TBAF lowered the yield of the desired alcohols and led to PhSeCF_2H as a by-product. This phenomenon was attributed to the electron-donating property of PhSe moiety, which resulted in more nucleophilic PhSeCF_2- than the CF_3- anion. Addition of molecular sieves to the reaction mixture could successfully overcome this problem. The resulting α,α-difluoro-β-hydroxy selenides could be further transferred to difluoromethyl alcohol **5** via treatment with Bu_3SnH/AIBN or $NaBH_4$/$InCl_3$ [29]. Besides aldehydes, the *gem*-difluoromethylenation of ketones with PhSeCF_2TMS under the same reaction conditions resulted in somewhat low yields (Scheme 54.3). In addition, it was also reported that when $Cu(OAc)_2$/dppe was used as the catalyst in DMF, treatment of aldehydes with 2.5 equiv of PhSeCF_2TMS gave the corresponding difluoro(phenylseleneyl)methyl adducts in high yields (Scheme 54.4) [30].

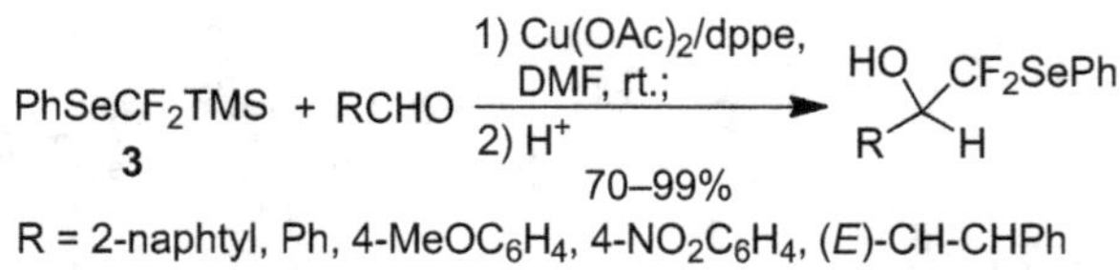

SCHEME 54.4 Nucleophilic addition of **3** with aldehydes in the presence of $Cu(OAc)_2$/dppe.

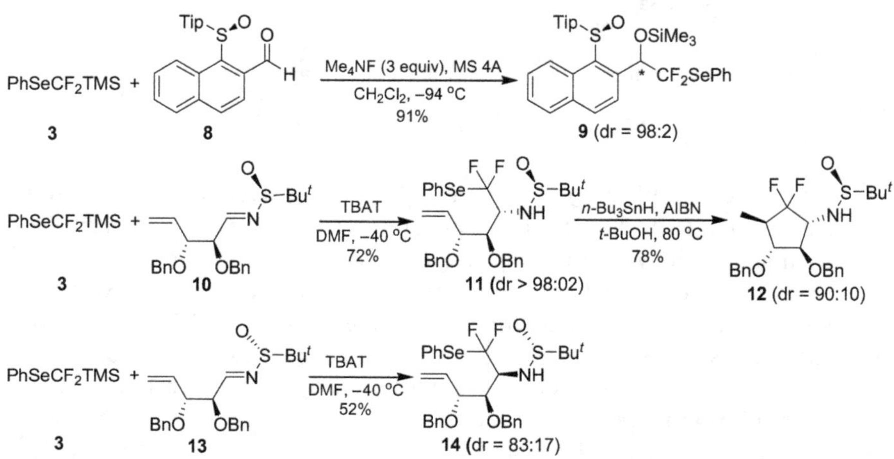

SCHEME 54.5 Asymmetric nucleophilic addition of **3**.

With $PhSeCF_2TMS$ as fluorination reagent, the remote asymmetric induction in the nucleophilic difluoromethylation of aldehyde with the sulfinyl group as a chiral auxiliary was developed [31]. For example, *gem*-difluoromethylenation of 1-[(2,4,4-triisopropylphenyl)sulfinyl]-2-naphthaldehyde (**8**) with $PhSeCF_2TMS$ at −94 °C gave adduct **9** in high yield with high diastereoselectivity (Scheme 54.5). Furthermore, the addition of $PhSeCF_2TMS$ to carbohydrate derived *tert*-butanesulfinylimines 10 and 13 proceeded smoothly using TBAT as a promoter [32]. The matched diastereoselectivity (dr > 98:02) and mismatched case (dr = 83:17) were observed. *n*-Bu_3SnH/AIBN-mediated reductive radical cyclization of adduct **11** afforded 5-deoxypentofuranose analog **12** in good yield.

Preparation of Ethyl (2*Z*)-4,4,4-trifluoro-3-iodo-2-butenoate (Scheme 54.6) [33–35]

Apparatus A 100-mL flask with septum, magnetic stirrer, syringe, safety glasses, sealed tube, laboratory coat, protective gloves.

Chemicals Hydroiodic acid, ethyl 4,4,4-trifluorobut-2-ynoate, diethyl ether, $Na_2S_2O_3$, HOAc, NaI, K_2CO_3, $MgSO_4$.

Attention! Safety glasses and protective gloves must be used at all times.

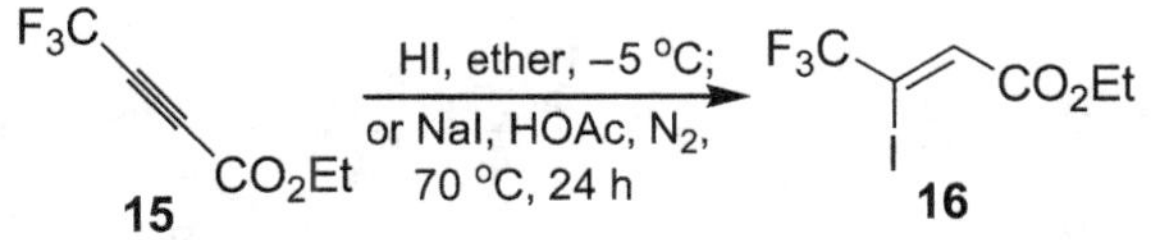

15 16

SCHEME 54.6 Preparations of trifluoromethylated building block **16**.

Caution! Because of its toxicity and volatility, care should be taken to avoid inhalation of diethyl ether and hydroiodic acid, and avoid contact of ethyl 4,4,4-trifluorobut-2-ynoate and $Na_2S_2O_3$. All reactions should be carried out in a well-ventilated hood.

Experimental Procedure (A) In a 100-mL flask, 14.6 mL (60 mmol) of an aqueous solution of hydroiodic acid (57%) was added drop by drop to a solution of ethyl 4,4,4-trifluorobut-2-ynoate (8.3 g, 50 mmol) in diethyl ether (50 mL). After stirring for 5 h at −5 °C, a 5% solution of $Na_2S_2O_3$ was added and the reaction mixture was then washed with brine and extracted with diethyl ether. The organic phases were dried over magnesium sulfate and concentrated, yielding (2*Z*)-4,4,4-trifluoro-3-iodo-2-butenoate (12.5 g, 85%), which was sufficiently pure to be used without purification.

(B) A mixture of ethyl 4,4,4-trifluorobut-2-ynoate (4.0 g, 24 mmol), NaI (4.5 g, 30 mmol), and HOAc (12 mL) was placed in a sealed tube. The reaction was carried out with magnetic stirring under nitrogen at 70 °C for 72 h. Then, water (60 mL) was added, and the mixture was cautiously neutralized with solid potassium carbonate, added in portions. Then, the aqueous solution was extracted with diethyl ether (3 × 50 mL), and the organic layer was dried over $MgSO_4$, concentrated, and purified by flash chromatography (petroleum ether/EtOAc = 50:1) to yield (2*Z*)-4,4,4-trifluoro-3-iodo-2-butenoate (5.3 g, 75%).

Characterization Data ^{1}H NMR (400 MHz, $CDCl_3$): δ 7.18 (q, *J* = 1.2 Hz, 1H), 4.34 (q, *J* = 7.2 Hz, 2H), 1.37 (t, *J* = 7.2 Hz, 3H) ppm. ^{13}C NMR (100 MHz, $CDCl_3$): δ 163, 132 (q, *J* = 5.3 Hz), 120.7 (q, *J* = 273 Hz), 98.4 (q, *J* = 37 Hz), 61.7, 14 ppm. ^{19}F NMR (378 MHz, $CDCl_3$): δ −69.2 (s, 3F) ppm. IR (cm^{-1}): $\bar{\upsilon}$ 2991, 1735, 1638, 1267. MS (*m/z*): 294 (75%), 266 (68%), 249 (100%), 221 (48%), 127 (26%), 103 (11%), 94 (10%), 75 (54%).

Application Subjecting (2*Z*)-4,4,4-trifluoro-3-iodo-2-butenoate (**16**) to Sonogashira reaction would afford a lots of conjugated (2*Z*)-en-4-ynoic acid derivatives containing trifluoromethyl group in high yield [33]. The reaction was catalyzed by $PdCl_2(PPh_3)_2$/CuI (Scheme 54.7). The configuration of the double bonds remained intact and the resulting stereodefined conjugated 2-en-4-ynoic derivatives (**18**) represented a class of important synthetic intermediates. For example, trifluoromethylated retinoic acid analog was prepared in a straightforward manner from the corresponding intermediate [36].

Under Suzuki cross-coupling reaction conditions, facile stereoselective synthesis of aryl-substituted α,β-unsaturated ester (**20**), stereodefined dienoate (**22**), and cyclopropylenoate (**24**), all containing a trifluoromethyl group,

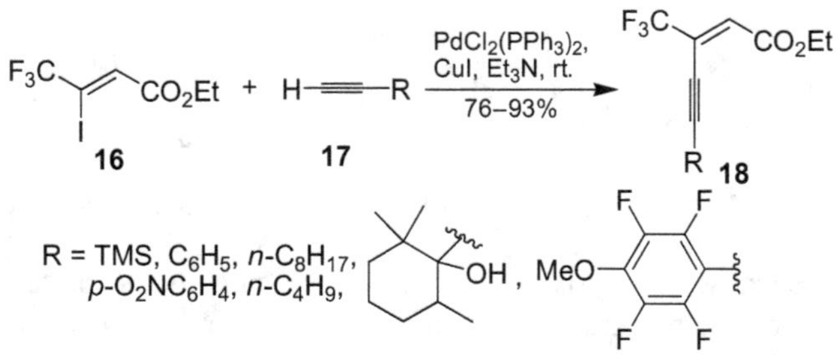

SCHEME 54.7 The Sonogashira reaction of **16**.

was achieved in moderate to high yield via reaction of compound **16** with arylboronic acid (**19**), *trans*-alkenylboronic acid (**21**), and *trans*-cyclopropylboronic acid (**23**), respectively (Scheme 54.8) [37, 38]. All the configurations, including double bonds and cyclopropyl stereoconfiguration, were retained in the reactions.

Stille cross-coupling of compound **16** with alkenyltin or alkynyltin provided a stereoselective access to functional diene **26** containing a trifluoromethyl group (Scheme 54.9) [34,35,39]. The reactions used $PdCl_2(MeCN)_2$ as catalyst. With a complete retention of the double bond configuration, the Stille cross-coupling resulted in good yield of dienes and no polymerization products were detected.

Addition of compound **16** to the solution of Bu_3SnCl in DMAC in the presence of activated zinc dust afforded the building block **27**, which was subjected to palladium-catalyzed cross-coupling of cyclic carbonate **28** to give the alcohol **29** in 51% yield [40]. Alcohol **29** could be further transferred to β-trifluoromethylgoniodiol (**30**) in two steps (Scheme 54.10).

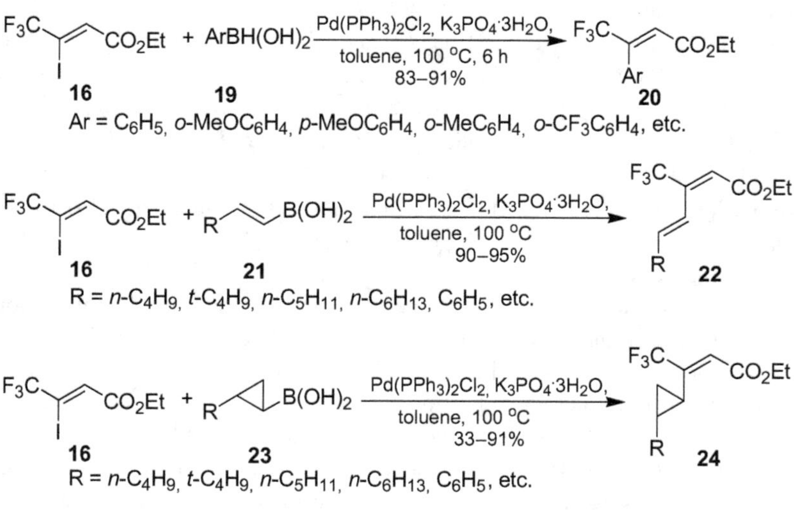

SCHEME 54.8 The Suzuki cross-coupling reaction of **16**.

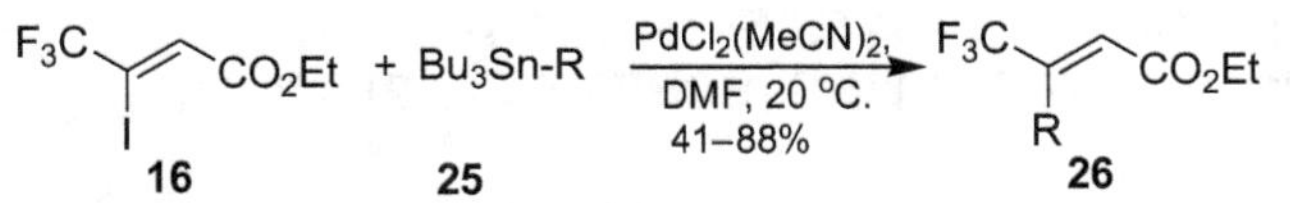

R = CH_2=CH; (*E*)-PhCH=CH; (*E*)-TMSCH=CH; Me_2C=CH; (*E*)-Bu_3SnCH=CH; $(EtO)_2CHCH_2C{\equiv}C$; $(EtO)_2CHC{\equiv}C$

SCHEME 54.9 The Stille cross coupling reaction of **16**.

Compound **16** could be converted to (*Z*)-ethyl-3-trifluoromethyl-3-magnesiated crotonate (**31**) by iodine–magnesium exchange reaction with isopropylmagnesium bromide in THF at −78 °C (Scheme 54.11) [41]. Transmetalation of **31** with THF-soluble copper salt CuCN·2LiCl provided a functionalized copper reagent, which reacted with acryl chloride and allyl bromide to afford the 1,4-ketoester **33** and allylated product **32**, respectively. In addition, reaction of **31** with aldehyde under Lewis acid $BF_3 \cdot OEt_2$ at −78 °C gave the butenolide **34** in good yield.

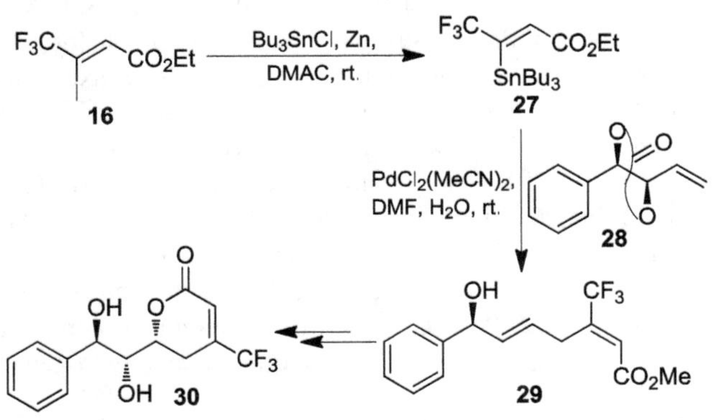

SCHEME 54.10 Synthesis of β-trifluoromethylgoniodiol **30** from **16**.

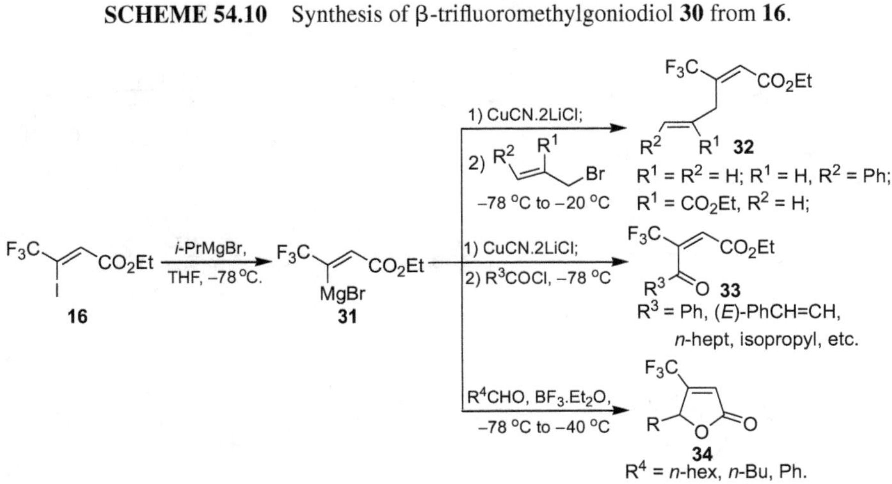

SCHEME 54.11 Tansformation of (Z)-ethyl-3-trifluoromethyl-3-magnesiated crotonate **31**.

54.2 PREPARATION OF (2*S*,4*S*)-*N*-BOC-4-TRIFLUOROMETHYLPROLINE (SCHEME 54.12) [42, 43]

Apparatus Flask with septum, magnetic stirrer, syringe, safety glasses, sealed tube, laboratory coat, protective gloves.

Chemicals Thionyl chloride, benzyl (2*S*,4*S*)-*N*-Boc-4-hydroxy-4-trifluoromethyl-L-prolinate, diethyl ether, pyridine, 1 N HCl, saturated aqueous $NaHCO_3$, Na_2SO_4, Pd/C (10% Pd), EtOH, silica gel, hexane, ethyl acetate (only for last two steps of method A).

Attention! Safety glasses and protective gloves must be used at all times.

Caution! Because of its toxicity and volatility, care should be taken to avoid inhalation of $SOCl_2$, pyridine, diethyl ether, and hydroiodic acid, and avoid contact of $SOCl_2$ and pyridine. In addition, care should be taken to use Pd/C and avoid fire or contact oxygen. All reactions should be carried out in a well-ventilated hood.

Experimental Procedure (Method A, last two steps) A mixture of benzyl (2*S*,4*S*)-*N*-Boc-4-hydroxy-4-trifluoromethyl-L-prolinate (**36**) (123 mg, 0.32 mmol), dry pyridine (4 mL), and $SOCl_2$ (300 μL) was refluxed under nitrogen for 20 min. H_2O (1 mL) was added to quench the reaction. The aqueous layer was extracted with Et_2O (3 × 20 mL). The combined organic phases were washed with 1 N HCl (20 mL), saturated aqueous $NaHCO_3$ (10 mL), H_2O (10 mL), and brine, and finally dried over anhydrous Na_2SO_4. After removal of the solvent in vacuo, the residue was purified by silica gel chromatography (hexane/EtOAc = 30:1) to give compound **37** as a clear oil (92 mg, 78%).

Pd/C (490 mg, 10% Pd) was added to a solution of compound **37** (394 mg, 1.39 mmol) in EtOH (25 mL). The solution was hydrogenated at room

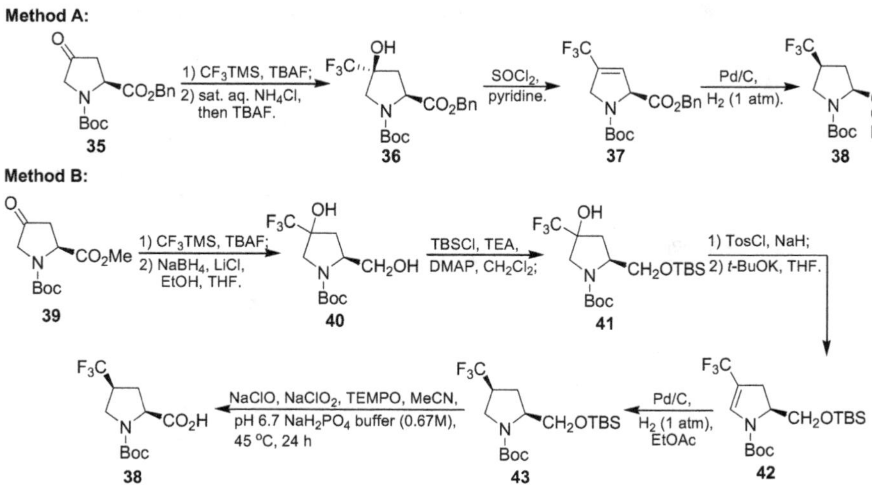

SCHEME 54.12 Preparation of (2S, 4S)-N-Boc-4-trifluoromethylproline **38**.

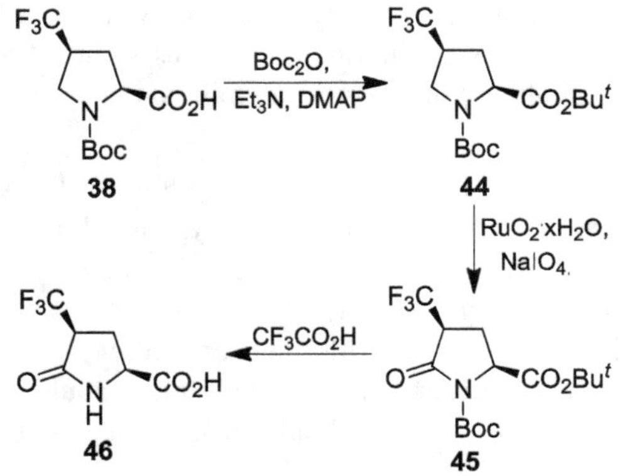

SCHEME 54.13 Preparation of (2*S*, 4*S*)-4-trifluoromentylpyroglutamic acid **46** from **38**.

temperature overnight. After filtration and removal of the solvent in vacuo, the residue was purified by silica gel chromatography (hexane/EtOAc = 1:1) to give (2*S*,4*S*)-*N*-Boc-4-trifluoromethylproline (**38**) as a white solid (360 mg, 90%).

Method B was reported by group of Goodman [43]. However, no detailed procedure was described.

Characterization Data ^{1}H NMR (300 MHz, $CDCl_3$): δ 8.98–8.91 (bs, 1H), 4.48–4.33 (dt, $J = 29.0$, 7.8 Hz, 1H), 3.96–3.82 (m, 1H), 3.53–3.45 (m, 1H), 3.04–2.90 (m, 1H), 2.67–2.53 (m, 1H), 2.41–2.19 (m, 1H), 1.49, 1.43 (2s, 9H) ppm. ^{19}F NMR (282 MHz, $CDCl_3$): δ −71.2 (d, $J = 4.8$ Hz, 3F) ppm. IR (thin film, cm^{-1}): $\bar{\upsilon}$ 3000–2500, 1726, 1647, 1481, 1445, 1407, 1280, 1164, 699. MS (*m*/*z*): 238 (3%), 182 (23%), 138 (37%), 57 (100%).

Application (2*S*,4*S*)-*N*-Boc-4-trifluoromethylproline (**38**) could be converted to (2*S*,4*S*)-4-trifluoromentylpyroglutamic acid (**46**) in a straightforward manner. The key step involved $RuO_2{\cdot}xH_2O$-catalyzed oxidation of protected prolinate **44** (Scheme 54.13) [44].

54.3 PREPARATION OF (*E*)-4,4,4-TRIFLUORO-2-BUTEN-1-OL (49) (SCHEME 54.14) [45, 46]

Apparatus Flask with septum, magnetic stirrer, syringe, safety glasses, sealed tube, laboratory coat, protective gloves.

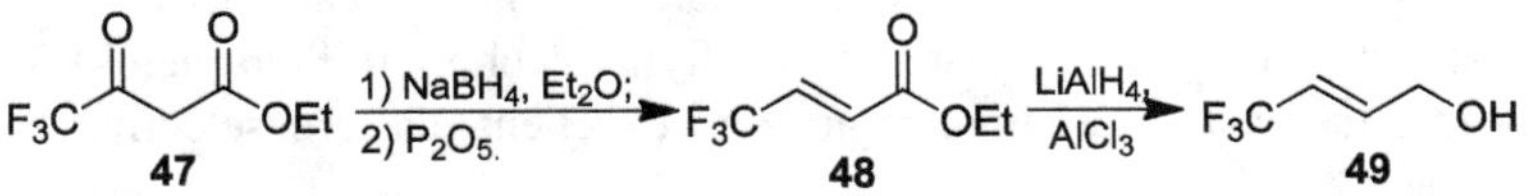

SCHEME 54.14 Preparation of (*E*)-4,4,4-trifluoro-2-buten-1-ol **49**.

Chemicals 4,4,4-Trifluoro-3-oxo-butyric acid ethyl ester, ethyl ether, sodium borohydride, aqueous hydrochloride, anhydrous sodium sulfate, phosphorus pentoxide, lithium aluminum hydride, aluminum chloride, anhydrous magnesium sulfate.

Attention! Safety glasses and protective gloves must be used at all times.

Caution! Because of its toxicity and volatility, care should be taken to avoid inhalation of 4,4,4-trifluoro-3-oxo-butyric acid ethyl ester and ethyl ether, and avoid contact of sodium borohydride, phosphorus pentoxide, lithium aluminum hydride, and aluminum chloride. In addition, phosphorus pentoxide, lithium aluminum hydride, and aluminum chloride should not be directly contacted with water. All reactions should be carried out in a well-ventilated hood.

Experimental Procedure To a solution of 4,4,4-trifluoro-3-oxo-butyric acid ethyl ester (55.2 g, 0.30 mmol) in Et_2O (600 mL), sodium borohydride was added in several portions over 30 min at 0 °C. The mixture was stirred at 0 °C for 1 h and at room temperature overnight. Then, a solution of aqueous HCl (10%, 300 mL) was carefully added to the reaction mixture and the resulting solid was removed by filtration. The aqueous layer was extracted with Et_2O (2 × 150 mL), and the combined organic phases were dried over anhydrous sodium sulfate, filtered, and evaporated to give a clear oil (45.7 g). Then, phosphorus pentoxide (16.9 g, 0.12 mmol) was added to the above resulting oil and the mixture was stirred at 100 °C for 1 h. After cooling to room temperature, the mixture was distilled to yield ethyl 4,4,4-trifluorocrotonate (33.3 g).

To a suspension of anhydrous $AlCl_3$ (2.27 g, 17.0 mmol) in Et_2O (30 mL) at 0 °C, a suspension of $LiAlH_4$ (1.90 g, 50.0 mmol) in Et_2O (40 mL) was added. The resulting mixture was stirred at this temperature for 15 min. A solution of ethyl 4,4,4-trifluorocrotonate (3.36 g, 20.0 mmol) in Et_2O (10 mL) was then added at 0 °C and stirring was continued for a further 2 h. The reaction mixture was subsequently quenched with aqueous saturated sodium sulfate at 0 °C. The resulting suspension was dried with anhydrous magnesium sulfate, filtered, and the solids were washed with diethyl ether. The combined organic phases were then concentrated in vacuo, and the residue was purified by distillation to afford pure (*E*)-4,4,4-trifluoro-2-buten-1-ol as a clear oil (2.27 g, 90%).

Characterization Data ^1H NMR (300 MHz, $CDCl_3$): δ 6.54–6.46 (m, 1H), 6.00–5.91 (m, 1H), 4.33–4.30 (m, 2H), 1.88 (br, 1H) ppm. ^{19}F NMR (282 MHz, $CDCl_3$): δ −64.47 (d, $J = 9.0$ Hz, 3F) ppm. IR (thin film, cm^{-1}): $\bar{\upsilon}$ 3427, 1692, 1638, 1300, 1268, 1132, 1086, 1019.

Application Bromination of the (*E*)-4,4,4-trifluoro-2-buten-1-ol (**49**) via treatment with PBr_3 in ether gave 4-bromo-1,1,1-trifluorobut-2-ene (**50**), which was subjected to the tin- and indium-mediated allylation reactions of aldehydes to afford β-trifluoromethylated homoallylic alcohols in high yields (Scheme 54.15). In most cases, high regio- and excellent diastereoselectivities were obtained [46, 47].

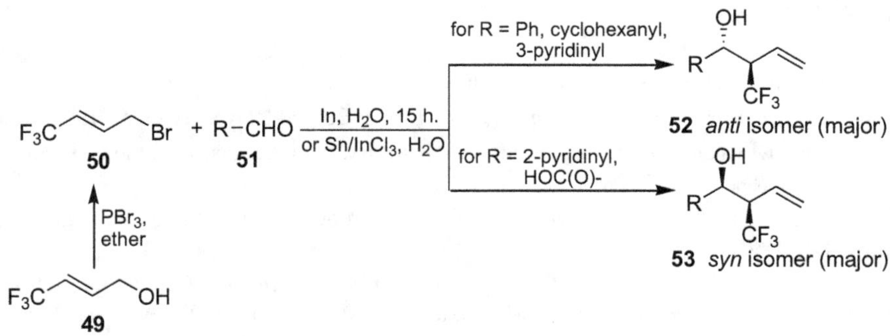

SCHEME 54.15 Synthesis of β-trifluoromethylated homoallylic alcohols from **49**.

The Sharpless AD reaction of benzylated compound **54** afforded the chiral building block **55**, which was successfully used for the preparation of trifluoromethylated β-L-fucofuranose (**56**) [45], trifluoromethylated β-L-4,6-dideoxyxylohexopyrnose (**57**) [45], (2*S*,3*R*)-4,4,4-trifluorothreonine (**58**) [48], (*S*)-2-(*tert*-butoxycarbonyl)amino-4,4,4-trifluorobutanoic acid (**59**) [48], (2*S*,3*S*)-*N*-benzoyl-3-(trifluoromethyl)isoserine (**60**) [49], and trifluoromethylated macrosphelid A (**61** and **62**) [50] (Figure 54.2).

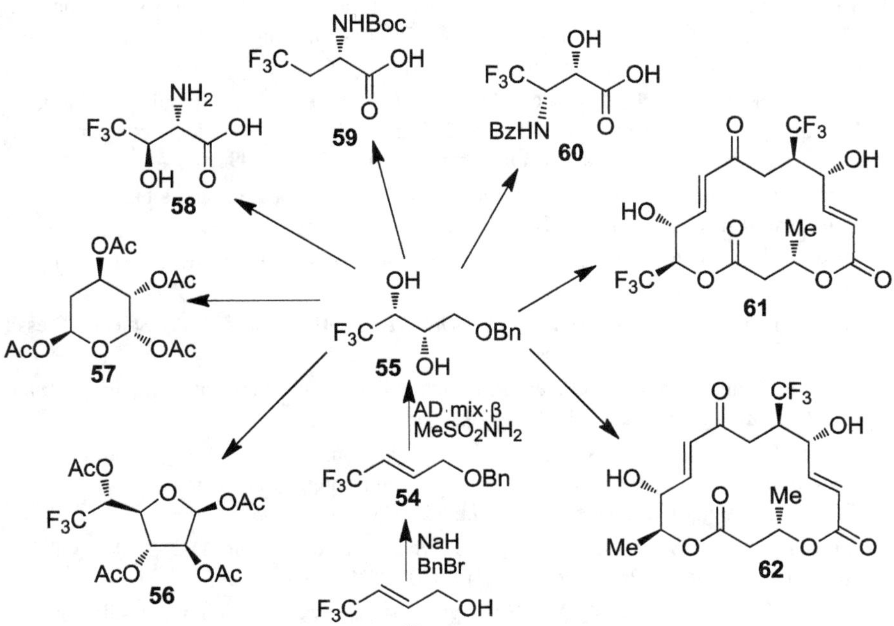

FIGURE 54.2 Transformations of the trifluoromethylated chiral building block **55**.

REFERENCES

1. Bondi A. van der Waals volumes and radii. *J Phys Chem* 1964;68:441–451.
2. Pauling L. *The Nature of the Chemical Bond.* Ithaca: Cornell University Press; 1960.
3. O'Hagan D. Understanding organofluorine chemistry. An introduction to the C–F bond. *Chem Soc Rev* 2008;37:308–319.
4. Clader JW. The discovery of ezetimibe: a view from outside the receptor. *J Med Chem* 2004;47:1–9.
5. Smith DH, van de Waterbeemd H, Walker DK. *Pharmacokinetics and Metabolism in Drug Design, Methods and Principles in Medicinal Chemistry.* Weinheim: Wiley-VCH; 2001.
6. Böhm H-J, Banner D, Bendels S, Kansy M, Kuhn B, Müller K, Obst-Sander U, Stahl M. Fluorine in medicinal chemistry. *ChemBioChem* 2004;5:637–643.
7. Müller K, Faeh C, Diederich F. Fluorine in pharmaceuticals: looking beyond intuition. *Science* 2007;317:1881–1886.
8. Guschlbauer W, Jankowski K. Nucleoside conformation is determined by the electronegativity of the sugar substituent. *Nucleic Acids Res* 1980;8:1421–1434.
9. Cheng DM, Kan L-S, Ts'o POP, Uesugi S, Takatsuka Y, Ikehara M. Multinuclear magnetic resonance studies of monomers and dimers containing 2′-fluoro-2′-deoxyadenosine. *Biopolymers* 1983;22:1427–1444.
10. Uesugi S, Kaneyasu T, Imura J, Ikehara M, Cheng DM, Kan L, Ts'o POP. ^{1}H-NMR studies on the dinucleoside monophosphates containing 2′-halogeno-2′-deoxypurinenucleosides: effects of 2′-substitutes on conformation. *Biopolymers* 1983;22:1189–1202.
11. Van der Boogaart JE, Kalinichenko EN, Podkopaeva TL, Mikhailopulo IA, Altona C. Conformation analysis of 3′-fluorinated A(2′-5′)A(2′-5′)A fragments. Relation between conformation and biological activity. *Eur J Biochem* 1994;221:759–768.
12. Barchi JJ, Karki RG, Nicklaus MC, Siddiqui MA, George C, Mikhailopulo IA, Marquez VE. Comprehensive structural studies of 2′,3′-difluorinated nucleosides: comparison of theory, solution, and solid state. *J Am Chem Soc* 2008;130:9048–9057.
13. Leroux F, Jeschke P, Schlosser M. α-Fluorinated ethers, thioethers, and amines: anomerically biased species. *Chem Rev* 2005;105:827–856.
14. Swain C, Rupniak NMJ. Progress in the development of neurokinin antagonists. *Ann Rep Med Chem* 1999;34:51–60.
15. Wermuth CG. *Rational Approaches in Drug Design.* Barcelona: Prous Science Press; 2001.
16. Park BK, Kitteringham NR, O'Neill PM. Metabolism of fluorine-containing drugs. *Annu Rev Pharmacol Toxicol* 2001;41:443–470.
17. Zhang X, Xia H, Dong X, Jin J, Meng W-D, Qing F-L. 3-Deoxy-3,3-difluoro-d-arabinofuranose: first stereoselective synthesis and application in preparation of gem-difluorinated sugar nucleosides. *J Org Chem* 2003;68:9026–9033.
18. Xu X-H, Qiu X-L, Zhang X, Qing F-L. Synthesis of L- and D-2′-3′-Deoxy-3′,3′-difluoronucleosides. *J Org Chem* 2006;71:2820–2824.
19. Wu Y-Y, Zhang X, Meng W, Qing F-L. Synthesis of new 2′,3′-dideoxy-6′,6′-difluoro-3′-thionucleoside from gem-difluorohomoallyl alcohol. *Org Lett* 2004;6:3941–3944.
20. Yue X, Qiu X-L, Qing F-L. Synthesis of 2′,3′-Dideoxy-6′,6′-difluoro-3′-azanucleosides. *J Fluorine Chem* 2008;129:866–874.

21. Wang R-W, Qiu X-L, Bols M, Ortega-Caballero F, Qing F-L. Synthesis and biological evaluation of glycosidase inhibitors: gem-difluoromethylenated analogues of deoxynojirimycin. *J Med Chem* 2006;49:2989–2997.
22. You Z-W, Wu Y-Y, Qing F-L. Synthesis of gem-difluoromethylenated massoialactone by ring-closing metathesis. *Tetrahedron Lett* 2004;45:9479–9481.
23. Zheng F, Zhang X, Qiu X-L, Zhang X, Qing F-L. Synthesis of L -β-3′-deoxy-3′,3′-difluoro-4-thionucleosides. *Org Lett* 2006;8:6083–6086.
24. Xu X-H, You Z-W, Zhang X, Qing F-L. Synthesis of 3-deoxy-3,3-difluoro-D-ribohexose from gem-difluorohomoallyl alcohol. *J Fluorine Chem* 2007;128:535–539.
25. X-H. Xu, Trunkfield AE, Bugg TDH, Qing F-L. Synthesis of gem-difluorinated nucleoside analogues of the liposidomycins and evaluation as MraY inhibitors. *Org Biomol Chem* 2008;6:157–161.
26. Qin YY, Qiu X-L, Yang YY, Meng W-D, Qing F-L. Synthesis and transformation of [Difluoro(phenylseleno)methyl]trimethylsilane ($PhSeCF_2TMS$). *J Org Chem* 2005;75:9040–9043.
27. Krishnamurti R, Bellew DR, Prakash GKS. Preparation of trifluoromethyl and other perfluoroalkyl compounds with (perfluoroalkyl)trimethylsilanes. *J Org Chem* 1991;56:984–989.
28. Singh RP, Shreeve JM. Nucleophilic trifluoromethylation reactions of organic compounds with (trifluoromethyl)trimethylsilane. *Tetrahedron* 2000;56:7613–7632.
29. Qin Y-Y, Yang Y-Y, Qiu X-L, Qing F-L. Reductive radical reaction of gem-difluorinated organoselenium compounds with indium(III) chloride-sodium borohydride system. *Synthesis* 2006;1475–1479.
30. Mizuta S, Shibata N, Ogawa S, Fujimoto S, Nakamura S, Toru T. Lewis acid-catalyzed tri- and difluoromethylation reactions of aldehydes. *Chem Commun* 2006;2575–2577.
31. Sugimoto H, Nakamura S, Shibata Y, Shibata N, Toru T. Remote asymmetric trifluoromethylation induced by chiral sulfinyl group: synthesis of enantiomerically pure 1-(2-naphthyl)-2,2,2-trifluoroethanol. *Tetrahedron Lett* 2006;47:1337–1340.
32. Fourriere G, Hijfte NV, Lalot J, Dutech G, Fragnet B, Coadou G, Quirion JC, Leclerc E. Synthesis of difluorinated carbocyclic analogues of 5-deoxypentofuranoses and 1-amino-5-deoxypentofuranoses: en route to fluorinated carbanucleosides. *Tetrahedron* 2010;66:3963–3972.
33. Qing F-L, Zhang Y. Ethyl 3-Iodo-4,4,4-trifluoro-2(Z)-butenoate: regio-and stereospecific preparation and palladium-catalyzed reaction with alkynes. *Tetrahedron Lett* 1997;38:6729–6732.
34. Prié G, Thibonnet J, Abarbri M, Duchêne A, Parrain J-L. A stereoselective access to functional dienes containing a trifluoromethyl group via stille cross coupling of ethyl 4,4,4-trifluoro-3-iodobutenoate. *Synlett* 1998;839–840.
35. Prié G, Abarbri M, Thibonnet J, Parrain J-L, Duchêne A. Efficient preparation of β-trifluoromethyl acrylates and derivatives via palladium cross-coupling reactions. *New J Chem* 2003;27:432–441.
36. Qing F-L, Ying J, Zhang Y. A stereoselective synthesis of ethyl 9-cis-7.8- didehydro-19-trifluromethylretinoate. *J Fluorine Chem* 2000;101:31–33.
37. Pan R-Q, Wang P-A, Deng M-Z. Facile stereoselective syntheses of aryl substituted α,β-unsaturated esters containing a trifluoromethyl group. *Heteroatom Chem* 2002;13:287–290.

38. Wang P-A, Deng M-Z, Pan R-Q, Zhang S-Y. Facile access to stereodefined dienoates and cyclopropylenoates containing a trifluoromethyl group. *J Fluorine Chem* 2003;124:93–97.

39. Thibonnet J, Prié G, Abarbri M, Duchêne A, Parrain J-L. Synthesis of ethyl (13e)-trifluoromethylretinoate and its analogues by palladium-catalysed cross-coupling. *Tetrahedron Lett* 1999;40:3151–3154.

40. Chen J, Lin G-Q, Wang Z-M, Liu H-Q. A short and general approach to the synthesis of styryllactones: (+)-goniodiol, its acetates and β-trifluoromethyl derivative, (+)-7-epi-goniodiol and (+)-9-deoxygoniopypyrone. *Synlett* 2002;1265–1268.

41. Thibonnet J, Duchêne A, Parrain J-L, Abarbri M. New stereoselective preparation of (z)-3-perfluoroalkyl-3-magnesiated enoates by an iodine–magnesium exchange reaction. *J Org Chem* 2004;69:4262–4264.

42. Qiu X-L, Qing F-L. Practical synthesis of boc-protected cis-4-trifluoromethyl and cis-4-difluoromethyl-l- prolines. *J Org Chem* 2002;67:7162–7164.

43. DelValle JR, Goodman M. Stereoselective synthesis of boc-protected cis and trans-4-trifluoromethylprolines by asymmetric hydrogenation reactions. *Angew Chem, Int Ed* 2002;41:1600–1602.

44. Qiu X-L, Qing F-L. Synthesis of cis-4-trifluoromethyl- and cis-4-difluoromethyl-lpyroglutamic acids. *J Org Chem* 2003;68:3614–3617.

45. Wang B-L, Yu F, Qiu X-L, Jiang Z-X, Qing F-L. Synthesis of trifluoromethylated analogues of β-l-6-deoxyfucofuranose and β-l-4,6-dideoxyxylohexopyranose. *J Fluorine Chem* 2006;127:580–587.

46. Loh T-P, Li X-R. Tin- and indium-mediated allylation reactions in water: highly stereoselective synthesis of β-trifluoromethylated homoallylic alcohols. *Eur J Org Chem* 1999;1893–1899.

47. Loh T-P, Li X-R. A highly stereoselective synthesis of β-trifluoromethylated homoallylic alchols in water. *Angew Chem, Int Ed Engl* 1997;36:980–982.

48. Jiang Z-X, Qin YY, Qing F-L. Asymmetric synthesis of both enantiomers of anti-4,4,4-trifluorothreonine and 2-amino-4,4,4-trifluorobutanioc acid. *J Org Chem* 2003;68:7544–7547.

49. Jiang Z-X, Qing F-L. Regioselective and stereoselective nucleophilic ring opening of trifluoromethylated cyclic sulfates: asymmetric synthesis of both enantiomers of syn-(3-trifluoromethyl)isoserine. *J Org Chem* 2004;69:5486–5489.

50. Wang B-L, Jiang Z-X, You Z-W, Qing F-L. Total synthesis of trifluoromethylated analogs of Macrosohelide A. *Tetrahedron* 2007;63:12671–12680.

CHAPTER 55

1,3-Dipolar Cycloaddition Reactions to Fluoroalkenes

HANNA WÓJTOWICZ-RAJCHEL AND HENRYK KORONIAK

1,3-Dipolar cycloaddition reactions are among the most important synthetic routes for the construction of five-membered heterocyclic systems [1, 2]. Dipoles can be classified as either allyl anion type or propargyl/allenyl anion type according to their structure. Particularly, nitrone cycloadditions are versatile for the construction of five-membered heterocycles [3]. Nitrones are very effective 1,3-dipoles, and they can readily undergo cycloaddition with electron-deficient olefins to produce substituted isoxazolidines with high degrees of regio- and stereocontrol.

Although there is a growing interest in development of new methods for the introduction of a fluorine atom or fluorinated groups to molecules, only some reports have been published on the reactions of fluoroalkenes with nitrones [4]. Recently, the reactions of 1,3-dipolar cycloaddition of nitrones to pentafluoropropene (PFP) and hexafluoropropene (HFP) have been analyzed [5]. Nitrones derived from aromatic or aliphatic aldehydes and ketones react with HFP to give the respective fluorinated isoxazolidine derivatives in good yields with complete regioselectivity and moderate diastereoselectivity. Dipolar cycloaddition of nitrones to electron-deficient alkenes such as HFP is probably a type I cycloaddition and is controlled by the interaction of highest occupied molecular orbital (HOMO) of the dipole and lowest unoccupied molecular orbital (LUMO) of the alkene. The obtained isoxazolidines are stable and can be further transformed under reductive conditions into α-trifluoromethyl-β-lactams and esters of β-amino acids (Scheme 55.1).

We have found that 5-fluorovinyl-substituted pyrimidines obtained by the Michael addition–elimination reaction [6] or by the Suzuki–Miyaura coupling [7] can be easily transformed into corresponding isoxazolidynyl derivatives (Scheme 55.2). Also, fluorinated enamines of nucleobases [8] can serve as precursors of isoxazolidynyl nucleoside analogs. The direct 1,3-dipolar cycloaddition of 5-fluorovinyl pyrimidines

Efficient Preparations of Fluorine Compounds, First Edition. Edited by Herbert W. Roesky.

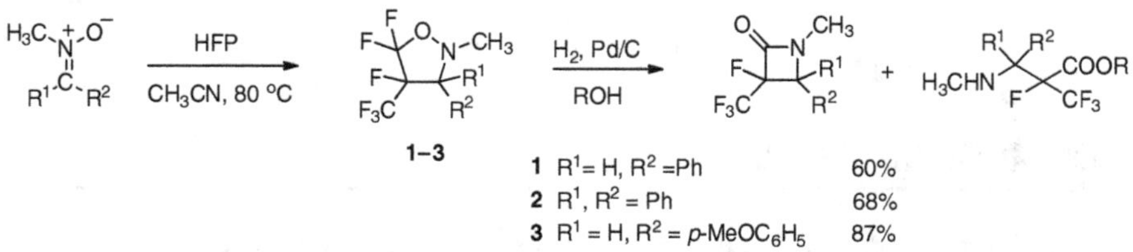

SCHEME 55.1 Cycloaddition of nitrones to HFP and the reductive cleavage of the N–O bond.

as dipolarophiles and model *N*-methyl and *N*-benzylnitrone leads to two diastereoisomers of the product with significant dominance of one of them. Any formation of regiostereoisomers was not observed; the cycloaddition reactions were completely regiospecific. This reaction followed by nonstandard hydrolysis of protective groups leads to a fluorinated isoxazolidine analog of pseudouridine.

55.1 PREPARATION OF 4,5,5-TRIFLUORO-2-METHYL-3-PHENYL-4-(TRIFLUOROMETHYL)ISOXAZOLIDINE

Apparatus Glass pressure tube, magnetic stirrer, oil bath, dry ice, Dewar flask, vacuum pump, safety glasses, laboratory coat, protective gloves.

Chemicals HFP, *N*-benzylidene-*N*-methylamine oxide, CH_2Cl_2, MeCN, hexanes, AcOEt, Et_2O, silica gel.

Attention! Safety glasses, protective gloves, sufficient ventilation.

Experimental Procedure HFP (0.5 mL, 4 mmol) was condensed in a glass pressure tube at −78 °C under argon atmosphere. MeCN (2.8 mL) and *N*-benzylidene-*N*-methylamine oxide (128 mg, 0.95 mmol) were introduced and the pressure tube was closed with a Teflon valve. The contents of the tube were stirred vigorously at 80 °C for 24 h. After opening the tube, the reaction mixture was poured into water (10 mL), and the products were extracted with CH_2Cl_2 (3 × 5 mL). Combined organic layers were washed with water (5 × 10 mL), dried over anhydrous Na_2SO_4, and concentrated. The product was purified by column chromatography on silica gel, using 10:1 or 5:1 mixtures of hexanes/AcOEt or hexanes/Et_2O as eluent. The product was obtained as colorless oil. Yield: 171 mg, 60%.

Characterization Data IR (film, ν_{max}/cm^{-1}): 2887, 1458, 1347, 1306, 1283, 1238, 1212, 1166, 1128, 1036, 1008, 729, 699. *Major diastereoisomer*: ^{1}H NMR: δ 2.79 (3H, d, $^5J_{HF} = 1.2$ Hz), 4.15 (1H, dd, $^3J_{HF} = 23.2$ Hz, $^4J_{HF} = 3.0$ Hz), 7.37–7.47 (5H, m) ppm. ^{13}C NMR: δ 43.8, 75.4 (d, $^2J_{CF} = 19.0$ Hz), 96.2 (dm, $^1J_{CF} = 220.7$ Hz, $^2J_{CF} = 27.6$ Hz), 120.3 (qdm, $^1J_{CF} = 285.4$ Hz, $^2J_{CF} = 30.2$ Hz), 122.8 (ddd, $^1J_{CF} = 283.6$ Hz, 261.2 Hz, $^2J_{CF} = 23.7$ Hz), 128.8, 128.8, 129.8, 130.2 ppm. ^{19}F NMR: δ −183.44 (1F, m), −85.20 (1F, dq, $^2J_{FF} = 144.5$ Hz, $^4J_{FF} = 12.5$ Hz), −82.18 (1F, d, $^2J_{FF} = 143.8$ Hz), −76.75 (3F, m) ppm. *Minor diastereoisomer*: ^{1}H NMR: δ 2.86 (3H, d, $^5J_{HF} = 1.5$ Hz),

4.25 (1H, dm, $^3J_{HF} = 29.7$ Hz), 7.37–7.47 (5H, m) ppm. ^{13}C NMR: δ 44.1, 79.3 (d, $^2J_{CF} = 26.7$ Hz), 96.2 (dm, $^1J_{CF} = 220.7$ Hz, $^2J_{CF} = 27.6$ Hz), 120.0 (qdm, $^1J_{CF} = 290.5$ Hz, $^2J_{CF} = 32.8$ Hz), 122.2 (ddd, $^1J_{CF} = 278.5$ Hz, 264.7 Hz, $^2J_{CF} = 21.5$ Hz), 128.3 (d, $^3J_{CF} = 2.6$ Hz), 128.8, 129.8, 130.0 ppm. ^{19}F NMR: δ −165.79 (1F, dm, $^3J_{FH} = 30.8$ Hz), −86.80 (1F, dm, $^2J_{FF} = 145.3$ Hz), −81.72 (1F, dm, $^2J_{FF} = 143.8$ Hz), −74.29 (3F, dd, $^4J_{FF} = 16.1$ Hz, $^3J_{FF} =$ 8.8 Hz) ppm. MS (EI): $m/z = 285$ (M^+). Anal. calcd for $C_{11}H_9F_6NO$: C, 46.32; H, 3.18; N, 4.91; F, 39.97. Found: C, 45.87; H, 2.79; N, 4.84; F, 39.98.

Application Numerous uses of isoxazolidines, obtained in 1,3-dipolar cycloaddition of nitrones to alkenes, have been reported in synthesis of biologically attractive compounds. This group of heterocycles shows very high ability to be transformed into polyfunctional amines, including aminoalcohols, aminoketones, and esters. The catalytic cleavage of comparatively weak N–O bond of described fluorinated isoxazolidines leads to fluorides of β-amino acids that undergo cyclization to α-trifluoromethylated β-lactams or form esters of α-trifluoromethylated β-amino acids.

55.2 PREPARATION OF 5-(2,4-DI-*TERT*-BUTOXYPYRIMIDIN-5-YL)-4,5-DIFLUORO-2-METHYL-4-(TRIFLUOROMETHYL)ISOXAZOLIDINE

Apparatus Carius tube, magnetic stirrer, oil bath, safety glasses, laboratory coat, protective gloves.

Chemicals *N*-Methylhydroxylamine hydrochloride, paraformaldehyde, 2,4-di-*tert*-butoxy-5-(perfluoroprop-1-enyl)pyrimidine, triethylamine, anhydrous toluene, $CHCl_3$, CH_2Cl_2, CH_3OH, hexane.

Attention! Use protective gloves and eye protection.

Experimental Procedure A mixture of paraformaldehyde (36 mg, 1.2 mmol), triethylamine (152 mg, 1.5 mmol), and *N*-methylhydroxylamine hydrochloride (100 mg, 1.2 mmol) in anhydrous toluene (10 mL) was placed in a Carius tube and stirred at 70 °C for 1 h. The solution was cooled to room temperature, the 2,4-di-*tert*-butoxy-5-(perfluoroprop-1-enyl)pyrimidine [1] (354 mg, 1 mmol) was then added and the solution was heated at 80 °C for 48 h. The solvent was evaporated and the residue was washed a few times with $CHCl_3$. The combined organic layers were dried with Na_2SO_4 and evaporated under reduced pressure to afford a crude mixture that was purified by column chromatography (silica gel, hexane, a gradient of hexane/CH_2Cl_2, and a gradient of CH_2Cl_2/CH_3OH). Yield: 219 mg, 53%.

Characterization Data White crystals, mp 111–113 °C; diastereoisomeric ratio 4.7:1. ^{1}H NMR ($CDCl_3$, 300 MHz): δ 1.62 *major diastereoisomer* and 1.66 (s, 9H), 1.64 *major diastereoisomer* and 1.67 (s, 9H), 3.00 (bs, 3H), 3.15 (bs, 1H), 4.01 (bs, 1H), 8.37 *major diastereoisomer* and 8.44 (bs, 1H) ppm. ^{19}F NMR ($CDCl_3$, 282 MHz): δ −75.06 and −77.26 *major diastereoisomer* (bs, 3F), −99.52 and −113.63 *major diastereoisomer* (bs, 1F), −165.15 *major*

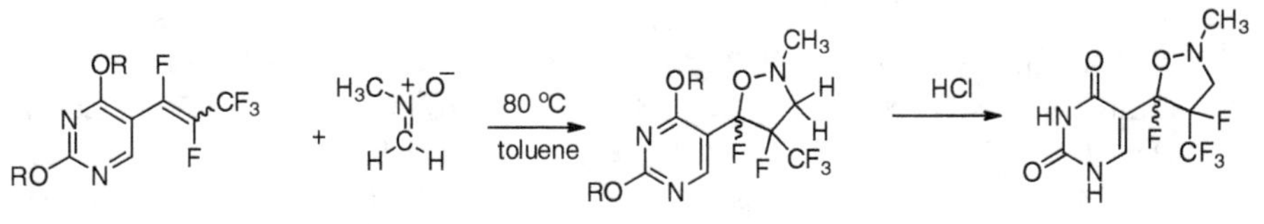

SCHEME 55.2 The synthesis of fluoroisoxazolidynyl derivatives of pyrimidines.

diastereoisomer and −168.57 (bs, 1F) ppm. ^{13}C NMR ($CDCl_3$, 101 MHz): δ 28.3, 28.4, 45.7, 63.4 (bs), 81.1, 83.0, 102.5 (m), 105.2 (d, $^2J_{CF} = 29$ Hz), 114.6 (dd, $^1J_{CF} = 238$ Hz and $^2J_{CF} = 30$ Hz), 121.0 (dq, $^1J_{CF} = 285$ Hz and $^2J_{CF} = 31$ Hz), 158.4 (bs), 165.0, 167.7 (d, $^3J_{CF} = 3$ Hz) ppm. MS (EI): m/z = 413 (M^+). Anal. calcd for $C_{17}H_{24}F_5N_3O_3$: C, 49.39; H, 5.85; N, 10.16. Found: C, 49.44; H, 6.64; N, 10.01.

Application A very suitable method for synthesis of pyrimidines with the fluorinated isoxazolidynyl ring at the 5-position. The di-*tert*-butoxy derivatives can be easily deprotected to the corresponding uracil bearing the heterocyclic ring instead of a sugar moiety. The fluorinated isoxazolidine analogs of C-nucleosides related to pseudouridine, which have a carbon–carbon linkage instead of a hydrolyzable carbon–nitrogen bond between the sugar and the aglycon, are very promising molecules with potential biological activities.

REFERENCES

1. Padwa A, Pearson WH, editors. *Synthetic Applications of 1,3-Dipolar Cycloaddition Chemistry Toward Heterocycles and Natural Products*. Hoboken: John Wiley and Sons; 2003.
2. Stanley LM, Sibi MP. Enantioselective copper-catalyzed 1,3-dipolar cycloadditions. *Chem Rev* 2008;108:2887–2902.
3. Gothelf KV, Jorgrnsen KA. Asymmetric 1,3-dipolar cycloaddition reactions. *Chem Rev* 1998;98:863–909.
4. (a) Bigdeli MA, Tipping AE. Reaction of nitrones with internal fluoroalkenes and with tetrafluorobenzyne. *J Fluorine Chem* 1992;58:101–110. (b) Loupy A, Petit A, Bonnet-Delpon D. Improvements in 1,3-dipolar cycloaddition of nitrones to fluorinated dipolarophiles under solvent-free microwave activation. *J Fluorine Chem* 1995;75:215–216. (c) Nguyen TB, Martel A, Dhal R, Dujardin G. N-(β,β-Difluorovinyl)oxazolidin-2-ones: first synthesis and application in [3 + 2]- and [4 + 2]-cycloaddition-type reactions. *Synlett* 2009;15:2492–2496. (d) Bigotti S, Malpezzi L, Molteni M, Mele A, Panzeri W, Zanda M. Functionalized fluoroalkyl heterocycles by 1,3-dipolar cycloadditions with γ-fluoro-α-nitroalkenes. *Tetrahedron Lett* 2009;50:2540–2542.

5. Jakowiecki J, Loska R, Mąkosza M. Synthesis of α-trifluoromethyl-β-lactams and esters of β-amino acids via 1,3-dipolar cycloaddition of nitrones to fluoroalkenes. *J Org Chem* 2008;73:5436–5441.
6. Wójtowicz-Rajchel H, Migas M, Koroniak H. Simple synthesis of some pentafluoropropenyl derivatives of pyrimidine and purine based on addition-elimination reaction. *J Org Chem* 2006;71:8842–8846.
7. Wójtowicz-Rajchel H, Koroniak H. Synthesis of 5-fluorovinyl derivatives of pyrimidines via Suzuki-Miyaura coupling and their 1,3-dipolar cycloaddition reactions with nitrones. *J Fluorine Chem* 212;135:225–230.
8. Wójtowicz-Rajchel H, Pasikowska M, Olejniczak A, Kartusiak A, Koroniak H. Fluorinated enamines of nucleobases as precursors of nucleoside analogues. Synthesis, spectroscopic and structural studies. *New J Chem* 2010;34:894–902.

CHAPTER 56

Synthesis of Fluorinated Vinyl Derivatives of Nucleic Acid Bases

HANNA WÓJTOWICZ-RAJCHEL, DONATA PLUSKOTA-KARWATKA, AND HENRYK KORONIAK

Since its beginning in the later part of the 19th century when Henri Moissan isolated the elemental fluorine, fluoroorganic chemistry has become the field of great significance. In the current period, the role of fluorine-containing compounds in all aspects of chemical industry (pharmaceuticals, agrochemicals, components of polymers) can be considered as authentically distinctive. Therefore, it is not surprising that there is a constant need to search for novel methodologies for preparation of such molecules.

Our interest has been focused on introduction of fluorine atoms or fluorinated groups into important biomolecules such as nucleic acid bases. This can influence their biological properties and convert them into potentially attractive compounds for medicinal chemistry.

We have used a general procedure for palladium(0)-catalyzed coupling of organometallic, unsaturated fragments with aromatic species [1, 2] to develop efficient method for the preparation of 5-perfluoroalkenyl-substituted uracils [3–5].

1,3-Dimethyl-5-iodouracil reacts smoothly with perfluoroalkenylzinc iodides in the presence of $Pd^0(PPh_3)_4$ catalyst, resulting in the formation of the appropriate derivatives at high and moderate yield (Scheme 56.1).

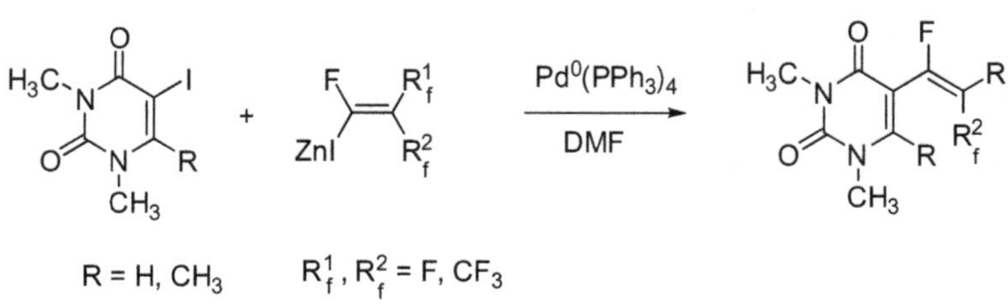

SCHEME 56.1 Synthesis of 5-perfluoroalkenyl derivatives of uracil.

Efficient Preparations of Fluorine Compounds, First Edition. Edited by Herbert W. Roesky.

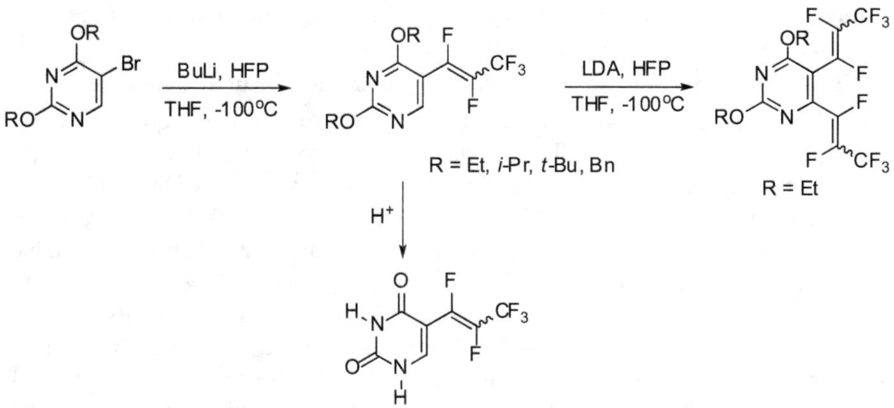

SCHEME 56.2 Synthesis of pentafluoropropenyl pyrimidines based on addition–elimination reaction.

We have also found that various pentafluoropropenyl-substituted pyrimidine and purine bases can be obtained by the method that involves the reaction of appropriate lithium derivatives prepared from both electron-rich and electron-poor pyrimidines with hexafluoropropene (HFP) via an addition–elimination process [6] (Scheme 56.2).

Organolithiums of pyrimidine and purine bases yield the addition–elimination products as *E*/*Z* mixtures.

Unsaturated, fluorinated substituents can also be introduced at N^1 and N^9 positions of appropriate nucleobases. Treatment of various nucleic acid bases and their derivatives with HFP or 1,1,3,3,3-pentafluoropropene in the presence of sodium hydride results in the formation of the corresponding N^1 and N^9 fluorine-containing enamines [7]. The products are obtained with high regioselectivity and without distinct *E*/*Z* stereoselectivity in moderate or high yields (Figure 56.1).

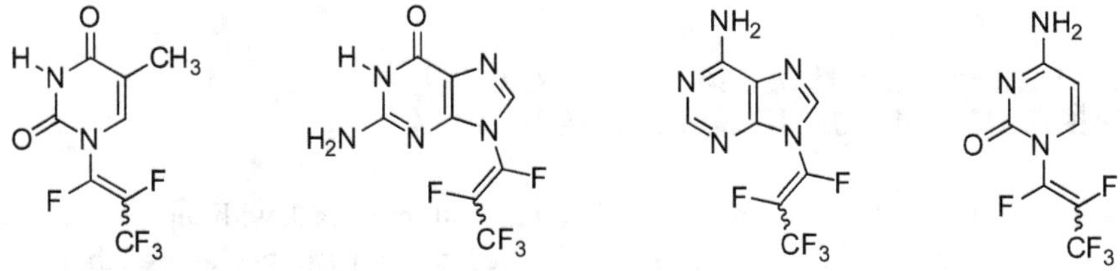

FIGURE 56.1 Examples of synthesized fluorinated enamines of nucleobases.

56.1 PREPARATION OF 1,3-DIMETHYL-5-PERFLUOROVINYLURACIL

Apparatus A 250 mL, round-bottomed flask, reflux condenser with septum, argon inlet, syringe (30 mL), magnetic stirrer, oil bath, safety glasses, laboratory coat, protective gloves.

Chemicals 1,3-Dimethyl-5-iodouracil, trifluorovinylzinc iodide, $Pd^0(PPh_3)_4$, anhydrous DMF, hexane, ethyl acetate, silica gel.

Attention! Safety glasses and protective gloves must be used at all time.

Experimental Procedure In a 250 mL, round-bottomed flask equipped with a magnetic stirrer and a reflux condenser with a septum and argon inlet, 1,3-dimethyl-5-iodouracil (3.2 g, 12 mmol) was dissolved in dry DMF (120 mL). Then $Pd^0(PPh_3)_4$ (350 mg, 0.3 mmol) and CuCl (a few milligrams) were added to the flask. The reactant was kept under argon atmosphere. Perfluorovinylzinc iodide (20 mmol, 1.5 molar excess) as a solution in dry DMF was next added to the reaction mixture by means of a syringe. Under stirring, the reaction mixture was then heated to about 100–110 ° C for 10–20 h. The progress of the reaction was monitored by thin layer chromatography (TLC) checking of the loss of starting 1,3-dimethyl-5-iodouracil. After cooling to ambient temperature, the oily reaction mixture was extracted with hexane/ethyl acetate (3:7 v/v) and after removing the excess of solvent was purified by column chromatography (silica gel, ethyl acetate/hexane, 7:3, v/v). The product was recrystallized from water. Yield: 60%.

Characterization Data Mp 69–71 °C. ^1H NMR (300 MHz, $CDCl_3$): δ 3.39 (s, 3H, N^1-CH_3), 3.47 (s, 3H, N^3-CH_3), 7.42 (s, 1H, C^6-H) ppm. ^{19}F NMR (300 MHz, $CDCl_3$): δ −98 (dd, 1F, C=CF*F*, $^3J_{cis\text{-FF}} = 30$ Hz, $^2J_{gem\text{-FF}} = 68$ Hz), −114.2 (dd, 1F, C=C*F*F, $^3J_{trans\text{-FF}} = 116$ Hz, $^2J_{gem\text{-FF}} = 68$ Hz), −165.5 (dd, 1F, C*F*=CFF, $^3J_{trans\text{-FF}} = 116$ Hz, $^3J_{cis\text{-FF}} = 30$ Hz) ppm.

Application It is an efficient method for introducing unsaturated perfluorinated substituents at 5-position of uracil. The *E* or *Z* geometry of the reacting pentafluoropropenylzinc iodides is fully preserved.

This methodology cannot be applied to the preparation of the analogous 6-substituted uracil derivatives most likely due to the nucleophilic character of the 6-position.

56.2 PREPARATION OF 2,4-DIBENZYLOXY-5-(PERFLUOROPROP-1-ENYL)PYRIMIDINE

Apparatus A 100 mL, three-necked, round-bottomed flask with septum, syringe (5 mL), Carius tube, magnetic stirrer, N_2/hexane bath, safety glasses, laboratory coat, protective gloves.

Chemicals HFP, 2,4-dibenzyloxy-5-bromopyrimidine, *n*-BuLi, anhydrous THF, CH_2Cl_2, hexane, silica gel.

Attention! Safety glasses and protective gloves must be used at all time.

Experimental Procedure To a stirred solution of *n*-BuLi (1.5 mmol) in anhydrous THF (20 mL) at −100 °C, a solution of 2,4-dibenzyloxy-5-bromopyrimidine (1 mmol) was added dropwise. The reaction mixture was kept for 30 min at this temperature before a solution of HFP (approximately

3 mmol) in 5 mL THF was added using a Carius tube. This mixture was kept at the above temperature for an additional 30 min and then allowed to warm to room temperature over a period of 3 h. The solution was evaporated to dryness under reduced pressure, and 10 mL water was added. The aqueous solution was extracted with CH_2Cl_2 (2 × 30 mL), and the combined extracts were dried over Na_2SO_4. The solvent was removed, and the crude product was separated by column chromatography (silica gel, hexane, a gradient of hexane/CH_2Cl_2, CH_2Cl_2) to give a mixture of *E*/*Z* alkenes. Yield: 0.35 g (82%).

Characterization Data Mp 48–51 °C. ^{1}H NMR (300 MHz, $CDCl_3$): δ 5.46–5.50 (m, 4H), 7.36–7.48 (m 10H), 8.28 (s, 1H, isomer *Z*), 8.41 (s, 1H, isomer *E*) ppm. ^{19}F NMR (300 MHz, $CDCl_3$), isomer *E*: δ −66.6 (dd, 3F, $^3J = 11$ Hz, $^4J = 22$ Hz), −135.1 (dq, 1F, $^3J = 141$ Hz, $^4J = 22$ Hz), −164.0 (dq, 1F, $^3J = 141$ Hz, $^3J = 11$ Hz); isomer Z: −66.0 (dd, 3F, $^3J = 8$ Hz, $^4J = 14$ Hz), −110.6 (m, 1F), −149.0 (m, 1F) ppm. Anal. calcd for $C_{21}H_{15}N_2O_2F_5$: C, 59.72; H, 3.58; N, 6.63. Found: C, 59.49; H, 3.49; N, 6.60. IR (KBr) :1720 m (CF=CF), 1685 m, 1648 m, 1598 s, 1553 s, 1435 s, 1361 s, 1207 s, 1145 s cm^{-1}. MS (EI) (*m*/*z*): (M^+) 422. HRMS: calcd. for $C_{21}H_{15}F_5N_2O_2$, 422.10538; found, 422.10845.

Application This is a simple, efficient, and very convenient method for synthesis of various pentafluoropropenyl derivatives of pyrimidine substituted at the 5- or 6-position as well as 5,6-disubstituted. This method can also be applied for the preparation of purine derivatives bearing a pentafluoropropenyl group.

56.3 PREPARATION OF N^1-(PERFLUOROPROP-1-ENYL)URACIL

Apparatus A 100 mL, three-necked, round-bottomed flask with septum, argon inlet, Carius tube, magnetic stirrer, reflux condenser with septum, oil bath, safety glasses, laboratory coat, protective gloves.

Chemicals Uracil, DMF, NaH (60% oil suspension), HFP, CH_3OH, CH_2Cl_2, hexane, silica gel.

Attention! Safety glasses and protective gloves must be used at all time.

Experimental Procedure Uracil (0.224 g, 2 mmol) was dissolved in DMF (5–15 mL) at 60 °C. The solution was cooled to room temperature, and NaH (60% oil suspension; 2.5–4 mmol) was added under argon. The mixture was stirred until evolution of H_2 had ceased. The reaction mixture was once more heated to 70 °C, and HFP (approximately 6 mmol) was added through a Carius tube. The mixture was kept at 70 °C for 15–30 min. After cooling to room temperature, the crude reaction mixture was poured into water (30 mL) and extracted with CH_2Cl_2 (3 × 20 mL). The combined extracts were washed with brine and dried with Na_2SO_4. After removal of solvent under reduced pressure, the residue was purified by silica gel column chromatography (silica gel, a gradient of hexane/CH_2Cl_2, CH_2Cl_2, a gradient of CH_2Cl_2/MeOH) to give a mixture of *E*/*Z* enamine. Yield: 0.343 g (71%).

Characterization Data Mp 135–140 °C. ^{1}H NMR (300 MHz, [D$_6$] DMSO): δ 5.95 (m, 1H, C^5-H), 7.96 (m, 1H, C^6-H), 12.06 (br s, 1H, N^3-H) ppm. ^{19}F NMR (282 MHz, [D$_6$] DMSO), isomer *E*: δ −67.51 (dd, 3F, $J = 11$ and 22 Hz, CF_3), −114.41 (dq, 1F, $J = 22$ and 132 Hz, F^α), −164.98 (dd, 1F, $J = 11$ and 132 Hz, F^β); isomer Z: −68.08 (dd, 3F, $J = 10$ and 12 Hz, CF_3), −94.45 (m, 1F, F^α), −154.65 (m, 1F, F^β) ppm. MS (EI): $m/z = 242$ $[M]^+$ (12%), 199 (100%). Anal. calcd. for $C_7H_3F_5N_2O_2$ (242.12): C, 34.73; H, 1.25; N, 11.57; found: C, 34.55; H, 1.33; N, 11.36.

Application Fluorinated enamines of nucleobases are easily prepared under very mild conditions in moderate to high yields by using readily available HFP, 1,2,3,3,3-pentafluoropropene, or 1,1,3,3,3-pentafluoropropene. They represent versatile intermediates for the synthesis of biologically interesting molecules.

REFERENCES

1. Heinze PL, Burton DJ. Palladium-catalyzed cross-coupling of perfluoroalkenylzinc reagents with aryl iodides. A new, simple synthesis of α,β,β-trifluorostyrenes and the stereoselective preparation of 1-arylperfluoropropenes . *J Org Chem* 1988;53:2714–2720.
2. Morken PA, Burton DJ. Preparation of β,β-difluoro-α-(trifluoromethyl)styrenes by palladium-catalyzed coupling of aryl iodides with pentafluoropropen-2-ylzinc reagent. *J Org Chem* 1993;58:1167–1172.
3. Koroniak H, Karwatka P, Pluskota D, Fiedorow P, Jankowski A. Synthesis of some 5-perfluoroalkenyl derivatives of uracil. *J Fluorine Chem* 1995;71:135–137.
4. Wójtowicz-Rajchel H, Bednarczyk I, Katrusiak A, Koroniak H. Palladium-catalysed cross coupling of perfluoroalkenyl zinc reagents with 5-halo-6-azauracils. *Mendeleev Commun* 2004;14:63–65.
5. Koroniak H, Karwatka P, Cytlak T. Photochemical behaviour of 5-perfluoroalkenyl uracils. *Tetrahedron Lett* 2004;45:5767–5769.
6. Wójtowicz-Rajchel H, Migas M, Koroniak H. Simple synthesis of some pentafluoropropenyl derivatives of pyrimidine and purine based on addition-elimination reaction. *J Org Chem* 2006;71:8842–8846.
7. Wójtowicz-Rajchel H, Koroniak H, Katrusiak A. A convenient method for the synthesis of stable α-fluoro enamines of nucleobases. *Eur J Org Chem* 2008;368–376.

CHAPTER 57

Synthesis of CF_3-Substituted Aziridine Ring by the Gabriel Reaction

TOMASZ CYTLAK, BARTOSZ MARCINIAK, AND HENRYK KORONIAK

The aziridine, three-membered ring system that is the smallest possible saturated nitrogen-containing heterocycle, is one of the most important compounds in modern synthetic chemistry. Its tremendous synthetic utility as a significant building block has expanded enormously in the past decades. Furthermore, the great interest in functionalized aziridines is connected with their potential biological activity. A number of synthetic compounds with aziridine ring in their structure exhibit useful biological properties. In addition, there are several classes of natural compounds containing aziridine framework in their structure.

There are plenty of methods in the literature for the synthesis of substituted aziridines. They can be classified into several approaches as cyclization reactions, transfer of nitrogen to olefins, transfer of carbon to imines, addition across the carbon–nitrogen double bond of azirines, reaction of ylides, aza-Darzens approaches, ring contraction, and functional group transformations [1].

One of these methods is described next. It is based on Gabriel reaction, simple intramolecular nucleophilic substitution of β-bromoamine moiety (Scheme 57.1) [2]. The β-bromoamine **3a** (**3b**) was obtained by addition of amine **2a** (**2b**) to 2-bromo-3,3,3-trifluoropropene (**1**). The reaction occurred in sealed glass tube (Carius tube) during approximately 10 days. Then, the product **3a** (**3b**) was isolated by column chromatography. In the second step, β-bromoamine **3a** (**3b**) was heated in dimethylformamide (DMF) in the presence of triethylamine for 3 h. The aziridine **4a** (**4b**), after removing the solvent, was isolated by column chromatography.

Efficient Preparations of Fluorine Compounds, First Edition. Edited by Herbert W. Roesky.

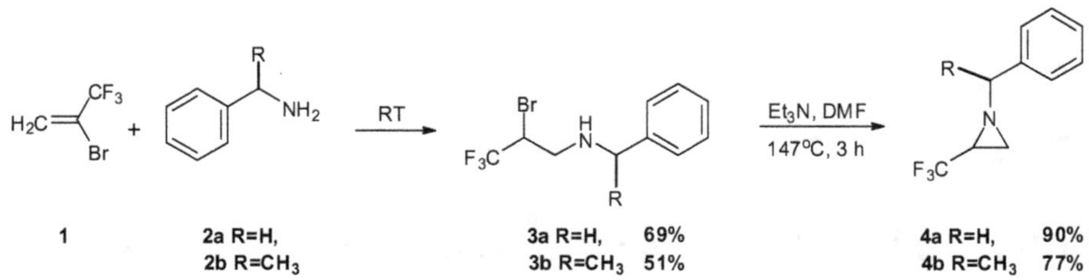

SCHEME 57.1 Synthesis route of obtaining aziridines 4a and 4b

57.1 PREPARATION OF *N*-BENZYL-2-BROMO-3,3,3-TRIFLUOROPROPAN-1-AMINE (3a) AND 2-BROMO-3,3,3-TRIFLUORO-*N*-(1-PHENYLETHYL)PROPAN-1-AMINE (3b)

Apparatus A 50-mL Carius tube with magnetic stirrer, safety glasses, laboratory coat, protective gloves.

Chemicals Benzylamine (**2a**), α-phenylethylamine (**2b**), 2-bromo-3,3,3-trifluoropropene (**1**).

Attention! Safety glasses and protective gloves must be used at all times.

Caution! Because of its toxicity and volatility, care should be taken to avoid inhalation or contact with skin of benzylamine (**2a**), α-phenylethylamine (**2b**), and 2-bromo-3,3,3-trifluoropropene (**1**). All reactions should be carried out in a well-ventilated hood.

Experimental Procedure Amine **2a** (**2b**) (2.5 mmol) was placed in a Carius tube equipped with a magnetic stirrer. Then the tube was connected to a vacuum pump and simultaneously cooled in a dry ice/isopropanol bath. After 5 min, the tube was closed, weighed, and connected to a cylinder with gaseous 2-bromo-3,3,3-trifluoropropene (**1**). Then the valve at the cylinder was opened and next a valve at the tube was gently opened to transfer 2-bromo-3,3,3-trifluoropropene (**1**) (2.5 mmol) to the tube. After closing the tube, the reaction was left at room temperature for approximately 10 days. After this, the content of the tube was dissolved in $CHCl_3$ and washed with saturated aqueous solutions of $NaHCO_3$ and brine, dried over Na_2SO_4, filtered, and evaporated under reduced pressure. The residue was purified by column chromatography with hexane/ethyl acetate (from 95:5 to 90:10, v/v). β-Bromoamines (**3a**, **3b**) are formed as colorless oils with 69% and 51% yield, respectively.

Characterization Data *N*-Benzyl-2-bromo-3,3,3-trifluoropropan-1-amine (**3a**): ^{1}H NMR (300 MHz, $CDCl_3$): δ 7.38–7.26 (m, 5H, C_6H_5), 4.31 (ddq, 1H, $J = 9.3, 7.1, 3.9$ Hz, C*H*Br), 3.87 (d, 1H, $J = 13.3$ Hz, CH_2Ph), 3.81 (d, 1H, $J = 13.3$ Hz, CH_2Ph), 3.21 (dd, 1H, $J = 13.6, 3.9$ Hz, CH_2NH), 3.05 (dd, 1H, $J = 13.5, 9.3$ Hz, CH_2NH) ppm. ^{19}F NMR (282 MHz, $CDCl_3$): δ −71.26 (d, $J = 7.2$ Hz) ppm. GC-MS (EI) (*m*/*z*): 280/282 [M-H], 120, 91 (100%).

2-Bromo-3,3,3-trifluoro-*N*-(1-phenylethyl)propan-1-amine (**3b**): ^{1}H NMR (300 MHz, $CDCl_3$): δ 7.36–7.31 (m, 5H, C_6H_5), 4.19 (ddq, $J = 9.0, 7.1,$

4.2 Hz, 1H, C*H*Br), 3.83 (q, J = 6.6 Hz, 1H, C*H*Ph), 3.05 (dd, J = 13.5, 4.2 Hz, 1H, CH_2NH), 2.95 (dd, J = 13.5, 9.0 Hz, 1H, CH_2NH), 1.39 (d, J = 6.6 Hz, 3H, CH_3) ppm. ^{19}F NMR (282 MHz, $CDCl_3$): δ −71.34 (d, $J = 7.3$ Hz) ppm. GC-MS (*m/z*): 280/282 [M-CH_3] (100%).

57.2 PREPARATION OF 1-BENZYL-2-(TRIFLUOROMETHYL)AZIRIDINE (4a) AND 1-(1-PHENYLETHYL)-2-(TRIFLUOROMETHYL)AZIRIDINE (4b)

Apparatus A 10-mL, round-bottomed flask with magnetic stirrer equipped with condenser, argon inlet, safety glasses, laboratory coat, protective gloves.

Chemicals Previously obtained respective β-bromoamines (**3a**, **3b**), triethylamine, dry DMF.

Attention! Safety glasses and protective gloves must be used at all times.

Caution! Because of its toxicity and volatility, care should be taken to avoid inhalation or contact with skin of β-bromoamines (**3a**, **3b**), triethylamine, and DMF. All reactions should be carried out in a well-ventilated hood.

Experimental Procedure Triethylamine (2 mmol) was added to the solution of β-bromoamine **3a** (**3b**) (1 mmol) in dry DMF (2 mL). The mixture was treated for 3 h at 150 °C in the presence of dry argon. After cooling down to room temperature, DMF was evaporated. The residue was purified by column chromatography eluting with hexane/ethyl acetate (from 90:10 to 50:50, v/v). Pure aziridines (**4a**, **4b**) are isolated as white solids with 90% and 77% yield, respectively.

Characterization Data 1-Benzyl-2-(trifluoromethyl)aziridine (**4a**): ^1H NMR (300 MHz, DMSO): δ 7.40–7.22 (m, 5H, C_6H_5), 3.50 (s, 2H, CH_2Ph), 2.74 (dquin, $J = 6.0, 3.1$ Hz, 1H, $CHCF_3$), 2.04 (d, $J = 3.0$ Hz, 1H, CH_2CH), 1.86 (d, $J = 6.3$ Hz, 1H, CH_2CH) ppm. ^{19}F NMR (282 MHz, DMSO): δ −68.95 (d, $J = 5.7$ Hz) ppm. GC-MS (EI) (*m/z*): 200 [M-H], 132, 110, 91 (100%).

1-(1-Phenylethyl)-2-(trifluoromethyl)aziridine (**4b**): ^1H NMR (300 MHz, $CDCl_3$): δ 7.42–7.32 (m, 5H, C_6H_5), 4.22 (dd, $J = 5.9, 3.7$ Hz, 1H, $CHCH_2$), 2.58 (q, $J = 6.5$ Hz, 1H, C*H*Ph), 2.15 (ddq, $J = 6.3, 5.3, 3.0$ Hz, 1H, $CHCF_3$), 2.04 (dd, $J = 3.7, 3.1$ Hz, 1H, $CHCH_2$), 1.48 (d, $J = 6.6$ Hz, 3H, CH_3) ppm. ^{19}F NMR (282 MHz, $CDCl_3$): δ −71.69 (d, $J = 5.3$ Hz) ppm. GC-MS (EI) (*m/z*): 200 [M-CH_3], 105 (100%).

Application The aziridine ring is one of the most valuable systems in modern synthetic chemistry, because of its widely recognized versatility as tremendously useful synthetic intermediate, significant building block for chemical bond elaborations and functional group transformation [3,4]. Due to the strain associated with the three-membered ring, aziridines undergo highly regio- and stereoselective ring-opening reactions involving nucleophilic substitution, with or without rearrangement, functional group transformations at the ring, etc. [5]. Furthermore, aziridines serve as versatile intermediates in natural product and pharmaceutical synthesis. Reactions with a broad range of nucleophiles proceed cleanly with excellent regioselectivity and/or stereoselectivity, furnishing products that bear useful amino group [6, 7].

REFERENCES

1. Singh GS, D'Hooghe M, De Kimpe N. Synthesis and reactivity of C-heteroatom-substituted aziridines. *Chem Rev* 2007;107:2080–2135.
2. Ignatova YuL, Karimova NM, Kil'disheva OV, Knunyants IL. Synthesis of 1-benzyl-2-trifluoromethylaziridine by the Gabriel reaction. *Russ Chem Bull* 1986;35:675.
3. Osborn HMI, Sweeney J. The asymmetric synthesis of aziridines. *Tetrahedron: asymmetry* 1997;8(11):1693–1715.
4. Sweeney JB. Aziridines: epoxides' ugly cousins? *Chem Soc Rev* 2002;31:247–258.
5. Hu XE. Nucleophilic ring opening of aziridine. *Tetrahedron* 2004;60:2701–2743.
6. Yudin AK, editor. *Aziridines and Epoxides in Organic Synthesis*. Weinheim: Wiley-VCH; 2006.
7. Padwa A, Murphree SS. *Epoxides and aziridines: a mini review. Arkivoc* 2006;(iii):6–33.

CHAPTER 58

Preparation of α-Fluoro Amino and α-Fluoro Enamino Reagents

JUSTYNA WALKOWIAK AND HENRYK KORONIAK

The steadily growing interest in organofluorine chemistry both for the design of new biologically active compounds and new functional materials stimulated considerable interest in the development of new methodologies for the introduction of a fluorine atom into molecules in highly selective ways [1–3]. The nucleophilic replacement of an oxygen function (hydroxyl, carbonyl groups) with fluorine is one of the most straightforward routes.

Fluorinating agents such as 70% hydrogen fluoride (HF)·pyridine (Olah's reagent) [4], sulfur tetrafluoride (SF_4) [5, 6], dialkylaminosulfur trifluoride (e.g., DAST) [1], 4-tert-butyl-2,6-dimethylphenylsulfur trifluoride (FLUOLEAD) [7], and fluoroamino reagents [9–11] (FARs, α-fluoroamines) [*N*,*N*-diethyl-1,1,2,3,3,3-hexafluoropropylamine (Ishikawa reagent) [9], and *N*,*N*-dimethyl-1,1,2,2-tetrafluoroethylamin (Petrov reagent) [10]] are widely employed in organic synthesis for the conversion of alcohols and carbonyl compounds into corresponding fluorides.

Relatively stable fluorinating agents useful for the replacement of hydroxyl group by fluorine are 1,1,3,3,3-pentafluoropropene (PFP) secondary amine adducts [12]. These reagents can be readily prepared by nucleophilic addition of secondary amine [(dimethylamine (DMA), diethylamine (DEA), pyrrolidine (Pyr), piperidine (Pip), morpholine (Mor)] across the double bond of PFP. Irrespective of applied synthetic method (in an autoclave at high pressure or atmospheric pressure), these reactions give a mixture of two products, the desired α-fluoroamine **1a-e**, and α-fluoroenamine **2a-e** in different molar ratios (Scheme 58.1).

These reaction mixtures, however, appear to be efficient fluorinating agents for replacing hydroxyl groups in alcohols by fluorine. In general, they react with alcohols yielding the corresponding fluorides, equimolar amounts of appropriate 3,3,3-trifluoropropionamide, and hydrogen fluoride. Aliphatic primary alcohols including

Efficient Preparations of Fluorine Compounds, First Edition. Edited by Herbert W. Roesky.

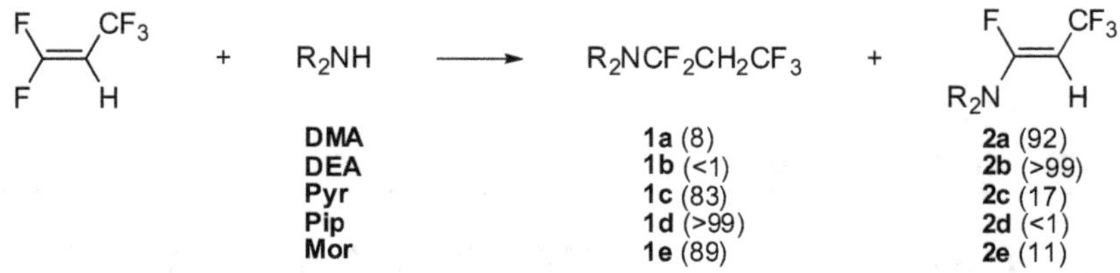

SCHEME 58.1 Synthesis of PFP Secondary Amine Adducts.

octanol and benzylic alcohol yield only alkyl fluorides. The secondary and tertiary alcohols, besides the desired fluorides, give usually considerable amount of alkenes.

The fluorinating mixtures are assumed to be the equilibrium between α-fluoroamine **1a-e** and α-fluoroenamine **2a-e**, where only the HF elimination/addition reaction determines the ratio of products. This also suggests that the presence of even traces of HF will trigger the fluorination. All presented fluorinating mixtures are readily hydrolyzed on being exposed to moist air. They react vigorously with water, forming the *N,N*-diethyl-3,3,3-trifluoropropionamide and hydrogen fluoride.

58.1 PREPARATION OF 1,1,3,3,3-PENTAFLUOROPROPENE SECONDARY AMINE ADDUCTS

Preparation of PFP DEA Adduct

Apparatus Carius tube, magnetic stirrer, 20-mL syringe, safety glasses, laboratory coat, protective gloves

Chemicals Dry DEA, PFP, dry diethylether

Attention Safety glasses and protective gloves must be used at all times.

Caution Because of volatility and toxicity, care should be taken to avoid inhalation of PFP, DEA, and diethyl ether, or contact of their solution with the skin. Fluorinating agent readily hydrolyzed being exposed to moist air forming the *N,N*-diethyl-3,3,3-trifluoropropionamide and hydrogen fluoride. Therefore, any contact with eyes, skin, or respiratory system should be avoided. All reactions should be carried out in a well-ventilated fume hood.

Experimental Procedure Diethylamine (10.5 g, 15 mL, 144 mmol) and dried diethylether (30 mL) were placed in an oven-dried Carius tube. Then the tube was cooled to –78 °C by a dry ice/isopropanol bath. The excess of PFP (22.7 g, 172 mmol) was transferred to the tube from a balloon via a needle (over 3 h). The mixture was stirred overnight while warming to room temperature. The solvent was removed in vacuum. The resulting orange residue was purified by distillation in a vacuum to yield 22.6 g (100.5 mmol, 70%) of pale yellow liquid [mixture of 1-diethylamino-1,3,3,3-tetrafluoropropene, **2b** and *N,N*-diethyl-1,1,3,3,3-pentafluoropropylamine, **1b** (less than 1%)].

Characterization Data bp 95 °C/72 mm Hg. ^{1}H NMR: δ 1.15 (t, $^3J_{H,H} = 7.1$ Hz, 6H), 3.14 (q, $^3J_{H,H} = 7.1$ Hz, 4H), 3.73 (dq, $^3J_{H,F} = 34.0$ Hz, $^3J_{H,F} = 7.2$ Hz, 1H)

ppm. ^{13}C NMR: δ 12.4 (s, 2C), 42.1 (s, 2C), 65.5 (dq, $^{2}J_{C,F}$ = 18.2 Hz, $^{2}J_{C,F}$ = 38.0 Hz, 1C), 125.8 (q, $^{1}J_{C,F}$ = 266.5 Hz, 1C), 160.7 (d, $^{1}J_{C,F}$ = 256.3 Hz, 1C) ppm. ^{19}F NMR: δ –51.8 (dd, $^{3}J_{F,H}$ = 7.1 Hz, $^{4}J_{F,F}$ = 15.2 Hz, 3F), –93.3 (dq, $^{3}J_{F,H}$ = 34.0 Hz, $^{4}J_{F,F}$ = 15.2 Hz, 1F) ppm.

58.2 PREPARATION OF PFP MORPHOLINE ADDUCT (PFPMor)

Apparatus Carius tube, magnetic stirrer, 5-mL syringe, safety glasses, laboratory coat, protective gloves

Chemicals Dry Mor, PFP, dry diethylether

Attention Safety glasses and protective gloves must be used at all times.

Experimental Procedure Morpholine (0.27 g, 0.27 mL, 3 mmol) and dried diethylether (3 mL) were placed in an oven-dried Carius tube, and then the tube was cooled to –78 °C by a dry ice/isopropanol bath. The excess of PFP (22.7 g, 172 mmol) was transferred to the tube portionwise. The mixture was stirred overnight while warming up to room temperature. The resulting white precipitate [mixture of 4-(1,1,3,3,3-pentafluoropropyl)morpholine, **1e** and 4-(1,3,3,3-tetrafluoroprop-1-enyl)morpholine, **2e**] is unstable under high vacuum distillation. Therefore, crude mixture of products **1e** and **2e** is used for the nucleophilic fluorination of hydroxyl compounds.

Characterization Data ^{19}F NMR: δ –52.5 (dd, $^{3}J_{F,H}$ = 7.0 Hz, $^{4}J_{F,F}$ = 15.2 Hz, 3F), –61,9 (tt, $^{3}J_{F,H}$ = 9.6 Hz, $^{4}J_{F,F}$ = 19.6 Hz, 3F), –87.3 (tq, $^{3}J_{F,H}$ = 11.0 Hz, $^{4}J_{F,F}$ = 19.6 Hz, 2F,), –91.9 (dq, $^{3}J_{F,H}$ = 31.6 Hz, $^{4}J_{F,F}$ = 15.2 Hz, 1F) ppm.

58.3 APPLICATION OF PFP SECONDARY AMINE ADDUCTS IN FLUORINATION REACTIONS

Tables 58.1 and 58.2 compare the reactivity of compounds possessing primary, secondary, and tertiary hydroxyl groups. The general procedure for each fluorination reaction is as follows: in case of PFPDEA (Scheme 58.2, reaction 1) alcohol solution is mixed with fluorinating reagent in molar ratio of 1:1.2, while for the reactions of PFPMor (Scheme 58.2, reaction 2) alcohol solution is added drop by drop directly into the crude fluorinating mixture (alcohol : Mor, molar ratio of 1:3). Yields are determined by ^{19}F NMR. Two internal references were used: α,α,α-tri-fluorotoluene ^{19}F NMR ($CDCl_3$, $CFCl_3$) δ –63.2 (s, 3F, CF_3) and *m*-fluorotoluene ^{19}F NMR ($CDCl_3$, $CFCl_3$) δ –114.1 (m, 1F, CF).

In general, the yields of reactions with α-fluoroamino and α-fluoroenamino mixtures are slightly lower in comparison with Ishikawa's reagent. Studies on optimization of the effect of stoichiometry on the fluorination of alcohols with PFP secondary amine adducts proved that these fluorinating mixtures are viable alternative reagents useful for the replacement of hydroxyl groups by fluorine. Table 58.3 compares the reactivity of the least reactive alcohols, such as 5-cholesten-3β-ol, (*R*)-(-)-2-octanol *tert*-butanol with PFPDEA.

TABLE 58.1 Fluorination of Alcohols with PFPDEA

		^{19}F NMR		
Alcohol	Yields %	δ, ppm	$^2J_{F,H}$, Hz	$^2J_{F,H}$, Hz
$CH_3(CH_2)_6CH_2OH$	74	–218.3	48	25
$PhCH_2OH$	64	–206.9	48	–
5-cholesten-3β-ol	33	–167.7	50	–
(*R*)-(–)-2-oktanol	54	–174.5	–	–
tert-Butanol	38	–130.7	–	21

Yields are determined by ^{19}F NMR.

TABLE 58.2 Fluorination of Alcohols with PFPMor

		Yield, %		^{19}F NMR		
Alcohol	Reaction condition, °C	1e	Ishikawa's reagent	δ, ppm	$^2J_{F,H}$, Hz	$^3J_{F,H}$, Hz
$CH_3(CH_2)_6CH_2OH$	25	18	87	–218.3	48	25
	65	**89**				
$PhCH_2OH$	25	82	60	–206.9	48	–
	65	**88**				
$PhCH_2CH_2OH$	25	30	89	–215.0	48	23
	65	**89**				
$CH_3CH(OH)(CH_2)_2CH_3$	25	**69**	62	–172.0	–	–
5-Cholesten-3β-ol	25	77	83	–167.7	50	–
	65	**89**				
Cyclohexanol	25	7	0	–172.6	–	–
	65	7				
tert-Butanol	25	**48**	78	–130.7	–	21
	65	2				

Yields are determined by ^{19}F NMR. Best yields are presented in bold.

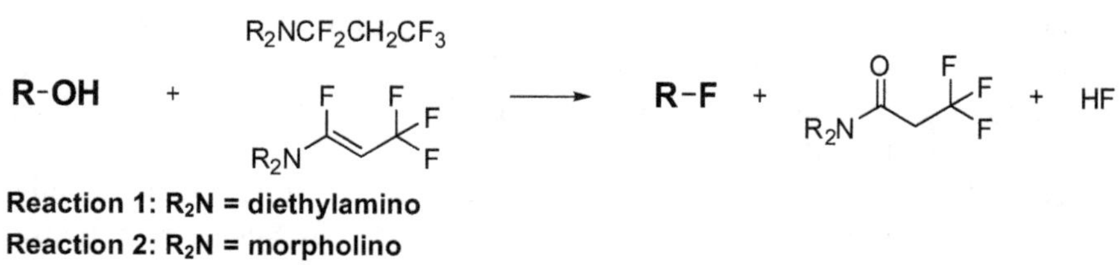

SCHEME 58.2 Fluorination of Alcohols with PFP Secondary Amine Adducts

TABLE 58.3 Optimization of the Fluorination Reactions with PFPDEA

	Fluoride yield, %		
Molar ratio ROH : 2b	(*R*)-(−)-2-octanol	5-cholesten-3β-ol	*tert*-butanol
1:1	54	33	38
1:2	75	79	**84**
1:3	**85**	**93**	71
	Fluoride yield with Ishikawa's reagent, %		
1:1.5	62	83	78

Yields are determined by ^{19}F NMR. Best yields are presented in bold.

Table 58.4 represents stoichiometry investigation on reactivity of PFPMor in fluorination reactions of benzyl alcohol.

TABLE 58.4 Optimization of the Fluorination Reactions of Benzyl Alcohol with PFPMor

	Benzyl fluoride yield, %	
Molar ratio ROH : Mor	25 °C	65 °C
1:1	48	46
1:2.5	99	77
1:3	88	92
	Benzyl fluoride yield with Ishikawa's reagent, %	
1:1.5	60	–

Yields are determined by ^{19}F NMR.

The yields of fluorination reactions studied in competitive studies using Ishikawa's reagent are comparable with those already reported in the literature.

REFERENCES

1. Hudlicky M, Pavlath AE. *Chemistry of Organic Fluorine Compounds*. Volume II. Washington, DC: American Chemical Society; 1995.
2. Olah GA, Chambers RD, Prakash GKS, editors. *Synthesis Fluorine Chemistry*. New York: Wiley; 1992. p 163.
3. Chambers RD. *Fluorine in Organic Chemistry*. Oxford: Blackwell; 2004.
4. Olah GA, Welch JT, Vankar YD, Nojima M, Kerekes I, Olah JA. Synthetic methods and reactions. 63. Pyridinium poly(hydrogen fluoride) (30% pyridine - 70% hydrogen fluoride): a convenient reagent for organic fluorination reactions. *J Org Chem* 1979;44:3872–3881.

5. Dmowski W, Koliński R. Reactions of sulfur tetrafluoride with carboxylic acids. Part II*. Reactions with acetic acid, propionic acid and the halo-acids. *Roczniki Chem* 1973;47: 1211–1221.
6. Hasek WR, Smith WC, Engelhardt VA. Chemistry of sulfur tetrafluoride.[1] III. Organoiminosulfur difluorides. *J Am Chem Soc* 1960;82:543–551.
7. (a) Umemoto T, Xu Y. *Substituted Phenylsulfur Trifluoride and Other Like Fluorinating Agents*. US patent 7,265,247 B1; (b) Umemoto T, Singh RP. *Substituted Phenylsulfur Trifluoride and Other Like Fluorinating Agents*. US patent 7,381,846 B2.
8. German LS, Zemskow S. *New Fluorinating Agents in Organic Synthesis*. New York: Springer; 1989. p 254.
9. Ishikawa N, Iwakiri H, Takaoka A. F-propene-dialkylamine reaction products as fluorinating agents. *Bull Chem Soc Jpn* 1979;52:3377–3380.
10. Petrov WA, Swearingen S, Hong W, Petersen WC. 1,1,2,2-Tetrafluoroethyl-*N,N*-dimethylamine: a new selective fluorinating agent. *J Fluorine Chem* 2001;109: 25–31.
11. Dmowski W, Kamienski M. Dialkyl-α,α-difluorobenzylamines and dialkyl(trifluoromethyl)amines novel fluorinating reagents. *J Fluorine Chem* 1983;23:219.
12. Koroniak H, Walkowiak J, Grys K, Rajchel A, Alty A, Du Boisson R. 1,1,3,3,3-Pentafluoropropene secondary amine adducts new selective fluorinating agents. *J Fluorine Chem* 2006;127:1245–1251.

CHAPTER 59

Synthesis of Original Fluoromonomers and Their Radical Copolymerization with Vinylidene Fluoride

BRUNO AMEDURI

Fluoropolymers are advanced materials endowed with remarkable properties that involved them in many High Tech applications [1–6]. One of the strategies to produce them is to get original fluorinated monomers that can copolymerize with commercially available fluoroolefins. Among various known pathways, the telomerization of various fluoroolefins is an interesting tool and was summarized several years ago [6].

The objective of this chapter is to supply relevant functional fluoropolymers from commercially available fluoroolefins. The involved strategy lies on the radical addition of a chain transfer agent onto a fluoroalkene, followed by straightforward organofluorine pathways to achieve functional monomers. The latters are further copolymerized with commercially available fluoroolefins, especially vinylidene fluoride (VDF).

59.1 RADICAL ADDITION OF IODINE MONOCHLORIDE ONTO FLUORINATED ALKENES

A typical example concerns the telomerization of various fluoroalkenes with iodine monochloride. Indeed, the radical addition of ICl onto different F-olefins (such as chlorotrifluoroethylene [7, 8], trifluoroethylene [9, 10], VDF [10], hexafluoropropylene (HFP) [11, 12], 1,1-difluorodichloroethylene [13], and perfluoromethyl vinylether [14]) was achieved by many scientists. However, in the case of asymmetric alkenes, these researchers paid a few attention to the presence of non-expected by-product arising from the reverse addition. Taking into account that the polarity of iodine monochloride is $\delta + \mathrm{I} \ldots \mathrm{Cl}\delta +$ and regarding polar, ionic, and steric effect, some reversed isomer cannot be neglected.

Efficient Preparations of Fluorine Compounds, First Edition. Edited by Herbert W. Roesky.

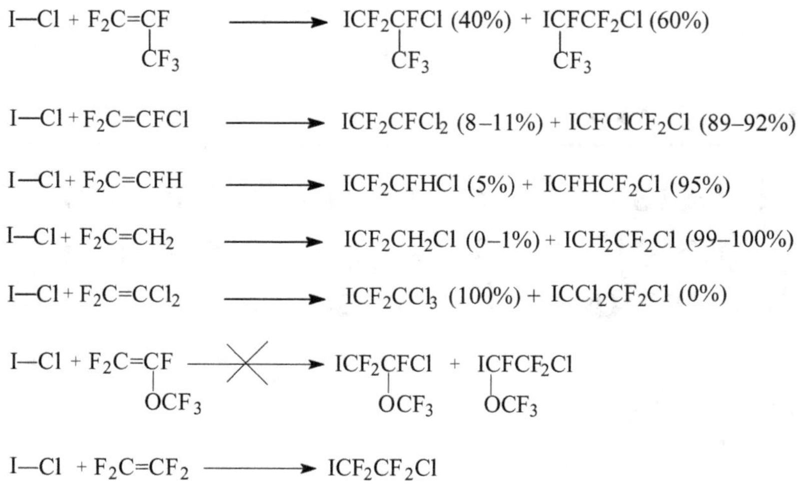

SCHEME 59.1 Radical addition of iodine monochloride onto various commercially available fluoroolefins [7–14].

In most cases, the different radical initiations (photochemical, thermal, or the use of redox catalysts or radical initiators) were investigated and the results are summarized in Scheme 59.1.

Interestingly, it is observed that the lower the fluorine content, the higher the selectivity.

The monoadduct was obtained in all cases, leading to original activated telogens containing CF2I or, better, CFClI end group [7, 8]. The addition of ICl onto 1,1-dichlorodifluoroethylene was found to be bidirectional but to a lesser extent [7] than in the case of CTFE [8]. The use of a catalyst (e.g., iron) affects the yield and the amount of by-product. HFP requires heat, unlike other monomers [7]. Hence, the following decreasing reactivity series can be suggested:

$$\text{CTFE} > \text{VDF} > \text{CF}_2 = \text{CCl}_2 > \text{HFP} > \text{PMVE}$$

Perfluoroalkyl iodides have been used as telogens with most fluoroalkenes. An extensive kinetic research on the synthesis of monoadducts was performed by Tedder and Walton. From various results of the literature [7–14], the following decreasing reactivity scale may be proposed:

$$\text{PMVE} > \text{VDF} > \text{TFE} > \text{F}_2\text{C} = \text{CFH} > \text{HFP} > \text{CTFE}$$

For less reactive monomer the more electrophilic the telogen radical, the less easier the reaction.

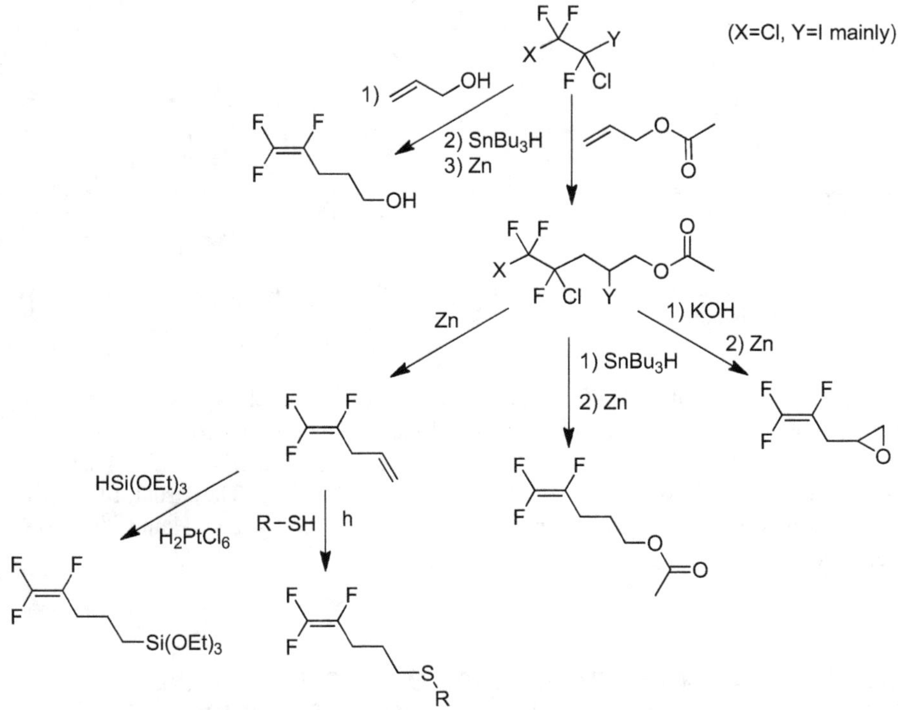

SCHEME 59.2 Various pathways to synthesize trifluorovinyl functional monomers from C2 telomers [15–20].

59.2 USE OF THESE FLUORINATED SYNTHONS TO PREPARE ORIGINAL FLUOROFUNCTIONAL MONOMERS

The above intermediates were interesting precursors of trifluorovinyl monomers as summarized in Scheme 59.2.

An example of ^{19}F NMR spectrum (Figure 59.1) shows the characteristic doublet of doublets or doublets of doublets of triplets assigned to the multiplicity of the coupling between the fluorine and fluorine atoms or hydrogen atoms.

More recently, a difluorocontaining monomer could be synthesized from a three-step procedure starting from the radical addition of methanol onto CFCl=CFCl (Scheme 59.3). The third and final step consists of a zinc-mediated dechlorination to obtain α,β-difluoroacrylic acid in ca. 29% of overall yield [21].

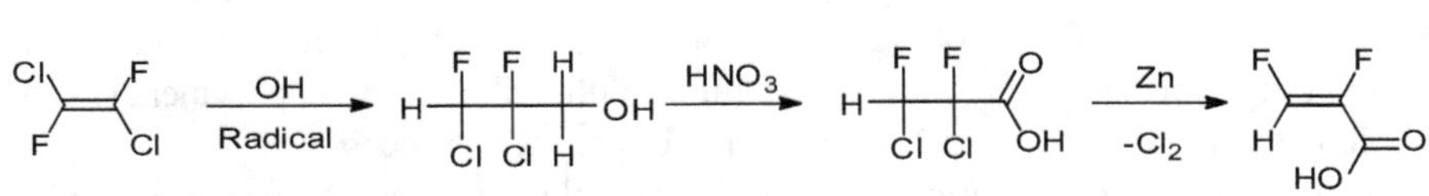

SCHEME 59.3 Strategy of synthesis of α,β-difluororacrylate from 1,2-dichloro-1,2-difluoroethylene [21].

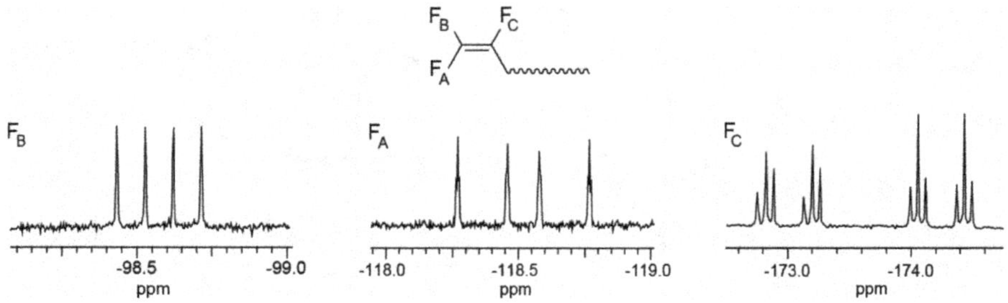

FIGURE 59.1 Expansion of the appropriate zones in the ^{19}F NMR spectrum of trifluorovinyl functional monomers listed in Scheme 59.2.

Only (*Z*) isomer of α,β-difluoroacrylic acid was obtained as this isomer is the thermodynamically favored compound [21]. Indeed, the coupling constant $^{19}F_{\alpha}$-$^{19}F_{\beta}$ from 60 to 70 Hz (see Experimental Procedure) is characteristic of (*Z*) isomer, whereas that of (*E*) isomer is much higher ($J_{F\text{-}F}$ = 250 Hz). The structure of (*Z*)-α,β-difluoroacrylic acid was confirmed by elemental analysis and ^{1}H and ^{19}F NMR spectroscopy [21].

59.3 RADICAL COPOLYMERIZATION OF TRIFLUOROVINYL FUNCTIONAL MONOMERS WITH VINYLIDENE FLUORIDE

The radical copolymerizations of VDF with various commercially available or synthesized monomers were investigated and recently reviewed [22; Scheme 59.4).

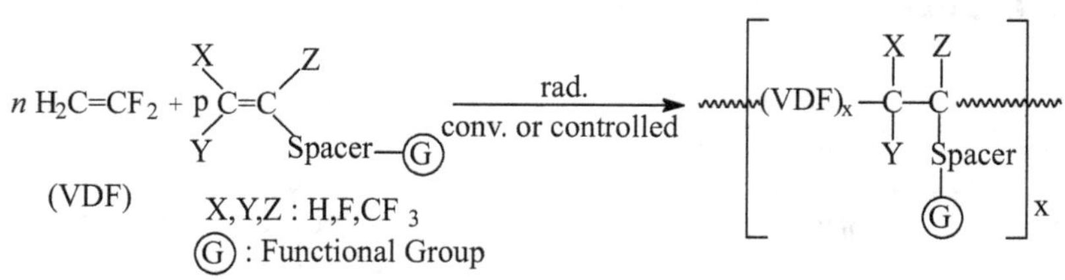

SCHEME 59.4 Radical copolymerization of VDF with fluorofunctional comonomers.

Actually, a very interesting comonomer is α-trifluoromethacrylic acid (that propagates under anionic initiation [23] but does not homopolymerize under radical conditions [24]) since it copolymerizes with VDF very well [25,26]. The nature of the process enables to obtain various structures, either alternating copolymers (from solution copolymerization [25]) or random (from aqueous process [26]).

In addition, the condensation of an ω-hydroxy oligo(ethylene oxide), oligo(EO), with α-trifluoromethacrylic acid led to original EO macromonomer that yielded PVDF-*g*-POE graft copolymers, as follows [27; Scheme 59.5):

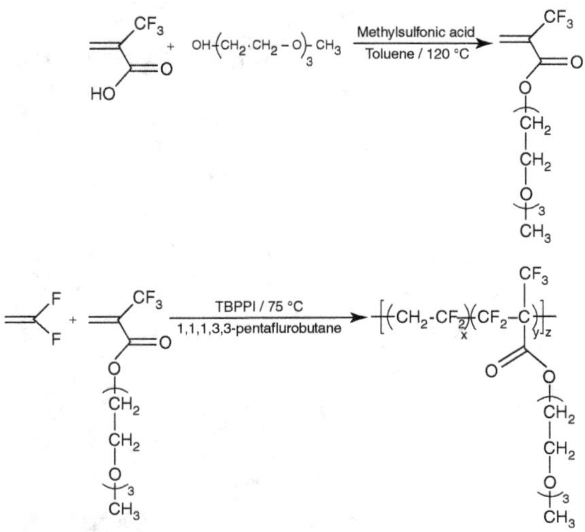

SCHEME 59.5 Synthesis of original α-trifluoromethacrylic monomers bearing EO oligomers and their radical copolymerization with VDF to obtain PVDF-*g*-PEO graft copolymers [27].

Controlled Radical Copolymerization of VDF with Various Comonomers

Few controlled radical (co)polymerizations of fluoromonomers (involving VDF) have been investigated in the literature, and this was recently reviewed [28]. An example in Scheme 59.6 proposes the iodine transfer copolymerization of VDF with perfluoromethyl vinylether that was successfully controlled by the presence of iodoperfluoroalkanes [14].

Radical copolymerization of VDF with trifluorovinyloxy comonomers bearing sulfonyl fluoride end-group (PFSVE) [29] was also investigated and the resulting purified copolymers were hydrolyzed in sulfonic acid, and further processed into membranes (Figure 59.2) for fuel cell applications.

A more recent study deals with the radical terpolymerization of VDF with PFSVE and vinyl tri(ethoxysilane) [30] for original crosslinked membranes (via a sol-gel procedure), the conductivities of which reached 43 mS/cm at 120 °C. Other methods of crosslinking have also been achieved by a photochemical irradiation of telechelic diacrylates [31] or under acidic conditions for a wide variety of telechelic bis(diethoxymethylsilane)s [32].

Experimental Procedure

Preparation of Trifluorovinyl Functional Monomers

Apparatus One hundred (or 250) milliliter Schlenk flask with septum, magnetic stirrer, T-shaped N_2 inlet, syringe (50 mL), safety glasses, laboratory coat, protective gloves

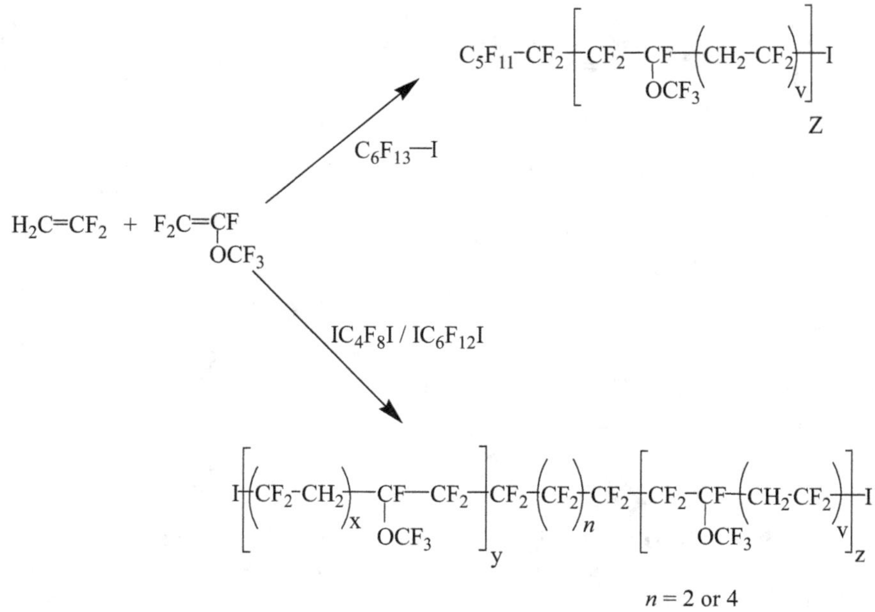

SCHEME 59.6 Iodine transfer radical copolymerization of VDF with perfluoromethyl vinyl ether controlled by iodoperfluoroalkanes [14].

Chemicals Vinylidene fluoride, radical initiators such as *tert*-butyl peroxypivalate, or *tert*-butyl cyclohexyl peroxybicarbonate, dry acetonitrile, iodine monochloride, tributyltin hydride, and 1,1,1,3,3-pentafluorobutane

Attention! Safety glasses and protective gloves must be used at all times.

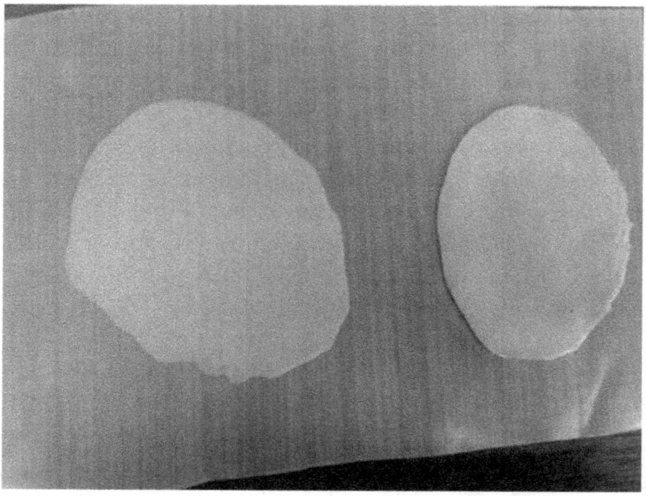

FIGURE 59.2 Original membranes from copolymers based on VDF and sulfonic acid containing fluorinated comonomers. (For a color version of the figure, please see color plates.)

Caution! Because of its toxicity and volatility, care should be taken to avoid inhalation of chlorotrifluoroethylene, iodine monochloride, and tributyltin hydride or contact of their solution with the skin. All reactions should be carried out in a well-ventilated hood.

Characterization The purity of trifluorovinyl monomers and the total product mixtures were analyzed by gas chromatography (GC) using a Delsi apparatus (model 330) equipped with an SE-30 column, 3 m × 0.3125 mm (internal diameter). The nitrogen pressure at the entrance to the column was maintained at 0.6 bar and the detector and injector temperatures were 260 and 255 °C, respectively. The temperature program started from 50 °C and reached 250 °C at a heating rate of 15 °C/min. The GC apparatus was connected to a Hewlett–Packard integrator (model 3390), which automatically calculated the area of each peak on the chromatogram. The products were characterized by ^{1}H and ^{19}F NMR spectroscopy at room temperature (ca. 23 °C). These spectra were recorded on Bruker WM-360 or 400 MHz instruments, using deuterated acetone or dimethyl formamide and tetramethyl silane or $CFCl_3$ as solvents and internal references, respectively.

Synthesis of 1-Iodo-1,2,2-Trifluoro-1,2-Dichloroethane A Carius tube (CT) containing iodine monochloride (113.73 g, 0.70 mmol), benzophenone (0.45 g, 2.5 mmol), and methylene chloride (200 mL) was frozen in an acetone/liquid nitrogen mixture (ca. −80 °C) where chlorotrifluoroethylene (82.2 g, 0.704 mol) was condensed. Then, the CT was sealed with the flame, and UV irradiated at room temperature for 14 h. After filtration and work-up, 1-iodo-1,2,2-trifluoro-1,2-dichloroethane was distilled under reduced pressure and bp 98–100 °C. The yield was 59%. ^{19}F NMR ($CDCl_3$), δ −54.9 (d, $^{3}J_{FF}$ = 14 Hz, Cl_2**CF**); −63.5 (AB system, $^{3}J_{FFgem}$ = 13.9 Hz, $^{3}J_{FFgem}$ = 15.7 Hz, $^{2}J_{FFvic}$ = 164 Hz, $\mathbf{CF_2}Cl$); −67.5 (t, $^{3}J_{FF}$ = 13.9 Hz, $I\mathbf{CF_2}$); −72.0 (t, $^{3}J_{FF}$ = 14.0 Hz, **CF**ClI) ppm.

Synthesis of α, β-Difluoroacrylic Acid [21] ^{1}H NMR ($CDCl_3$), δ 7.4 (dd, $^{2}J_{H(F\text{-}3)}$ = 69.6 Hz, $^{3}J_{H(F\text{-}2)}$ = 13.8 Hz, **HFC**, 1H); 9.0 (s, COO**H**) ppm.

^{13}C NMR ($CDCl_3$), δ 138.4 (dd, $^{1}J_{C(F\text{-}2)}$ = 255.4 Hz, $^{2}J_{C(F\text{-}3)}$ = 10.2 Hz, **C**F); 145.4 (dd, $^{1}J_{C(F\text{-}3)}$ = 280.3 Hz, $^{2}J_{C(F\text{-}2)}$ = 10.2 Hz, HF**C**); 158.0 (dd, $^{2}J_{C(F\text{-}2)}$ = 30 Hz, $^{3}J_{C(F\text{-}3)}$ = 9.5 Hz, **C**OOH) ppm.

^{19}F NMR ($CDCl_3$), δ −139.4 (dd, $^{3}J_{F(F\text{-}2)}$ = 6.3 Hz, $^{2}J_{F(H\text{-}3)}$ = 69.9 Hz, H**F**C); −152.4 (dd, $^{3}J_{F(H\text{-}3)}$ = 13.5 Hz, $^{3}J_{F(F\text{-}3)}$ = 6.7 Hz, C**F**) ppm.

Radical Copolymerization of Trifluorovinyl Functional Monomers or α,β-Difluoroacrylic Acid with VDF

The investigations (even for kinetics) of copolymerization of VDF with fluorofunctional monomers (or α,β-difluoroacrylic acid) were performed in thick borosilicate

FIGURE 59.3 Picture of manifold used to prepare the sealed CTs for the radical copolymerization of VDF with fluorofunctional comonomers. (For a color version of the figure, please see color plates.)

CTs in a batch process (length, 130 mm; internal diameter, 10 mm; thickness, 2.5 mm; for a total volume of 8 mL). After the initiator, $F_2C{=}CF$-spacer-G or α,β-difluoroacrylic acid and the solvent were added, the tube was connected to a manifold (Figure 59.3) that was linked to a vacuum line and purged several times by evacuating and flushing with helium. After at least five thaw–freeze cycles to remove oxygen, VDF was trapped under vacuum in the tube frozen in liquid nitrogen, after release in an intermediate metallic container calibrated in pressure. The required amount of VDF (ranging between 0.385 and 1.300 $\pm$ 0.008 g) introduced into the tube was assessed by the relative drop of pressure in this release container, initially fed by a cylinder of 300 g of VDF. A beforehand calibration curve—weight of trapped VDF (in g) versus drop of pressure (in bar)—was determined (for 0.385 and 0.900 g of VDF, differences of pressure of 0.26 and 0.60 bar were required, respectively). The tube, under vacuum and immersed in liquid nitrogen, was sealed under vacuum and placed into a shaking oven at the required temperature. After reaction (ca. 5 times higher than the half-life of the initiator at the required temperature), the tube was frozen in liquid nitrogen and then opened (Figure 59.4). The (co)polymers were isolated by removal of the solvent. The (co)polymers were further dried under high vacuum at 70 °C until constant weight. The obtained powders were then weighed to determine the yield, and characterized by the usual spectroscopic techniques.

FIGURE 59.4 Picture of CTs after the radical copolymerization of VDF with α,β-difluoroacrylic acid. (For a color version of the figure, please see color plates.)

59.4 CONCLUSION

Novel strategies to synthesize fluorinated functional comonomers for the conventional or controlled radical copolymerization with VDF are still a challenge. The function brought by the comonomer is crucial to enable the targeted properties: conductivity, crosslinking, and the kinetic of copolymerization deserves to be investigated to compare the reactivities of both comonomers. That also induces to assess the content of VDF in the copolymer to offer either elastomeric or thermoplastic morphologies. Hence, possible applications such as piezoelectric devices, surfactants, fuel cell membranes, coatings, and polymer electrolytes for lithium ion batteries can be obtained. These applications are still key challenges to face and should attract the interest of many academic and industrial researchers.

REFERENCES

1. Wall LA. *Fluoropolymers*. New York: Wiley Interscience; 1972.
2. Feiring AE. In: Banks RE, Smart BE, Tatlow JC, editors. *Organofluorine Chemistry: Principles and Commercial Applications*. Volume 15. New York: Plenum Press; 1994. p 339–372.
3. Scheirs J. *Modern Fluoropolymers*. New York: John Wiley & Sons; 1997.
4. Hougham G, Cassidy PE, Johns K, Davidson T. *Fluoropolymers 2: Properties*. New York: Kluwer/Plenum; 1999.
5. Ameduri B, Boutevin B, Kostov G. Fluoroelastomers: synthesis, properties and applications. *Prog Polym Sci* 2001;26:105–187.

6. Ameduri B, Boutevin B. *Well-Architectured Fluoropolymers: Synthesis, Properties and Applications*. Amsterdam: Elsevier; 2004.
7. Haupstchein M, Braid M, Fainberg A. Addition of iodine halides to fluorinated olefins. I. The direction of addition of iodine monochloride to perhaloölefins and some related reactions. *J Am Chem Soc* 1961;83:2495–2500.
8. Ameduri B, Boutevin B, Kostov GK, Petrova P. Novel fluorinated monomers bearing reactive side groups. Part 1. Preparation and use of ClCF2CFClI as the telogen. *J Fluor Chem* 1995;74:261–267.
9. Ameduri B, Boutevin B, Kharroubi M, Kostov G, Petrova P. Radical addition of iodine monochloride to trifluoroethylene. *J Fluor Chem* 1998;91:41–48.
10. Kharroubi M, Manseri A, Ameduri B, Boutevin B. Radical addition of iodine monochloride to vinylidene fluoride. *J Fluor Chem* 2000;103:145–153.
11. Manseri A, Ameduri B. Unpublished results.
12. (a) Chambers RD, Musgrave WKR, Savory J. The addition of "halogen monofluorides" to fluoroolefins. *Proc Chem Soc* 1961;113–118; (b) Chambers RD, Musgrave WKR, Savory J. Mixtures of halogens and halogen polyfluorides as effective sources of the halogen monofluorides in reactions with fluoroolefins. *J Chem Soc* 1961;3779–3786.
13. Bissel J. Some isomers of chloroiodotrifluoroethane. *J Chem Soc* 1963;227–231.
14. Boyer C, Ameduri B, Hung MH. Telechelic Diiodopoly(VDF-*co*-PMVE) copolymers by iodine transfer opolymerization of vinylidene fluoride (VDF) with perfluoromethyl vinyl ether (PMVE). *Macromolecules* 2010;43:3652–3663.
15. Ameduri B, Boutevin B, Fruchier A, Kostov G, Petrova P. Synthesis and polymerization of fluorinated monomers bearing a reactive lateral group. Part 2. Synthesis of nonconjugated dienes. *J Fluor Chem* 1998;89:167–172.
16. (a) Ameduri B, Boutevin B, Kostov G, Petrova P. Synthesis and polymerization of fluorinated monomers bearing a reactive lateral group. Part 9. Bulk copolymerization of vinylidene fluoride with 4,5,5-trifluoro-4-ene pentyl acetate. *Macromolecules* 1999;32:4544–4550; (b) Petrova P, Ameduri B, Boutevin B, Kostov G. US patent 6,753,392 (assigned to Solvay). 2004.
17. Ameduri B, Boutevin B, Kostov G, Petrova P. Synthesis and polymerization of fluorinated monomers bearing a reactive lateral group. Part 6. Synthesis of trifluorovinyl epoxide and its 1,2-diol. *J Fluor Chem* 1999;93:139–144.
18. Ameduri B, Boutevin B, Kostov G, Petrova P. Synthesis and polymerization of fluorinated monomers bearing a reactive lateral group. Part 5. Radical addition of iodine monobromide to chlorotrifluoroethylene to form a useful intermediate in the synthesis of 4,5,5-trifluoro-4-ene-pentanol. *J Fluor Chem* 1999;93:117–127.
19. Guiot J, Ameduri B, Boutevin B. Synthesis and polymerization of fluorinated monomers bearing a reactive lateral group. XII. Copolymerization of vinylidene fluoride with 2,3,3-trifluoroprop-2-enol. *J Polym Sci A Polym Chem* 2002;40:3634–3643.
20. Guiot J, Ameduri B, Boutevin B, Lannuzel T. Original crosslinking of poly(vinylidene fluoride) via trialkoxysilane-containing cure-site monomers. *J Polym Sci A: PolymChem* 2006;44:3896–3910.
21. Boschet F, Cracowski J-M, Montembault V, Ameduri B. Radical copolymerization of a,b-difluoroacrylic acid with vinylidene fluoride. *Macromolecules* 2010;43:4879–4888.

22. Ameduri B. From vinylidene fluoride (VDF) to the applications of VDF-containing polymers and copolymers: recent developments and future trends. *Chem Rev* 2009;109:6632–6686.

23. Narita T. Stimulation on the addition reactivity of fluorinated vinyl monomers-Facile carbon-carbon bond formation by the aid of fluorine substituents. *J Fluor Chem* 2010;131:812–833.

24. (a) Ito H, Giese B, Engelbrecht R. Radical reactivity and Q-e values of methyl α-(trifluoromethyl)acrylate. *Macromolecules* 1984;17:2204–2210; (b) McElroy KT, Purrington ST, Bumgardner CL, Burgess JP. Lack of polymerization of fluorinated acrylates. *J Fluor Chem* 1999;95:117–125.

25. Souzy R, Ameduri B, Boutevin B. Radical copolymerization of -trifluoromethylacrylic acid with vinylidene fluoride and vinylidene fluoride/hexafluoropropene. *Macromol Chem Phys* 2004;205:476–485.

26. (a) Ameduri B, Boyer C. Japanese patent 2008/214,420 (assigned to Tosoh F-Tech Inc.) (Chem. Abstr. 2008:1122917). (b) Boyer C, Ameduri B. Iodine transfer copolymerization of vinylidene fluoride and a-trifluoromethacrylic acid in emulsion process without any surfactants. *J Polym Sci A Polym Chem* 2009;47:4710–4722.

27. Galiano H, Ameduri B, Vergnaud J. French patent FR 2010/1055964 (assigned to CEA and CNRS).

28. Ameduri B. Controlled radical (co)polymerization of fluoromonomers. *Macromolecules* 2010;43:10163–10184.

29. Sauguet L, Ameduri B, Boutevin B. Boutevin fluorinated copolymers and terpolymers based on vinylidene fluoride and bearing sulfonic acid side-group. *J Polym Sci A: Polym Chem* 2007;45:1814–1834. DOI: 10.1002/pola.21978.

30. Sel O, Soules A, Ameduri B, Boutevin B, Laberty-Robert C, Gebel G, Sanchez C. Original fuel-cell membranes from crosslinked terpolymers via a "Sol–gel" strategy. *Adv Funct Mater* 2010;20:1090–1098.

31. Hung MH, Ameduri B, Kostov GK. US patent 2011/0015359 (assigned to DuPont Performance Elastomers).

32. Hung MH, Ameduri B. US patent 2011/0136999 (assigned to DuPont Performance Elastomers).

CHAPTER 60

Convergent ^{18}F Radiosynthesis

V. GOUVERNEUR AND O. LOZANO

The availability of ^{18}F-radiolabeled probes is important for *in vivo* studies by positron emission tomography (PET) [1]. Among the frontier challenges in ^{18}F-radiochemistry are the interconnected goals of increasing synthetic efficiency and diversity in the construction of ^{18}F-labeled radiotracers. ^{18}F-radioretrosynthetic strategies implemented to date are typically linear sequences of transformations designed with the aim of introducing the ^{18}F-label ideally in the last step or at least as late as possible. Conceptually, convergent ^{18}F-radiochemistry offers the possibility to rapidly assemble functionalized ^{18}F-radiotracers from readily accessible ^{18}F-labeled prosthetic groups. Using multicomponent reactions [1,3] for proof of concept, Ugi-4CR, Passerini-3CR, Biginelli-3CR, and Groebke-3CR were performed successfully using ^{18}F-benzaldehyde derivatives; these highly convergent reactions delivered in high decay-corrected radiochemical yield, structurally complex ^{18}F-radiotracers with the ^{18}F-label positioned on an aryl motif not responsive to direct nucleophilic fluorination. These data establish an unprecedented connection between radiochemistry for PET and the field of multicomponent chemistry and demonstrate that convergent retroradiosynthesis is a powerful strategy expanding dramatically the scope of ^{18}F-prosthetic group radiochemistry. The preparation of ^{18}F-labeled prosthetic groups from [^{18}F]fluoride ion is commonly performed in many labeling laboratories, so the concept of convergent ^{18}F-radiosynthesis can easily be applied [4].

Apparatus Glass vial, magnetic stirrer, nitrogen inlet, microwave, heating plate, 75-mm hot cells designed for working with radioactivity, SCINTOMICS (customized automated radiochemical apparatus), high-performance liquid chromatography (HPLC) instrument with an NaI-radiodetector, Mini-Scan thin-layer chromatography (TLC) scanner system (LabLogic Systems Ltd.), safety glasses, laboratory coat, protective gloves.

Chemicals *For the radiolabeling*. [^{18}F]fluoride solution in water [supplied by PETNET solutions at Mont Vernon Hospital (UK)], Kryptofix® 222, K_2CO_3,

Efficient Preparations of Fluorine Compounds, First Edition. Edited by Herbert W. Roesky.

anion exchange cartridges (Synthra), C-18 Sep-Pak cartridges (Waters Corporation), dry acetonitrile.

For the Radio-Biginelli reaction. 4-[^{18}F]fluorobenzaldehyde, urea, ytterbium (III) trifluoromethanesulfonate, ethyl acetoacetate.

For the Radio-Ugi reaction. 4-[^{18}F]fluorobenzaldehyde, benzoic acid, benzyl isocyanide, 1-propylamine, ethanol.

For the Radio-Groebke-Bienaymé-Blackburn reaction. 4-[^{18}F]fluorobenzaldehyde, scandium (III) trifluoromethanesulfonate, 2-aminopyridine, benzyl isocyanide, 3-methyl-1-butanol.

For the Radio-Passerini reaction. 4-[^{18}F]fluorobenzaldehyde, benzoic acid, cyclohexyl isocyanide in 4 M lithium chloride (aq.), methanol.

Attention! Follow all the safety regulations necessary when performing experiments in a radiochemistry laboratory.

Experimental Procedure *Preparation of 4-[^{18}F]fluorobenzaldehyde.* K^{18}F/Kryptofix® 222 in dimethyl sulfoxide (0.5 mL) was added to a sealed reaction vial containing 4-formyl-*N,N,N*-trimethylbenzenaminium trifluoromethane sulfonate (10 mg) and heated under 50 W microwave irradiation for 100 s. The crude reaction mixture was diluted with water (10 mL). Analysis by HPLC (CH_3CN/H_2O) and radio-TLC (5% water in CH_3CN) indicated a radiochemical yield of 83% ($n = 3$).

Radio-Biginelli reaction. A solution of 4-[^{18}F]fluorobenzaldehyde in 3-methyl-1-butanol (200 μL, 5–15 MBq) was dispensed into a sealed vial containing urea (3 mg), ytterbium(III) trifluoromethanesulfonate (3 mg) and ethyl acetoacetate (6 μL). The mixture was heated under vigorous stirring at 130°C for 20 min before being allowed to cool to room temperature and diluted with methanol (1 mL). Analysis by HPLC (CH_3CN/H_2O) and by radio-TLC (66% ethyl acetate in hexane, $R_f = 0.3$) indicated a radiochemical yield of 66% ($n = 3$).

Radio-Ugi reaction. A solution of 4-[^{18}F]fluorobenzaldehyde in ethanol (200 μL, 5–15 MBq) was dispensed into a sealed vial containing benzoic acid (6 mg), benzyl isocyanide (6 μL), and 1-propylamine (4 μL). The mixture was heated to 100°C and stirred for 30 min before being allowed to cool to room temperature. Analysis by HPLC (CH_3CN/H_2O) and radio-TLC (33% ethyl acetate in hexane, $R_f = 0.3$) indicated a radiochemical yield of 62% ($n = 3$).

Radio-Groebke-Bienaymé-Blackburn reaction. A solution of 4-[^{18}F]fluorobenzaldehyde in 3-methyl-1-butanol (200 μL, 5–15 MBq) was dispensed into a sealed vial containing scandium(III) trifluoromethanesulfonate (5 mg) and 2-aminopyridine (3 mg). A solution of benzyl isocyanide (3 μL) in 3-methyl-1-butanol (300 μL) was then added via syringe. The reaction was heated at 170°C for 15 min before being allowed to cool to room temperature. The mixture was then diluted with CH_3CN (1 mL). Analysis by radio-HPLC [CH_3CN/phosphate buffer (10 mM, pH 7.0)] and by radio-TLC (50% ethyl acetate in hexane, $R_f = 0.2$) indicated a radiochemical yield of 85% ($n = 4$).

TABLE 60.1 ^{18}F-Radio-Multicomponent Reaction (4-[^{18}F]Fluorobenzaldehyde as the Prosthetic Group)

^{18}F-labeled product	MCR	RCY (%)[a]
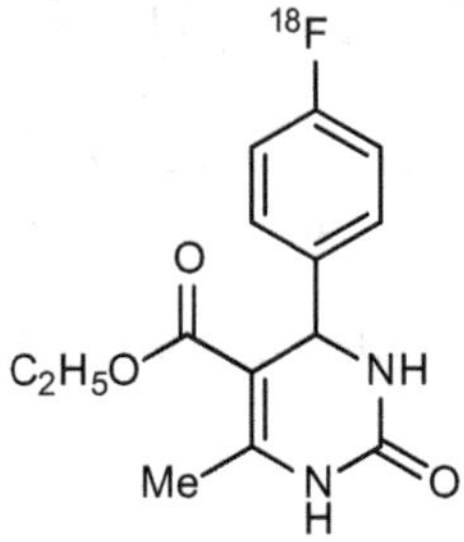	Biginelli	66 ($n = 3$)
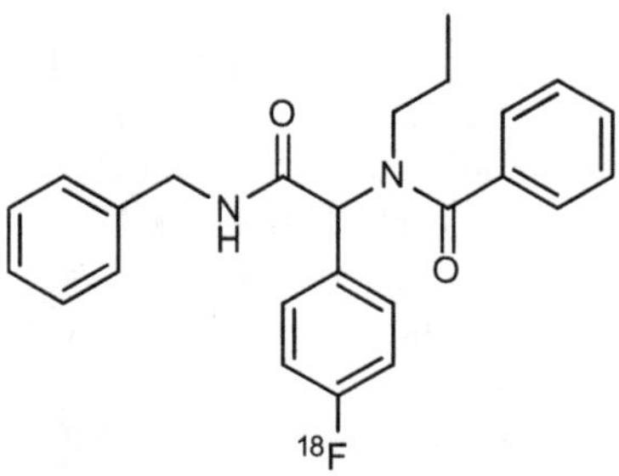	Ugi	62 ($n = 3$)
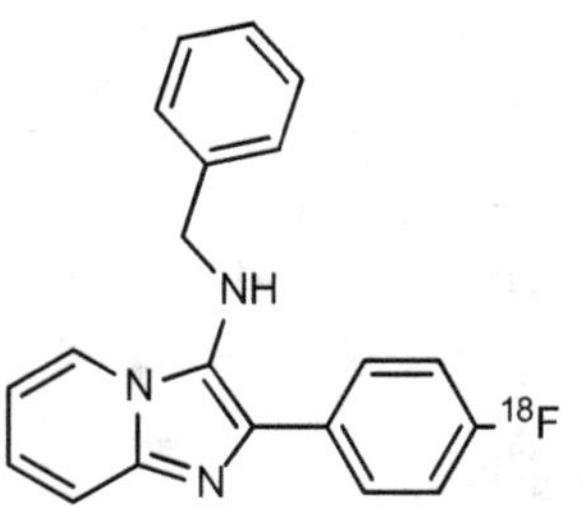	Groebke	85 ($n = 4$)
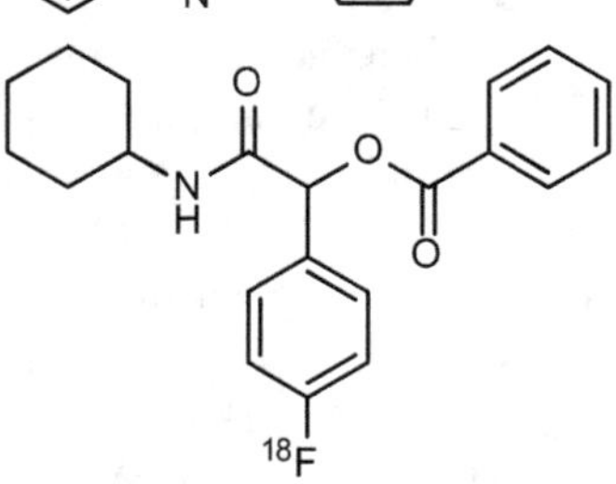	Passerini	65 ($n = 3$)

[a] Crude RCY based on n experiments (decay-corrected).
MCR, multicomponent reaction; RCY; radiochemical yield.

Radio-Passerini reaction. A solution of 4-[^{18}F]fluorobenzaldehyde in methanol (100 μL, 5–15 MBq) was dispensed into a sealed vial containing benzoic acid (7 mg) and cyclohexyl isocyanide (4 μL) in an aqueous solution of LiCl (4 M, 600 μL). The mixture was heated at 100°C for 30 min before being allowed to cool to room temperature and diluted with CH_3CN (300 μL). Analysis by radio-HPLC (CH_3CN/H_2O) indicated a radiochemical yield of 65% ($n = 3$).

Application Convergent ^{18}F-radiochemistry allows for the rapid assembly of functionalized ^{18}F-radiotracers from readily accessible ^{18}F-labeled prosthetic groups. Table 60.1 illustrates how multicomponent reactions allowed for the preparation of various heterocyclic and peptidic-like compounds (e.g., ethyl 4-(4-[^{18}F]fluorophenyl)-6-methyl-2-oxo-1, 2,3,4-tetrahydropyrimidine-5-carboxylate, *N*-[2-(benzylamino)-1-(4-[^{18}F]fluorophenyl)-2-oxoethyl]-*N*-propylbenzamide, *N*-benzyl-2-(4-[^{18}F]fluorophenyl)imidazo[1,2-*a*]pyridin-3-amine and 2-(cyclohexylamino)-1-(4-[^{18}F]fluorophenyl)-2-oxoethylbenzoate), thereby highlighting the potential of this approach in the preparation of complex ^{18}F-radiotracers.

REFERENCES

1. For selected recent reviews, see (a) Phelps ME, Mazziotta JC. Positron emission tomography: human brain function and biochemistry. *Science* 1985;228:799–801; (b) Schirrmacher R, Wängler C, Schirrmacher E. Recent developments and trends in 18F-radiochemistry: syntheses and applications. *Mini-Rev Org Chem* 2007;4:317–329; (c) Ametamey SM, Honer M, Schubiger PA. Molecular imaging with PET. *Chem Rev* 2008;108:1501–1516; (d) Townsend DW. Dual-modality imaging: combining anatomy and function. *J Nucl Med* 2008;49:938–955; (e) Hooker JM. Modular strategies for PET imaging agents. *Curr Opin Chem Biol* 2010;14:105–111.
2. For selected recent reviews, see (a) Dömling A, Ugi I. Multicomponent reactions with isocyanides. *Angew Chem Int Ed* 2000;39:3168–3210; *Angew Chem* 2000;112:3301–3344; (b) Weber L. Multi-component reactions and evolutionary chemistry. *Drug Discov Today* 2002;7:143–147; (c) Zhu J, Bienaymé H, editor. *Multicomponent Reactions*. Weinheim: Wiley-VCH; 2005; (d) Bembenek SD, Tounge BA, Reynolds CH. Ligand efficiency and fragment-based drug discovery. *Drug Discov Today* 2009;14:278–283.
3. For multicomponent reactions based on fluorinated building blocks, see (a) Li D, Song L, Song S, Zhu S. Regioselective synthesis of 6-trifluoromethyl-1,4,5,6-tetrahydropyrazolo[3,4-b]pyran derivatives. *J Fluor Chem* 2007;128:952–957; (b) Gulevich AV, Shevchenko NE, Balenkova ES, Röchenthaler G-V, Nenajdenko VG. The Ugi reaction with CF3-carbonyl compounds: effective synthesis of α-trifluoromethyl amino acid derivatives. *Tetrahedron* 2008;64:11706–11712; (c) Wu J, Cao S. Multicomponent reactions based on fluoro-containing building blocks. *Curr Org Chem* 2009;13:1791–1804; (d) Yanai H, Taguchi T. Novel defluorinative alkylation of trifluoroacetaldehyde N,O-acetal derivatives and its application to multi-component reaction. *Chem Commun* 2009;9:1034–1036; (e) Faggi C, Neo AG, Marcaccini S, Menchi G, Revuelta J. Ugi four-component condensation with two cleavable components: the easiest synthesis of 2,N-diarylglycines. *Tetrahedron Lett* 2008;49:2099–2102.
4. Li L, Hopkinson MN, Leuma Yona R, Bejot R, Gee AD, Gouverneur V. Convergent ^{18}F radiosynthesis: a new dimension for radiolabellin. *Chem Sci* 2011;2:123–131.

CHAPTER 61

Asymmetric Fluorocyclization Reactions

V. GOUVERNEUR AND O. LOZANO

Despite recent progress in the field of asymmetric halogen-promoted cyclizations, only few reactions belonging to this category are known that lead to fluorinated carbo- or heterocycles. Fluorocyclizations typically involve two processes, a fluorination and a cyclization, the order of events depending on the nature of the substrate and the fluorinating reagent [1]. Catalytic asymmetric fluorocyclizations are rare with only two elegant transformations reported to date. A representative fluorinated indanone was prepared via an enantioselective Nazarov cyclization/electrophilic fluorination using 10 mol% of $Cu(OTf)_2$/(*R*)-Ph-bis(oxazoline) in up to 95% yield and *trans/cis* ratio superior to 49:1. In essence, this process is a catalytic asymmetric conrotatory electrocyclic ring closure followed by a diastereoselective fluorination [2]. A unique recent example of electrophilic fluorination/cyclization was recently validated using prochiral indoles as starting material, *N*-fluorobenzenesulfonamide (NFSI) as the fluorinating reagent, and a cinchona alkaloid as the chiral catalyst. This reaction afforded a range of enantioenriched fluorinated analogs of natural products featuring the pyrrolo[2,3-*b*]indole and furo[2,3-*b*]indole skeleton [3].

61.1 FLUORINATED INDANONES VIA COPPER-CATALYZED TANDEM NAZAROV-FLUORINATION [2]

Apparatus Round-bottom flask with septum, magnetic stirrer, argon inlet, safety glasses, laboratory coat, protective gloves.

Chemicals $Cu(OTf)_2$, substituted alkylidene β-ketoesters, NFSI, 1,2-dichloroethane (DCE), $Cu(OTf)_2$/(*R*)-Ph-bis(oxazoline).

Attention! Safety glasses and protective gloves must be used at all times.

Efficient Preparations of Fluorine Compounds, First Edition. Edited by Herbert W. Roesky.

Experimental Procedure Substituted alkylidene β-ketoesters can be prepared using described procedures [2]. Under a positive pressure of argon at room temperature, $Cu(OTf)_2$ (8.0 mg, 0.022 mmol), methyl (2*Z*)-2-(1,3-benzodioxol-5-ylcarbonyl)-3-(4-methoxyphenyl)prop-2-enoate (75.3 mg, 0.22 mmol) and NFSI (83.7 mg, 0.27 mmol) were dissolved in 1,2-dichloroethane (3 mL) in a 15-mL Schlenk flask. The resulting reaction mixture was stirred at 80 °C. After 8 h, the reaction mixture was cooled to room temperature and purified by silica gel column chromatography [gradient increasing from 10% to 20% ethyl acetate in petroleum ether (60–90 °C)] to furnish methyl (5*R*,6*S*)-6-fluoro-5-(4-methoxyphenyl)-7-oxo-6,7-dihydro-5*H*-indeno[5,6-*d*][1,3]dioxole-6-carboxylate as a white solid; yield: 75.3 mg (95%).

Application Cu(II) complexes proved to be effective for a series of tandem Nazarov carbocyclizations followed by electrophilic fluorination with NFSI, a process leading to α-fluoroindanones with high level of enantio- and diastereocontrol. When no chiral ligand is present, the *trans* diastereomers were formed preferentially (dr > 49:1) in good yields (Table 61.1). Extension of this protocol to a catalytic enantioselective tandem transformation proved successful with a representative enantioenriched fluoro indanone made accessible with up to 95% *ee* in the presence of $Cu(OTf)_2$/(*R*)-Ph-bis(oxazoline).

TABLE 61.1 Catalytic Diastereoselective Tandem Nazarov-Fluorination

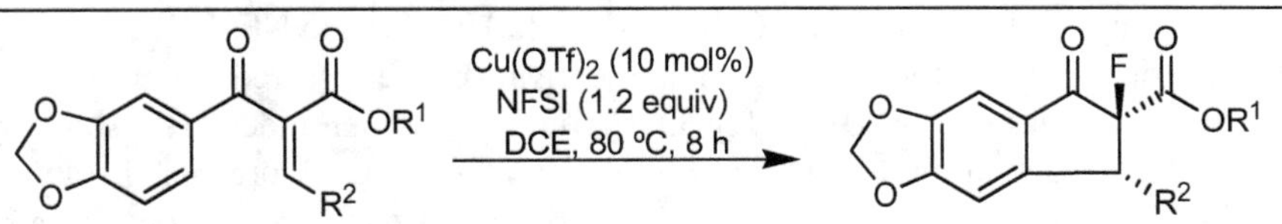

1a-d **2a-d**

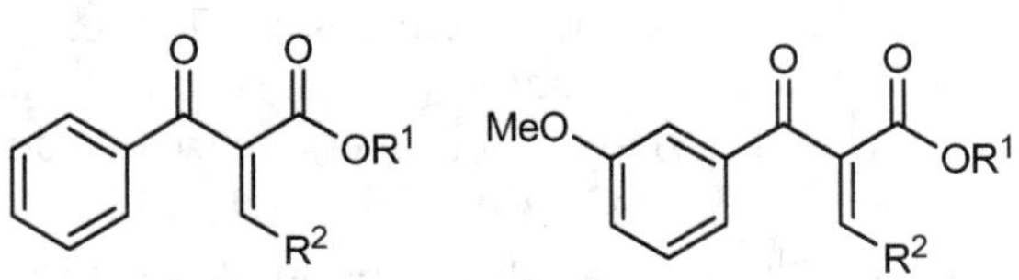

1e-f **1g**

1	R^1	R^2	*Trans* /*cis*[a]	Yield, %
1a	Me	$4\text{-}MeOC_6H_4$	32/1	95
1b	Me	$3,4,5\text{-}MeOC_6H_2$	>49/1	82
1c	Me	$2,4,6\text{-}MeOC_6H_2$	32/1	78
1d	Me	$2\text{-}MeOC_6H_4$	24/1	67
1e	Me	$4\text{-}ClC_6H_4$	24/1	82
1f	Me	$4\text{-}NO_2C_6H_4$	19/1	84
1g	Et	$2,4,6\text{-}MeOC_6H_2$	19/1	85

[a] Determined by ^{1}H-NMR or ^{19}F-NMR.

Catalytic Enantioselective Tandem Nazarov-Fluorination

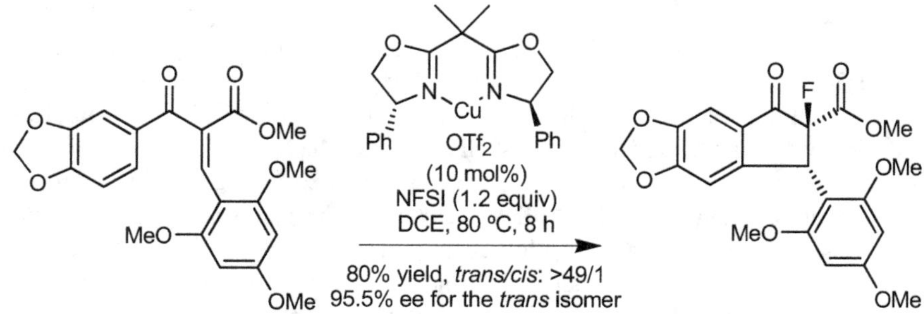

61.2 ORGANOCATALYZED ENANTIOSELECTIVE FLUOROCYCLIZATION OF INDOLES [3]

Apparatus Ten-milliliter round-bottom flask with septum, magnetic stirrer, argon inlet, safety glasses, laboratory coat, protective gloves, cryocool.

Chemicals NFSI, 1-chloromethyl-4-fluoro-1,4-diazoniabicyclo[2.2.2]octane (Selectfluor), dihydroquinine phthalazine-1,4-diyl diether [$(DHQ)_2PHAL$], 3-substituted indole derivatives, $NaHCO_3$, K_2CO_3, acetone.

Attention! Safety glasses and protective gloves must be used at all times.

Experimental Procedure 3-Substituted indole derivatives can be prepared using described procedures. General procedure for the catalytic electrophilic fluorocyclization reactions: $(DHQ)_2PHAL$ (20 mol%) and NFSI (1.2 equiv) in acetone (1.5 mL for 20 mg of indole derivative) were stirred under argon at room temperature for 30 min. K_2CO_3 (6.0 equiv) was then added to the solution, and the reaction mixture stirred for 30 min at −78 °C. A precooled solution (−78 °C) of the indole derivative (1 equiv) in acetone (0.5 mL for 20 mg of indole derivative) was added drop by drop to the catalyst suspension and the reaction mixture was stirred at the same temperature overnight. The reaction mixture was evaporated and the residue was purified through neutral alumina (eluting with hexane : ethyl acetate, 6 : 4) to give the fluorocyclized product. The *ee* were determined by chiral HPLC on CHIRALCEL OJ-H, AD, or OD.

Application Table 61.2 summarizes the first examples of organocatalyzed asymmetric fluorocyclizations mediated by cinchona alkaloids. Two sets of reaction conditions were considered for comparative purpose. The first approach, noncatalytic, involves the premixing of equimolar amount of the cinchona alkaloid and Selectfluor prior to the addition of the starting material. The second strategy uses catalytic amount of the catalyst (20 mol%) as well as NFSI instead of Selectfluor as the electrophilic source of fluorine. The reaction delivered enantioenriched tetrahydrofuroindoles and hexahydropyrroloindoles with different substitution patterns in the aromatic ring and various protecting groups on the pending nitrogen nucleophile. The fluorocyclized products were obtained with good yields and good to excellent enantioselectivities (up to 92% *ee*).

TABLE 61.2 Fluorocyclization of Indole Derivatives

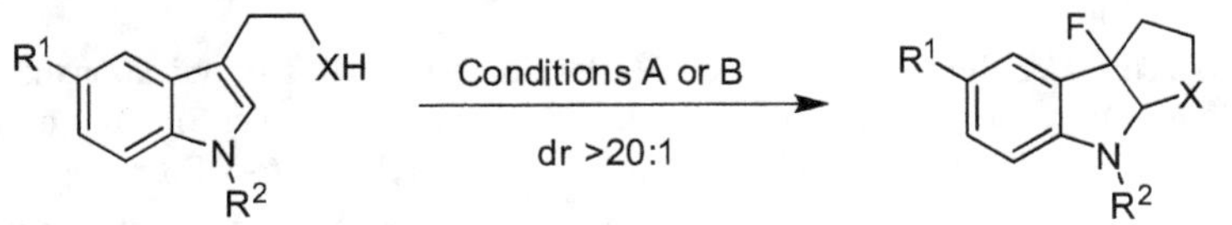

Conditions[a]	R^1	R^2	XH	Yield, %[b]	*ee*, %[c]
A	H	Me	OH	56	74
B	H	Me	OH	72	66
A	OMe	Me	OH	90	86
B	OMe	Me	OH	65	74
A	OBn	Me	OH	69	84
B	OBn	Me	OH	78	74
A	Oallyl	Me	OH	53	86
B	Oallyl	Me	OH	65	78
A	Mes	Me	OH	57	90
B	Mes	Me	OH	55	84
A	H	Me	NHTs	54	76
B	H	Me	NHTs	59	64
A	OMe	Me	NHTs	55	78
B	OMe	Me	NHTs	51	70
A	Ph	Me	NHTs	50	82
B	Ph	Me	NHTs	70	70
A	Mes	Me	NHTs	60	92
B	Mes	Me	NHTs	80	84
A	Mes	Me	NHCOMe	38[d]	92
B	Mes	Me	NHCOMe	65	92
A	H	Me	$NHCO_2Me$	56	78
B	H	Me	$NHCO_2Me$	76	74
A	H	Me	$NHCO_2Bn$	40	78
B	H	Me	$NHCO_2Bn$	47	77
A	H	Me	NHBoc	67	86
B	H	Me	NHBoc	70	78

[a] Conditions A: 1.2 equiv Selectfluor, 1.2 equiv $(DHQ)_2PHAL$, 1.2 equiv $NaHCO_3$, acetone, −78 °C; Conditions B: 1.2 equiv NFSI, 0.2 equiv $(DHQ)_2PHAL$, 6 equiv K_2CO_3, acetone, −78 °C.

[b] Yield of isolated products.

[c] *ee* determined by chiral stationary phase HPLC.

[d] Conversion.

Mes, mesityl.

REFERENCES

1. (a) Wilkinson SC, Salmon R, Gouverneur V. Developments in fluorocyclization methodologies. *Future Med Chem* 2009;1:847–863; (b) Castellanos A, Fletcher SP. Current methods for asymmetric halogenation of olefins. *Chem Eur J* 2011;17:5766–5776.
2. Nie J, Zhu H-W, Cui H-F, Hua M-Q, Ma J-A. Catalytic stereoselective synthesis of highly substituted indanones via Tandem Nazarov cyclization and electrophilic fluorination trapping. *Org Lett* 2007;9:3053.
3. Lozano O, Blessley G, Martínez Del Campo T, Thompson AL, Giuffredi GT, Bettati M, Walker M, Borman R, Gouverneur V. Organocatalyzed enantioselective fluorocyclizations. *Angew Chem Int Ed*. 2011;50:8105–8109.

CHAPTER 62

Preparation of Allylic Fluorides

V. GOUVERNEUR AND O. LOZANO

Allylic fluorides are important building blocks, which can be manipulated through the double bond [1]. This structural motif is also found unmodified in important life science compounds such as insecticides, herbicides, fungicides, and pharmaceutical drugs. Retrosynthetic strategies based on the functionalization of a fluorinated precursor have been reported to access allylic fluorides (e.g., Wittig olefination of fluorinated aldehydes) [2] but more recently catalytic processes based on the use of either a nucleophilic or electrophilic fluorine source were successfully implemented with exquisite level of regio- and stereocontrol. Allyl *p*-nitrobenzoates were found to be responsive to nucleophilic fluorination under Pd catalysis delivering a series of linear allylic fluorides in moderate to excellent yields [3, 4]. This remarkably fast reaction is suitable for ^{18}F-radiolabeling. In the same vein, the fluorination of cyclic allylic chlorides with AgF using a Pd(0) catalyst and Trost bisphosphine ligand gave access to enantioenriched allylic fluorides in up to 96% *ee* [5, 6]. The asymmetric electrophilic fluorination of allylsilanes is an alternative route to produce enantioenriched branched allylic fluorides [7]. Catalytically, this is achieved combining a cinchona alkaloid with *N*-fluorobenzenesulfonimide (NFSI), an electrophilic fluorinating reagent of tamed reactivity with respect to Selectfluor [8]. Rewardingly, this strategy complements the Pd-catalyzed fluorination process as it allows access to allylic fluorides featuring the fluorine substituent on a quaternary stereogenic center with *ee* up to 95%.

62.1 PD-CATALYZED FLUORINATION OF ALLYL *P*-NITROBENZOATES [3, 4]

Apparatus Glass vial with septum, magnetic stirrer, argon inlet, safety glasses, laboratory coat, protective gloves.

Efficient Preparations of Fluorine Compounds, First Edition. Edited by Herbert W. Roesky.

Chemicals Reaction: [Pd(dba)$_2$], PPh$_3$, substituted 2-phenylprop-2-en-1-yl 4-nitrobenzoates, dry THF, TBAF(*t*-BuOH)$_4$.

Preparation of TBAF(*t*-BuOH)$_4$: TBAF·3H$_2$O, *t*-BuOH, hexane [5].

Attention Safety glasses and protective gloves must be used at all times.

Experimental Procedure Substituted 2-phenylprop-2-en-1-yl 4-nitrobenzoates can be prepared using described procedures [3]. For the preparation of TBAF(*t*BuOH)$_4$ [5]: Commercially available TBAF trihydrate (1.0 g, 3.17 mmol) was added to *t*BuOH (88 mL) and *n*-hexane (22 mL). The mixture was stirred for 30 min at 90 °C. During this time TBAF dissolved completely. The solution was cooled to room temperature, and a white crystalline solid precipitated. The crystalline solid was filtered and washed rapidly with 40 mL of 70% *t*BuOH/hexane. The filtrate was kept in vacuum for 15–20 min to remove residual solvent, and TBAF(*t*BuOH)$_4$ (1.63 g, 2.92 mmol) was obtained as a white crystalline solid in 92% yield. ^{1}H-NMR (600 MHz, CDCl$_3$): δ 1.01 (t, J = 9.0 Hz, 12H), 1.27 (s, 36H), 1.42–1.48 (m, 8H), 1.66–1.71 (m, 8H), 3.44 ppm (t, J = 9.0 Hz, 8H) ppm; ^{13}C-NMR (150 MHz, CDCl$_3$): δ 13.6, 19.6, 24.0, 31.1, 58.5, 68.2 ppm.

General procedure for the palladium-catalyzed allylic fluorination: [Pd(dba)$_2$] (6 mg, 0.01 mmol) and PPh$_3$ (8 mg, 0.03 mmol) were added to a solution of 2-(4-*tert*-butylphenyl)prop-2-en-1-yl-4-nitrobenzoate (68 mg, 0.2 mmol) in THF (2 mL). TBAF(*t*BuOH) (279 mg, 0.5 mmol) was then added in one portion. The reaction was stirred at room temperature for 1 h. The reaction was quenched by the addition of saturated NH$_4$Cl solution. The aqueous layer was extracted with Et$_2$O (2–5 mL), and then the combined organic extracts were washed with NH$_4$Cl (2–10 mL), H$_2$O (1–5 mL), and dried (Na$_2$SO$_4$). The solvent was removed *in vacuo* and the crude reaction mixture purified by silica gel column chromatography with 100% petroleum ether (bp 30–40 °C) eluent.

Application The scope of this palladium-catalyzed allylic fluorination is exemplified in Table 62.1. The *p*-nitrobenzoate leaving group proved to be most suitable for this reaction; when acetate, trifluoroacetate, or benzoate was used instead of *p*-nitrobenzoate, no reaction or low conversion was observed with, in some cases, undesired allylic alcohol being formed preferentially. Some aryl-substituted propenyl fluorides were obtained in high yields (>95%) under the reaction conditions. This mild procedure also allowed for the preparation of linear allylic fluorides although the yields were generally lower.

62.2 ASYMMETRIC PALLADIUM-CATALYZED FLUORINATION OF ALLYLIC CHLORIDES [6]

Apparatus Round-bottom flask with septum, magnetic stirrer, safety glasses, laboratory coat, protective gloves.

TABLE 62.1 Palladium-Catalyzed Allylic Fluorination

Ester (R=$COC_6H_4pNO_2$)	Allyl fluoride	Yield, %[a]
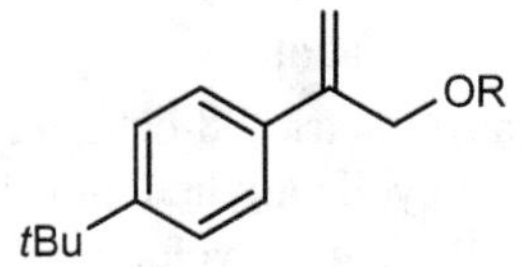	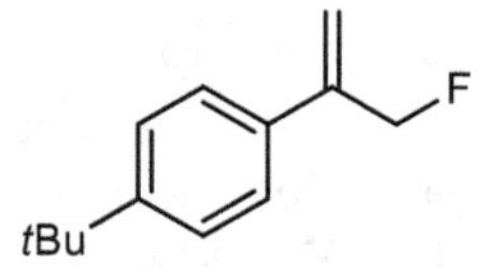	>95
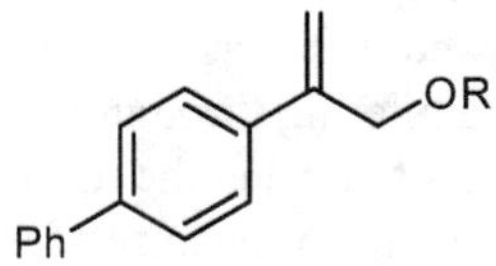	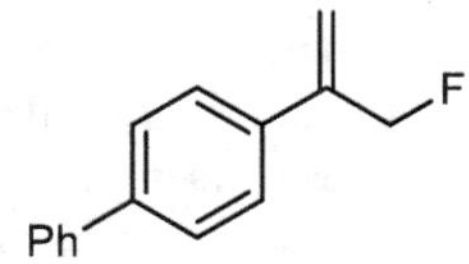	95
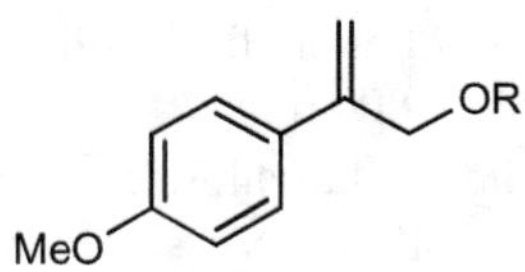	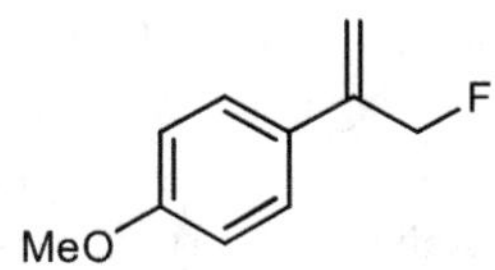	85
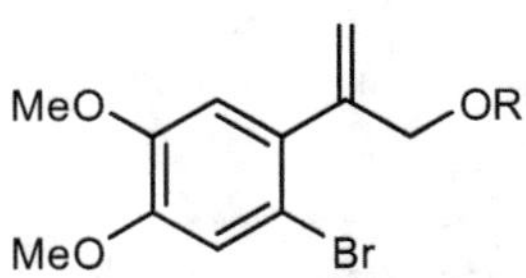	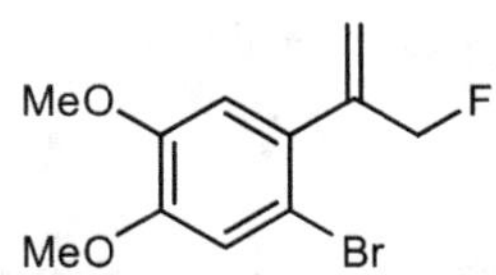	84
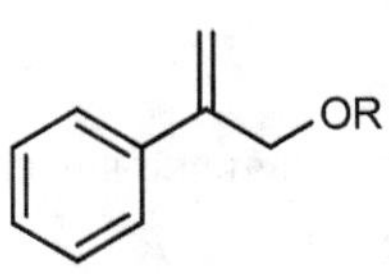	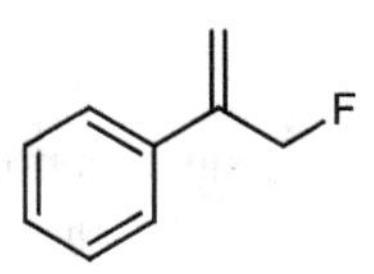	66[b]
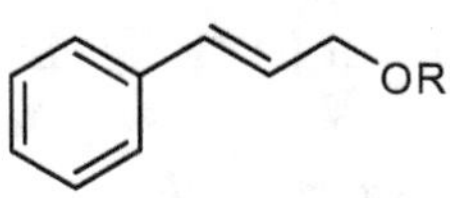	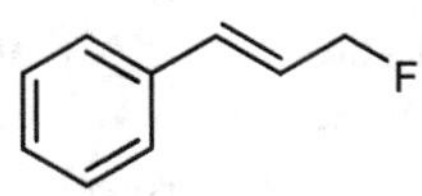	>95
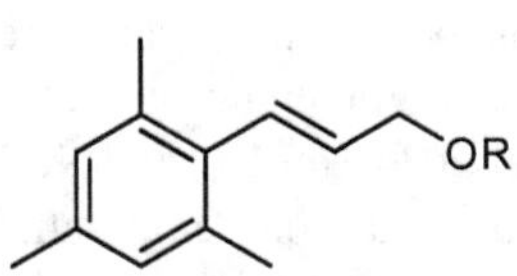	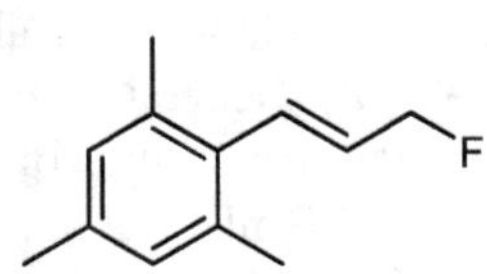	65
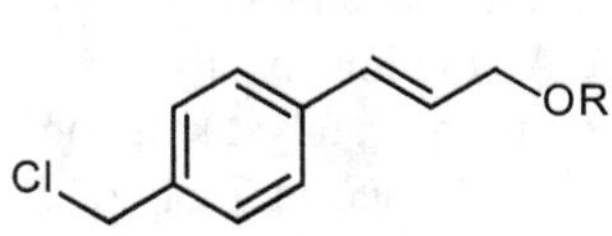	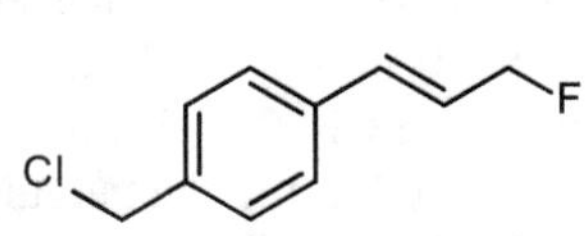	60

[a] Yield of the isolated product.

[b] The volatility of the allylic fluoride attenuates the yield of isolated product.

Chemicals $[Pd_2(dba)_3]$, AgF, (1*R*,2*R*)-(+)-1,2-diaminocyclohexane-*N*,*N*′-bis(2′-diphenylphosphinobenzoyl), substituted cyclic allylic chlorides, dry THF.

Attention Safety glasses and protective gloves must be used at all times.

Experimental Procedure General procedure for the asymmetric Pd-catalyzed fluorination of allylic chlorides [6]: to a solution of the cyclic allylic chloride (1 equiv) in THF (0.1 M) was added (1*R*,2*R*)-(+)-1,2-diaminocyclohexane-*N*,*N*′-bis(2′-diphenylphosphinobenzoyl) (10 mol%), AgF (1.1 equiv), and $[Pd_2(dba)_3]$ (5 mol%). The reaction mixture was allowed to stir at room temperature, wrapped with aluminum foil, for 24 h before filtering through a pad of Celite. The filtrate was concentrated under reduced pressure and the residue purified by silica gel column chromatography. *Note*: The heterogeneity of the reaction was found to have a large impact on the outcome (rate, yield, *ee*, and dr). Ideally, fluorinations should be conducted in a round-bottom flask with an egg-shaped stir bar, magnetically stirring at a rate of 600–800 rpm.

Application The enantioselective fluorination of cyclic allylic chlorides with AgF with Pd(0) catalyst and a Trost bisphosphine ligand allows access to enantioenriched cyclic allylic fluorides. The reaction was found to be compatible with a selection of functional groups (Table 62.2).

62.3 ORGANOCATALYTIC ASYMMETRIC FLUORINATION OF ALLYLSILANES [7, 8]

Apparatus Ten-milliliter round-bottom flask with septum, magnetic stirrer, argon inlet, safety glasses, laboratory coat, protective gloves.

Chemicals NFSI, dihydroquinine 2,5-diphenylpyrimidine-4,6-diyl diether $[(DHQ)_2PYR]$, dihydroquinine phthalazine-1,4-diyl diether $[(DHQ)_2PHAL]$, substituted allylsilanes, K_2CO_3, dry acetonitrile.

Attention Safety glasses and protective gloves must be used at all times.

Experimental Procedure Substituted allylsilanes can be prepared using described procedures [7, 8]. General procedure for the catalytic enantioselective fluorination of allylsilanes: a solution of cinchona alkaloid (10 mol%) and NFSI (1.2 equiv) in CH_3CN (1.0 mL) was stirred under nitrogen at room temperature for 30 min. K_2CO_3 (6.0 equiv) was then added to the solution, and the reaction mixture stirred for 30 min at −20 °C or −40 °C. A solution of allylsilane (0.084–0.131 mmol) in CH_3CN (1.0 mL) was added to the catalyst solution. The reaction was stirred at the temperature and time indicated in Table 62.3 with monitoring by thin-layer chromatography (TLC). The reaction mixture was filtrated through alumina and the solvent was evaporated. The residue was purified by alumina preparative TLC or column chromatography on alumina eluting with hexane. The *ee* were determined by chiral HPLC on CHIRALCEL OD-H or OJ and CHIRALPAK AD-H columns.

TABLE 62.2 Palladium-Catalyzed Asymmetric Allylic Fluorination

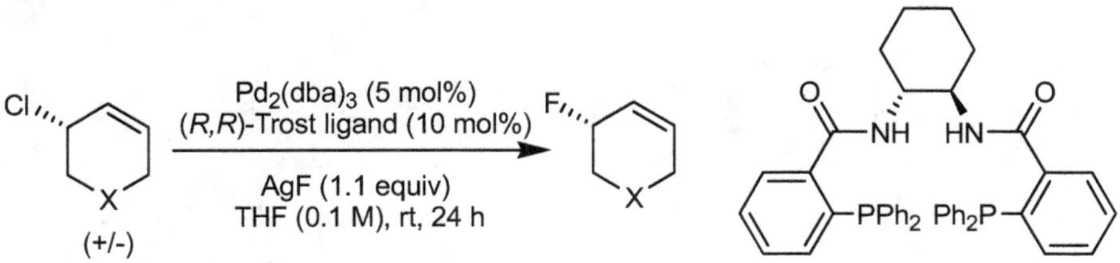

(*R,R*)-Trost ligand

Product	Yield, %	*ee*, %[a]	dr[b]
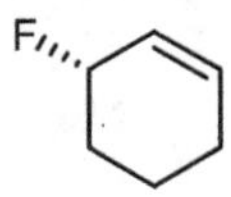	85[c]	88	N/A
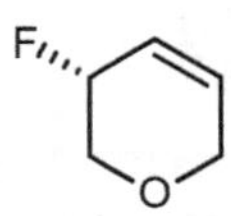	62[c,d]	90	N/A
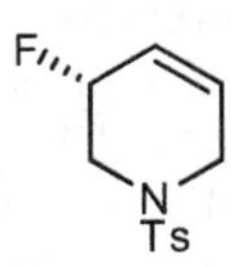	74[d,e]	96	N/A
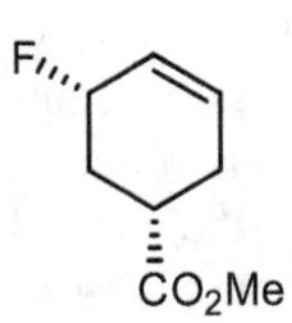	70	91	3 : 1
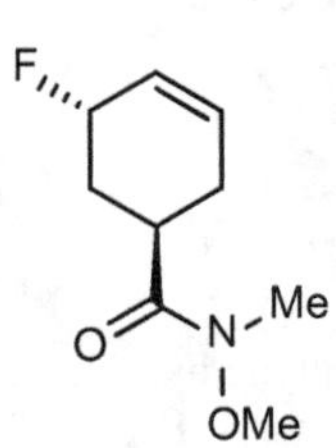	85	85	>20 : 1

[a] Determined by chiral GC or HPLC.
[b] Determined by ^{1}H NMR; drs were unchanged from starting material to product.
[c] Determined by GC using octane as an internal quantitative standard.
[d] Used (1*R*,2*R*)-(+)-1,2-diaminocyclohexane-*N*,*N*′-bis(2-diphenylphosphino-1-naphthoyl).
[e] Reaction conducted in toluene with 2.0 equiv of AgF.

TABLE 62.3 Enantioselective Fluorodesilylation of Allylsilanes

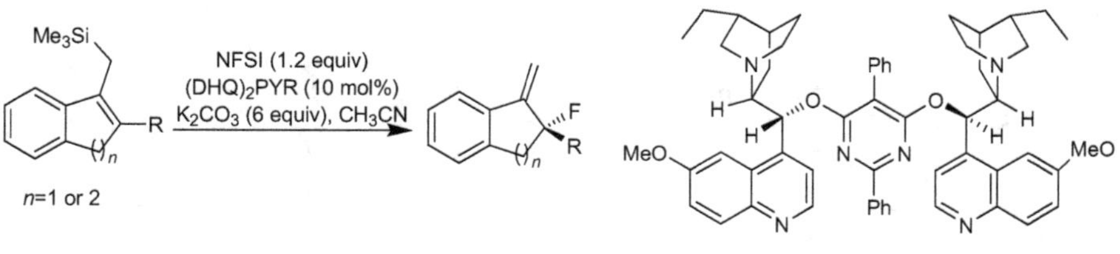

$(DHQ)_2PYR$

R	*n*	T, °C	*t*	*ee*, %	Yield, %
$CH_2C_6H_5$	1	–40	3 d	94	75
4-$MeCH_2C_6H_4$	1	–20	12 h	95	75
4-$ClCH_2C_6H_4$	1	–20	18 h	94	81
2-naphtylmethyl	1	–20	34 h	91	69
4-$MeCH_2C_6H_4$	2	–20	36 h	81	71

Application The first examples of reagent-controlled asymmetric electrophilic fluorodesilylation leading to enantioenriched allylic fluorides were reported by Gouverneur and co-workers in 2003 [7]. In 2008, Shibata and co-workers [8] developed the catalytic variant of this transformation using NFSI in combination with a catalytic amount of either $(DHQ)_2PYR$ or $(DHQ)_2PHAL$. These reactions provide the desired allylic fluorides in good yield and with moderate to high enantioselectivities (up to 95% *ee*). The presence of a large R substituent on the silylated indenes is a prerequisite for synthetically useful level of enantiocontrol. The reaction typically requires several days to go to completion. Table 62.3 illustrates with some representative examples the scope of this catalytic asymmetric fluorodesilylation reaction. Significantly, the method allows access to various cyclic enantioenriched allylic fluorides inclusive of those featuring the fluorine substituent on a quaternary stereogenic center.

REFERENCES

1. Pacheco MC, Purser S, Gouverneur V. The chemistry of propargylic and allylic fluorides. *Chem Rev* 2008;108:1943–1981.
2. Jiang H, Falcicchio A, Jensen KL, Paixão MW, Bertelsen S, Jørgensen KA. Target-directed organocatalysis: a direct asymmetric catalytic approach to chiral propargylic and allylic fluorides. *J Am Chem Soc* 2009;131:7153–7157.
3. Hollingworth C, Hazari A, Hopkinson MN, Tredwell M, Benedetto E, Huiban M, Gee AD, Brown JM, Gouverneur V. Palladium-catalyzed allylic fluorination. *Angew Chem Int Ed* 2011;50:2613–2617; *Angew Chem* 2011;123:2661–2665.
4. Hazari A, Gouverneur V, Brown JM. Palladium-catalyzed substitution of allylic fluorides. *Angew Chem Int Ed* 2009;48:1296–1299; *Angew Chem* 2009;121:1322–1325.

5. Kim DW, Jeong H-J, Lim ST, Sohn M-H. Tetrabutylammonium tetra(tert-butyl alcohol)-coordinated fluoride as a facile fluoride source. *Angew Chem Int Ed* 2008;47:8404–8406; *Angew Chem* 2008;120:8532–8534.
6. Katcher MH, Doyle AG. Palladium-catalyzed asymmetric synthesis of allylic fluorides. *J Am Chem Soc* 2010;132:17402–17404.
7. Greedy B, Paris J-M, Vidal T, Gouverneur V. Regio- and enantioselective synthesis of allylic fluorides by electrophilic fluorodesilylation of allyl silanes. *Angew Chem Int Ed* 2003;42:3291–3294; *Angew Chem* 2003;115:3413–3416.
8. Ishimaru T, Shibata N, Horikawa T, Yasuda N, Nakamura S, Toru T, Shiro M. Cinchona alkaloid catalyzed enantioselective fluorination of allyl silanes, silyl enol ethers, and oxindoles. *Angew Chem Int Ed* 2008;47:4157–4161; *Angew Chem* 2008;120:4225–4229.

CHAPTER 63

Dehydroxyfluorinations of Primary or Secondary Alcohols Using Perfluoro *n*-Butylsulfonyl Fluoride (Nonaflyl Fluoride) in Combination with 1,8-Diaza-bicyclo[5.4.0]undec-7-ene

ORLIN PETROV, MATTHIAS SCHNEIDER, ROLF BOHLMANN,
STEPHAN VETTEL, AND HELMUT VORBRÜGGEN

The combination of the stable and nontoxic nonaflyl fluoride (NfF), (bp 65 °C), which is readily obtained in 50–60% yield in one reaction step on a technical scale by anodic fluorination of sulfolene [1], with the strong base 1,8-diazabicyclo[5.4.0]undec-7-ene (DBU) converts readily primary and secondary alcohols at −35 °C in high yields with inversion into the corresponding primary or secondary fluorides, without affecting other functional groups such as ketones or α,β-unsaturated ketones. As an example, the secondary steroid alcohol **1** undergoes readily dehydroxyfluorination at −35 °C into the inverted fluorinated steroid **2**, which is converted *in situ* in 86% overall yield into the nicely crystallizing dienol acetate **3**, an intermediate in the synthesis of powerful anti-estrogens [2–5].

63.1 CONVERSION OF 11α-HYDROXYESTRA-4-EN-3,17-DIONE 1 VIA 11β-FLUORO-ESTRA-4-ENE-3,17- DIONE 2 INTO 11β-FLUORO-3-ACETOXY-ESTRA-3,5-DIEN-17-ONE 3

Apparatus One-liter round-bottom flask connected to an addition funnel and a distillation apparatus, a powerful stirrer, as well as an efficient cooling bath.

Chemicals 11α-hydroxyestra-4-en-3,17-dione, NfF, and DBU.

Attention Safety glasses and protective glasses should be used all the times.

Efficient Preparations of Fluorine Compounds, First Edition. Edited by Herbert W. Roesky.

Caution The contact with the volatile NfF as well as the strongly basic DBU as well as the biologically potent steroids should be avoided.

Experimental Procedure A mixture of 11α-hydroxyestra-4-en-3,17-dione **1** (50 g, 173.4 mmol) and DBU (85.6 mL, 572.2 mmol) in dry EtOAc (250 mL) was cooled to −35––40 °C and NfF (50 mL, 277.4 mmol) in EtOAc (100 mL) added slowly within 60 min with vigorous stirring in order to keep the temperature of the highly exothermic reaction below −30 °C. A clear solution was obtained, which was further stirred at −35 °C for 90 min. High-performance liquid chromatography control of the reaction showed complete conversion of the starting alcohol after 30 min. The reaction mixture was warmed to −10 °C, stirred for 1 h at this temperature, and quenched by adding water (20 mL). Next, 2 N sulfuric acid (100 mL) was added, and the reaction mixture was stirred at 10 °C for 90 min. The phases were separated and 2 N sulfuric acid (22 mL) was added to the organic phase to adjust a pH value of 2. After separation, the organic phase was washed with saturated $NaHCO_3$ solution (50 mL) followed by saturated NaCl solution (50 mL) and concentrated at reduced pressure to 160 mL. EtOAc (200 mL) was added and the mixture concentrated again to 160 mL. The procedure was repeated twice, whereupon a crystalline slurry was obtained. The slurry was cooled at 0 °C and acetic anhydride (147.1 mL, 1.56 mol) and methanesulfonic acid (2.24 mL, 34.5 mmol) were added under stirring. Stirring was continued for 44 h at 0 °C and the slurry was filtered and washed four times with ice-cold isopropyl acetate (40 mL each portion). The obtained 11β-fluoro-3-acetoxy-estra-3,5-dien-17-one **3** was dried at 40 °C *in vacuo* to give pure **3** (49.6 g, 86.1%), mp 175–177 °C, $[\alpha]_D = -32.3°$ in $CHCl_3$. ^1H NMR (δ in $CDCl_3$): 5.82 [d, 1H, $J = 2$ Hz], 5.50 [m, 1H], 5.08 [d(br), 1H, $J = 49$ Hz], 2.15 [s, 3H], 1.04 [d, 3H, $J = 1.5$ Hz] ppm.

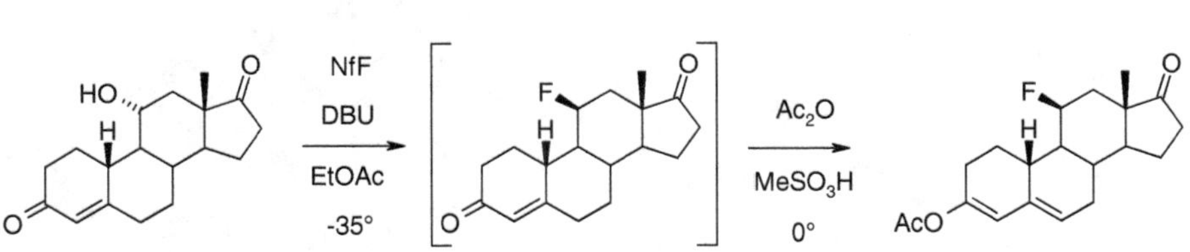

1 **2** **3** (86%)

REFERENCES

1. Zimmer R, Webel M, Reissig H-U. Nonafluoro-1-butanesulfonyl fluoride: more than a fluorinating reagent. *J Prakt Chem* 1998;340:274–277.

2. Bennua-Skalmowski B, Vorbrüggen H. A facile conversion of primary or secondary alcohols with n-perfluorobutanesulfonyl fluoride/1,8-diazabicyclo[5.4.0]undec-7-ene into their corresponding fluorides. *Tetrahedron Lett* 1995;36:2611–2614.

3. Vorbrüggen H. The conversion of primary or secondary alcohols with nonaflyl fluoride into their corresponding inverted fluorides,. *Synthesis* 2008;8:1165–1174.
4. Vorbrüggen H. On novel fluorine reagents in preparative organic chemistry. *Helv Chim Acta* 2011;94:947–965.
5. Petrov O, Schneider M, Bohlmann R, Vettel S, E.P. 2053055, C.A. 150, 472959.

CHAPTER 64

Preparation of Rare Earth Fluorosulfides and Oxyfluorosulfides

A. DEMOURGUES AND A. TRESSAUD

Rare earth-based mixed-anion compounds containing sulfur atoms have been largely investigated for their outstanding optical properties, particularly in absorption and luminescence fields. Rare earth sesquisulfides Ln_2S_3 and ALn_2S_4 (A = Ca, Sr) have been proposed as inorganic color pigments, because they exhibit pronounced colors ranging from yellow to red, depending on the rare earth [1, 2]. Rare earth-based oxide sulfides have also been largely investigated for their specific properties in luminescence fields [3]. These materials are used, for instance, as cathode-ray phosphors present in color TV screens or as X-ray phosphors in conventional intensifying screens in radiography [4]. It has been recently shown that the presence of fluorine in the lattice leads to remarkable behavior: On the one hand, LnSF-type rare earth fluorosulfides exhibit good chromatic properties and high stability in acidic medium and are, therefore, good candidates as inorganic color pigments [5]. On the other hand, complex La-based fluorosulfides and oxyfluorosulfides display noteworthy absorption in the ultraviolet (UV) range [6–8].

Because the size of F^- and O^{2-} anions is similar, they are often randomly distributed in the same crystallographic sites, creating mainly ionic or ionocovalent bonds. F^- and O^{2-} anions can be considered as hard Lewis bases, whereas S^{2-} anion is a soft base. The resulting structures, in which the three anions are involved with electrostatic/ionic and covalent interactions, often exhibit strong anisotropy, leading to two-dimensional (2D) network due to the difference in polarizabilities of the anions. Moreover, as recently discovered in oxypnictides with high-T_C superconductivity properties [9, 10], fluorine and oxygen atoms are engaged in ionic bonds with rare earth and alkaline earth (hard acid) metals, whereas the more electronegative transition metals (borderline or soft acid) are surrounded by sulfur atoms in tetrahedral sites (sp^3 hybridization), which exhibit a lower electronegativity than fluorine or oxygen. From a chemistry point of view, the coexistence of $F^-/O^{2-}/S^{2-}$ anions in the

Efficient Preparations of Fluorine Compounds, First Edition. Edited by Herbert W. Roesky.

vicinity of rare earth, alkaline earth, and transition metal leads to prepare separately the starting binary fluorides, oxides, oxyfluorides, or fluorocarbonates, which can be either processed in a gas–solid reaction under sulfur atmosphere (CS_2/H_2S) under various reductive conditions or in a solid–solid reaction by mixing together binary fluorides and oxides under an inert atmosphere (sealed quartz tube).

This chapter is related to the preparation methods of CeSF, SmSF, LaSF, $Ln_2AF_4S_2$, $Ln_3OF_3S_2$, $La_2O_{1.5}FS$, $Ln_4O_2F_2S_3$ fluorosulfides and oxyfluorosulfides and to their subsequent morphological and optical properties.

64.1 SYNTHESIS ROUTES OF RARE EARTH FLUOROSULFIDES AND OXYFLUOROSULFIDES

Route A: CeSF, SmSF, and LaSF (via Solid-State Syntheses)

Rare earth fluorosulfides α-LnSF, with Ln = La, Ce, and Sm, are synthesized by reaction of stoichiometric quantities of high-purity rare earth sulfides in their α forms and rare earth fluorides. α-Ln_2S_3 sulfides are prepared from reactions involving pure rare earth metal and sulfur, slowly heated in sealed quartz tubes at 400 °C during 24 h and then at 700 °C for 24 h. Before use, these phases are heated under dynamical vacuum for 3 h at 150 °C to remove traces of moisture. They should be kept under an Ar atmosphere.

Fluorination Apparatus for Preparing Starting Rare Earth Fluorides Synthesis methods for obtaining inorganic solid fluorides have been described in several review articles, in particular by Grannec and Lozano [11]. A fluorination line designed to obtain metal fluorides, including rare earth fluorides, is given in Figure 64.1.

Pure fluorine or 10% F_2 diluted in N_2 is stored in commercial cylinders. The cylinder must be located in a well-ventilated, independent container and all pieces of the apparatus must be built under hoods. Just next to the F_2 gas cylinders, a container is filled under 5 bars with the fluorinating gas, in order the main fluorine source could be turned out for safety reasons during the experiment. Any traces of HF are trapped by passage through a column filled with NaF pellets. The reaction tube in the horizontal electrical furnace is made of Ni or Monel. Mass flow controllers allow checking the amount of gas that is passed during the experiment. Generally, the samples for fluorination are placed in a Ni boat. The appropriate temperature is obtained by a resistance that is wound around the horizontal Ni tube. It is controlled by a thermocouple located inside a Ni sheath placed just under the sample. Unreacted fluorine is destroyed in columns containing anhydrous alumina, calcium oxide, or soda lime. Note that fluorination may be carried out either in a dynamic way or in a static way in Ni reactors.

LaF_3 is obtained by fluorination of lanthanum oxide (Rhodia 99.99%). The oxide powder is placed in a nickel boat inside an electrical furnace. In a first

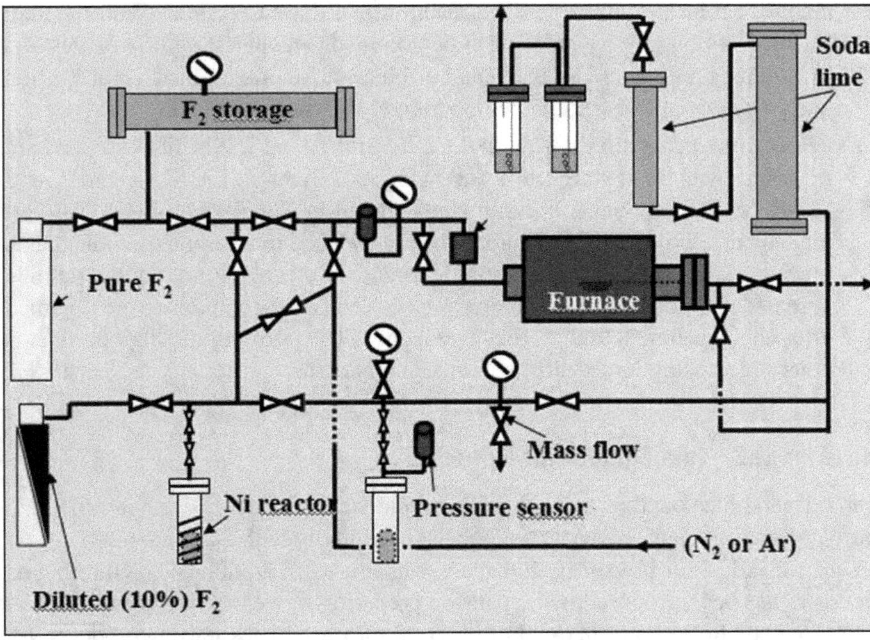

FIGURE 64.1 Scheme of a fluorination apparatus using F_2 gas. (For a color version of the figure, please see color plates.)

step, the product is heated under a dynamic vacuum at 500 °C to remove traces of water. The temperature is lowered to room temperature and F_2 gas is then introduced in the same furnace up to 1 bar pressure. In order not to have a strong exothermic reaction, F_2 gas is first introduced as 20% diluted in Ar. While the temperature is increased up to 500 °C, F_2 rate in Ar is concomitantly increased up to pure F_2 concentration. Fluorination for 10–12 h in pure F_2 gas at 500 °C is required to obtain LaF_3.

In the case of Ce and Sm trifluorides, starting oxides CeO_2 and Sm_2O_3 are treated with $NH_4F{\cdot}HF$ in HF solution (40% in water). The mixture is heated on a sand bath in a Pt crucible at 150–200 °C. This step is repeated several times. The powder is finally purified under an anhydrous HF stream for 5 h at 800 °C to remove remaining traces of starting oxide and to obtain the pure trifluorides.

Chemicals F_2 gas containers are commercialized by Comurhex Co.; rare earth oxides La_2O_3, CeO_2, and Sm_2O_3 (99.99%) are commercial products from Rhodia Co.; and $NH_4F{\cdot}HF$ and HF solution (40% in water) are commercial products available in several chemicals companies. Anhydrous hydrogen fluoride is commercialized by Praxair, Alfa Aesar, or other manufacturers.

Caution The synthesis of starting fluoro-materials requires the use of fluorinating substances such as F_2 gas, anhydrous HF, and NH_4HF_2, which should be used very cautiously because of their high reactivity and corrosive power. Contact of F_2 with any organic substance, grease, etc., may lead to uncontrolled chain reaction or explosion. The reaction of F-based products with the skin may lead to burns very difficult to be confined. The handling procedures and drastic precautions to be taken can be found in Reference 11. Eye protection and protective gloves should be imperatively worn.

Experimental Procedure The final CeSF, SmSF, and LaSF fluorosulfides are prepared by solid-state reaction following the equation: $Ln_2S_3 + LnF_3 \rightarrow 3$ LnSF. Both components are intimately mixed in a glove box under an inert atmosphere. The ground mixture is then introduced in a carbon crucible that is placed in a quartz tube. After being evacuated during 40 min, the quartz tube is sealed and placed in an electric furnace. The LnSF compounds are obtained after an annealing at 900 °C for several days. The purity of the final product is determined using X-ray diffraction (XRD) patterns.

Route B: LaSF (via Gas–Solid Route)

It should be pointed out that, using the above solid-state method, it is rather difficult to control the granulometry of the final products. In addition, such a way cannot be easily adapted for industrial processes. For these reasons, sulfurization of oxyfluorinated precursors has been proposed, which allows preparing rare earth fluorosulfides with submicronic particle size from oxyfluorides or fluorocarbonates starting products [8].

Starting Ln-based compounds, LnOF and $LnFCO_3$ (bastnaesite) (except Ce, because of the occurrence of the two oxidation states Ce^{4+}/Ce^{3+}), can be prepared by precipitating rare earth chlorides or nitrates at room temperature in acidic medium with stoichiometric amounts of HF and NH_4HCO_3. The precipitates are annealed at 700 °C during 12 h under Ar to obtain pure LnOF and $LnFCO_3$.

Sulfurization Apparatus The sulfurization of LnOF and $LnFCO_3$ is carried out in equipment illustrated in Figure 64.2.

Two gas lines (1) allow using either CS_2 and/or H_2S. Whereas H_2S is a gas at room temperature, CS_2 is a liquid with a boiling point of 46 °C. Therefore, a bath with controlled temperature (2) allows having a constant vapor pressure of CS_2 in the line. The ratios between the amounts of the two gases are controlled by two mass-flow controllers (3). The starting F-based Ln compounds (either LnOF or $LnFCO_3$) are set in a boat inside an alumina tube (4), in a tubular electrical furnace (5). The unreacted gases are neutralized in containers filled with soda solution (6).

Chemicals Ln chlorides or nitrates, NH_4HCO_3, and HF solution (40% in water) are commercial products available from several chemical companies. H_2S is commercialized by ALPHAGAZ and CS_2 by Alfa Aesar.

Caution See above, for handling fluorine-based products. When inhaled, both CS_2 and H_2S are poisonous gases, the latter causing lethal accidents. All the experiments involving fluorine-based chemicals, CS_2, and H_2S should be manipulated in specific laboratories in well-ventilated hoods. Eye protection and protective gloves should be imperatively worn.

FIGURE 64.2 Sulfurization apparatus for preparing lanthanum fluorosulfide LaSF. (For a color version of the figure, please see color plates.)

Experimental Procedure Fluorosulfides and oxyfluorosulfides are obtained by annealing LnOF and $LnFCO_3$ at 500 °C<*t*<700 °C under CS_2/H_2S gas mixture (100:0, 30:70, 50:50, 70:30, 0:100 ratios), depending on the F/O and (F + O)/S anionic ratio to be reached. CS_2 is a liquid (vapor pressure of 350 Torr at 25 °C), whereas H_2S is a gas phase at room temperature. CS_2 liquid is slowly heated at 85 °C to reach a sufficient sulfur partial pressure. The sulfur partial pressure is higher for CS_2 at <700 °C. Above 700 °C, the sulfur partial pressure is higher for H_2S and the conditions become more reductive with a competition between hydrogen and sulfur gases. LaSF compound with small crystallite sizes (ϕ = 50 nm) is obtained by treating LaOF at 500 °C in the tubular electrical furnace under pure CS_2 during 12 h. The phase is also synthesized under H_2S steam diluted in Ar for 12 h at 600 °C.

It should be noted that the XRD patterns of LaSF prepared using the gas–solid routes with H_2S or CS_2 exhibit a line widening, associated with an increase of the background. The lower crystallinity and smaller particle size of the final products can be related to lower reaction temperature at which the sulfurization takes place. These results are illustrated in Figure 64.3. The particle size is around 3 μm for the solid-state method, whereas it is submicronic for the gaseous sulfurization route.

Route C: Preparation of Complex Fluorosulfides and Oxyfluorosulfides $Ln_2AF_4S_2$, $Ln_3OF_3S_2$, $La_2O_{1.5}FS$, and $Ln_4O_2F_2S_3$ (via Solid-State Syntheses)

Apparatuses The experimental procedures are similar to those presented in Section 64.1: starting rare earth fluorides are prepared by fluorinating oxide

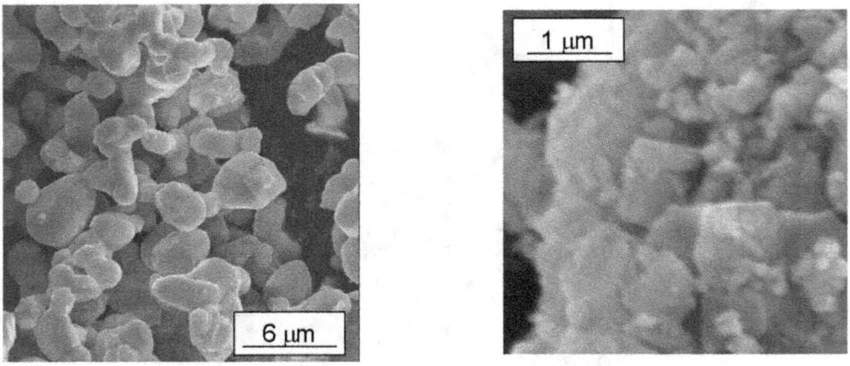

FIGURE 64.3 Comparison of scanning electron microscope photos of LaSF prepared via two different syntheses (solid-state and sulfurization routes).

precursors as stated earlier and rare earth sulfides by sulfurization of pure rare earth metals (see earlier).

Chemicals Alkaline earth fluorides are generally commercial products (>99.9%). They are dehydrated from water traces at 150 °C in dynamical vacuum for 4 h before any experiments and stored under Ar atmosphere.

Caution See earlier, for handling fluorine-based products. Eye protection and protective gloves should be imperatively worn.

Experimental Procedure $Ln_2AF_4S_2$ phases with A = Ca and Sr are obtained by mixing rare earth sulfides, rare earth fluorides, and alkaline earth fluorides in a glove box, following the reaction: $2/3Ln_2S_3 + 2/3LnF_3 + AF_2 \rightarrow Ln_2AF_4S_2$. The components are intimately mixed in a glove box and introduced in a Pt tube that is placed in a quartz tube. After being evacuated during 40 min, the quartz tube is sealed and placed in an electric furnace. The thermal process is the following: temperature increasing rate: 1 °C/min up to 1100 °C; this temperature is maintained for 24 h and the system is then cooled down to room temperature at 2.5 °C/min. To get a good homogeneity, the powder is ground in a glove box and the heating process is repeated one or even two times. The purity of the final product is determined using XRD patterns.

Rare earth oxyfluorosulfides $Ln_3OF_3S_2$, $La_2O_{1.5}FS$, and $Ln_4O_2F_2S_3$ (Ln = La, Ce) are obtained by mixing the corresponding rare earth sulfides, oxides, and fluorides in a glove box, following the reactions:

$$\frac{2}{3}Ln_2S_3 + LnF_3 + \frac{1}{3}Ln_2O_3 \rightarrow Ln_3OF_3S_2$$

TABLE 64.1 Sintering Procedure for Obtaining Complex Rare Earth Fluorosulfides and Oxyfluorosulfides (Solid-State Route)

(Oxy)fluorosulfides	*T* increasing rate, °C/min	Reaction duration	*T* cooling rate, °C/min
$Ln_2AF_4S_2$	1	24 h at 1100 °C	2.5
$Ln_3OF_3S_2$	1	24 h at 1200 °C	2.5
$La_2O_{1.5}FS$	1	24 h at 900 °C	2.5
$Ln_4O_2F_2S_3$	1	48 h at 1250 °C	2.5

$$\frac{1}{3}\,Ln_2S_3 + \frac{1}{3}\,LnF_3 + \frac{1}{2}\,Ln_2O_3 \rightarrow Ln_2O_{1.5}FS$$

$$Ln_2S_3 + \frac{2}{3}\,LnF_3 + \frac{2}{3}\,Ln_2O_3 \rightarrow Ln_4O_2F_2S_3$$

The synthesis conditions are similar to those described for $Ln_2AF_4S_2$ earlier and the thermal processes are grouped in Table 64.1.

Additional annealing can be carried out depending on the purity of the final phases. Note that single crystals of $Ln_3OF_3S_2$ (Ln = La, Ce) for structural determination purpose can be grown through the following procedure: the thoroughly ground mixture was heated up to 1200 °C at a 15 °C/h speed and slowly cooled down to room temperature (5 °C/h).

64.2 ANALYTICAL AND PHYSICAL–CHEMICAL CHARACTERIZATIONS OF RARE EARTH FLUOROSULFIDES AND OXYFLUOROSULFIDES (A, B, AND C ROUTES)

The amount of fluorine and sulfur in the compounds was determined by chemical analysis. For fluorine determination, the samples were dissolved in a flux of $CaCO_3/K_2CO_3$ at 840 °C and the amount of the fluorine measured using a LaF_3 specific electrode. For sulfur determination, the samples were heated at 1300 °C in air to ensure the formation of SO_2. The quantity of SO_2 was measured by infrared spectroscopy taking into account absorption bands associated with S–O vibrations at 1360 and 1151 cm^{-1}, and their integrated areas were compared with references. Particles size and distribution of these phases prepared via solid-state reactions are quite identical, that is, around 5–10 μm.

The absence of impurities was controlled by powder XRD (Philips PW 1050, CuK_α radiation, in Bragg–Brentano geometry). The Rietveld method was used for structure refinement of powder XRD data using a pseudo-Voigt profile function. When single crystals were obtained, as in the case of $Ln_2CaF_4S_2$ or $Ce_3OF_3S_2$, intensity data were collected on an Enraf-Nonius CAD-4 form circle diffractometer, using graphite monochromated Mo-$K\alpha$ radiation. Weissenberg and precession photographs of $Ce_3OF_3S_2$ single crystal showed an orthorhombic *Pnnm* symmetry. Neutron diffraction patterns of $La_2O_{1.5}FS$ were obtained on the D1B diffractometer of the Institut Laue-Langevin (ILL) at Grenoble, France. Whereas $La_2O_{1.5}FS$ crystallizes with hexagonal symmetry in the *P*-3*m*1 space group ($a = 4.111(7)$ Å, $c = 6.917(9)$ Å), $La_3OF_3S_2$ adopts the orthorhombic *Pnnm* space group ($a = 5.804(2)$ Å, $b = 5.784(2)$ Å, $c = 19.482(5)$ Å). Diffuse reflectance spectra were recorded on a Cary 17 spectrophotometer in the visible region [5–8].

64.3 PROPERTIES/APPLICATIONS

LnSF phases crystallize in the *P*4/*nmm* space group related to the PbFCl-type structure (LaSF: $a = 4.0398(3)$ Å, $c = 6.9697(6)$ Å). The structures of α-LnSF and $Ln_2AF_4S_2$ (A = Ca, Sr) fluorosulfides can be considered as formed of stacking of various sheets of rare earth or alkaline earth, fluorine, and sulfur. The rare earth and fluorine atoms form $[Ln_2F_2]^{4+}$ or $[Ln_2AF_4]^{4+}$ fluorite-type blocks, which alternate with double $[S_2]^{4-}$ layers along the *c*-axis. $Ln_3OF_3S_2$ (Ln = La, Ce) and $La_2O_{1.5}FS$ oxyfluorosulfides are also related to α-LnSF and Ln_2O_2S networks.

Within the LnSF family, SmSF and CeSF powders exhibit strong yellow and red color (Figure 64.4), with absorption edges at 490 and 597 nm, respectively.

These materials have been proposed as industrial inorganic pigments [12]. The chromatic properties have been correlated to the structural features [6–8]. The involved electronic transitions, except Ce (4f → 5d), are the result of charge transfer mechanisms from sulfur 3p band to rare earth (4f, 5d) bands [13]. In La-based fluorosulfides and oxyfluorosulfides, the absorption threshold has been correlated with the structural features. It has been deduced that the value of the absorption edge increases with the number of sulfur atoms surrounding the rare earth. For compounds with double sulfur layers, that is, LaSF, $La_2CaF_4S_2$, and $La_3OF_3S_2$, the absorption

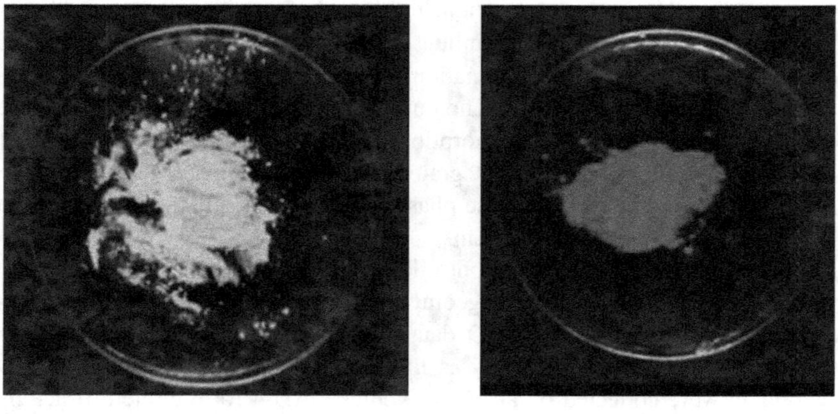

FIGURE 64.4 Inorganic pigments SmSF and CeSF with yellow and red color, respectively. (For a color version of the figure, please see color plates.)

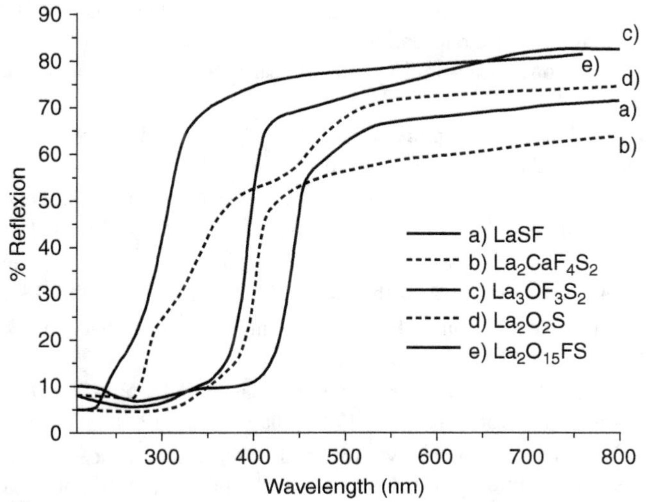

FIGURE 64.5 Diffuse reflectance spectra of LaSF, $La_2CaF_4S_2$, $La_3OF_3S_2$, La_2O_2S, and $La_2O_{1.5}FS$.

takes place in the 380–440 nm range and some of these materials can be used as UV absorbers (Figure 64.5).

ACKNOWLEDGMENTS

H. Laronze (PhD thesis, 1999) and D. Pauwels (PhD thesis, 2003) are deeply acknowledged for their involvement in these topics, and E. Durand for setting up the fluorination and sulfurization equipments. P. Maestro and Rhodia Co. are thanked for financial support.

REFERENCES

1. Flahaut J, Domange L, Patrie M. Combinaisons formées par les sulfures des éléments du groupe des terres rares. Etude cristallographique des phases ayant le type structural du phosphure de thorium Th_3P_4. *Bull Soc Chim* 1962;2048.
2. Mauricot R, Gressier P, Evain M, Brec R. Comparative study of some rare earth sulfides: doped γ-[A]M_2S_3 (M = La, Ce and Nd, A = Na, K and Ca) and undoped γ-M_2S_3 (M = La, Ce and Nd). *J Alloys Comp* 1995;223:130–138.
3. Ozawa L, Forest H. Determination of trivalent europium concentration in yttrium oxysulfide phosphor. *Anal Chem* 1973;45:978.
4. Blasse G, Grabmaier BC. *Luminescent Materials*. Berlin, Heidelberg: Springer-Verlag; 1994. p 161.
5. Demourgues A, Tressaud A, Laronze H, Macaudière P. Rare earth fluorosulfides LnSF and $Ln_2AF_4S_2$ as new colour pigments. *J Alloys Compd* 2001;223:323–324.

6. Pauwels D, Demourgues A, Laronze H, Gravereau P, Guillen F, Isnard O, Tressaud A. Structural features of new rare earth-based mixed anions (O, S, F) compounds: relationships between optical absorption and rare earth environment. *Solid State Commun* 2002;4:1471.

7. Pauwels D. Relations entre structure et absorption UV-visible dans les (oxy)fluorosulfures de terres rares. PhD thesis. University Bordeaux 1; 2003.

8. Pauwels D, Weill F, Tressaud A, Demourgues A. Rare earth oxyfluorosulfides: a new class of compounds with modulated structure. *Chem Mater* 2006;18:6121–6131.

9. Kamihara Y, Watanabe T, Hireno M, Hosono H. Iron-based layered superconductor $La[O_{1-x}F_x]FeAs$ ($x = 0.05-0.12$) with $T_c = 26$ K. *J Am Chem Soc* 2008;130:3296–3297.

10. Chen XH, Wu T, Wu G, Liu RH, Chen H, Fang DF. Superconductivity at 43 K in $SmFeAsO_{1-x}F_x$. *Nature* 2008;453:761–762.

11. Grannec J, Lozano L. Preparative methods. In: Hagenmuller P, editor. *Inorganic Solid Fluorides*. New York: Academic Press; 1985. Chapter 2, pp 17–46.

12. Macaudière P, Morros J, Tourré JM, Tressaud A, inventors; Procédé de traitement de pigments à base de sulfures de terres rares, nouveaux pigments ainsi obtenus et leurs utilisations. Rhône-Poulenc Chimie, assignee. *Patent 9306899*. 1993 June 15; Demourgues A, Laronze H, Macaudière P, Tressaud A, inventors; Nouveaux fluorosulfures et oxyfluorosulfures de terres rares, leurs procédés de préparation et leur utilisation comme pigments colorants. Rhodia Chimie International, assignee; *Patent WO 0073210*. 2000 July 15.

13. Goubin F, Rocquefelte X, Pauwels D, Tressaud A, Demourgues A, Jobic S, Montardi Y The dielectric function of LnSF rare-earth fluorosulfides: experiment and theory. *J Solid State Chem* 2004;177:2833–2840.

CHAPTER 65

Preparation of Carbon–Fluorine Compounds and Fluoride or Oxide Fluoride-Intercalated Graphites

TSUYOSHI NAKAJIMA

Fluorination of carbon materials by elemental fluorine (F_2) yields several types of compounds depending on reaction temperatures and crystallinity of carbon materials. Fluorination of high-crystallinity graphite such as natural graphite between 350 and 600 °C gives graphite fluorides, $(CF)_n$, $(C_2F)_n$, or their mixtures [1,2]. Graphite fluorides have covalent C–F bonds, i.e. puckered carbon layers with hybridized sp^3 orbital. The $(C_2F)_n$-type graphite fluoride is prepared by the fluorination of graphite between 350 and 400 °C, and $(CF)_n$ is prepared between 550 and 600 °C. In the intermediate temperatures, mixtures of $(CF)_n$ and $(C_2F)_n$ are obtained. Therefore, chemical composition successively changes depending on the fluorinated products. Low-crystallinity carbons such as petroleum cokes are often used for the preparation of graphite fluoride. Fluorination of petroleum coke gives only $(CF)_n$-type graphite fluoride between 350 and 550 °C. However, those graphitized between 2600 and 2800 °C yield $(C_2F)_n$ between 350 and 400 °C. Fluorination of graphite at temperatures below ca. 100 °C gives fluorine-graphite intercalation compound, C_xF, in the presence of a Lewis acid such as HF. The C_xF yields stage structures according to the degree of fluorination as conventional graphite intercalation compounds. It maintains planar carbon layers with sp^2 orbital, i.e. has ionic bonding except highly fluorinated stage 1 compound which partly contains puckered carbon layers with sp^3 bonds. Fluorination rate of graphite is low between 100 and 300 °C, where only surface regions of graphite are fluorinated. At temperatures higher than 600 °C, main products are fluorocarbon gases such as CF_4 and C_2F_6. Fluoride or oxide fluoride-graphite intercalation compounds are conveniently synthesized using F_2 gas and simple substances or oxides in the same manner. High-crystallinity graphite such as natural graphite

Efficient Preparations of Fluorine Compounds, First Edition. Edited by Herbert W. Roesky.

and highly oriented pyrolytic graphite (HOPG) are normally used as host materials. Chemical bonds between host graphite and guest species are ionic, showing higher electrical conductivities than that of host graphite.

65.1 PREPARATION OF GRAPHITE FLUORIDES, $(CF)_n$ AND $(C_2F)_n$

Apparatus Fluorination apparatus with fluorine cylinder, pressure gauge for F_2, nickel reactor, electric furnace, stainless-steel column containing porous activated alumina to consume unreacted F_2, rotary pump (see Figure 65.1) [1,2].

Caution Pure nickel without welding should be used for the reactor because fluorination of graphite by F_2 is made at high temperatures between 350 and 600 °C. Stainless steel is used for flange of the reactor and fluorination line. Copper pipe can be also used for fluorination line. A special pressure gauge for F_2 gas should be used. Aluminum metal packing is used for the connection of fluorine cylinder with pressure gauge [1,2].

Chemicals F_2 cylinder, carbon materials such as natural graphite flake or powder and petroleum cokes.

$(CF)_n$ is prepared from both high- and low-crystallinity carbon materials. However, high-crystallinity natural or synthetic graphite should be used for the preparation of $(C_2F)_n$.

Experimental Procedure A batch-type reaction is normally adopted to prepare a small amount of graphite fluoride on a laboratory scale [1,2]. F_2 gas flow method is also possible. A nickel boat containing a carbon material (<300 mg) is placed in a nickel reactor and then the reactor is evacuated by a rotary pump for 12 h. Temperature of the reactor is gradually raised and adjusted at a fluorination temperature under vacuum. After that, F_2 gas is slowly introduced into the reactor ($\sim 1.0 \times 10^5$ Pa). Graphite is allowed to react with F_2 for

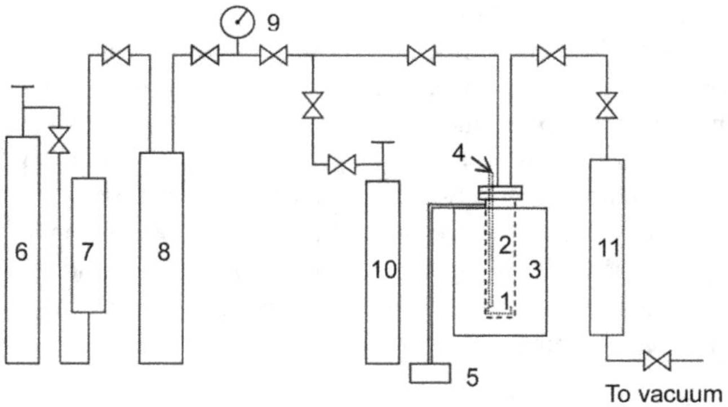

FIGURE 65.1 An example of a fluorination apparatus. 1. Ni boat. 2. Ni reactor. 3. Electric furnace. 4. Thermocouple. 5. Temperature controller. 6. F_2 cylinder. 7. HF absorber (NaF pellets heated at ca. 120 °C). 8. Buffer tank. 10. N_2 cylinder. 11. Activated alumina column.

1–10 days. Preparation of $(C_2F)_n$ needs 1 week to 10 days because the reaction temperature is limited between 350 and 400 °C. With increasing fluorination temperature, the duration of the fluorination is shortened. After the fluorination is completed, residual F_2 is allowed to react with porous activated alumina by slow evacuation using a rotary pump. F_2 easily reacts with activated alumina at room temperature. Since this reaction is highly exothermic, we can find which part of an activated alumina column is reacting with F_2 by touching the alumina column. When F_2 flow method is taken, F_2 flow rate should be low to avoid the rapid increase in temperature of activated alumina column. Unreacted F_2 is completely removed by the reaction with activated alumina at room temperature.

Characterization Data for $(CF)_n$ Color: gray~white; chemical composition: F/C $\approx$ 1.0; X-ray diffraction: d_{001} $\approx$0.6 nm; IR absorption: 1220 cm^{-1} (vs), 1350 cm^{-1} (w), 1075 cm^{-1} (w), 940 cm^{-1} (vw); XPS: C1s: 290 eV (C–F bond), F1s: 689–690 eV [1,2].

Characterization Data for $(C_2F)_n$ Color: black, chemical composition: F/C $\approx$0.6; X-ray diffraction: d_{001} $\approx$0.9 nm; IR absorption: 1220 cm^{-1} (vs), 1350 cm^{-1} (s), 940 cm^{-1} (m); XPS: C1s: 288 eV (C–C bond), 290 eV (C–F bond), F1s: 689–690 eV [1,2].

The color of graphite fluoride changes from black, gray to white, and F/C ratio continuously increases from $\approx$0.6 to $\approx$1.0 with increasing fluorination temperature. Thin $(CF)_n$ layers are formed in the surface disordered region of graphite particle when $(C_2F)_n$ is prepared. Therefore, the lowest F/C value of $(C_2F)_n$ is approximately 0.6. The F/C of $(C_2F)_n$ prepared from graphitized petroleum coke is $\approx$0.7 because such synthetic graphite has the lower crystallinity than natural graphite. The d_{001} of $(CF)_n$ prepared from a low-crystallinity carbon, for example, petroleum coke, is $\approx$0.7 nm larger than $\approx$0.6 nm.

Caution The reaction of a carbon material with F_2 is highly exothermic. Fluorination of more than 1 g of carbon should be carefully performed because rapid increase in local temperature often causes explosion. It is also difficult to fluorinate a low-crystallinity carbon such as activated carbon with a large surface area by F_2 of 1×10^5 Pa even at room temperature. Under such condition, activated carbon vigorously reacts with F_2, yielding fluorocarbon gases such as CF_4 and C_2F_6. It is necessary to use F_2/Ar mixture with low F_2 partial pressure ($<5 \times 10^4$ Pa) for the fluorination of low-crystallinity carbons.

Application Graphite fluorides are hydrophobic materials due to their covalent C–F bonds, being stable under usual environment. However, electrochemical reduction of graphite fluorides easily takes place in organic electrolyte solution. The most important application of $(CF)_n$ is the use as cathode material of primary Li battery. Li/$(CF)_n$ battery shows excellent properties such as high voltages, high energy densities, long shelf life, high thermal stability, good pulse discharge characteristics, and so on (see Figure 65.2) [2]. Graphite fluorides are also applied to solid lubricant and water repellent.

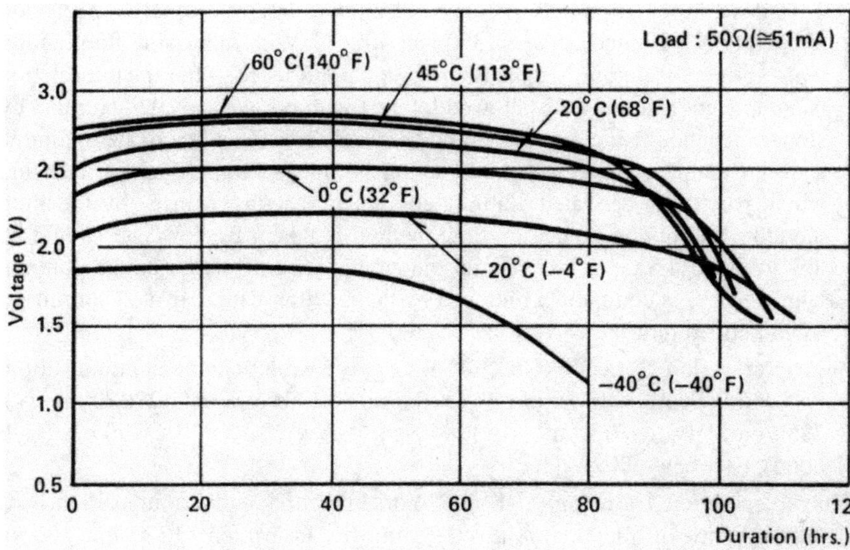

FIGURE 65.2 Discharge curves of Li/(CF)$_n$ battery (BR-C type).

65.2 PREPARATION OF FLUORINE–GRAPHITE INTERCALATION COMPOUND, C$_x$F

Apparatus (1) Lewis acids such as HF are often used as catalysts for fluorine intercalation into graphite, which occurs at temperatures below ca. 100 °C. Preparation of C_xF is normally accomplished at room temperature. When a mixture of a Lewis acid such as HF and F_2 gas or only F_2 gas is used for fluorination of graphite, the apparatus shown in Figure 65.1 can be employed [3–5].

(2) When fluorine intercalation is made in anhydrous liquid HF (aHF), fluoropolymer such as Teflon perfluroroalkoxy (PFA) tube is used for fluorination line and reactor. Since the boiling point of HF is 19.5 °C, fluorination is performed at temperatures lower than the boiling point of HF, very often between 0 and 19 °C. Figure 65.3 shows an example of fluorination line for using aHF [6–8].

Chemicals High-purity F_2 cylinder, carbon materials such as natural graphite flake or powder, HOPG, high-oxidation-state-complex fluorides such as K_2NiF_6 and $KAgF_4$, MoF_6 and SbF_5 cylinders, Sb, W, and Ag powder.

aHF is purified by bubbling high-purity F_2 gas through it. High-oxidation-state-complex fluorides such as K_2NiF_6 and $KAgF_4$ are used for purification of aHF and/or as oxidizer for graphite. Lewis acids such as MoF_6 and SbF_5 are sometimes used as initiators of fluorine intercalation in graphite for opening the edge of graphite (oxidizer for graphite).

Experimental Procedure (1) When only F_2 gas or a mixture of F_2 and HF is used for fluorination of graphite, the apparatus shown in Figure 65.1 can be

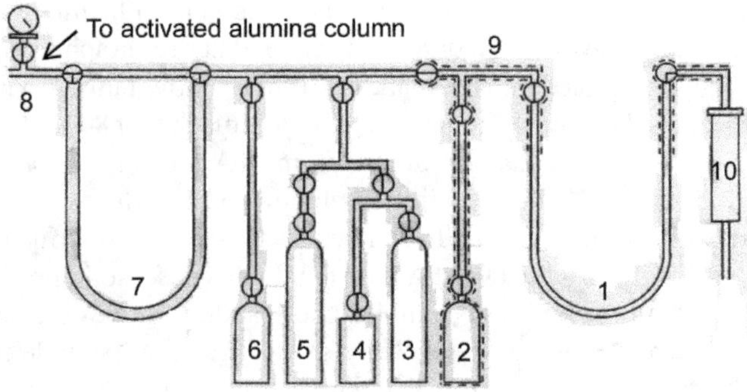

FIGURE 65.3 An example of fluorination apparatus using aHF. 1. Reaction tube (Teflon PFA). 2. MoF_6 or SbF_5 cylinder. 3. F_2 cylinder. 4. F_2 buffer tank. 5. Ar cylinder. 6. HF cylinder. 7. Distillation tube (Teflon PFA). 8. Pressure gauge. 9. Ribbon heater. 10. Activated alumina column.

used. Only F_2 gas is often employed because F_2 yields a trace of HF by the reaction with adsorbed moisture in a nickel reactor. A nickel boat containing graphite (<300 mg) is placed in a nickel reactor and then the reactor is evacuated by a rotary pump for 12 h. Next day, F_2 gas or F_2 and HF are slowly introduced into the reactor ($\sim 1.0 \times 10^5$ Pa). Graphite is allowed to react with F_2 for 1–10 days. C_xF samples (x: 2.0–8.0) with stages 1–3 or mixed stages are usually obtained [3, 5].

(2) Another method of C_xF preparation is to add a trace of Sb or W (<10 mg) to graphite sample when graphite is fluorinated by F_2. Sb or W easily gives SbF_5 or WF_6 as an oxidizer for graphite, respectively, by fluorination at 150–200 °C. After Sb or W is fluorinated to SbF_5 or WF_6, the reactor is cooled down to room temperature to facilitate fluorine intercalation into graphite. The apparatus shown in Figure 65.1 is used for this method [3, 5].

(3) C_xF is also prepared under highly purified condition without HF using the apparatus shown in Figure 65.1. Graphite is placed in a nickel boat closely with Ag powder. Nickel reactor is purified at 200 °C by repeating the introduction of high-purity F_2 gas (HF: <0.01%) and evacuation by rotary and diffusion pumps. After confirmation of no IR absorption by HF at 3800–4000 cm^{-1}, graphite is allowed to react with F_2 for 5~34 days. Metallic Ag powder is fluorinated by F_2, giving AgF_2 as an oxidizer for graphite. Weak X-ray diffraction peaks indicating AgF_2 are detected with those for C_xF [4].

(4) When aHF is used as a reaction medium, fluorination line made by Teflon PFA tube is used. Figure 65.3 shows an example of fluorination apparatus. The aHF is transferred from HF cylinder (6 in Figure 65.3) to Teflon PFA tube (7 in Figure 65.3) using liquid N_2 bath. A trace of water in HF is removed by bubbling high-purity F_2 gas through it for 6–12 h. The PFA tube (7) is cooled by ice bath during F_2 bubbling. Graphite powder (50–150 mg) is placed in the PFA tube (1 in Figure 65.3), followed by pumping. The purified aHF is

then transferred from PFA tube (7) to PFA tube (1). To facilitate the fluorine intercalation, a trace of MoF_6 or SbF_5 as an oxidizer for graphite is sometimes added. F_2 gas is bubbled into PFA tube (1) from F_2 buffer tank (4 in Figure 65.3) for several days, keeping the temperature of PFA tube (1) at 8–25 °C. During the fluorination, aHF is gradually removed from PFA tube (1). Unreacted F_2 and HF are allowed to react with activated alumina. C_xF samples (x: 2.0–2.4) are obtained by this procedure [6]. To synthesize more highly fluorinated graphite, K_2NiF_6 or $KAgF_4$ is added to PFA tube (1). In this case, solid product NiF_2(s) or AgF_2(s) is mixed with C_xF sample. These solid fluorides are separated by the following reactions: $2AsF_5 + NiF_2(s) \rightarrow Ni(AsF_6)_2$ (soluble in aHF) and $2AsF_5 + AgF_2(s) \rightarrow Ag(AsF_6)_2$ (soluble in aHF). Highly fluorinated C_xF samples (x: 1.2–2.3) with stage 1 structure are obtained by this method. A trace of HF is usually contained in the products. It is removed by pumping at room temperature or slightly higher temperature than room temperature [7, 8].

Characterization Data color: black, X-ray diffraction: I_c (repeat distance along c-axis): ~0.6 nm (stage 1); I_c: 0.94 nm (stage 2); I_c: 1.28 nm (stage 3); IR absorption: 1084–1134 cm^{-1} (vs) (C–F bond vibration), 1196–1230 cm^{-1} (s) (C–F bond vibration), 1250–1257 cm^{-1} (s) (C–C bond vibration), 1523–1575 cm^{-1} (vs) (C–C bond vibration, A_{2u}); XPS: C1s: 284–285 eV (C–C bond), 287–288 eV (C–F bond), and F1s: 687 eV for C_xF (x: 2.0–7.0), C1s: 287–288 eV (C–C bond), 290–292 eV (C–F bond), 294 eV (CF_2 bond), and F1s: 689–691 eV for C_xF (x: <2.0) [3–8].

Caution Handling of aHF should be carefully done since the boiling point of HF is 19.5 °C close to room temperature with the consequence that vapor pressure of HF is always high. During experiment, cooling of aHF by ice bath is necessary to reduce its high vapor pressure.

Application C–F bond of C_xF changes from ionic to covalent with increasing fluorine content, i.e. with decreasing stage number. Electrical conductivity of stage 1 C_xF is similar to or lower than that of host graphite because stage 1 compound has covalent C–F bonds. However, stage 2 or higher stage compounds than stage 2 have nearly ionic C–F bonds. Electrical conductivity is much higher than that of host graphite because a large number of positive holes is created in graphene layers by charge transfer. C_xF can be used as an additive for $(CF)_n$ cathode of primary lithium battery because $(CF)_n$ is an electric insulator, having a high resistance at the beginning of discharge. Electro-conductive C_xF facilitates the discharge of $(CF)_n$ cathode because of its higher discharge potential than that of $(CF)_n$.

65.3 PREPARATION OF FLUORIDE- AND OXIDE FLUORIDE-GRAPHITE INTERCALATION COMPOUNDS

Apparatus The same apparatus as shown in Figure 65.1 is used for the preparation of fluoride or oxide fluoride-graphite intercalation compounds.

Chemicals F_2 cylinder, simple substances such as Ti, V, Nb, Ta, Mo, W, As, Sb, I, oxides such as V_2O_5, WO_3, carbon materials such as natural graphite flake or powder, HOPG [9–13].

Experimental Procedure Nickel boats separately containing graphite and powder of simple substance or oxide are placed in a nickel reactor. Temperature of the reactor is raised to a temperature between 160 and 310 °C under vacuum. After 6 h, F_2 gas is introduced into the reactor at the same temperature. Simple substance or oxide is fluorinated by F_2, yielding fluoride such as TiF_4, VF_5, NbF_5, or oxide fluoride such as VOF_3, WOF_4, WO_2F_2. Fluorination of simple substance or oxide by F_2 is exothermic reaction, easily giving corresponding fluoride or oxide fluoride. Fluoride or oxide fluoride is intercalated into graphite with fluorine at the same temperature between 160 and 310 °C. Handling of volatile fluorides is normally difficult because most of them are hygroscopic. If F_2 gas is employed, preparation of fluoride or oxide fluoride-graphite intercalation compounds is easily made by using simple substance or oxide [9–13].

Characterization Data color: deep blue; X-ray diffraction: I_c (repeat distance along c-axis): 0.82–0.83 nm (stage 1), I_c: 1.14–1.17 nm (stage 2), I_c: 1.49–1.51 nm (stage 3) [9–13].

Caution Chemical bond between intercalated fluoride or oxide fluoride and host graphite is ionic. Some intercalation compounds are hygroscopic, which are unstable in air.

Application Fluoride or oxide fluoride-graphite intercalation compounds have much higher electrical conductivities than host graphite because positive holes are created in graphene layers by electron transfer from graphite to intercalated fluoride or oxide fluoride. Figure 65.4 shows electrical conductivities of C_xTiF_5 [9]. Potential application is materials for shielding of electromagnetic wave.

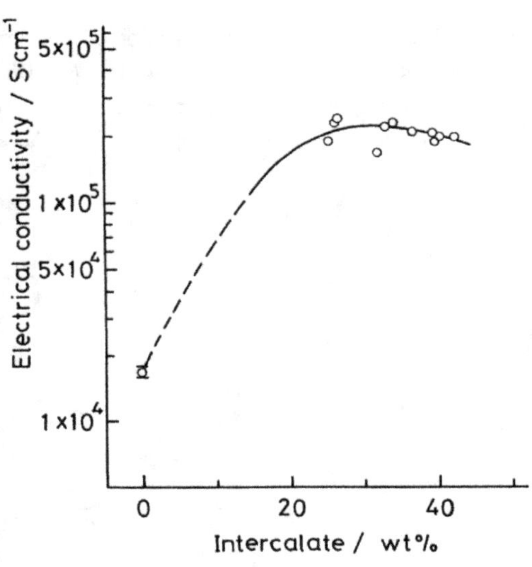

FIGURE 65.4 Electrical conductivities of C_xTiF_5 prepared from HOPG.

REFERENCES

1. Kita Y, Watanabe N, Fujii Y. Chemical composition and crystal structure of graphite fluoride. *J Am Chem Soc* 1979;101:3832–3841.
2. Nakajima T, Watanabe N. *Graphite Fluorides and Carbon-Fluorine Compounds*. Boca Raton, FL: CRC Press; 1991.
3. Nakajima T, editor. *Fluorine-Carbon and Fluoride-Carbon Materials – Chemistry, Physics, and Applications*. New York, NY: Marcel Dekker; 1995.
4. Nakajima T, Molinier M, Motoyama M. Structure of fluorine-graphite intercalation compounds. *Carbon* 1991;29:429–437.
5. Amine K, Nakajima T, Motoyama M. Structure of fluorine-graphite intercalation compounds synthesized in gaseous and anhydrous liquid hydrogen fluoride. *Carbon* 1994;32:1067–1071.
6. Nakajima T, Namba M. Synthesis of fluorine-graphite intercalation compounds in anhydrous liquid hydrogen fluoride containing a Lewis acid. *TANSO* 1995;170:255–258.
7. Nakajima T, Koh M, Gupta V, Žemva B, Lutar K. Electrochemical behavior of graphite highly fluorinated by high oxidation state complex fluorides and elemental fluorine. *Electrochim Acta* 2000;45:1655–1661.
8. Nakajima T, Gupta V, Ohzawa Y, Groult H, Mazej Z, Žemva B. Influence of cointercalated HF on the electrochemical behavior of highly fluorinated graphite. *J Power Sources* 2004;137:80–87.
9. Nakajima T, Nakane K, Kawaguchi M, Watanabe N. Preparation, structure and electrical conductivity of graphite intercalation compound with titanium fluoride. *Carbon* 1987;25:685–689.
10. Nakajima T, Okahara K, Padma DK, Watanabe N. Preparation and electrical conductivity of niobium fluoride-graphite intercalation compound. *Carbon* 1988;26:825–829.
11. Nakajima T, Matsui T, Motoyama M, Mizutani Y. Preparation, structure and electrical conductivity of vanadium fluoride-graphite intercalation compound. *Carbon* 1988;26:831–836.
12. Nakajima T, Molinier M. Preparation and characterization of tungsten oxide fluoride and tungsten fluoride-graphite intercalation compounds. *Synth Metals* 1989;34:103–108.
13. Nakajima T, Nagai Y, Motoyama M. Synthesis, structure and electrical conductivity of vanadium oxide fluoride-graphite intercalation compound. *Eur J Solid State Inorg Chem* 1992;29:919–928.

CHAPTER 66

Safe Synthesis of Superstoichiometric Mesoporous Fluorocarbons

VALENTIN N. MITKIN

Over the last 40 years it had been a challenging task for creating a new generation of chemical current sources (CCSs) with optimal characteristics. New CCSs should operate not below 3 V, they should have specific electric capacity, and energy not below 250 Ah and 700 Wh per kg of current cell, small weight, and the compactness combined with inherent reliability and working capacity under arctic and tropical conditions. Moreover, they should exhibit small self-discharge and a long-term operation without maintenance and optimal safety. This task has been solved at the end of the seventies by the technical development and marketing issues of several electrochemical systems based on lithium anode.

As can be seen from Table 66.1 , the leading systems are Li cells with the electrochemical system "Li–CF_x" [1,2]. The current forming reaction is mainly due to the high C–F bond energy (sp^3–C–F $\sim$ 480 kJ$\cdot$mol^{-1}) given in reaction 66.1.

$$x\text{Li}_{(\text{solid})} + \text{CF}_{x(\text{solid})} \rightarrow x\text{LiF}_{(\text{solid})} + \text{C}_{(\text{solid})} \qquad (66.1)$$

From reaction 66.1 and Table 66.1 it follows that one of the most perspective directions on the development are fluorocarbon materials and lithium metal for chemical power generation with new superstoichiometric fluorocarbons of composition CF_{1+x} ($x \geq 0$). Moreover, it is important to develop safe synthetic methods without explosion during the fluorination of CF_{1+x} materials [2]. As an interesting precursor a new mesoporous carbon material was used (called sibunite; Figure 66.1) [3]. Fluorination of sibunite carbon material allowed to prepare a new generation of superstoichiometric fluorocarbons (fluorosibunites) under mild conditions with the atomic ratio of C : F 1.18–1.33 [4].

Efficient Preparations of Fluorine Compounds, First Edition. Edited by Herbert W. Roesky.

TABLE 66.1 Main Characteristics of Electrochemical Systems Based on Lithium Anode

Electrochemical system	Voltage, V: Open circuit voltage	Voltage, V: Nominal exploitation	Specific capacity of cathode, Ah/kg	Theoretical energy of cathode, Wh/kg	Maximal current density, mA/cm^2
$Li–F_2$	6.05	Non-realized	1413	6254	Non-realized
$Li–O_2$	2.90	Non-realized	1678	5900	Non-realized
Li–S[a]	2.52	2.2–2.4	1167	2560–2800	10–100
$Li\text{-}(CF_{1.0})_n$[b]	3.20	2.4–2.7	861	2000–2320	0.1–0.2
$Li–(C_2F)_n$[b]	3.25	2.9–3.1	624	1800–1930	0.5–1.0
$Li–SOCl_2$	3.61	2.9–3.3	605	1750–2000	2–5
$Li–MnO_2$	3.69	2.5–2.8	309	770–862	0.5–2.0
$Li–SO_2$	2.95	2.2–2.7	420	900–1100	2–5
$Li–FeS_2$	2.85	1.2–1.4	895	1075–1250	0.01–0.05
Li–CuO	2.24	1.2–1.4	575	1288	0.1–0.2

[a] High-temperature system.
[b] "Matsushita Electric Co" (Panasonic).

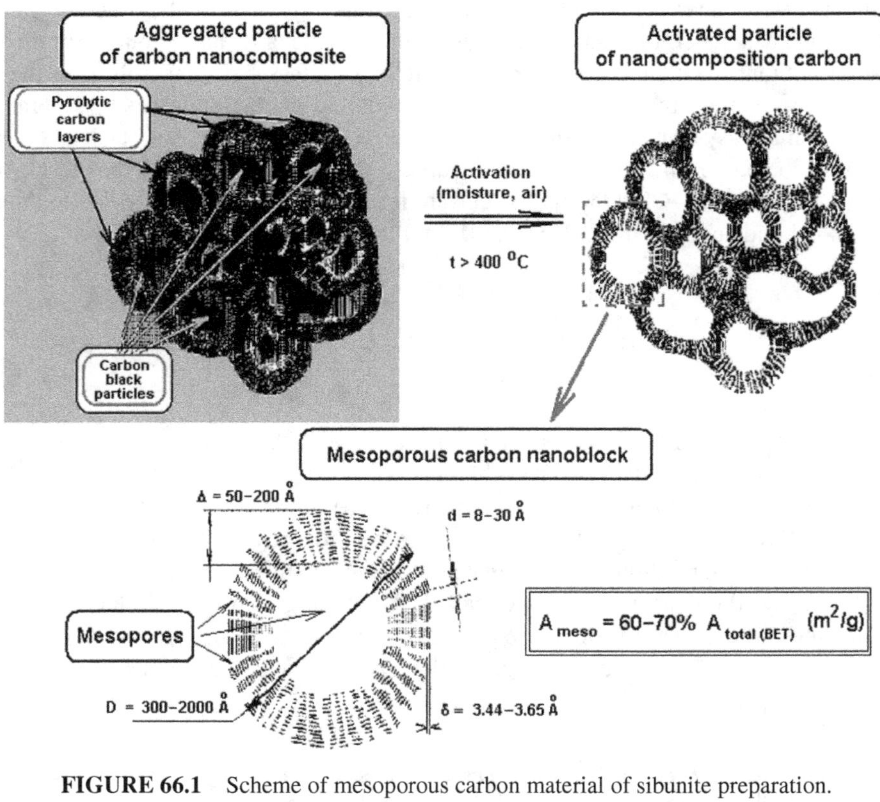

FIGURE 66.1 Scheme of mesoporous carbon material of sibunite preparation.

66.1 GASEOUS FLUOROXIDANTS

Fluorine gas was produced in a steel F-cell supplied with a nickel anode on a large laboratory scale. The range of current was 0–100 A with productivity of up to 70 g of F_2 per hour. As an electrolyte KF∗nHF ($n = 1.83$–1.95) has been used. The purity of HF was better than 99.97 wt%. Electrolysis was working at a current of 10–80 A and working temperature of 85–115 °C. Anode gas fluorine was passed through traps with tablets of sodium fluoride at 100–130 °C for elimination of HF. Chlorine trifluoride was obtained from the Siberian Group of Chemical Enterprises (Seversk, Tomsk). Content of ClF_3 in this commercial product was better than 98 wt%.

66.2 INITIAL CARBON MATERIALS FOR PREPARATION OF SUPERSTOICHIOMETRIC FLUOROCARBONS

A series of mesoporous carbon materials (sibunites M-19, M-36, M-85P) were produced by Omsk Institute of Hydrocarbon Processing SB RAS, according to the technical specification TU 38 41538-94 [3] and main scheme of Figure 66.1. For the synthesis of CF_{1+x}, various sibunite fractions with particle sizes that range from <40 to 2000 μm were used. A series of sibunites have a specific surface area of 100–600 m^2/g (method BET, N_2), a density of 2.01–2.11 g/cm^3, a bulk density of 0.61–0.81 g/cm^3, and a volume of free pores of 57–69 rel%. The content of absorbed H_2O was 0.45–0.60 wt%. Carbon in sibunite series was found to be 98.8–99.1 wt% and H was 0.05–0.12 wt%. The total amount of impurities was detected by atomic emission spectral analysis as 120 to 340 ppm (Fe, Si, Al, Ca, Mg, Ni, Mn, etc.). Oxygen content was calculated as 0.44–1.03 wt% as difference [$100\% - C - H - H_2O - \sum$impurities]. Sizes of carbon nanoblocks in sibunite particles, determined as coherent diffraction area on 002 and 004 XRD peaks, were found to be about ~25–45 Å. Main interlayer distances measured from 002 XRD peak were 3.49–3.52 Å.

From data of Figure 66.1 it follows that all sibunites have a mesoporous architecture, where the volume of mesopores reaches 60–70 rel% of the total porosity. The macro- and microstructure units of sibunite are the hollow shells with diameters of 500–2000 Å and thicknesses of the shell layers of 50–500 Å. These shells are formed by graphite-like blocks whose sizes are in the range of 50–100 Å. The structure of sibunite with its high specific area and high porosity of carbon particles provides the right carbon material for effective removal of the heat (Q_r) during the fluorination process by taking off gaseous products during reaction 66.2.

$$(1 + y)C^{+}_{(solid)}[0.5\,(1 + x) + 2y]F_{2\,(gas)} \rightarrow CF_{1+x(solid)} + y\,CF_{4\,(gas)} + Q_r \quad (66.2)$$

66.3 LABORATORY SYNTHESES OF THE SUPERSTOICHIOMETRIC MESOPOROUS FLUOROCARBONS

Syntheses of CF_{1+x} were accomplished in two basic types of laboratory Monel reactors, which are shown in Figure 66.2. The horizontal reactor MIG-1 had a volume

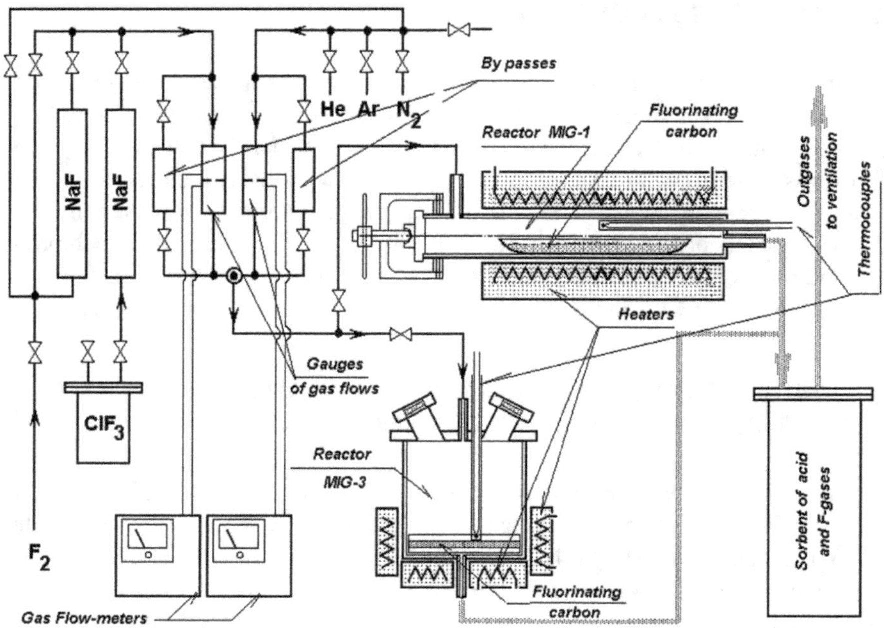

FIGURE 66.2 Laboratory pilot installation for studies and preparation of superstoichiometric fluorocarbon materials.

of 350 mL that allowed a maximum loading of carbon material of 6–8 g. The vertical reactor MIG-3 had a volume of 1800 mL that allowed a fluorination of a maximum loading of 10–12 g of carbon. Additional advantage of the vertical reactor was the possibility to see the progress of the reaction through a transparent sapphire window in the shell of MIG-3 reactor.

The process of fluorination in the horizontal reactor MIG-1 was conducted in nickel or vitrous carbon (glassy carbon) boats. The fluorination in the vertical reactor was done on a fixed nickel net supported by a nickel framework holder (diameter 120 mm), where a thin layer (0.8–2.0 mm) of the carbon material was placed. The working gas was passed through the carbon layer instead of flowing over as in a horizontal reactor.

Data on experimental conditions for CF_{1+x} syntheses of the horizontal and vertical reactors are shown in Tables 66.2 and 66.3. These data indicate that fluorination of sibunites leads mainly to the formation of white powders of composition $CF_{1.18\text{-}1.33}$, and this is accompanied by partial loss of the initial carbon through the formation of gaseous products. Thus, this loss decreases using F_2 or ClF_3 diluted with an inert gas (He, Ar, N_2).

It was noted that in the horizontal reactor MIG-1 in almost all cases (270–280 °C) there was a substantial loss of carbon materials, and yields of targeted fluorocarbon seldom exceeded 50%.

The experiments with lower temperatures (250–260 °C) resulted in less volatile products (10–15%), although the product formation was less favorable.

TABLE 66.2 Fluorination of Sibunite (M-36) in the Horizontal Reactor MIG-1 (fraction size 70–200 μm, A_{BET} 350 m^2/g)

Initial load of carbon, g	Final weight of CF_{1+x}, g	Flow of F_2 (ClF_3), mL/min	Flow of inert gas, mL/min	Temperature, °C	Time, min	A_{BET} of final CF_{1+x}, m^2/g	Kind of fluorination product and its chemical composition
4.48	7.72	80 ClF_3	240 N_2	270	170	373	White $CF_{1.18}$ with small black points
3.90	2.74	50 ClF_3	150 He	260	360	308	White $CF_{1.28}$
4.48	7.19	50 ClF_3	300 N_2	280	150	283	White $CF_{1.33}$
3.12	3.61	100 F_2	100 He	280	240	422	Snow white powder $CF_{1.33}$
7.63	10.36	170 F_2	150 He	270	185	380	White $CF_{1.30}$

TABLE 66.3 CF_{1+x} Formation in the Vertical Reactor MIG-3

Type of sibunite	Fraction of initial carbon, mcm	Initial load of carbon, g	Final weight of CF_{1+x}, g	Flow of F_2 (ClF_3), mL/min	Flow of inert gas, mL/min	Temperature, °C	Time, min	Kind of fluorination product and its chemical composition
M-19	70–100	5.0	12.4	160 F_2	800 Ar	275	2	White powder $CF_{1.32}$. Dusty
						300	15	
						330	45	
						350	30	
	100–200	5.0	11.7	275 F_2	800 He	280	45	Snow white powder $CF_{1.33}$
						360	155	
	40–70	7.0	13.7	140	715	300	20	White powder $CF_{1.18}$ Dusty
					a3OT	330	36	
						345	60	
	200–250	10.0	19.3	160 F_2	800 He	315	45	White granules $CF_{1.24}$ Not dusty
				120 F_2		350	60	
				100 F_2		360	120	
	70–100	7.5	17.0	250 F_2	800 He	275	2	Snow white powder $CF_{1.33}$ Dusty. Maximal yield
				160 F_2		300	15	
				120 F_2		330	45	
				100 F_2		350	30	
	20–40	5.0	11.3	150 ClF_3	520 He	280	60	White powder $CF_{1.19}$ Dusty
						350	60	
M-85P	40–70	5.0	10.7	130 ClF_3	400 He	310	120	White powder $CF_{1.14}$ Dusty
	40–70	5.1	8.0	210 ClF_3	400 He	300	45	Gray powder $CF_{1.12}$ Dusty
	40–70	5.0	10.8	300 ClF_3	1000 He	285	180	Gray powder $CF_{1.09}$
	40–70	5.0	11.4	155 F_2	480 He	310	160	White powder $CF_{1.25}$. Dusty

The process conducted in the vertical reactor MIG-3 is more favorable and therefore the preparation of relatively large amounts of fluorinated carbon materials (up to 18 g in one load) and a yield of about 75% was obtained according to Equation 66.3.

$$4C^{+}_{(solid)}4F_{2(gas)} \rightarrow 3\,CF_{1.33(solid)} + CF_4(gas) \qquad (66.3)$$

From the highlighted data it follows that both in isothermal (Table 66.2) and in polythermal (Table 66.3) modes of fluorination it is possible to have process optimum in a temperature range of 300–350 °C. When the fluorination processes are carried out using non-diluted fluorine, the reaction of thin powdered sibunites became uncontrollable in some "temperature–time ranges" (variable for each type of sibunite), which often led to ignition and practically full burning of the carbon powder owing to the sharp starting initiation process.

In case of diluted fluorine gas the reactions (entry 2–3) proceed smoother. The yield and properties of the obtained CF_{1+x} poorly depend on the F_2 partial pressure in the studied range of F_2 to inert gas of 1:3 to 1:10. However, the reaction increases with a big excess of fluorine, although in the temperature range of 200–250 °C the reaction proceeds slowly and only the surface of the carbon species is fluorinated.

On the basis of the analysis of obtained data on fluorination of sibunite materials in a vertical reactor it has been possible to select optimal synthetic conditions for any initial carbon material. In most of the cases the synthesis of CF_{1+x} can be finished within 17–20 g/h/dm^2, which allows to enter into safe non-explosion reaction of about 7–10 g/dm^2 of sibunite materials. Moreover, the system is ready for operation after heating for 60–80 min and with the appropriate mixture of a stream of fluorine and inert gas.

66.4 BRIEF DESCRIPTION OF PHYSICAL-CHEMICAL PROPERTIES FOR MESOPOROUS FLUOROCARBON MATERIALS CF_{1+x}

All properties of the superstoichiometric fluorocarbon materials and their physical-chemical characteristics have been well described in the technical specification [5] and previous publications [6–9]. The texture of CF_{1+x} is identical to initial sibunite. Usually CF_{1+x} samples have a specific surface area of 280–600 m^2/g (method BET, N_2), a density of 2.57–2.67 g/cm^3, a bulk density of 0.85–0.95 g/cm^3, a volume of free pores 57–69 rel%, where 60–70 rel% are mesopores.

The content of absorbed H_2O usually does not exceed 0.05–0.08 wt%. Chemical formula is usually $CF_{1.14\text{-}1.33}$. Total amount of impurities, detected by atomic emission spectral analysis, do not exceed 500 ppm (Fe, Si, Al, Ca, Mg, Ni, Mn, etc.). The sizes of fluorocarbon nanoblocks in $CF_{1.14\text{-}1.33}$ particles were determined as CDA on 001 and 003 XRD peaks and analyzed to be about ~20–25 Å. The main interlayer distances were measured on 002 XRD peaks of 6.5–6.9 Å [4,7]. The characteristic electron micrographs for laboratory samples of $CF_{1.33}$ and its structure are shown in Figure 66.3. The most characteristic bands in the Fourier transform infrared spectra for CF_{1+x} samples are strong bands between 1200 and 1230 cm^{-1} (sp^3–C–F bonds

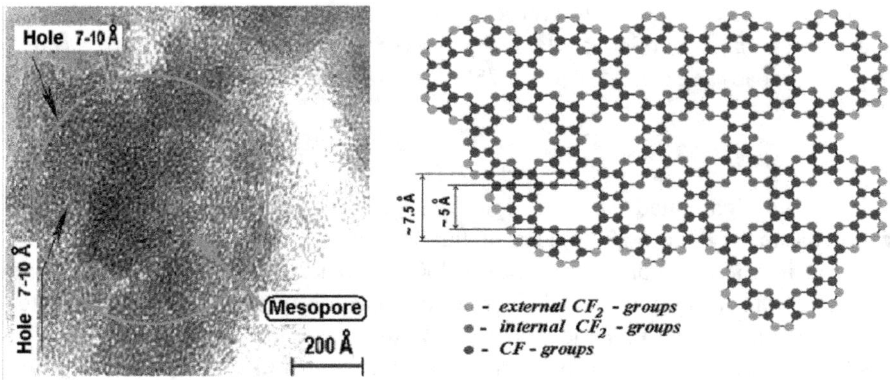

FIGURE 66.3 Image of microparticle surface and assumed structure for $CF_{1.33}$ [7,8]. (For a color version of the figure, please see color plates.)

in C–F groups), and also strong asymmetric vibrations of sp^3–C–F bonds in CF_2 groups between 1320 and 1340 wave numbers [4,6].

Independent tests of the superstoichiometric fluorocarbons as cathode materials were carried out in BR-2325 Li cells [8,9] and these data are given in Table 66.4, and compared with those results of comparative tests of common industrial CF_x materials in the same Li cells. Results of these tests indicate that CF_{1+x}-based cathodes have a specific capacity in a range of 580–680 mAh/g (30–50% higher compared to the best analogs), and the current discharge density is 10–30 times better than market analogs with long shelf-life of up to 25 years.

TABLE 66.4 Characteristics of CF_{1+x}-Based Cathodes for Li Cells of BR-2325 Coin Type [8,9]

Technical properties	New BR-2325[a]	Market BR-2325
Open circuit voltage, V	3.0–3.3	3.0–3.3
Short current, mA	20–100	10–50
Nominal capacity, mAh (warranted)	240	160
Nominal load, kΩ	10-30	30–100
Nominal voltage, V (dependent on load)	2.4–2.9	2.4–2.7
Maximal continual current to 2 V cutoff, mA	0.2	0.03–0.1
Maximal possible shelf-life, years	15–25	10
Warranted shelf-life, years	5–7	5
Material of cathode	Novel $CF_{1.25}$	Market $CF_{0.92\text{-}0.98}$
Theoretical capacity of real cathode, mAh/g	798	680
Specific capacity at 0.5 mA/cm^2, mAh/g	644	446
Specific capacity at 0.125 mA/cm^2, mAh/g	650	511
Specific capacity at 0.03 mA/cm^2, mAh/g	680	535

[a] BR-2325—Type size according to the International Classification; BR is "CF_x–Li" system; 23—diameter of cell in millimeter; 25—thickness of cell is 2.5 mm.

REFERENCES

1. Ishikawa N, editor. *Modern Application Technology of Fluorine Compounds.* Tokyo: CCMC; 1981 (Russian Edition - Mir, Moscow, 1984, 592 p., cited p. 134).
2. Watanabe N, Nakajima T, Touhara. H. *Graphite Fluorides.* Amsterdam, Oxford, New-York: Elsevier, Science Publishers B.V.; 1988. –p 263.
3. Technical carbon. Technical specification TU 38 41538-94, Boreskov Institute of Catalysis SB RAS, Novosibirsk. 1994, p 34; see also Likholobov VA, Surovikin VF, Plaksin GV, Tsekhanovich MS, Surovikin YV, Baklanova ON. Catal Ind 2009;1:11–16. DOI: 10.1134/S2070050409010024
4. Zemskov SV, Gornostaev LL, Mitkin VN, Ermakov YI, Lisitsyn AS, Likholobov VA, Kedrinskij IA, Pogodaev VP, Plaksin GV, Surovikin VF. C.A. 126:10026. Carbon fluoride and its manufacture for battery electrodes. Russian patent RU No 2054375 C1 20 Feb 1996 from Izobreteniya 1996; 5:152. http://www.lookchem.com/cas-571/57158-31-3.html
5. Mitkin VN, Levchenko LM, Galkin PS, et al. Polycarbonofluoride powdered of FS, FT types. Technical specification TU 34 9735-0003-03533984-97, in Russian.Institute of Inorganic Chemistry SB RAS, Angarsk, Novosibirsk; Angarsk Electrochemical Combinate, JS Novosibirsk Chemical Concentrates Plant, Angarsk, Novosibirsk, 1997. p 94.
6. Mitkin VN, Oglezneva IM. *J Appl Chem (Russ)* 2000;73:1595–1603.
7. Mitkin VN. *J Struc Chem* 2003;44:99–138.
8. Mitkin VN. Novel Electrode Materials for Lithium Chemical Energetics – in Russian (Novejshiye electrodnye materialy dlya litievoj khimicheskoj energetiki). Novosibirsk: JS NCCP Publishers; 2001. p 162.
9. Mitkin VN. Mechanochemical synthesis and physical-chemical properties of carbon-fluorocarbon nanocomposition materials. A review. *J Fluor Chem* 2011;132:1047–1066. Special issue (in press), DOI: 10.1016/j.jfluchem.2011.06.039.

CHAPTER 67

Preparation of Fluorinated γ-Alumina

ERHARD KEMNITZ, TOMAŽ SKAPIN, AND JOHN M. WINFIELD

The importance of the high-surface-area oxides, chromia or γ-alumina, as heterogeneous catalysts or catalyst supports is well established in large-scale preparations of carbon–halogen compounds. Where anhydrous HF is used as the fluorinating agent (usually under flow conditions) the oxide surfaces become fluorinated; therefore, the active catalytic sites are in fluorinated or partially fluorinated environments. These sites are highly acidic, having Lewis or Brønsted character depending on the synthesis conditions. Both types of fluorinated oxide have been extensively investigated under laboratory conditions to obtain fundamental information [1–4] but syntheses of model fluorinated materials from the defect spinel, γ-alumina, are rather better defined than those from chromia. Work with the latter material is complicated by the possibility of oxidation states of Cr $>$ III and by an oxide precursor in which long-range order is absent.

Fluorinated aluminas are often the model compounds of choice for laboratory catalysts, catalytic support materials, or for solid acids, but it should be remembered that in applications where fluorination and chlorination processes are both involved, there is an important thermodynamic difference between chromia and alumina. The enthalpy difference between fluorination and chlorination of Cr_2O_3 is small, whereas fluorination of Al_2O_3 is enthalpically far more favorable than its chlorination. A wide variety of fluorine-containing compounds can be used to achieve the fluorination of an alumina surface under laboratory conditions, illustrated here with the fluorinating agents, sulfur tetrafluoride, SF_4 [5–9], and trifluoromethane, CHF_3 [10,11]. The latter reagent is preferred over other carbon halogen compounds in view of its availability and the absence of Cl, which can complicate the resulting surface.

Efficient Preparations of Fluorine Compounds, First Edition. Edited by Herbert W. Roesky.

67.1 PREPARATIONS OF FLUORINATED ALUMINAS

Preparation of Fluorinated Alumina using SF_4 under Static Conditions

Apparatus Simple vacuum line fabricated from Monel or stainless steel, Monel metal pressure vessels (90–100 mL) for fluorination reactions and storage of volatile materials; metal equipment should be well passivated before use, e.g. by multiple exposures to ClF_3; the line should be provided with a metal Bourdon pressure/vacuum gauge and each section calibrated before use; volumes of the various sections of the vacuum line and reagent quantities should be chosen to ensure that pressures during manipulation of volatile species are contained below 0.5 atm; safety glasses, face shield, laboratory coat, protective gloves.

Chemicals Sodium-free, γ-alumina, lecture bottle containing SF_4 and equipped with a main valve and reducing valve; if these are not available, SF_4 can be prepared following a reliable literature synthesis [8,9] using 1/10th scale; SF_4 can be purified via its adduct with BF_3 decomposing the latter with dry Et_2O; liq. N_2.

Attention! Eye protection (safety glasses and/or face mask) and protective gloves must be used at all times.

Caution! SF_4 is very toxic and hydrolyses readily to give HF, which is dangerous and must be handled with the appropriate precautions Transfers involving γ-alumina can produce harmful dust and should be carried out in a fume cupboard. All equipment should be housed in a well-ventilated hood or similar facility.

Experimental Procedure Sodium-free γ-alumina (1.5 g, BET, Brunauer-Emmett-Teller area 110 $m^2\ g^{-1}$), which has been previously caked and sieved to produce particles in the range 500–1000 μm, is added to a pressure vessel, attached to a metal vacuum line by a high vacuum metal valve and evacuated. An aliquot (9.0 mmol) of purified SF_4 is condensed into the vessel, the temperature allowed to rise from 77 K to ambient, and the mixture allowed to stand for 2 h. Volatile products, a mixture of OSF_2 and SO_2 and identified using Fourier transform infrared spectroscopy, are removed by distillation. This sequence is repeated twice using two further aliquots of SF_4. For *in situ* use, the procedure can be carried out on a reduced scale (0.5 g with 3×3 mmol of SF_4 in FEP or even Pyrex ampoules). Fluorinated γ-alumina contains some adsorbed HF (see below), so it is stable in Pyrex for only very short periods.

Characterization Data Brunauer–Emmett–Teller (BET) area 80–90 $m^2\ g^{-1}$, F content ca. 22% (value calculated from a radiotracer study of the reaction)

Preparation of Fluorinated Alumina using SF_4 under Flow Conditions

Apparatus Flow system built from stainless-steel tubing (outside diameter, (o.d.) ca. 65 mm) incorporating a vertical tube reactor (o.d ca. 100 mm, gas flow top to bottom) and mass flow control valves; safety glasses, face shield, laboratory coat, protective gloves.

Chemicals γ-Alumina and SF_4 as specified above; O_2 free, dry N_2.

Attention! Eye protection (safety glasses and/or face mask) and protective gloves must be used at all times.

Caution! SF_4 is very toxic and hydrolyses readily to give HF (see above). Transfers involving γ-alumina can produce harmful dust and should be carried out in a fume hood. All equipment should be housed in a well-ventilated hood or similar facility.

Experimental Procedure This procedure is designed particularly for *in situ* catalytic experiments.

γ-Alumina (300–500 mg) is placed on a pad of steel mesh supported within the reactor and calcined at 523 K under flow of dry N_2 (2.5 L h^{-1}) for 2 h. The N_2 stream is replaced by a mixed SF_4/N_2 stream (20% SF_4, 2.5 L h^{-1}) for 2 h maintaining the reactor temperature at 523 K. After this period, the reactor is allowed to cool to ambient temperature under N_2 flow.

Characterization Data BET area 67 m^2 g^{-1}, F content 47.1%. X-ray diffraction (XRD) analysis indicates the presence of γ-alumina with, possibly also, an amorphous phase.

Preparation of Fluorinated Alumina using CHF_3 under Flow Conditions

Apparatus Inlet flow system built from copper tubing (inside diameter, i.d. ca. 2 mm) incorporating a vertical tube reactor from nickel or Monel (i.d. ca. 22 mm, gas flow top to bottom) and mass flow control valves; furnace with programmable temperature regulator, plastic exhaust tubing connected to a plastic wash bottle with 5–10% NaOH or KOH solution, safety glasses, face shield, laboratory coat, protective gloves.

Chemicals Commercial γ-alumina in the form of grains or pellets with particles in the range 500–1000 μm, CHF_3 container equipped with a main valve and reducing valve.

Attention! Eye protection (safety glasses and/or face mask) and protective gloves must be used at all times.

Caution! Manipulation of CHF_3 is in general much safer when compared with other conventional fluorinating agents, like HF, because of its low toxicity, relatively high inertness, and non-flammability. However, gaseous products from the fluorination with CHF_3 contain water, HF, and CO, and must be handled with caution. In the exhaust tubing highly corrosive and toxic aqueous HF may condense. Transfers involving γ-alumina and the fluorinated products can produce harmful dust and should be carried out in a fume hood. All equipment should be housed in a well-ventilated hood or similar facility.

Experimental Procedure This procedure is convenient for the preparation of larger laboratory batches (several grams) of fluorinated γ-alumina.

γ-Alumina (10 g, BET area 240 m^2 g^{-1}) is placed on a nickel perforated holder plate, positioned in the middle of the vertical reactor and calcined at 523 K under flow of dry N_2 (25 L h^{-1}) for 2 h. The N_2 stream is firstly replaced by a mixed CHF_3/N_2 stream (20% CHF_3, 30 L h^{-1}) and the temperature is increased to 623 K in 2 h. The mixed CHF_3/N_2 stream is then replaced with pure CHF_3 (6 L h^{-1}) for 3 h maintaining the reactor temperature at 623 K. After this period, the reactor is allowed to cool to ambient temperature under N_2 flow.

Characterization Data BET area 34 m^2 g^{-1}, F content 58.4%. XRD analysis indicates the presence of α- and β-AlF_3 with, possibly also, an amorphous phase.

Applications/Properties Irrespective of the preparative method used, fluorinated γ-alumina exhibits significant surface acidity and is an effective heterogeneous catalyst for many reactions involving carbon–halogen compounds [2–4,6, 7,10–14]. Lewis acid surface sites are present in all cases but when SF_4 is used as the fluorinating agent under static conditions, there are additionally Brønsted surface sites, believed to be the result of incomplete removal of adsorbed HF formed during the fluorination process. The material formed by fluorination using CHF_3 has been used as a benchmark in comparative studies of catalytic activity involving isomerization and dismutation of C_2 chlorofluorocarbons. Examination of the surface of this material using various X-ray spectroscopic methods shows that the fluorine uptake is a slow process and is initiated at the surface of oxide particles; subsequently F is incorporated into sub-surface regions. The X-ray BE data indicate, however, that the Lewis acid sites have Al(III) in disordered O/F rather than fully fluorinated environments. They indicate in addition that oxidic species are present at the surface.

Events that occur during the fluorination of the alumina surface by SF_4 under static conditions can be inferred from radiotracer experiments. Although the treatment described above is carried out nominally at ambient temperature, the surface hydrolysis reaction is strongly exothermic. The initial step involves replacement of surface –OH and some Al–O–Al groups by Al–F with the concurrent formation of OSF_2 followed by SO_2. Although there is no evidence from radiotracer studies for strong adsorption of OSF_2 or SO_2, retention of HF appears to be essentially complete.

The chemical reactivity of the surface that results from the fluorination of γ-alumina by SF_4 under static conditions suggests that the Lewis acid sites formed may be stronger than those resulting from the other treatments described. For example, dehydrochlorination of CH_3CCl_3 and *t*BuCl are both observed *at room temperature* and are accompanied by oligomerization of the olefins formed. In the former case the resulting surface catalyses F-for-Cl halogen exchange, *also at room temperature*, in hydrochlorocarbons that themselves undergo dehydrochlorination.

REFERENCES

1. Winfield JM. In: Cornils B, Herrmann WA, Schlögl R, Wong C-H, editors. *Catalysis from A to Z*. Wiley-VCH, Weinheim; 2000. p 222.
2. Kemnitz E, Winfield JM. In: Nakajima T, Žemva B, Tressaud A, editors. *Advanced Inorganic Fluorides: Synthesis, Characterization and Applications*. Chapter 12. pp 367–401. Amsterdam: Elsevier Science S. A; 2000.
3. Bozorgzadeh H, Kemnitz E, Nickkho-Amiry M, Skapin T, Winfield JM. *J Fluor Chem* 2001;110:181–189.
4. Winfield JM. *J Fluor Chem* 2009;130:1069–1079.
5. Bendada A, Webb G, Winfield JM. *Eur J Solid State Inorg Chem* 1996;33:907–916.
6. Bozorgzadeh H, Kemnitz E, Nickkho-Amiry M, Skapin T, Winfield JM. *J Fluor Chem* 2001;107:45–52.
7. Barclay CH, Bozorgzadeh H, Kemnitz E, Nickkho-Amiry M, Ross DEM, Skapin T, Thomson J, Webb G, Winfield JM. *J Chem Soc Dalton Trans* 2002;40–47.
8. Fawcett FS, Tullock CW. *Inorg Synth* 1963;7:119–124.
9. Kolta GA, Webb G, Winfield JM. *J Fluor Chem* 1981/82;19:89–90.
10. Skapin T. *J Mater Chem* 1995;5:1215–1222.
11. Skapin T, Mazej Z, Makarowicz A, Jesih A, Nickkho-Amiry M, Schroeder SLM, Weiher N, Žemva B, Winfield JM. *J Fluor Chem* 2011; 132: 703–712.
12. Hess A, Kemnitz E. *J Catal* 1994;149:449–457.
13. Boese O, Unger WES, Kemnitz E, Schroeder SLM. *Phys Chem Chem Phys* 2002;4:2824–2832.
14. Nickkho-Amiry M, Winfield JM. *J Fluor Chem* 2007; 128:344–352.

CHAPTER 68

High-Yield Synthesis of a Single Asymmetric Isomer of $C_{70}(CF_3)_{10}$ by High-Temperature Radical Trifluoromethylation

NATALIA B. SHUSTOVA, STEVEN H. STRAUSS, AND OLGA V. BOLTALINA

The first high-temperature preparation of 1,4,10,19,25,41,49,60,66,69-$C_{70}(CF_3)_{10}$ compound using CF_3I gas as a perfluoroalkylating reagent was reported by Boltalina and co-workers [1]. It resulted in a 27% yield of trifluoromethylated fullerenes based on converted C_{70}. The $C_{70}(CF_3)_{10}$ single isomer constituted about 40 mol% of all $C_{70}(CF_3)_n$ compounds in the crude mixture. Following attempts to optimize the synthetic conditions using the same procedure led to 55 mol% of the desired $C_{70}(CF_3)_{10}$ isomer [2].

Further significant improvement of yield and selectivity of the title compound is described in the present procedure. The distribution of trifluoromethylated fullerenes in the crude mixture is a function of the synthetic conditions, among which the crucial parameter is a residence time of $C_{70}(CF_3)_n$ compounds in the reactor hot zone. Decreasing the CF_3I flow rate and changing the reactor length from 5 to 40 cm resulted in the increase of the residence time and hence improved selectivity. Careful studies of the reaction parameters led to the preparation of 1,4,10,19,25,41,49,60,66,69-$C_{70}(CF_3)_{10}$ in 68 mol% yield and 89 mol% purity in the 40 cm long quartz tube at 530 °C for 90 min using 0.025 mmol/min flow rate of CF_3I. Such high selectivity toward the single isomer is rare in fullerene chemistry, especially for the derivatives with a high number of the attached groups on a fullerene cage [3]. The presented procedure allows one to prepare $C_{70}(CF_3)_{10}$ isomer with good purity on a gram-scale without time-consuming and labor-demanding HPLC procedure.

Efficient Preparations of Fluorine Compounds, First Edition. Edited by Herbert W. Roesky.

68.1 PREPARATION OF 1,4,10,19,25,41,49,60,66,69-$C_{70}(CF_3)_{10}$

Apparatus A 0.7 cm internal diameter quartz tube is placed in a 40 cm tube furnace and connected to a mineral-oil bubbler and a source of CF_3I gas. The length of the furnace hot zone is critical (e.g., the 5 cm shorter furnace leads to low yield and purity of the prepared compound). The Teflon connectors are used to ensure proper fitting of the quartz tube. Safety glasses, a laboratory coat, protective gloves are used during experiment.

Chemicals C_{70}, CF_3I, toluene, and heptane (solvents were used for HPLC purification).

Attention! Safety glasses and protective gloves are necessary.

Caution! CF_3I decomposes in air above 300 °C and produces toxic HF, COF_2, and I_2. It should be handled in a well-ventilated fume hood.

Experimental Procedure Finely ground C_{70} (0.02 g, 0.024 mmol) is loaded in a 0.7 cm internal diameter quartz tube. The central part of the tube containing C_{70} is placed in a 40 cm long tube furnace. The tube is connected to a gas handling system at one end and a mineral-oil bubbler at the other. After purging the system with nitrogen, the reactor is heated to 530 °C in the presence of steady flow of CF_3I gas (0.025 mmol/min) for 90 min. The orange-brown product and purple iodine condense on the cold end of the quartz tube. The reactor is then cooled to room temperature under CF_3I flow. The product is washed with heptane and the washings are evaporated to dryness. Excess of iodine is removed under vacuum at room temperature. The ^{19}F NMR of the crude mixture showed that the title compound constituted 89 mol% of all $C_{70}(CF_3)_{10-14}$ compounds present. If the higher purity is necessary (>89 mol%), the isolated $C_{70}(CF_3)_{10}$ compound can be purified by HPLC (5 mL/min eluent flow rate, 10 mm I.D. × 250 mm Cosmosil Buckyprep Column, Nacalai Tesque Inc., 300 nm UV detector) using a two-stage quick procedure. In the first stage, 3.6–3.9 min fraction is collected using toluene as an eluent. The fraction is then eluted with 60:40 (v:v) toluene/heptane mixture to separate $C_{70}(CF_3)_{10}$ compound from $C_{70}(CF_3)_8$ and $C_{70}(CF_3)_{12}$. The ^{19}F NMR spectrum of $C_{70}(CF_3)_{10}$ prepared by this procedure indicated 98 mol% purity.

Characterization Data ^{19}F NMR (376.5 MHz, chloroform-*d*, C_6F_6 internal standard (δ −164.9 ppm)): δ −60.9 (m), −62.9 (m), −63.3 (m), −63.3 (m), −63.6 (m), −63.6 (m), −64.9 (m), −65.5 (m), −67.6 (q, J_{FF} = 15.9 Hz), 70.7 (q, J_{FF} = 10.3 Hz). APSI-MS: *m/z* 1530.

Application Perfluroalkyl fullerenes, in particular $C_{70}(CF_3)_{10}$, exhibit high solubility in most organic solvents, thermal/air stability and possess better electron-accepting properties than the parent C_{70} due to the presence of electron-withdrawing groups on the fullerene cage. Such a combination of chemical properties with a tunable range of reduction potentials makes them attractive candidates for electron-accepting materials in organic photovoltaic devices.

REFERENCES

1. Kareev IE, Kuvychko IV, Lebedkin SF, Miller SM, Anderson OP, Seppelt K, Strauss SH, Boltalina OV. *J Am Chem Soc* 2005;127:8362–8375.
2. Popov AA, Kareev IE, Shustova NB, Lebedkin SF, Strauss SH, Boltalina OV, Dunsch L. *Chem Eur J* 2008;14:107–121.
3. Shustova NB, Kareev IE, Kuvychko IV, Whitaker JB, Lebedkin SF, Popov AA, Dunsch L, Chen Y-S, Seppelt K, Strauss SH, Boltalina OV. *J Fluor Chem* 2010;132:1198–1212.

INDEX

Acetylhypofluorite, 139, 141, 143, 144
N-Acetyl-trifluoromethylphenylsulfilimine, 263
$AgBF_4$, 9, 42, 43
Ag^+ BrF_4^-, 9
$Ag(CF_3SO_3)$, 50
$[Ag(CO)]_2B_{12}F_{12}$, 18
Ag_2CO_3, 9, 10, 88, 179, 260
AgF, 9, 10, 37–39, 405, 408, 409
AgF_2, 9, 38, 39, 429, 430
$AgNO_3$, 9, 10
Ag_2O, 9, 10, 38, 43
$AgSCF_3$, 26
aHF, 11–14, 36, 38–45, 64, 428–430
$Al(CF_3SO_3)_3$, 46
Alkanoyl vinylic ether, 300
Allylic fluoride, 405–408, 410, 411
Allylic fluorination, 406, 407, 409, 410
Antimony pentafluoride, 1, 51, 52
Aqueous HF, 2, 4, 9, 10, 73, 77, 444
AsF_5, 12, 82, 84–86, 334, 430
Asymmetric fluorocyclization, 400, 402
AuF_3, 94–97
1-(1-Phenylethyl)-2-(trifluoromethyl)aziridine, 377

$B(C_6F_5)_3$, 60–64
Bartlett, 11, 15, 107, 330
$BaSO_4$, 8
[[Difluoro(trimethylsilyl)methyl]seleno]benzene, 352, 353
1-Benzyl-2-(trifluoromethyl)aziridine, 377
N-Benzyl-2-bromo-3,3,3-trifluoropropan-1-amine, 376
Benzylperfluorobutanedithioate, 294
BF_3, 40–43, 60, 61, 67–69, 148, 335, 443
1,4-Bis(Z-2-iodo-1,2-difluoroethenyl)benzene, 246, 247
Bis[2-(2-fluoroethoxy)ethyl] ether, 130
$BMIM^+$ $CF_3SO_2NSO_2C_4F_9^-$, 259
(2*S*,4*S*)-*N*-Boc-4-trifluoromethylproline, 358, 359
BrF, 134–136
BrF_3, 70, 154–170
1-Bromo-2-pentafluorosulfanylheptyl acetate, 276
2-Bromo-3,3,3-trifluoro-*N*-(1-phenylethyl) propan-1-amine, 376
tert-Butyldifluoroamine, 127, 128
N-Butyl-*N*-(trifluoromethyl)hexaneamide, 162
N-Butyltrifluorodithioacetate, 294
Butyl-2-trifluoromethyl-2-*t*-butylacetate, 164

Calcium gluconate, 36, 37, 40–52, 80, 102, 104–106, 118
$C_{70}(CF_3)_{10}$, 447, 448
para-$C_{60}(CF_2C_6F_5)_2$, 19, 20
para-$C_{60}(CH_2Ph)_2$, 19
CeSF, 416, 418, 422
$(CF_3)_6P_3N_3$, 124, 125
$C_6F_5SiF_3$, 65, 66, 68–70
$CF_3SO_2N(Na)SiMe_3$, 257, 258
$CF_3SO_2N(Na)SO_2C_4F_9$, 257, 258
CF_3SO_2NHNa, 257, 258
$[C_6F_5Xe][B(C_6F_5)_2F_2]$, 60, 62
3-Chlorodifluoromethoxyaniline, 160, 161
2-Chloro-2-(trifluoromethyl)-3,6-dihydro-2H-thiopyran, 296
$(CH_3)_4NF$, 7
$(C_2H_5)_3N{\cdot}3HF$, 39, 55
ClF_3, 4, 95–97, 100 –106, 435–438, 443
$(\eta^5\text{-}C_5Me_5)TiF_3$, 271
Copper, 3, 19, 20, 25, 26, 34, 36, 38, 39, 47, 48, 58, 99, 136, 140, 153–155, 187, 241, 242, 244, 247, 268, 287, 357, 368, 400, 426, 444
$[CpFe(CO)_2(SO_2)]$ $[AsF_6]$, 83, 84
$CpTiF_3$, 271, 272
$Cs[B(C_6F_5)_4]$, 63, 64
$Cs_2B_{12}F_{12}$, 16, 17
CsF, 7, 8, 76, 88, 91, 92, 124, 125, 177, 179, 187, 346

Efficient Preparations of Fluorine Compounds, First Edition. Edited by Herbert W. Roesky.

$Cs[1\text{-}H\text{-}CB_{11}F_{11}]$, 22–24
Cs_2SO_4, 8
$Cu(CF_3SO_3)_2$, 47
CuF_2, 36, 37, 55, 179
$CuSCF_3$, 26
Cyclohexylacetyl fluoride, 158
Cyclotron, 11

Dehydrofluorination, 148, 275, 306, 315
Dehydroxyfluorination, 412
Dibenzosuberenone, 140, 141
S-(Trifluoromethyl)dibenzothiophenium-3-sulfonate, 280, 283, 284
S-(Trifluoromethyl)dibenzothiophenium tetrafluoroborate, 280
S-(Trifluoromethyl)dibenzothiophenium triflate, 280, 282
2,4-Dibenzyloxy-5-(perfluoroprop-1-enyl) pyrimidine, 372
1,1-Dichloropolyfluoroalkylsulfides, 292
Diethyl-β,β-difluoroglutarate, 170
Diethyl-1,1(difluoro)-3-hydroxybutylphosphonate, 211
α,β-Difluoroacrylic acid, 391
2,2-Difluoroadamantane, 155, 156
4,4-Difluoro-1-*t*-butylcyclohexane, 138
2-(Bromomethyl)-1,1-difluorocyclopropane, 338, 339, 341
2,2-Difluorocyclopropylmethanol, 339, 341
1,1-Difluoro-2,2-dibutylethane, 168
2,2-Difluoro-3,3-diiodo butane, 135
2,2-Difluoro-1,3-dioxolo-4,5-pyridine, 160
3-Difluoromethoxyaniline, 161
5-(2,4-Di-*tert*-butoxypyrimidin-5-yl)-4,5-difluoro-2-methyl-4-(trifluoromethyl) isoxazolidine, 367
1,1-Difluoro-1-methylthioundecane, 163
1,1-Difluorotridecane, 171
1,1-Difluoroundecane, 162, 168
1,1-Dihydropolyfluoroalkylsulfides, 291
(2*R*)-1,2-Di-*O*-isopropylidene-4,4-difluoro-5-hexen-1,2,3-triol, 350, 351
Dimethyl-2,2-dibromo-3,3-difluorosuccinate, 136
1,3-Dimethyl-5-perfluorovinyluracil, 371
3,7-Dinitro-*S*-(trifluoromethyl) dibenzothiophenium-triflate, 282
1,3-Dioxolo-4,5-pyridine-2-thione, 160
erythro-1,3-Diphenyl-1-fluoro-2-methoxy-3-propanone, 144
S-(Trifluoromethyl)diphenylsulfonium triflate, 267, 268
Diphenyltetrafluoroethane, 151
(DPTPT), 195

Electric discharge, 11
Elemental fluorine, 1, 4–6, 16, 37, 39, 55, 80, 94, 95, 100, 111–113, 127, 142, 148, 150
Enaminone, 300, 302
Enantioselective fluorination, 408, 411
Enantioselective fluorocyclization of indoles, 402
Ethyl-2,2-difluoro-3-methylundecanoate, 169
Ethyl(2*Z*)-4,4,4-trifluoro-3-iodo-2-butenoate, 354
Ethyl-2-trifluoromethylcycloheptanoate, 165

F_2, 1, 3–6, 8, 11–14, 16, 17, 23, 95, 135, 136, 139, 140, 142, 146–153, 155, 173, 273
^{18}F, 11, 12, 14, 144, 394–397, 403
$4\text{-}FC_6H_4IF_4$, 72, 73
Fluoride-intercalated graphite, 425
Fluorinated γ-alumina, 442–444
Fluorinated indanones, 400
Fluorinating agent, 3, 5, 11, 16, 37, 40, 66, 72, 74, 91, 94, 100, 108, 139, 140, 154, 271, 379, 380, 384, 442, 444, 445
Fluorination of aliphatics, 149
Fluorination of 4-alkylcyclohexanol, 147
Fluorination of bicyclic compounds, 149
Fluorination of steroids, 149
11β-Fluoro-3-acetoxy-estra-3,5-dien-17-one, 412
Fluoroalkenes, 74, 202, 203, 243, 315, 316, 365, 368, 369, 385, 386
4-[^{18}F]Fluorobenzaldehyde, 397, 398
Fluorobis(phenylsulfonyl)methane (FBSM), 191, 194
Fluorodesilylation, 411
α-Fluoro enamino reagents, 379
Fluoroethylene/perfluoropropylene copolymer, 3, 11
trans-1-Fluoro-2-iodomenthan, 135
3-Fluoro-3-methyloxindole, 128
^{18}F positron emission tomography, 11
(FPTPT), 196–199
^{18}F radiosynthesis, 396, 399
$FSO_2N{=}SF_4$, 91

Gabriel reaction, 375, 378
Gaseous HF, 9
Graphite fluorides, 425–427, 432, 441

Heptafluoropropyl trifluorovinyl ether, 152
Hexafluorophosphate, 35, 36, 44, 45, 57
HF, 2–4, 8–11, 13–16, 22–24, 38–44, 55, 62, 67, 68, 70, 72–74, 76, 77, 79–81, 92, 100–106
HF/H_2O, 8–10

$Hg(NSF_2)_2$, 89–91
HMGF, 220, 221
H_2O_2, 2, 17, 292
HPF_6, 43–46
$HSbF_6$, 51
Hydrogen fluoride, 13–15, 38, 39, 41, 42, 55, 64, 72, 81, 102, 105, 106, 111, 117

IF, 134–138, 155
IF_5, 70, 72–75, 78
(*Z*)-2-Iodo-1,2-difluorostyrene, 243, 244
1-Iodo-1,2,2-trifluoro-1,2-dichloroethane, 391
Ionic liquids, 46, 57, 200, 256, 260, 261, 284, 334, 337, 389

$K[AsF_5OH]$, 118
$K_2B_{12}F_{12}$, 16, 17
$K[C_6F_5BF_3]$, 66, 67
Kel-F, 12, 13, 80, 84, 91, 92
Kel-F valve, 12
Ketosulfones, 202, 295
KF, 2, 7, 16, 17, 95, 96, 179, 187, 188, 222, 241, 243, 244, 246, 249
KHF_2, 16
K_2MnF_6, 1, 2, 4
$KMnO_4$, 2, 277, 345
K_2NiF_6, 37, 80, 428, 430
KPF_6, 45, 46

$La_2O_{1.5}FS$, 416, 419–423
LaSF, 416, 418–420, 422, 423
$[LCaF(thf)]_2$, 272, 273
$LiBF_4$, 40, 41, 56
$Li(CF_3SO_3)$, 49, 58
$LiPF_6$, 44, 57

Magic acid, 51, 59
Me_3SnF, 270–273
Mesoporous fluorocarbon, 433, 435, 439
Metal vacuum, 12, 79, 100, 443
2-Methoxy-6-(trifluoromethyl)-3,4-dihydro-2H-thiopyran-3,4-diol, 299
2-Methoxy-6-trifluoromethyl-2H-thiopyran, 297, 298
(2,2-Difluorocyclopropyl)methyl benzoate, 340
Methyl-3,3-dimethyl-2-fluorobutyrate, 143
Methyl glutaroyl fluoride, 157
MnF_3, 1
Moissan, 1, 5, 146, 370
Monel, 4, 16, 37–39, 94–97, 100, 103–105, 111, 136, 140, 148, 153, 155, 309, 312, 416, 435, 443, 444
Monel U-tube, 3, 4
4-[2-Hydroxy-2-phenyl-2-(trifluoromethyl) acetyl]morpholine, 123, 124

$Na[SbF_5OH]$, 119
$NbCl_4F$, 111, 112, 115
NbF_5, 112, 336, 431
$[N(CH_3)_4]F$, 76–78
$[N(C_2H_5)_4][SbF_6]$, 118, 119
$[N(C_2H_5)_4][SbF_5OH]$, 119
$(NF_4)_2MnF_6$, 3
Nickel, 3, 6, 12, 13, 37, 39, 164, 166, 236, 313, 331, 416, 426, 429, 431, 435, 436, 444, 445
7-Nitro-S-(trifluoromethyl)dibenzothiophenium-3-sulfonate, 280, 284
$[NMe_4]SCF_3$, 27, 28
$[NMe_4]SeCF_3$, 27–29
$[NMe_4]TeCF_3$, 31, 33
NO_2BF_4, 41, 43
Noble-gas compounds, 11
NSF, 85, 86, 88, 90–92

Octafluorocyclooctatetraene, 308, 312–314
1,1,2,2,9,9,10,10-Octafluoroparacyclophane, 343
Octyltrifluoromethyl ether, 158

$[PBr_4][PF_6]$, 110, 115
PCl_4F, 114, 115
$[PCl_4]\ [PF_6]$, 109
trans-$[Pd(H)\text{-}(4\text{-}C_5NF_4)(P\textit{i}\text{-}Pr_3)_2]$, 238
5,5,6,6,6-Pentafluorohexane-3-ol, 207
2,2,3,3,3-Pentafluoro-1-phenylpropan-1-ol, 206
1,1,3,3,3-Pentafluoropropene (PFP), 379
1,1,3,3,3-Pentafluoropropenemorpholine adduct (PFPMor), 379
2-Pentafluorosulfanylheptanal, 277
2-Pentyl-β,β-difluoroacrylate, 166
Perfluoroalkyliodide, 386
Perfluoro *n*-butylsulfonyl fluoride (nonaflyl fluoride), 412
Perfluoromethyl vinyl ether, 385, 389, 390, 394
Perfluoro[2.2]paracyclophane, 343–348
Perfluoropropylene copolymer, 11
PF_5, 45, 109, 110, 112–115, 334
(*Z*)-*N*-Phenyl-2,3-difluoro-3-(triethylsilyl) acrylamide 249
$PhSCF_3$, 178, 180, 182
$PhSCH_2F$, 187, 188, 190
$PhSO_2CF_2H$, 181–186
$PhSO_2CF_2TMS$, 184–187
S-(Difluoromethyl)-*S*-phenyl-2,3,4,5-tetramethylphenylsulfonium tetrafluoroborate, 195

S-(Fluoromethyl)-*S*-phenyl-2,3,4,5-tetramethylphenylsulfonium tetrafluoroborate, 196
$PhSO_2CF_3$, 176, 178, 180–182
$PhSO_2CH_2F$, 187–195
Polyethylene beaker, 2
Potassium hexafluoromanganate, 1
4-[(*E*)-1-Chloro-2-(pentafluoro- λ^6-sulfanyl)vinyl]-4′-propyl-1,1′-bi(cyclohexyl), 215
4-[(Pentafluoro-λ^6-sulfanyl)ethynyl]-4′-propyl-1,1′-bi(cyclohexyl), 216
Pyrazoles, 301, 306, 307
Pyrimidines, 302, 307, 365, 368, 369, 371

Radical copolymerization, 385, 388–392, 394, 395
Raman, 13–15, 80, 97–99, 106, 107, 111, 335, 336
$[Re(CO)_5(NSF)][AsF_6]$, 85, 86
ReF_6, 101, 103, 104, 107
$ReOF_5$, 100–104
$ReSF_5$, 79
$[Rh(Bpin)(PEt_3]$, 235
Ruppert–Prakash reagent, 174–182, 186, 199, 200

$SbCl_4F$, 110, 111, 113, 115
SbF_5, 1–4, 51, 52, 82, 111, 120, 428–430
SF_6, 5, 107, 276
Sibunite, 433–435, 437–439
Silver monofluoride, 9
SmSF, 416, 418, 422
Stainless steel, 3, 8, 12, 45, 67, 80, 100–106, 335, 426, 443
S-(Phenyl)-*S*-(trifluoromethyl)sulfoximine, 263
Superstoichiometric fluorocarbon, 433, 435, 436, 439, 440

$TaCl_4F$, 112, 113, 115
TaF_5, 113, 334
Tandem Nazarov-fluorination, 400 – 402
Teflon, 2–5, 12, 16, 36, 37, 39–52, 61–63, 67, 72, 73, 76, 83, 85, 91, 92, 101–106, 154, 240, 257–259, 310, 366, 428, 429, 448
Tefloncoated, 2, 36, 37, 40–44, 46–51, 240, 243, 246, 250, 258
Teflon-FEP, 3–5
$TESCF_3$, 179
Tetrafluoroallene, 316–318, 330, 331
Tetrafluorocatechol, 252, 254, 255
1,1,2,2-Tetrafluoro-1,2-diphenylethane, 131
2,3,5,6-Tetrafluoropyridine, 237, 239
Thiazolidinone, 305
6-(Trifluoromethyl)-2H-thiopyran, 297
$TMSCF_3$, 174–182, 186, 199, 200
p-Tolyltrifluorodithioacetate, 293
Triazoles, 303, 304, 307
(*Z*)-1-(4-Heptafluorotolyl)-2-triethylsilyl-1,2-difluoroethene, 241
Triflate, 35, 36, 46 –51, 54, 57–59, 196, 198, 199, 267, 268, 280, 282, 285, 287
(*E*)-4,4,4-Trifluoro-2-buten-1-ol, 359, 360
2-(Ethoxyoxalyl-amino)-3,3,3-trifluoro-2-hydroxypropionic acid methyl ester, 229
(2*S*,4*S*)-4-Trifluoromentylpyroglutamic acid, 359
Trifluoromethane, 121, 122, 126
2-Trifluoromethylimidazoline, 306
4,5,5-Trifluoro-2-methyl-3-phenyl-4-(trifluoromethyl)isoxazolidine, 366
2-Trifluoromethylpyrrole, 122, 123
2-Trifluoromethylthiopyrylium tetrafluoroborate, 298
2-Trifluoromethyl-3,3,3-trifluoropropanal, 250
2-[*N*-(2-Ethoxyoxalyl)imino]-3,3,3-trifluoropropanoate, 230
Trifluorothioacetamide, 304–306
Trifluorothioacetylchloride, 293, 297
(3,3,3-Trifluoropropyl)trimethoxysilane, 232–234
Trimetylsilyl-2,2-difluoro-2-(fluorosulfonyl) acetate, 339
Tris(tetramethylguanido)sulfoniumchloride, 223, 224

UF_3, 94, 95, 97
UF_6, 5, 94, 96–99, 107, 336, 337
N^1-(Perfluoroprop-1-enyl)uracil, 373

van de Graff accelerator, 11
Vinylidene fluoride, 385, 388, 390, 394, 395
4-Vinyltetrafluoropyridine, 237–239

WSF_4, 79–81

Xe, 11–14, 39
$Xe(C_6F_5)_2$, 78
XeF_2, 11–14, 39, 60, 62, 68, 78, 142
XeF_4, 11, 68
XeF_6, 11
Xenon difluoride, 11–14, 39, 70, 75
Xenon tetrafluoride, 14

Zinc bars, 10

www.ingramcontent.com/pod-product-compliance
Lightning Source LLC
LaVergne TN
LVHW011553200926
848484LV00001B/310